S0-EMD-925

Asymmetric Organic Reactions

Prentice-Hall International Series in Chemistry

Boudart **Kinetics of Chemical Processes**
Ferguson **The Modern Structural Theory of Organic Chemistry**
Harvey **Introduction to Nuclear Physics, 2nd ed.**
Jolly **Synthesis and Characterization of Inorganic Compounds**
Kiser **Introduction to Mass Spectrometry and its Applications**
McGlynn, Azumi & Kinoshita **Molecular Spectroscopy of the Triplet State**
Morrison & Mosher **Asymmetric Organic Reactions**
Pasto & Johnson **Organic Structure Determination**
Sandorfy **Electronic Spectra and Quantum Chemistry**

PRENTICE-HALL, INC.
PRENTICE-HALL INTERNATIONAL, INC., UNITED KINGDOM AND EIRE
PRENTICE-HALL OF CANADA, LTD., CANADA

Asymmetric Organic Reactions

James D. Morrison
Department of Chemistry
University of New Hampshire

Harry S. Mosher
Department of Chemistry
Stanford University

Prentice-Hall, Inc., Englewood Cliffs, New Jersey

Printed in the United States of America

13-049551-4

Library of Congress Catalog Card No.: 76-118316

Current printing (last digit):

10 9 8 7 6 5 4 3 2 1

Prentice-Hall, Inc., London
Prentice-Hall of Australia, Pty. Ltd., Sydney
Prentice-Hall of Canada, Ltd., Toronto
Prentice-Hall of India Private Limited, New Delhi
Prentice-Hall of Japan, Inc., Tokyo

Preface

Asymmetric organic reactions have proven to be of value in the study of reaction mechanisms, in the determination of relative configurations and in the practical synthesis of optically active compounds. The time is propitious for a comprehensive review of the field. We have attempted to interpret the published research on asymmetric synthesis through 1968. Brief addenda of relevant references appearing in 1969 have been appended to each chapter. The early work has not been treated in as great depth as that from the more recent literature for two reasons. First, the material published prior to 1933 has been reviewed by P. D. Ritchie [*Asymmetric Synthesis and Asymmetric Induction* (London: Oxford University Press) 1933]. Secondly, the reliability of the results from much of this early work, especially the quantitative aspects, is difficult to ascertain.

It has been our goal to make this review so complete that others will not need to cover this ground again but can use this volume as a source and guide for the prior literature. It has been necessary to impose some boundaries on our subject matter. In particular the areas of asymmetric enzymatic reactions, polymerizations and heterogeneous catalytic reactions have been slighted. This was first planned as a result of space limitations but can be justified on the basis that others have now written authoritatively on these subjects [cf. R. Bentley, *Molecular Asymmetry in Biology*, (New York: Academic Press) Vol. I, 1969; Vol. II, 1970; G. Natta and F. Danusso, *Stereoregular Polymers and Stereospecific Polymerization* (New York: Pergamon Press) 1967; E. I. Klabunovskii, *Stereospecific Catalysis* (Moscow: Nauka) 1968].

We have endeavored to be critical in the evaluation of the work reviewed. The reader will perceive that we have often avoided consideration of highly speculative stereochemical analyses of transition states for asymmetric syntheses. It is not that we consider such speculation unprofitable but rather that we believe this type of conjecture is best handled in the confines of informal research seminars. Where sufficient data have been available we have presented configurational correlation models for the asymmetric synthesis under consideration. Rarely, however, is there enough evidence to warrant presentation of such models as valid representations of the transition state.

It is our intent that the reader should cover the first chapter as a necessary introduction and background to the viewpoints and definitions adopted for the subsequent chapters. The remaining chapters may be consulted independently of each other for specific treatments of narrow regions within the broad area of asymmetric organic reactions. Our treatment of the subject presumes that the reader is familiar with many of the fundamentals of stereochemistry which are dealt with in Eliel's *Stereochemistry of Carbon Compounds* [New York–London: McGraw-Hill Book Co. Inc.) 1962] and Mislow's *Introduction to Stereochemistry* [New York–Amsterdam: W. A. Benjamin, Inc.) 1966]. We not only acknowledge our indebtedness to these texts but also to these authors for many fruitful discussions.

We invite comments concerning all errors in the book, especially inadvertent errors of omission or misrepresentation.

Both of us are grateful for many opportunities which have made this volume possible. One of us (H.S.M.) is especially indebted to the Petroleum Research Fund of the American Chemical Society for a grant and to Stanford University for a sabbatical leave which enabled the preparation of a portion of this manuscript. The sabbatical leave was spent with Professor V. Prelog at the Eidg. Technische Hochschule in Zürich, Switzerland; we are grateful to Professor Prelog and his colleagues at the E.T.H. for the stimulating atmosphere and warm hospitality. One of us (J.D.M.) is particularly indebted to Professor J. W. Nowell and Dean E. G. Wilson of Wake Forest University who arranged a travel grant which aided in the early stages of the literature research for this volume. He also thanks the Stanford University Department of Chemistry for its hospitality on several occasions.

We are appreciative of the contributions made to this book by our students who read and criticized preliminary manuscript and suffered a certain amount of what may have been salubrious neglect during the writing and publication process. Finally, we are indebted to Carol W. Mosher who read and criticized early portions of the manuscript and proofread the final galleys.

James D. Morrison
Harry S. Mosher

Contents

3 Reactions of Achiral Reagents with Chiral Aldehydes and Ketones 84

4 Various Asymmetric Addition Reactions of Carbonyl Compounds 133

Asymmetric Organic Reactions

1

Introduction

1-1. History

In 1894, Emil Fischer clearly outlined the concept of asymmetric synthesis based upon his experiments in the conversion of one sugar to its next higher homolog via the cyanohydrin reaction, relating this process directly to the biochemical process for the production of optically active sugars in plants.[1] He made the assumption that carbon dioxide and water condensed to give formaldehyde under the influence of sunlight and chlorophyll. Furthermore, he assumed that the formaldehyde condensed with itself and with simple carbohydrates under the direction of the optically active substances in the chlorophyll-containing granules in the cells in such a way that the incorporation of each successive asymmetric carbon atom into the chain produced only one of the two possible stereoisomeric forms. As the reaction proceeded, a sugar molecule formed in close association with the chlorophyll. This formation was followed by separation of the optically active sugar and regeneration of the chlorophyll catalyst so that it was available to continue the cycle.[1] The now known chemical pathway of carbon in the photosynthesis process shows little correspondence to the details of this oversimplified scheme propounded at the end of the last century. Nevertheless,

[1] E. Fischer, *Ber.*, **27**, 3231 (1894). K. Freudenberg has summarized E. Fischer's contribution in this field [*Adv. in Carbohydrate Chem.*, **21**, 1 (1960)].

the concept of asymmetric synthesis as envisaged by Emil Fischer is, in its essence, valid today.

During the intervening years, the subject of asymmetric synthesis has had its share of controversies. Japp, in a published lecture[2] given in 1898 on the subject "Steoreochemistry and Vitalism," emphasized the role of the living cell in this process: "Only the living organism with its asymmetric tissue, or the asymmetric products of living organisms, or the living organism with its concept of asymmetry can produce this result. Only asymmetry can beget asymmetry." The ensuing series of published replies and rebuttals[3] concerning the details of this process proved how deeply embedded were the roots of the asymmetric synthesis concept in the vitalistic theory. In addition, from 1900 to 1935 when W. Marckwald, E. Erlenmeyer, Jr., A. McKenzie, and P. Ritchie were prominent figures in this field, another controversy was created over the subject of asymmetric induction.[3,4] This concept of asymmetric induction was never completely freed from the elements of some mysterious unsymmetrical intramolecular force acting at a distance and it was never completely liberated from the vitalistic heritage of optically active substances.

Progress was slow, but by 1949 the science of organic chemistry was ready for a succession of events which led to the interpretation of several key asymmetric reactions on a rational stereochemical basis founded upon conventional steric and electronic concepts.

Doering and Young[5] and Jackman, Mills, and Shannon,[6] in experiments foreshadowed by the publication of Baker and Linn,[7] carried out the asymmetric Meerwein–Ponndorf–Verley reduction of a ketone with an optically active alcohol in the presence of aluminum alcoholate to give a second partially optically active alcohol. They interpreted these results in terms of steric interactions in the activated complex. Vavon and his co-workers,[8,9] using a terpene-derived Grignard reagent, reported the first asymmetric Grignard reduction. Mosher and La Combe[10,11] had started independently

[2] F. R. Japp, *Nature*, **58**, 482 (1898).

[3] P. D. Ritchie, *Asymmetric Synthesis and Asymmetric Induction* (London: Oxford University Press, 1933). This historically interesting controversy is neatly summarized with complete references in Chapter I of Ritchie's book.

[4] G. Kortüm, "Neuer Forschungen über die optische aktivität chemischer Molekule," in *Sammlung Chimie und Chemische Technologie* (Stuttgart: F. Enke, 1932), Vol. 10. Kortüm in this review summarized this very interesting controversy and defined asymmetric induction as "the action of a force arising in all asymmetric systems, which influences certain adjacent systems originally of symmetrical configuration in such a way that they become asymmetric."

[5] W. Doering and R. W. Young, *J. Am. Chem. Soc.*, **72**, 631 (1950).

[6] L. M. Jackman, J. A. Mills, and J. S. Shannon, *J. Am. Chem. Soc.*, **72**, 4814 (1950).

[7] R. H. Baker and L. E. Linn, *J. Am. Chem. Soc.*, **71**, 1399 (1949).

[8] G. Vavon and B. Angelo, *Compt. rend.*, **224**, 1435 (1947).

[9] G. Vavon, C. Rivière, and B. Angelo, *Compt. rend.*, **222**, 959 (1946).

[10] H. S. Mosher and E. La Combe, *J. Am. Chem. Soc.*, **72**, 3994 (1950).

[11] H. S. Mosher and E. La Combe, *J. Am. Chem. Soc.*, **72**, 4991 (1950).

a systematic study of the asymmetric reduction of ketones such as pinacolone (**2**), using a chiral Grignard reagent (**1**) to give an optically active alcoholate (**3**). They also interpreted their results in terms of a model based on steric interactions in the transition state. During the same period, Prelog[12]

$$\underset{\mathbf{1}}{Me\text{—}C(CH_2MgCl)(C_2H_5)\text{—}H} + \underset{\mathbf{2}}{CH_3\text{—}C(=O)\text{—}C(CH_3)_3} \longrightarrow CH_2{=}C(Me)(C_2H_5) + \underset{\mathbf{3}}{H\text{—}C(CH_3)(C(CH_3)_3)\text{—}OMgCl}$$

reexamined the many examples compiled from 1904 to 1936 by McKenzie and co-workers on the addition reactions at the carbonyl group in optically active esters of α-ketoacids such as **4**. He showed that the configuration of the alcohol moiety of the benzoylformic ester (**4**) could be correlated with that of the partially active atrolactic acid (**6**) resulting from saponification of the

$$\underset{\mathbf{4}}{H\text{—}C^{*}(R)(R')\text{—}O\text{—}C(=O)\text{—}C(=O)\text{—}Ph} \xrightarrow[2)\ H_2O]{1)\ MeMgI} \underset{\mathbf{5}}{H\text{—}C^{*}(R)(R')\text{—}O\text{—}C(=O)\text{—}C^{*}(Me)(OH)\text{—}Ph} \xrightarrow[(KOH)]{H_2O} \underset{\mathbf{6}}{Me\text{—}C^{*}(COOH)(Ph)\text{—}OH} + H\text{—}C^{*}(R)(R')\text{—}OH$$

intermediate ester, **5**. The correlations were based upon eminently reasonable stereochemical assumptions concerning the nature of the reaction and the relative steric hindrance of R and R′.

Studies of Cram and Abd Elhafez,[13] published in 1952, on the stereochemistry of addition reactions of aldehydes or ketones having a chiral

$$\underset{\mathbf{7}}{(H)(Me)(Ph)C\text{—}C(=O)H} + MeMgI \longrightarrow \underset{\mathbf{8A}\ (67\%)}{(H)(Me)(Ph)C\text{—}C(OH)(H)(Me)} \text{ and } \underset{\mathbf{8B}\ (33\%)}{(H)(Me)(Ph)C\text{—}C(OH)(Me)(H)}$$

center adjacent to the carbonyl group, as in **7** → **8A** and **8B**, led to the now well-known rule of *steric control of asymmetric induction.* Thus in a 3-year period from 1949 to 1952, the foundations were laid for the rational interpretation of the stereochemical course of those organic reactions which had been generally classified as asymmetric syntheses. The period since the early 1950's has been one of exploring new systems and consolidating our knowledge in this area.

[12] V. Prelog, *Helv. Chim. Acta*, **36**, 308 (1953). See also V. Prelog and W. Dauben, *Abstr. XII Internatl. Congress Pure Appl. Chem.*, 401 (Sept., 1951).

[13] D. J. Cram and F. A. Abd Elhafez, *J. Am. Chem. Soc.*, **74**, 5828, 5851 (1952).

We may now reasonably attempt to analyze critically all such asymmetric syntheses on a rational thermodynamic and kinetic basis in terms of the interactions, steric and electronic, which are stereochemically controlling. These same forces operate in all reactions whether the substrates and reactants are optically active, racemic, or achiral. Numerous recent studies in this field have increased our knowledge of the details of asymmetric processes, and although our understanding is far from complete, especially in terms of the quantitative aspects of the problem, the time is propitious for a comprehensive review of the field. This review will evaluate the known asymmetric organic reactions with respect to their value in the study of reaction mechanisms, their reliability for the prediction of configuration, and their usefulness in the stereochemical control of synthetic processes. Several previous reviews on this overall subject[3,4,14–19] or on special aspects of asymmetric transformations[20–25] have been published. We especially urge the reader to consult these reviews for additional facets of historical,[3,18,19,22] biochemical,[21] heterogeneous catalytic,[15,20] and polymer[25] aspects of this problem which we have not attempted to treat fully in this present volume.

1-2. Asymmetric Synthesis

At this point we shall elaborate on what is to be understood by the term *asymmetric synthesis.* In its broadest interpretation, an asymmetric synthesis

[14] (a) J. Mathieu and J. Weil-Raynal, *Bull. Soc. Chim. Fr.*, 1211 (1968). (b) H. Pracejus, *Fortsch. Chem. Forsch*, **8**, 493–553 (1967). (c) L. Velluz, J. Valls, and J. Mathieu, *Angew. Chem.*, **79**, 774 (1967); *Internat. Edn.*, **6**, 778 (1967). (d) D. R. Boyd and M. A. McKervey, *Quart. Rev.*, **22**, 95 (1968).

[15] E. I. Klabunovskii, *Asymmetric Synthesis* (Moscow: Gosundart Nauchtekl, Izdatel Khim, 1960). German translation by G. Rudakoff (Berlin: Deutscher Verlag der Wissenschaften, 1963).

[16] W. Theilacker in *Heuben-Weyl Methoden der Organischen Chemie*, 4th Edition, ed. E. Müller (Stuttgart: Thieme Verlag, 1955), Vol. 4, Part 2, pp. 505–538.

[17] K. Bláha, *Chemie* (Prague), **9**, 845–870 (1957).

[18] E. E. Turner and M. M. Harris, *Quart. Rev.*, **1**, 299–330 (1947).

[19] F. Ebel in *Stereochemie*, ed. K. Freudenberg (Leipzig and Vienna: Franz Deuticke, 1933), pp. 580ff.

[20] M. M. Harris, *Progr. Stereochem.*, **2**, 157–197 (1958).

[21] P. D. Ritchie, *Advan. Enzymol.*, **7**, 65–106 (1947).

[22] P. Chanussot, *Chimia*, **15**, 17–28 (1946).

[23] E. L. Eliel, *Stereochemistry of Carbon Compounds* (New York: McGraw-Hill Book Company, 1962), pp. 81–87.

[24] K. Mislow, *Introduction to Stereochemistry* (New York: W. A. Benjamin, Inc., 1966), p. 122.

[25] (a) M. Goodman, *Topics in Stereochemistry*, ed. E. L. Eliel and N. L. Allinger (New York: Interscience Pub. Inc., 1967), Vol. 2, pp. 73–156. (b) R. C. Schulz and E. Kaiser, *Adv. Polymer Sci.*, **4**, 236 (1965).

is a reaction in which an achiral unit in an ensemble of substrate molecules* is converted by a reactant into a chiral unit in such a manner that the stereoisomeric products† are produced in unequal amounts. This is to say an asymmetric synthesis is a process which converts a prochiral unit into a chiral unit so that unequal amounts of stereoisomeric products result. For the purpose of this definition, reactant includes not only the usual chemical reagents but also solvents (Sec. **10-1**), catalysts (Sec. **6-7**), and physical forces such as a circularly polarized light (Sec. **10-4**).

This is a more inclusive definition of asymmetric synthesis than that which was formed by Marckwald. It is based on symmetry principles rather than on the observation of optical activity.§ The classical and narrower definition

* The substrate molecule must have either enantiotopic or diastereotopic groups or faces [K. Mislow and M. Raban, *Topics in Stereochemistry*, ed. E. L. Eliel and N. L. Allinger (New York: Interscience Pub. Inc., 1966), Vol. 1, p. 1].

† The stereoisomeric products may be enantiomeric or diastereomeric, and they will be chiral and nonracemic with rare exceptions. A few cases which give an achiral product must be called asymmetric synthesis, for instance, the reduction of the chiral hydroxy ketone A to give

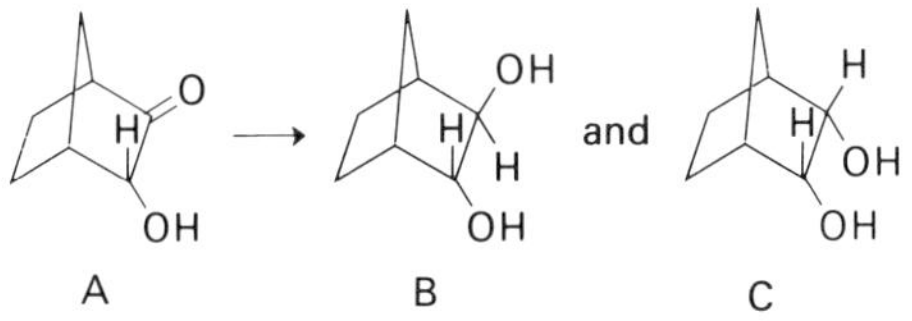

the chiral glycol B and the achiral glycol C. If the reaction were 100% stereoselective giving the achiral product C, the reaction would still be considered an asymmetric synthesis. Also, as developed on page 11 and in Sec. **3-1**, much stereochemical information may be obtained from the study of a reaction which starts with a racemate‡ and gives a mixture of diastereomeric products. Generally we shall be dealing with reactions involving optically active materials but occasionally we shall find it convenient to discuss examples involving racemates.

‡ A racemic mixture constitutes an achiral system even though it is composed of individual chiral molecules. The term *chiral and nonracemic* is redundant but unambiguously defines an ensemble of chiral molecules in which one enantiomer predominates.

§ Optical activity is an operational term meaning that the system rotates the plane of polarized light. All systems which are optically active must be chiral; but whether or not a chiral system is observably optically active depends upon the sensitivity of the polarimeter and the specific conditions of the determination. For instance if there were a collection of hydrocarbon molecules,

$$^{14}CH_3(CH_2)_5\overset{*}{C}H(CH_3)(CH_2)_5CH_3$$

of only one stereoisomeric form, most certainly this would have zero rotation within experimental error on the most sensitive polarimeter now available; yet it would be chiral by virtue of isotopic substitution. Furthermore many chiral systems will have zero optical rotation under specific conditions. Thus, if a solution has a (−) rotation at one wavelength and a (+) rotation at another, there must be an intermediate wavelength where the rotation is zero regardless of the sensitivity of the polarimeter. Temperature changes, concentration changes, and changes in ratios of mixed solvents can also lead to zero rotations.

which was enunciated by Marckwald[26] in 1904 and has an operational basis is as follows: "Asymmetric syntheses are those reactions which produce optically active substances from symmetrically constituted compounds with the intermediate use of optically active materials but with the exclusion of all analytical processes."* According to this historically important definition, an asymmetric synthesis must start with an achiral substance. The broader definition, which we shall use, specifies a symmetric but prochiral center or group which may be incorporated into either a chiral or an achiral molecule. However, most planned studies of asymmetric syntheses have concentrated on cases in which an achiral starting material is transformed into a chiral compound by the action of a chiral reagent under conditions whereby the original chiral group subsequently is removed, leaving behind a chiral product. The present broader definition differs from the classical Marckwald description in that it concentrates on the specific reaction where the new chiral unit is created instead of considering the sequence of reactions including the one whereby the "inducing" chiral reagent is removed from the product. The present review is not concerned as much with investigating the precise area outlined by Marckwald's historically important definition as with understanding the chemistry behind all asymmetric processes. These differences can be explained best by examples.

Terms such as chiral, achiral, prochiral, enantiotopic, diastereotopic, and diastereomer† for the most part follow the generally accepted usage of Eliel,[23] Mislow,[24] Mislow and Raban,[27a] and Hanson.[27b]

1-3. Illustrative Examples of Asymmetric Synthesis

Despite our intention to treat most thoroughly those examples which conform to Marckwald's classical definition, we shall first consider several diverse examples so that the boundaries of the subject may be defined and subsequent examples can be placed in their proper perspective.

Consider the addition of the phenyl Grignard reagent to the aldehyde group of the optically active sugar derivative **9**. Since in this reaction a new chiral center is being formed in a chiral environment, the two epimeric

[26] W. Marckwald, *Ber.*, **37**, 1368 (1904).

[27] (a) K. Mislow and M. Raban, *Topics in Stereochemistry*, ed. E. L. Eliel and N. L. Allinger (New York: Interscience Pub. Inc., 1966), Vol. 1, p. 1. (b) K. R. Hanson, *J. Am. Chem. Soc.*, **88**, 2731 (1968).

* Marckwald's discussion implied, but did not specify, that the chiral reagent should be recoverable and not destroyed in the process.

† Eliel and Mislow use different definitions for diastereomers in their texts, but Eliel has now adopted the definition of a diastereomer as "all stereoisomers which are not enantiomeric" in accord with Mislow's usage. (Private communication.)

products **10A** and **10B** will not be produced at the same rate and therefore not in the same amount. Bonner[28] found that these epimers were produced in 92% yield in the ratio of 65:35. Let us assume that the initial reaction

product containing a mixture of the two products **10A** and **10B** is quantitatively isolated and then carried through a series of steps so that, by methylation and oxidation, the carbohydrate moiety is degraded to the carboxyl group of *O*-methylmandelic acid (**11**). If there is no racemization, kinetic resolution, or analytical concentration of any isomer during this sequence, then the enantiomeric ratio, (*R*-**11**):(*S*-**11**), as revealed by optical rotation, will be an accurate measure of the ratio of the epimers **10A** and **10B**. In this process the symmetrical aldehyde group of the original sugar derivative has been converted into the asymmetric center of the mandelic acid and at the same time the original optically active portion of the molecule has been converted into the symmetrical carboxyl group. This example fulfills the conditions of our present concept of an asymmetric synthesis but fails to conform to the classical definition in that the symmetrical carbonyl group is a part of the original optically active molecule, not a separate entity, and the chiral inducing moiety is not recoverable as implied by Marckwald.

Another example of an asymmetric synthesis is the reduction of an optically active steroidal ketone to give the corresponding secondary alcohol with a new asymmetric center, in which the α- and β-epimers are not formed in equal amounts. Consider, as a specific example, the sodium borohydride reduction of cholestan-3-one (**12**), which gives an 80% yield of secondary alcohols consisting of 87% cholestanol (β-OH, β-**13**) and 13% epicholestanol (α-OH, α-**13**).[29] During this process, a new asymmetric center has been

[28] W. A. Bonner, *J. Am. Chem. Soc.*, **73**, 3126 (1951).

[29] (a) C. W. Shoppee and C. H. R. Summers, *J. Chem. Soc.*, 687 (1950). (b) O. H. Wheeler and J. L. Mateos, *Chem. Ind.* 395 (1957). (c) D. H. R. Barton, *J. Chem. Soc.*, 1027 (1953).

formed from the originally achiral carbonyl group. The resulting diastereomers are not formed in equal amounts because of the diastereomeric environments of the α and β faces of the carbonyl group.

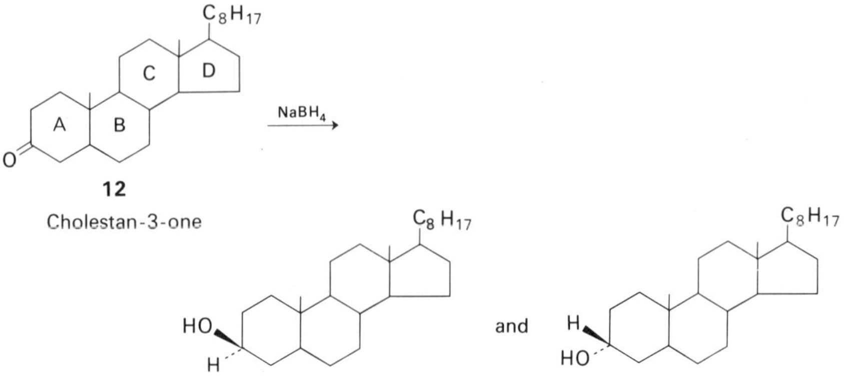

C_8H_{17} C D A B O $\xrightarrow{NaBH_4}$

12
Cholestan-3-one

C_8H_{17} HO H and H HO C_8H_{17}

β-**13** (87%)
Cholestanol
(OH equatorial)

α-**13** (13%)
Epicholestanol
(OH axial)

Barton[29c] recognized that steric factors were controlling in such metal hydride reactions and generalized the known results by stating that the equatorial alcohol will predominate when the carbonyl group is relatively unhindered and the axial carbinol will be the major product when the carbonyl group is hindered.

Let us assume that in this reaction we could take the total reduction product consisting of a mixture of α-**13** and β-**13** and by some hypothetical sequence of reactions we could degrade the steroid structure so that the C_3 carbinol carbon atom of ring A retained only carbon atoms C_1—C_2 and C_4 attached, respectively, as ethyl and methyl groups. By such a process we could have converted an achiral structure, namely the 3-keto group, under the intervening influence of other chiral centers in the molecule in such a way that an optically active product would be produced, namely a

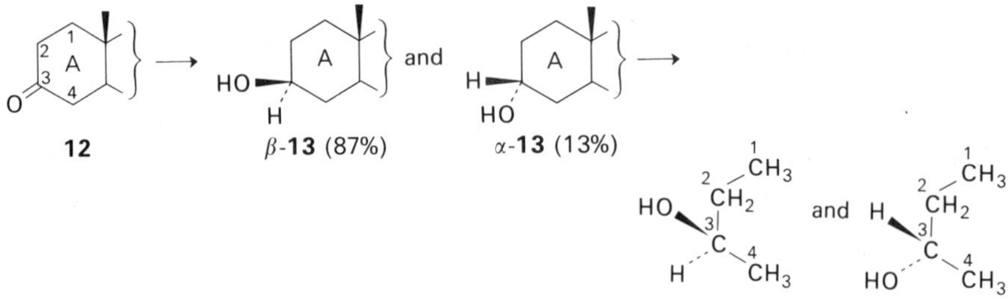

12 → β-**13** (87%) and α-**13** (13%) →

R-**14** (87%) and S-**14** (13%)

mixture of 87% (*R*)-2-butanol (*R*-**14**) and 13% (*S*)-2-butanol (*S*-**14**). One can also conceive of a total synthesis of cholestanone (**12**) which would involve the reaction of an achiral carbonyl substrate such as methyl vinyl ketone with a complex, optically active reagent incorporating rings B, C, and D of the cholestanone molecule. This postulated total process of synthesis, followed by reduction and subsequent degradation, would then have converted a symmetrically constituted compound, namely methyl vinyl ketone, into an optically active substance, namely 2-butanol (**14**), by the intervention of an optically active reagent. Such a hypothetical example fully meets Marckwald's operational definition.* Thus, the difference between the classical definition and the one now adopted is one of degree only. We shall consider that the asymmetric synthesis takes place during that reaction in which the new chiral center is formed from the previously achiral center and that it is *not* the total operational process of synthesis, reaction, and degradation outlined by Marckwald. In essence, the analytical procedures of organic chemistry are now so much better than they were in Marckwald's time that we do not need to carry through the complete sequence to demonstrate a valid asymmetric synthesis. At the same time this example illustrates one of the flaws of the definition of asymmetric synthesis which we have adopted. In this case the factors which have led to an asymmetric synthesis, namely a predominance of axial attack to give equatorial alcohol (β-**13**), are obviously unrelated to the chirality of the system† per se.

If the substrate is a flexible molecule rather than the rigid steroid system as in the example above, the interpretation of the results is not so obvious but the demonstration of an asymmetric synthesis is more direct and dramatic. Optically active (*S*)-(−)-2-octyl benzoylformate (**17**) can be synthesized readily from optically active (*S*)-(−)-2-octanol (**16**) and achiral benzoylformic acid via the acid chloride (**15**). When the Grignard reagent from methyl iodide is allowed to react with (*S*)-(−)-2-octyl benzoylformate (*S*-(−)-**17**), an unequal mixture of diastereomeric esters *S*,*S*-**18** and *R*,*S*-**19** is produced.

McKenzie and Ritchie[30] found that quantitative hydrolysis of the unpurified diastereomeric mixture of the resulting 2-octyl atrolactates (*S*,*S*-**18** and *R*,*S*-**19**) gave atrolactic acid (**20**), which had an optical rotation of $[\alpha]_D^{20}$ − 6.6° corresponding to a mixture of 59% *R*-(−)-**20** and 41% *S*-(+)-**20**. This corresponds to a difference in energies of activation ($\Delta\Delta G^{\ddagger}$) of the competing transition states of 0.22 kcal/mole at room temperature.

* Although the implied condition of recovering the chiral inducing agent is not met.

† Consider for instance the lithium aluminum hydride reduction of 4-*t*-butylcyclohexanone (Table **3-11**) which gives 92% *trans*- and 8% *cis*-4-*t*-butylcyclohexanol. Both the substrate and the product are achiral and thus the ratio of products is independent of the chirality of the system. Similar factors, however, control the extent of asymmetric synthesis in the example starting with cholestan-3-one.

[30] A. McKenzie and P. D. Ritchie, *Biochem. Z.*, **237**, 1 (1931).

We shall define the percent asymmetric synthesis as the extent to which one enantiomer (or epimer) is produced in excess over the other. In this case,

Ph—C(=O)—C(=O)—Cl (**15**) + Me—C(OH)(H)—C_6H_{13} (*S*-(−)-**16**) ⟶ Ph—C(=O)—C(=O)—O—C(H)(Me)—C_6H_{13} (*S*-(−)-**17**) $\xrightarrow{\text{1) MeMgI; 2) } H_2O}$

Me—C(Ph)(OH)—C(=O)—O—C(Me)(H)—C_6H_{13} (*S,S*-**18** (41%)) and HO—C(Ph)(Me)—C(=O)—O—C(Me)(H)—C_6H_{13} (*R,S*-**19** (59%)) $\xrightarrow[(OH^-)]{H_2O}$ Me—C(Ph)(OH)—COOH (*S*-(+)-**20** (41%)) and HO—C(Ph)(Me)—COOH (*R*-(−)-**20** (59%)) + Me—C(OH)(H)—C_6H_{13} (*S*-(−)-**16**)

the excess of *R*-(−)-**20** over *S*-(+)-**20** is 18%; i.e., 18% asymmetric synthesis or 18% enantiomeric excess (18% e.e.). Note that this is the same as the percent excess of epimer *R,S*-**19** over epimer *S,S*-**18**. In fact the asymmetric synthesis takes place during the reaction which produces **18** and **19** and the subsequent quantitative hydrolysis step [**18** and **19** → *S*-(+)-**20** and *R*-(−)-**20**] is an analytical device for determining the composition of the epimeric mixture. The percent asymmetric synthesis is obtained then from the polarimetrically determined enantiomeric composition[31] of the isolated atrolactic acid (**20**). "Optical purity" is an operational term which is determined by dividing the observed rotation by the rotation of the pure enantiomer both of which have been taken under identical conditions.

$$\text{Percent optical purity} = \frac{[\alpha]_{\text{obs}}}{[\alpha]_{\text{max}}} \times 100$$

Assuming a linear relationship between rotation and composition and no experimental error, percent "optical purity" is equated with the percent excess of one enantiomer over the other, which we shall designate percent enantiomeric excess (% e.e.): This is a term which we shall generally use.

$$\text{Percent enantiomeric excess} = \frac{[R] - [S]}{[R] + [S]} \times 100 = \%R - \%S$$

The terms *percent asymmetric synthesis* and *percent stereoselectivity*[23,24,31]

[31] M. Raban and K. Mislow, in *Topics in Stereochemistry*, ed. E. Eliel and N. Allinger (New York, N.Y.: Interscience Pub., 1967), Vol. 2, pp. 199–230, have reviewed and discussed the modern methods in addition to polarimetry for determining enantiomeric composition.

are closely related; the latter, however, is a broader term and would, for instance, include the percent excess production of one geometrical isomer (A) over another (B):

$$\text{Percent stereoselectivity} = \frac{[A] - [B]}{[A] + [B]} \times 100 = \%A - \%B$$

For the purpose of our present example, it is important to note that it is unnecessary to hydrolyze the mixed diastereomeric atrolactic esters **18** and **19** to the mixture of *R* and *S* atrolactic acids (**20**) if a suitable direct quantitative method for analysis of the diastereomers themselves is available. This is now possible[31,32] in suitable cases with nuclear magnetic resonance spectroscopy (nmr). Since the asymmetric synthesis has taken place during the Grignard addition step, i.e., **17** → **18** + **19**, a direct analysis of the epimeric mixture of **18** and **19** would constitute a direct determination of the percent stereoselectivity which is the percent asymmetric synthesis. Thus the complete Marckwald sequence from achiral benzoylformic acid (**15**) to optically active atrolactic acid (**20**) dramatically illustrates that an asymmetric synthesis has taken place and this process has served a very useful purpose historically; nevertheless, the study of any particular asymmetric synthesis need not involve the operational sequence laid down in the Marckwald definition if a suitable analytical procedure can be devised for studying the specific reaction in question.

Furthermore the *extent* of asymmetric synthesis can be determined without the use of optically active reagents! Assume that we start with racemic 2-octanol (*RS*-**16**) and prepare racemic 2-octyl benzoylformate (*RS*-**17**). When this ester is treated with the methyl Grignard reagent, four diastereomers will result as summarized in the following equations.

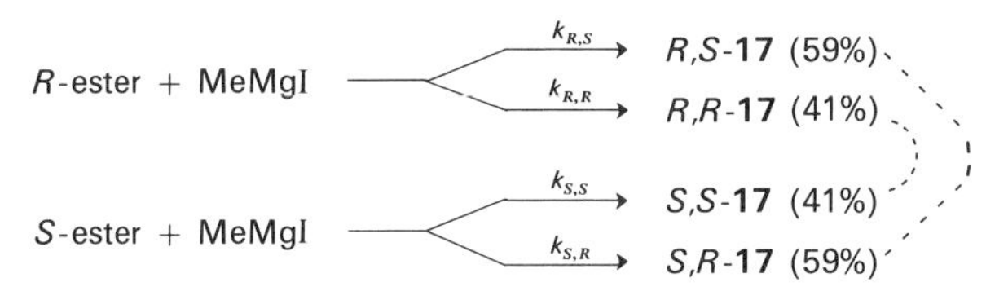

The *R,S* and *R,R* esters will be formed in the same 59:41 ratio already found for the pure isomer (assuming, of course, negligible effects for a slightly different solvent environment in the racemic case). Symmetry principles require that the rate $k_{R,S}$ for forming the *R,S* isomer will be identical to $k_{S,R}$ for forming the enantiomeric *S,R* product. The same will apply to $k_{S,S}$ and $k_{R,R}$. Therefore the ratio of the two enantiomeric pairs (*R,S* + *S,R*):(*R,R* + *S,S*) produced from racemic materials will be the

[32] (a) J. A. Dale, D. L. Dull, and H. S. Mosher, *J. Org. Chem.*, **24**, 2543 (1969). (b) B. Braman, M.S. thesis, Stanford University, Jan., 1967.

same as the ratio of the epimers (*R*,*S*):(*S*,*S*) obtained using optically active ester (*S*-(−)-**17**). It would of course be necessary to determine which diastereomeric pair was produced in excess by some independent means, since hydrolysis of the *R*,*R* + *S*,*S* mixture or the *R*,*S* + *S*,*R* mixture would yield only racemic alcohol and acid. Once the configuration of the predominant isomers has been determined, however, there would be no need to use optically active reagents if a direct method for analysis of the diastereomers were available.

McKenzie and Ritchie[30] in 1931 invoked the phenomenon of *asymmetric induction*, as they then understood it, to explain the optical activity which has been "induced" in the symmetrical α-keto portion of the benzoylformate moiety by the stable asymmetric center in the alcohol portion of the ester and which was then manifest in the unequal production of diastereomers **18** and **19**. Prelog[12] brilliantly demonstrated that this asymmetric induction could be rationalized in terms of steric interactions among the various possible reacting conformations. In terms of modern transition-state theory, there is no longer any necessity of invoking the effect of an "inductive force" acting upon the symmetrical carbonyl group either through space or through bonds.

The atrolactic asymmetric synthesis differs from that in the earlier example of cholestan-3-one (**12**) reduction by sodium borohydride in that the new asymmetric center can be separated easily by hydrolysis from the "inducing" chiral center and readily recovered, thereby establishing very clearly that an asymmetric synthesis has occurred. In the cholestanone-cholestanol case, it is quite impractical, although, theoretically possible, to degrade the product in such a manner that the asymmetric carbinol carbon (C-3) is left with no other optically active group attached. Fundamentally, however, these two reactions are closely related from a stereochemical standpoint. The fact that the atrolactate molecule has a flexible chain structure while the steroid molecule has a rigid cyclic structure and that in the former the original asymmetric portion of the molecule can be separated conveniently from the product, while in the latter case it cannot, corresponds to differences in degree, not in kind.

A fourth example is the reduction of isopropyl phenyl ketone (**21**) by the Grignard reagent from (+)-1-chloro-2-phenylbutane (**22**),[33] in ether solvent.

$$\underset{\mathbf{21}}{(i\text{-Pr})(\text{Ph})\text{C}{=}\text{O}} + \underset{\mathbf{22}}{\text{H}\blacktriangleright\text{C}(\text{CH}_2\text{MgCl})(\text{Ph})\blacktriangleleft\text{Et}} \xrightarrow[\text{2) } H_2O]{\text{1) } Et_2O} \underset{S\text{-}\mathbf{23}\ (9\%)}{\text{H}\blacktriangleright\text{C}(i\text{-Pr})(\text{Ph})\blacktriangleleft\text{OH}} \text{ and } \underset{R\text{-}\mathbf{23}\ (91\%)}{\text{HO}\blacktriangleright\text{C}(i\text{-Pr})(\text{Ph})\blacktriangleleft\text{H}} + \underset{\mathbf{24}}{\text{CH}_2{=}\text{C}(\text{Ph})\text{—Et}}$$

[33] J. S. Birtwistle, K. Lee, J. D. Morrison, W. A. Sanderson, and H. S. Mosher, *J. Org. Chem.*, **29**, 37 (1964).

The achiral ketone (**21**) is reduced in about 69% yield at the expense of the optically active Grignard reagent (**22**) which is oxidized concomitantly to achiral 2-phenyl-1-butene (**24**). The reduction product, isopropylphenylcarbinol (**23**), is optically active and is a mixture of 91% of the *R*-(+) isomer and 9% of the *S*-(−) isomer corresponding to an 82% excess of the *R* over the *S* enantiomer (82% e.e., $\Delta\Delta G^{\ddagger} = 1.5$ kcal/mole). The overall effect is the transfer of the asymmetry of the reducing agent to the product. This type of asymmetric synthesis in which an asymmetric center is created simultaneously with destruction of another has been termed "self-immolative" by Mislow[24] and "asymmetric transfer" by Pracejus.[14b]

No optically active diastereomeric intermediate is isolable in this asymmetric reduction by an optically active Grignard reagent in contrast to the case of the previous Grignard addition to (−)-2-octyl benzoylformate. To this extent, the two reactions differ from each other, but they possess the common feature that they proceed via two competing diastereomeric transition states. In Sec. **5-2** we shall interpret the observed stereochemistry in terms of established concepts of steric and electronic group interactions in the competing diastereomeric transition states.

A fifth example, the asymmetric Meerwein–Ponndorf reduction, is of the same type. Doering and Young[5] treated methyl isohexyl ketone (**26**) with an excess of (*S*)-(+)-2-butanol (**25**) in the presence of aluminum-2-butoxide (8 hr at 36°); optically active (+)-6-methyl-2-heptanol (**27**) containing about a 6% excess of the *S*-(+) isomer was isolated.

$$\underset{S\text{-}(+)\text{-}\mathbf{25}}{H\text{—}\overset{CH_3}{\underset{C_2H_5}{C}}\text{—}OH} + \underset{\mathbf{26}}{\overset{CH_3}{\underset{i\text{-}C_6H_{13}}{C}}{=}O} \xrightarrow{Al(OR)_3} \overset{CH_3}{\underset{C_2H_5}{C}}{=}O + \underset{S\text{-}(+)\text{-}\mathbf{27}}{H\text{—}\overset{CH_3}{\underset{i\text{-}C_6H_{13}}{C}}\text{—}OH} \text{ and } \underset{R\text{-}(-)\text{-}\mathbf{27}}{HO\text{—}\overset{CH_3}{\underset{i\text{-}C_6H_{13}}{C}}\text{—}H}$$

Streitwieser and co-workers[34] in a similar manner have used the magnesium alcoholate of (−)-isoborneol (**29**) to reduce benzaldehyde-1-*d* (**28**) to benzyl-α-*d* alcohol (**30**) which was optically active (α_D^{25} −0.715°, neat, $l = 1$) by virtue of deuterium substitution. In this case the chiral reducing agent is oxidized to (+)-camphor (Sec. **5-1**).

$$\underset{\mathbf{28}}{D\text{—}\overset{O}{\underset{Ph}{C}}} + \underset{\mathbf{29}}{[\text{isobornyl}]\text{—}OMgX} \longrightarrow\longrightarrow \underset{R\text{-}(-)\text{-}\mathbf{30}\,(73\%)}{H\text{—}\overset{OH}{\underset{Ph}{C}}\text{—}D} \text{ and } \underset{S\text{-}(+)\text{-}\mathbf{30}\,(27\%)}{D\text{—}\overset{OH}{\underset{Ph}{C}}\text{—}H} + \underset{\mathbf{31}}{[\text{camphor}]}$$

Gerlach[35] has shown by ingenious experiments, based upon the proton magnetic resonance analysis of diasteromeric derivatives of optically active

[34] A. Streitwieser, Jr., J. R. Wolfe, Jr., and W. D. Schaeffer, *Tetrahedron*, **6**, 338 (1959).
[35] H. Gerlach, *Helv. Chim. Acta*, **49**, 2481 (1966).

benzyl-1-*d* amine and the stereospecific conversion of benzyl-1-*d* alcohol to benzyl-1-*d* amine, that the product from Streitwieser's asymmetric reduction was approximately 46% enantiomerically pure, i.e., 73% *R*-**30** and 27% *S*-**30**. When benzaldehyde-1-*d* (**28**) was reduced with actively fermenting yeast,[36] instead of isobornyloxymagnesium bromide (**29**) the resulting benzyl-1-*d* alcohol was the enantiomerically pure *R*-(−)-**30** isomer.[35] This quantitative difference in stereoselectivity between the "chemical" versus the enzymatic asymmetric synthesis is one of the most challenging aspects of the subject of asymmetric synthesis.

From these limited examples, we see that asymmetric synthesis, taken in its broad meaning, encompasses a wide spectrum of reactions which vary greatly in character and the study of these reactions can serve many useful purposes. Where the mechanism of the reaction involved is established and the substrate is a molecule of known, rigid orientation, the asymmetric synthesis process may be well understood. Where the mechanism is in doubt, the stereochemical course of the asymmetric synthesis may help to clarify the mechanism. Where flexible species are involved and the mechanism known, a study may offer a unique opportunity for establishing the interplay of forces, both steric and electronic, controlling the topologies and relative energies of the diastereomeric transition states. Furthermore, as shown in the following chapters, asymmetric synthesis is often a practical method for obtaining certain optically active compounds. Finally, there is the challenge to employ "chemical" asymmetric syntheses to better understand the biochemical counterparts and the challenge to develop chemical systems as efficient (i.e., 100% stereoselective) as the enzymatic ones.

1-4. Methods of Producing Optically Active Compounds

Optically active substances in solution can be obtained only through the intervention of some chiral reagent to give diastereomeric transition states, products, or complexes (including solvates). In the sense of this statement, "chiral reagent" must include a chiral physical force, such as a circularly polarized light, which would lead to an *absolute asymmetric synthesis*. The general processes whereby optically active substances are obtained from optically inactive materials can be grouped into four broad categories: physical separation via enantiomeric crystalline forms, resolutions based upon separation of diastereomeric forms, thermodynamically controlled asymmetric transformations of stereochemically labile diastereomers, and kinetically controlled asymmetric transformations. This classification is not perfect; borderline and combination situations are obviously

[36] V. E. Althouse, D. M. Feigl, W. A. Sanderson, and H. S. Mosher, *J. Am. Chem. Soc.*, **88**, 3595 (1966).

possible. We are concerned primarily with kinetically controlled processes in the discussion of asymmetric synthesis; the subsequent chapters deal essentially with reactions of this fourth class. We do not intend to discuss specifically or in detail all the ways of producing optically active compounds, but the different methods are outlined and examples given so that asymmetric synthesis can be seen in its proper relationship to these other methods.

METHODS OF PRODUCING OPTICALLY ACTIVE COMPOUNDS

(1) Physical Separation via Enantiomeric Crystalline Forms
- (a) Physical sorting of enantiomeric crystals.
- (b) Selective seeding of a solution of racemate with crystals of one enantiomer (or isomorphous crystal).
- (c) Preferential incorporation of one enantiomer in an inclusion compound.

(2) Resolution Based upon Separation of Diastereomeric Forms
- (a) Classical resolution. Stable diastereomer formation (including molecular complexes) followed by physical separation based upon crystallization, chromatography, distillation, etc.
- (b) Selective association on a chiral adsorbant or selective solvent extraction using a chiral solvent or solution (equivalent to labile diastereomer formation).

(3) Thermodynamically Controlled Asymmetric Transformations of Stereochemically Labile Diastereomers.

(4) Kinetically Controlled Asymmetric Transformations
- (a) Reactions of racemic substrates with chiral reagents.
- (b) Reactions of achiral substrates with chiral reagents.
- (c) Absolute asymmetric degradation and synthesis.

1-4.1. Physical Separations via Enantiomeric Crystalline Forms. Resolution methods have been reviewed recently[37] and the following therefore will be a very brief summary. Pasteur's separation by physical sorting of enantiomeric crystals of racemic sodium ammonium tartrate was the first resolution. However, the cases in which this can be done are rare and unpredictable (unless the phase diagram has been determined) and the technique laborious,[37a] although this may not always be so (cf. heptahelicene, **36C**, Sec. **1-4.3b**). It is also possible that, from a solution of a racemate,

[37] (a) R. M. Secor, *Chem. Rev.*, **63**, 297 (1963). (b) D. R. Buss and T. Vermeulen, *Ind. Eng. Chem.*, **60**, [8], 12 (Aug., 1968). (c) J. W. Westley, B. Halpern, and E. L. Karger, *Anal. Chem.*, **40**, 2046 (1968).

crystals will grow which have no detectable hemihedral facets and yet are made up of one chiral form. In this case each crystal must be tested, either as the crystal or in solution for optical activity. Tri-*o*-thymotide is such an example[38] (see also tri-*o*-carvacrotide, **37**, Sec. **1-4.3b**).

One need not have a chiral molecule in order to have a chiral crystal (e.g., SiO_2 and quartz). Urea crystallizes from solution in a helical pattern (C_2 symmetry) leaving chiral interstices which can occlude solvent.[39] When such a clathrate or inclusion compound is formed from racemic diheptyl malate, the malate ester recovered from selected crystals is optically active.[39a] Inclusion compounds of cyclodextrin have been studied thoroughly.[39d] The ethyl esters of α-methyl-, α-chloro-, α-bromo-, and α-hydroxyphenylacetic acids are preferentially occluded by the crystals of cyclodextrin from which 3–12% excess of the *R*-(−) enantiomer can be recovered.

The seeding of a solution of a racemate by the crystals of one pure enantiometer (or an isomorphic crystal) can, under the proper circumstances, lead to the separation of one form and the retention of the other in solution. Separation can then be achieved by filtration or centrifugation. Additional racemate may be dissolved in the mother liquor which is then seeded with the other enantiomer and this alternating process repeated indefinitely.[37]

Exclusive spontaneous crystallizations of one chiral form of a crystal have been reported,[19] but repetition of such spontaneous crystallizations under conditions which rigorously exclude accidental seeding must give, on the average, one chiral form as often as the other. This does not mean that in one crystallization container there will be as many (+) crystals as (−) crystals, since self-seeding may take place. Soret[40] carried out the spontaneous crystallization of sodium chlorate 938 times in sealed ampoules. The molecules of sodium chlorate are achiral but the crystal is chiral. Dextrorotatory crystals formed 433 times; levorotatory crystals, 411 times; and a mixture of both, 94 times.

1-4.2. Resolutions Based upon Separations of Diastereomeric Forms. Resolution of a racemic pair is most generally carried out by formation of a diastereomeric derivative, which can then be separated by virtue of some difference in physical properties. Most generally this is a difference in solubilities of a crystalline derivative such as an alkaloid salt, but differences in boiling point, in chromatographic adsorption, as well as in gas chromatographic retention times[31] can be employed. This process can

[38] A. C. D. Newman and H. M. Powell, *J. Chem. Soc.*, 3747 (1952).

[39] (a) W. Schlenk, Jr., in *100 Jahre* BASF, *Aus der Forschung* (Ludwigshafen Am Rhein, 1965), pp. 1–32. (b) *Fortschr. Chem. Forsch.*, **2**, 92 (1951). (c) *Experientia*, **8**, 337 (1952). (d) F. Cramer and W. Dietsche, *Chem. Ber.*, **92**, 378 (1959).

[40] C. H. Soret, *Ztsch. Krystallogr. Mineral*, **34**, 630 (1901).

be generalized as follows:

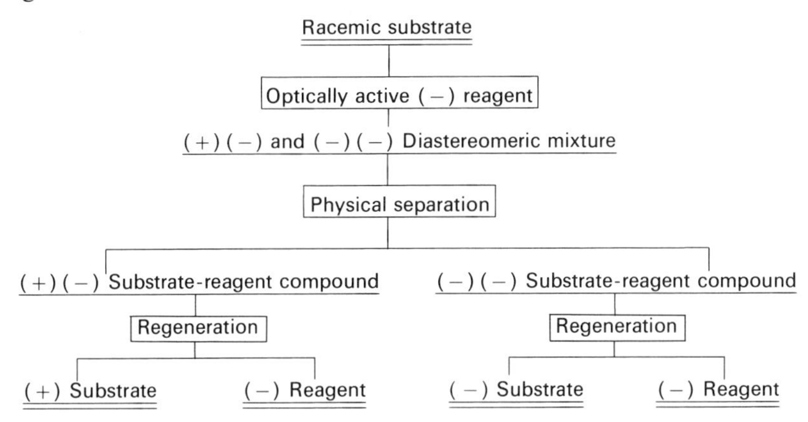

Such resolutions are not asymmetric syntheses, since they clearly involve the separation of members of a stereochemically stable diastereomeric pair by physical means. Classical resolutions of this type are the most general method of obtaining optically active substances other than isolation from nature. The theory[23] and technique[37] have been reviewed and will not be elaborated here.

Resolutions can also be achieved by formation of an unstable combination of a stable racemate and stable optically active resolving agent.[37b] The following are representative examples: the resolution of Troeger's base by chromatography on a lactose column,[41a] resolution of mandelic acid on amylose or starch columns,[41b] metallocenes on acetyl cellulose columns,[41c] amino acids by chromatography on cellulose paper,[41d] glycols by extraction with chiral solvents,[41e] amino acids on chiral ion exchange resins,[41f] and of trifluoroacetyl amino acids by gas chromatography using a chiral stationary phase.[42]

1-4.3. Thermodynamically Controlled Asymmetric Transformations of Stereochemically Labile Diastereomers. If the isomers being subjected to resolution are stereochemically labile under the conditions of the separation, there is no sharp division between the physical separation associated with resolution and that which has been termed an *asymmetric transformation.* Both processes, physical separation and *asymmetric transformation,* can occur at the same time and a clear distinction may

[41] (a) V. Prelog and P. Wieland, *Helv. Chim. Acta*, **27**, 1127 (1944). (b) M. Ohara, Chan-Yang Chen, and T. Kwan, *Bull. Chem. Soc. Japan*, **39**, 1440 (1966). (c) H. Falk and K. Schlögel, *Tetrahedron*, **22**, 3047 (1966). (d) C. E. Dalgliesch, *J. Chem. Soc.*, 137 (1952). (e) N. S. Bowman, G. T. McCloud, and G. K. Schweitzer, *J. Am. Chem. Soc.*, **90**, 3848 (1968). (f) J. A. Lott and W. Rieman, *J. Org. Chem.*, **31**, 561 (1966).

[42] E. Gil-Av and B. Feibush, *Tetrahedron Lett.*, 3345 (1967).

be difficult to establish. Such a case involving a stereochemically labile racemate does not constitute a true resolution, nor can it be classified as an asymmetric synthesis since the starting point is a racemic mixture, albeit one in which the enantiomers are equilibrated via a symmetrical intermediate. The subject of asymmetric transformations of stereochemically labile compounds has been critically reviewed.[18,20]

Thermodynamically controlled processes may be described by an energy-reaction coordinate diagram where reactants and products are connected by a pathway which permits equilibration under the conditions of the experiment (Fig. **1-1**). If the energy barrier of this pathway is in the range of 15–18 kcal/mole or less (E_{act}, Fig. **1-1**), the reactants and products will be

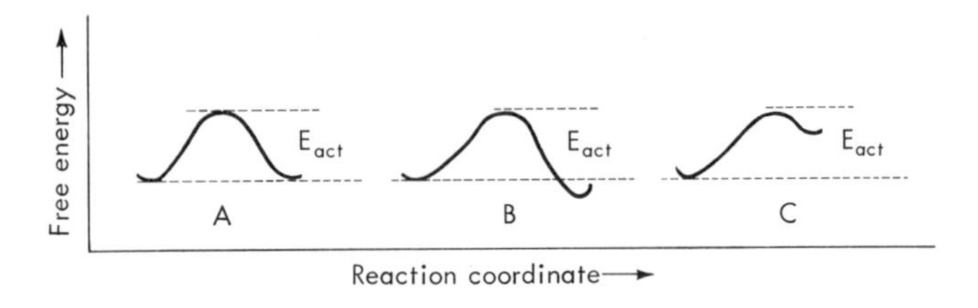

Fig. 1-1. Energy-reaction coordinate diagram for racemization (A) and epimerization (B or C).

in equilibrium at room temperature and not separable by conventional experimental procedures. In such processes, with a low energy barrier between reactants and products, the composition of a homogeneous reaction mixture depends only on the difference in free energies of the ground states of reactants and products ($\Delta G°$) and is independent of the pathway; i.e.,

$$\Delta G° = G°_{\text{products}} - G°_{\text{reactants}} = -RT \ln K$$

where K is the equilibrium constant for the process:

$$\text{Reactants} \rightleftarrows \text{Products}$$

If the reactants and products are enantiomeric, their free energies are identical in an achiral environment, i.e., $\Delta G° = 0$, and the process represents racemization (Fig. **1-1**A) where the mole fraction of reactants must equal that of products at equilibrium. But if the reactant and product are diastereomers, in which there is one stereochemically labile center, this process represents epimerization (Figs. **1-1**B and **1-1**C) where $\Delta G° \neq 0$. Both processes are fundamental to a discussion of the production of optically active compounds.

A variety of structural or mechanistic processes exist which permit thermodynamic equilibration of optically active molecules.

(a) Unimolecular inversion without bond breaking of an unsymmetrically trisubstituted tetrahedral atom (such as nitrogen, sulfur, phosphorus, or antimony, Sec. **1-4.3a**).
(b) Unimolecular rotation around a conformationally mobile bond so that one chiral form is converted into another (as in the atropisomerism of *ortho*-substituted biphenyls, Sec. **1-4.3b**).
(c) Interconversion of chiral centers via an achiral intermediate (for example, enol-keto isomerism, ring-chain tautomerism, formation of carbanions, etc., Sec. **1-4.3c**).
(d) Reversible isomerism about a double bond, C=N, N=N, or C=C, in such a way that chiral forms are interconverted (Sec. **1-4.3d**).

We shall now consider briefly specific examples of these various types of processes in asymmetric transformations of stereochemically labile enantiomers and diastereomers.

1-4.3a. *Unimolecular Inversion of an Unsymmetrically Trisubstituted Tetrahedral Atom.* When a chiral sulfoxide is heated, it undergoes unimolecular racemization.[43] This is illustrated in Fig. **1-2**, where a planar (or

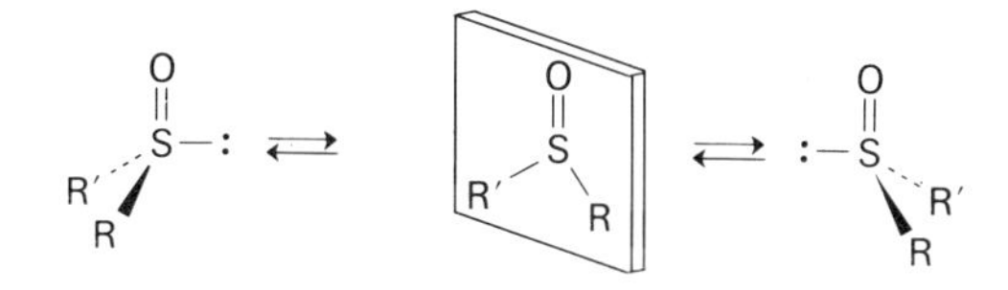

Fig. 1-2. Unimolecular thermal inversion of sulfoxides.

near planar) transition state is shown in which the electron pair "tunnels" through the sulfur atom of the molecule resulting in its inversion. The half-life for inversion at 210° in benzene solvent is about 6 hr when R and R′ equal phenyl and *p*-tolyl (E_{act} = 39 kcal/mole, $\Delta H^{\ddagger}$ = 36.2 kcal/mole; $\Delta S^{\ddagger}$ = 3.1 e.u.).[43a] Additional pathways have been demonstrated when R equals benzyl (radical)[43b] and when R equals allyl (concerted).[43c]

The inversion process for phosphorus compounds corresponding to Fig. **1-2** occurs at a comparable rate at about 130° (E_{act} = 30–31 kcal/mole in benzene)[43a] and for trisubstituted nitrogen compounds the inversion process is usually fast at room temperature.[44a] However, trisubstituted

[43] (a) D. R. Rayner, A. J. Gordon, and K. Mislow, *J. Am. Chem. Soc.*, **90**, 4854 (1968). (b) E. G. Miller, D. R. Rayner, H. T. Thomas, and K. Mislow, *ibid.*, **90**, 4861 (1968). (c) P. Bickart, F. W. Carson, J. Jacobus, E. G. Miller, and K. Mislow, *ibid.*, **90**, 4869 (1968).

[44] (a) D. L. Griffith and J. D. Roberts, *J. Am. Chem. Soc.*, **87**, 4089 (1965). (b) D. Felix and A. Eschenmoser, *Angew. Chem.*, **80**, 197 (1968); *Internat. Edn.*, **7**, 224 (1968). (c) S. J. Brois, *J. Am. Chem. Soc.*, **90**, 508 (1968). (d) F. Montanari, I. Moretti, and G. Torre, *Chem. Commun.*, 1694 (1968).

nitrogen compounds are stabilized to inversion by substitution of halogen on nitrogen and by incorporation of the nitrogen into a three-membered ring.[44b,44c,44d] 7-Chloro-7-azabicyclo[4.1.0] heptane (**32**) has been separated at room temperature into *endo* (**32A**) and *exo* (**32B**) forms:[44b] The latter was stable to distillation at 57–58° and 11 mm pressure. Similarly a simple aziridine has been separated into two diastereomeric racemic forms[44c] **32C** and **32D**. The chiral oxaziridine **32E** has been obtained by asymmetric peroxidation (see also Sec. **8-2**). When the nitrogen atom is so constrained in a ring system that inversion is precluded on structural grounds, the compound is stereochemically stable; two such examples which have been resolved are Troeger's base[41a] (**33A**) and 6,6-dimethyl-2-quinuclidone.[45]

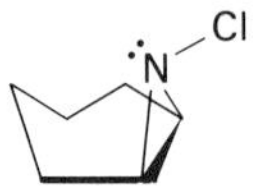

32A  **32B** 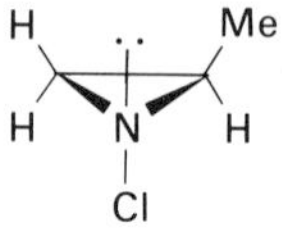 **32C** 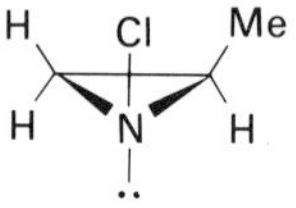 **32D** 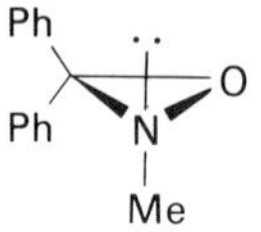 **32E**

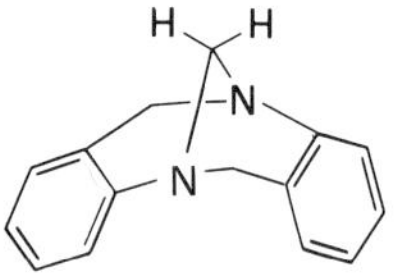

33A 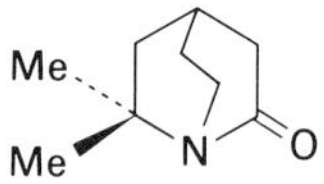 **33B**

1-4.3b. *Rotation Around a Conformationally Mobile Bond.* When steric hindrance or torsional strain associated with rotation about a single bond becomes low enough, enantiomeric conformations can be interconverted. The chair conformations of *cis*-1,2-disubstituted cyclohexane (e.g., **34** where the two R groups are the same) exist in enantiomeric forms **34A** and **34B**. Since **34A** and **34B** are in rapid equilibrium at room temperature, no conformational isomers corresponding to these two structures have as yet been isolated.* In principle, low temperature techniques, such as have been used

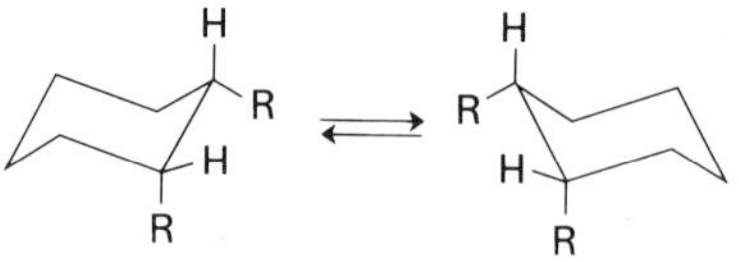

34A **34B** 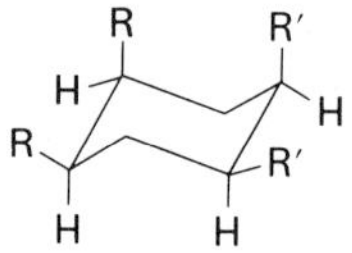 **34C**

[45] H. Pracejus, *Ber.*, **92**, 988 (1959).

* S. Wolfe and J. R. Campbell, *Chem. Commun.*, 874 (1967) have interpreted low temperature nmr studies on *cis*-1,2-dicarbomethoxy cyclohexane-5,3,4,5,6,6-d_6 to mean that the interconversion of the two forms of *cis*-1,2-dichlorocyclohexane takes place via the half-chair transition state of lowest energy. With all *cis*-1,2,4,5-tetra-substituted derivatives such as **34C** one of the half-chair transition states requires that the two R groups become coplanar and in in the other the two R′ groups. Thus resolution of suitably substituted compounds of type **34C** should be practical.

to isolate the equatorial form of chlorocyclohexane[46] at $-150°$, could be adopted for obtaining individual chiral forms such as **34A** and **34B**, or possibly low temperature chromatography on a chiral adsorbant.

The barrier to rotation about the single bond joining the two aromatic rings in *ortho*-substituted biphenyls has been studied extensively. The subject has been reviewed[23,24] and a quantitative treatment,[47] based on steric factors of the various *ortho* substituents and buttressing effects of *meta* substituents, has been developed. The activation energies for racemization of many biphenyls represented by **35A** → **35B** and related compounds have

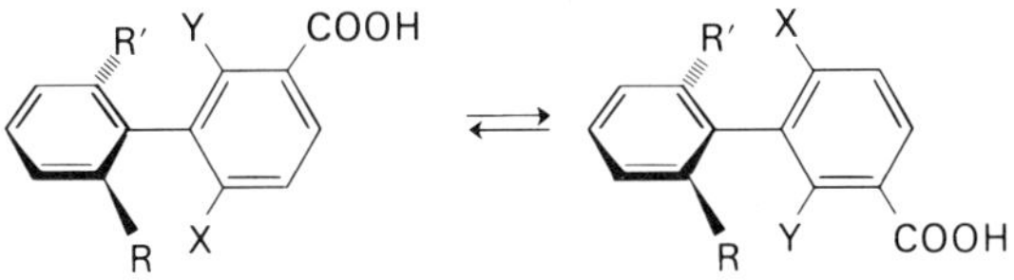

35A ⇄ **35B**

been determined. Some biphenyl derivatives are not racemized at elevated temperatures (**35**, R and X = *t*-butyl; R′ and Y = H, remains unchanged on melting at 191°), others cannot be resolved at 0° (any derivative of **35** with X, Y, and one R = hydrogen) and compounds of all intermediate degrees of stereochemical stability are known. The isolation or observation of enantiomeric rotomers (conformational isomers or atropisomers) is dependent not only upon the size and nature of the various groups, but also upon the isolation temperature and the time interval of the observation. Thus differentiation between conformational isomerism and the usual stereoisomerism is completely arbitrary and depends upon the time and temperature scale being employed or specified.

Other interesting examples of racemization by rotation about conformationally mobile bonds are shown in formulas **36–42**. For instance, 1-fluoro-12-methylbenzo[c]phenanthrene, which has been resolved by use of the optically active complexing agent (−)-α-2,4,5,7-tetranitro-9-fluorenylidinaminooxypropionic acid,[48] is a compound which owes its asymmetry to molecular distortions out of planarity, resulting in right- and left-handed helical configurations (**36A** and **36B**). (−)-1-Fluoro-12-methylbenzo[c]phenanthrene (**36A**) racemizes slowly by a first-order process upon heating to about 100° (E_{act} = 31 kcal/mole). Heptahelicene[49] (**36C**) gives chiral crystals

[46] F. R. Jensen and C. H. Bushweller, *J. Am. Chem. Soc.*, **88**, 4279 (1966); **91**, 3223 (1969).

[47] F. Westheimer, "Calculation of the Magnitude of Steric Effects," in *Steric Effects in Organic Chemistry*, ed. M. Newman (New York, N.Y.: John Wiley & Sons, Inc., 1956), Chapter 12.

[48] (a) M. S. Newman, R. G. Mentzer, and G. Slomp, *J. Am. Chem. Soc.*, **85**, 4018 (1963). (b) M. S. Newman and W. B. Lutz, *J. Am. Chem. Soc.*, **78**, 2469 (1956). (c) M. S. Newman and D. Lednicer, *J. Am. Chem. Soc.*, **78**, 4765 (1956).

[49] (a) R. H. Martin, M. Flammang-Barbieux, J. P. Cosyn, and M. Gelbcke, *Tetrahedron Lett.*, 3507 (1968). (b) H. Wynberg and M. B. Groen, *J. Am. Chem. Soc.*, **90**, 5339 (1968).

(belonging to the noncentric $P_{2,1}$ space group with two molecules per asymmetric unit) which can be separated mechanically. One such crystal

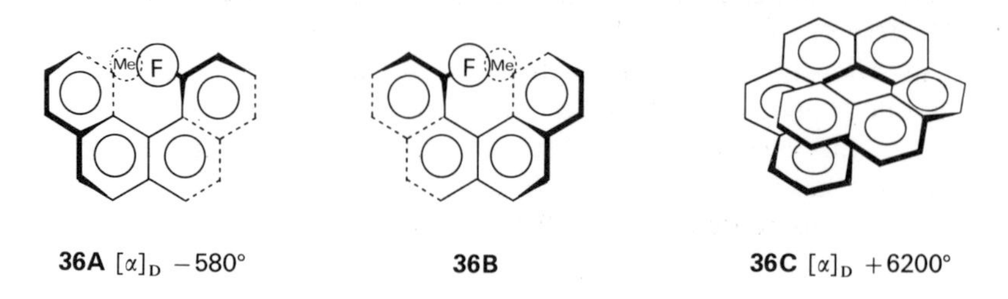

36A $[\alpha]_D$ −580° **36B** **36C** $[\alpha]_D$ +6200°

gave a solution with $[\alpha]_D + 6200 \pm 200°$ ($CHCl_3$). Hexa- and hepta-heterohelicenes[49b] have been prepared and show similar properties. Other types of helical compounds run the gamut of stereochemical stability.[50] Tri-*o*-carvacrotide[51] is an unusual compound which is reported to owe its optical activity not only to a right-handed (**37A**) versus left-handed (**37B**) three-bladed propeller type of dissymmetry (C_3 symmetry) but also to helical dissymmetry (C_2 symmetry). The free energy of activation for the interconversion of propeller to helix ($\Delta G^{\ddagger} = 20.6 \pm 0.2$ kcal/mole), helix back to propeller ($\Delta G^{\ddagger} = 20.3 \pm 0.2$ kcal/mole), and helix to enantiomeric helix ($\Delta G^{\ddagger} = 17.6 \pm 0.2$ kcal/mole) have been deduced from nmr studies.

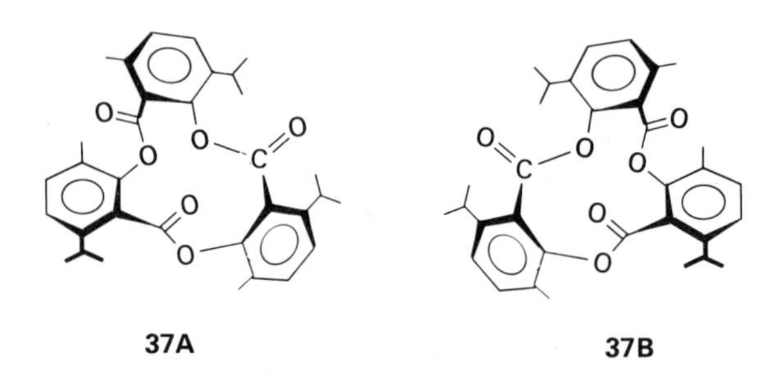

37A **37B**

Still another type of rotation around single bonds which can lead to the racemization of chiral forms is encountered in cyclophanes and related compounds. Rotation of the benzene ring containing the carboxyl group in **38A** leads to **38B**, its mirror image. The derivative in which $m = n = 2$ is stereochemically stable,[52] while that where $n = 3$, $m = 4$ racemizes at about 160°, and that with $m = n = 4$ could not be resolved above 0°.

[50] For citation to related compounds, see Reference 48. For the absolute configurational determination of **36A**, see C. M. Kemp and S. F. Mason, *Chem. Commun.*, 559 (1965).

[51] A. P. Downing, W. D. Ollis, and I. O. Sutherland, *Chem. Commun.*, 171 (1967).

[52] D. J. Cram and N. L. Allinger, *J. Am. Chem. Soc.*, **77**, 6289 (1955).

In connection with the problem of nonbonded electron interactions, the rate of ring "swiveling" in the *meta* cyclophanes **39**, **40**, and **41** has been

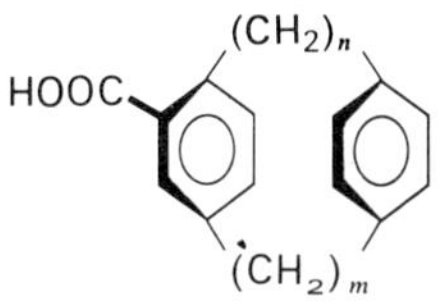

38A 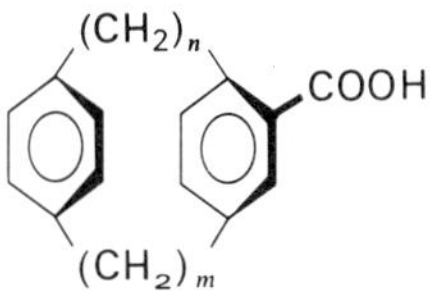 **38B**

determined by variable temperature nmr studies.[53] The 2,2′-dimethyl homolog of **39** has been resolved[54] and is stereochemically stable to 200°, while **39** itself shows no indication of ring swiveling up to 190° ($\Delta G^{\ddagger} > 27$ kcal/mole). But the nitrogen analog (**40**) shows coalescence of the nmr

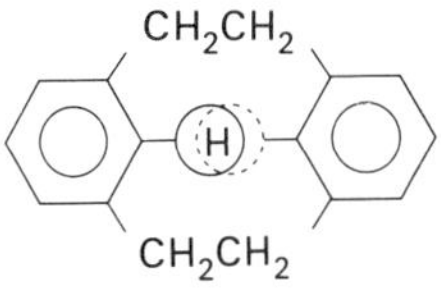

39A 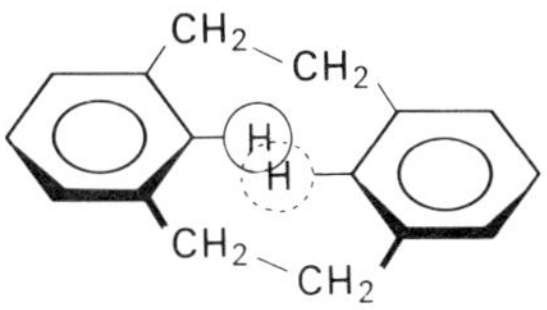 **39B**

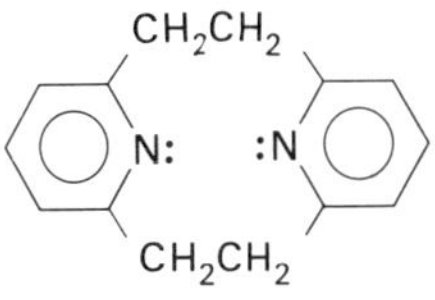

40 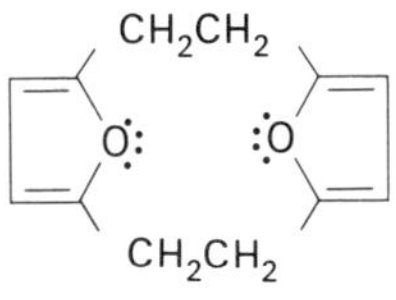 **41**

signals for the diastereotopic methylene protons at about 13.5° ($\Delta G^{\ddagger} = 14.8$ kcal/mole) and the furan derivative at 63° ($\Delta G^{\ddagger} = 16.8$ kcal/mole).

When reactant and product are diastereomers in which only one out of two or more asymmetric centers is stereochemically labile, then interconversion of isomers in solution will lead to unequal amounts of epimers which will differ in ground state free energies (i.e., Fig **1-1B**, $\Delta G^{\circ} \neq 0$).[55] Mutarotation is evidence that this process is occurring. In this respect the equilibration of epimers differs from that of enantiomers which at equilibrium (in an achiral environment) must always give a statistical 50:50 mixture (i.e., $\Delta G^{\circ} = 0$) and leads to racemization. The interconversion of epimers in solution has

[53] I. Gault, B. J. Price, and I. O. Sutherland, *Chem. Commun.*, 540 (1967).

[54] T. Sato, S. Akabori, M. Kainosho, and K. Hata, *Bull. Chem. Soc. Japan*, **39**, 856 (1966).

[55] These energy differences may be small and the ratio of epimers very close to 50:50. Under special but very rare circumstances, the ratio of epimers might be exactly 50:50. If we assume an equilibration of epimers which gives a ratio of 49:51 at one temperature but 51:49 at another temperature, then at some intermediate temperature the ratio must be exactly 50:50.

been termed "first-order asymmetric transformation."[18] When equilibration is accompanied by the separation of a crystalline phase from solution, the terms "second-order asymmetric transformation" or "optical activation"[56] have been applied. Molecules containing a stereochemically mobile structure such as those in formulas **35–41** might in addition possess a stereochemically stable chiral unit. Those processes which lead to racemization in molecules with a single mobile chiral center lead to mutarotation under proper circumstances in those molecules which possess both stereochemically labile and stable chiral centers. The stereochemically important and often practical occurrence of "second-order asymmetric transformation" or "optical activation" will take place if conditions are such that one epimer of a readily interconvertible diastereomeric pair is less soluble than the other and crystallizes from solution at a temperature where the equilibration occurs. The equilibrium will be disturbed by crystallization of this less soluble diastereomer and separation can continue until the original mixture is largely converted into one solid form.

This situation is most often encountered when alkaloid salt formation is used for resolution of a *stereochemically labile* substrate as generalized in the following equation:

$$\left\{\begin{matrix}(+)\text{-substrate}\\(+)\text{-alkaloid}\end{matrix}\right\} \underset{\longleftarrow}{\overset{k_1}{\longrightarrow}} \left\{\begin{matrix}(-)\text{-substrate}\\(+)\text{-alkaloid}\end{matrix}\right\} \underset{\longleftarrow}{\longrightarrow} \left\{\begin{matrix}(-)\text{-substrate}\\(+)\text{-alkaloid}\end{matrix}\right\}$$

Solution Solution Solid

Turner and Harris[18] have emphasized the unique aspect of such transformations; namely, the rate of crystal formation increases if the solution is warmed instead of cooled. The crucial test that such a process is not a simple resolution is the demonstration that more than 50% of the original racemate is obtained in one optically active form.

A specific and dramatic example taken from Mislow's work[57] on bridged biphenyls is the following: Racemic 1,2,3,4-dibenzocyclonona-1,3-diene-7-carboxylic acid (**42**) exists in *R* and *S* forms by virtue of the barrier to rotation of the two phenyl rings imposed by the fused carbocyclic system at the two

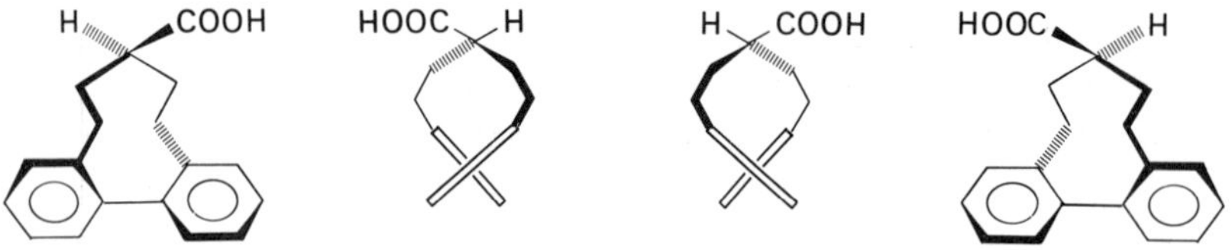

H COOH HOOC H H COOH HOOC H

R-(+)-**42** *S*-(−)-**42**

[56] See page 42 of Reference 23 and page 304 of Reference 18a for a discussion of this term.

[57] K. Mislow, S. Hyden, and H. Schaefer, *J. Am. Chem. Soc.*, **84**, 1449 (1962).

ortho positions. Refluxing an acetone solution of 0.500 g of racemic **42** and the equivalent amount of quinidine for 1 hr caused separation of crystals from the initially clear solution. These crystals of the (+)-quinidine salt, 0.824 g, 84% yield, $[\alpha]_D^{25}$ + 129° (c 1, $CHCl_3$), were treated with dilute hydrochloric acid at 0° and the product crystallized at −20° to give 0.370 g of recovered acid, $[\alpha]_D^{28}$ − 42.3° (c 1, $CHCl_3$). There is excellent reason to believe that this product represents one substance and that the original 0.500 g of racemic mixture gave 0.370 g or 74% of relatively pure *S*-(−) isomer. Since this is more than 50% of the original racemic mixture, a second-order asymmetric transformation must have taken place. This compound had a half-life of 53 min at 50° in xylene. Thus, in refluxing acetone, the two isomers are in rapid equilibrium and the quinidine salt of the (−) acid, which is rather insoluble, separates, shifting the equilibrium to the right. Steric repulsion to rotation and the excessive angle strain in the transition state are enough to prevent appreciable interconversion at 0°. The energy of activation for the racemization was found to be 24 kcal/mole.

1-4.3c. *Interconversion of Chiral Centers via an Achiral Intermediate Moiety.* The equilibrium of 1-decalone (**43**) represents the epimerization of one of two asymmetric centers in a molecule via a keto-enol tautomerism.[58] This process is of course catalyzed by either acid or base and could be classified in the latter case as being stereochemically labile because of the formation of a carbanion rather than enol. (+)-*cis*-1-Decalone in solution equilibrates with the (+)-*trans* isomer giving a 5:95 mixture in favor of the *trans*

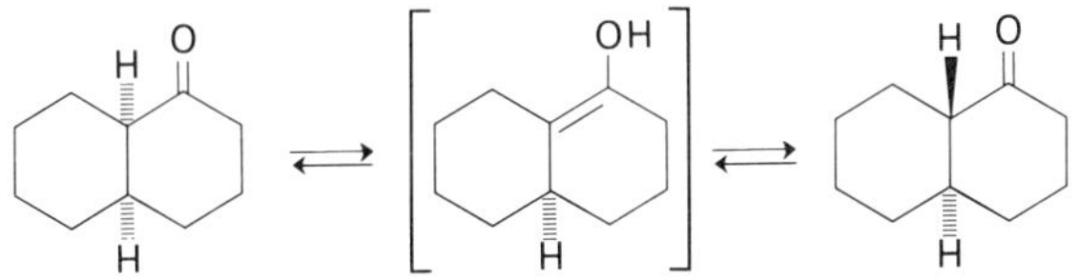

H O H OH H O H H H

43A (5%) $[\alpha]_D$ −71° **43B** **43C** (95%) $[\alpha]_D$ −10°

fused ring system. This interconversion is shown for the (9*R*)-(−)-*cis* isomer (**43A**) going to the 9(*S*)-(−)-*trans* isomer (**43C**). The intermediate enol (**43B**) no longer has a chiral center at C-9, and when it ketonizes it may give either isomer **43A** or **43C**. Since **43A** and **43C** do not possess the same ground state free energies, they will not be formed in equal amounts. If pure **43A** were observed in solution in the presence of a trace of acid or base catalyst, the rotation would change from an initial value of $[\alpha]_D$ − 71° to an equilibrium value of $[\alpha_D]$ − 16°, corresponding to the 5:95 mixture of **43A** and **43C**. During this isomerization, the stereochemistry at the second chiral center

[58] W. Acklin, V. Prelog, F. Schenker, B. Serdarević, and P. Walter, *Helv. Chim. Acta*, **48**, 1725 (1965).

at C-10 is unaffected. Pure **43C** can readily be isolated from the equilibrium mixture. The preference for the *trans* fused ring system is not always so great. The 10-angular methyl homolog gives a 60:40 *trans-cis* mixture at equilibrium.[59]

The most extensively studied example of ring-chain tautomerism leading to mutarotation is that of glucose. α-D-Glucose (**α-44**) is equilibrated via the aldehyde form (**44A**) with β-D-glucose (**β-44**) by either acid or base catalysis.

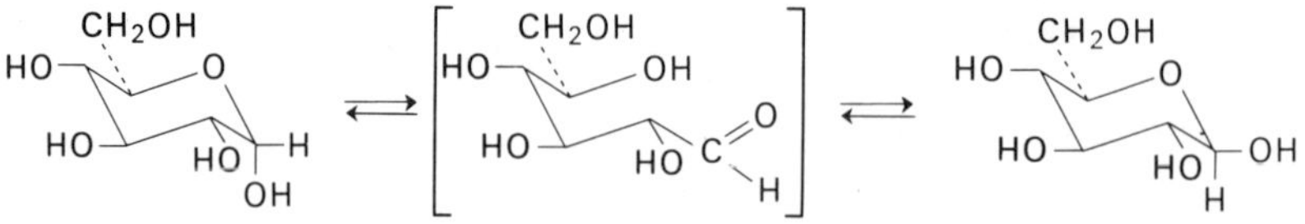

α-**44** (36%) $[\alpha]_D^{22}$ + 110° **44A** β-**44** (64%) $[\alpha]_D^{22}$ +19.7°

If the temperature of the aqueous glucose solution is held between 35° and 40° while the solvent is allowed to evaporate slowly, anhydrous α-D-glucose crystallizes. Thus, the thermodynamically less stable α-form separates from a solution containing an excess of the β-form and the equilibrium is shifted so that the mixed α- and β-forms in solution are almost completely converted to the crystalline α-D-glucose; but when glucose is crystallized from hot acetic acid, the β-form separates. The solution at equilibrium in water at room temperature has $[\alpha]_D^{20}$ + 52.6° corresponding to 36% α-D-glucose (α-**44**) and 64% β-D-glucose (β-**44**) representing a free energy difference ($\Delta G°$) of about 0.37 kcal/mole in favor of the all equatorial, β-form. In absolute methanol solution at room temperature, equilibrium is reached with 45% α- and 55% β-D-glucose. When the reactants constitute a flexible structural system, in contrast to the rigid ring systems of 1-decalone and D-glucose, the very same considerations obtain. For instance, McKenzie and Smith[60] studied the mutarotation of the menthyl α-chlorophenylacetates. The (−)-menthyl ester (**45A**) from the (+)-α-chlorophenylacetic acid ($[\alpha]_D$ + 155.8°) had $[\alpha]_D$ + 5.7 and ester (**45C**) from the (−) acid had $[\alpha]_D$ −150.1° while ester made from the racemic acid had $[\alpha]_D$ − 72.2°. Saponification of any of these esters gave recovered α-chlorophenylacetic acid

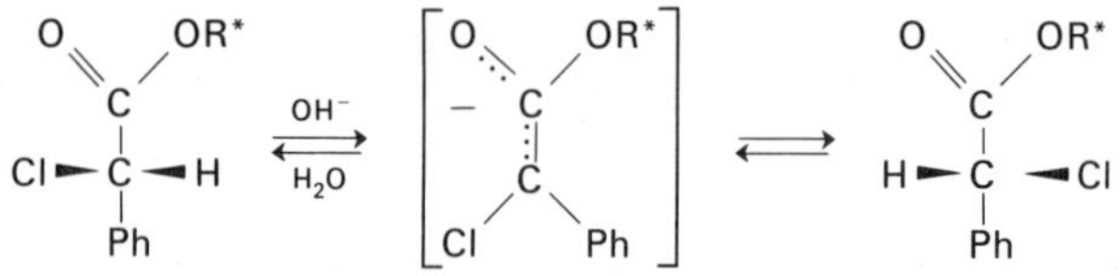

45A **45B** **45C**

R* = (−)-menthyl

[59] F. Sondheimer and D. Rosenthal, *J. Am. Chem. Soc.*, **80**, 3995 (1958).
[60] A. McKenzie and I. A. Smith, *J. Chem. Soc.*, **125**, 1582 (1924); *ibid.*, **123**, 1962 (1923).

with $[\alpha]_D$ − 1.6 to − 2.2°. These results were quite puzzling at that time. A few drops of base added to the solution of either pure diastereomer gave in a short time a solution with the same specific rotation ($[\alpha]_D$ − 86.5° corresponding to an 18% excess of the (−)-isomer). McKenzie and Smith postulated that there was an equilibration to give unequal amounts of the two diastereomers, **45A** and **45C** (through a form which we would now write as the enol or enolate anion, **45B**), and that the α-chlorophenylacetate moiety was largely racemized during the saponification.

The reason for the stereochemical lability of an asymmetric center is not always this apparent. (−)-Menthyl benzoylformate (**46B**) undergoes mutarotation when dissolved in ethanol (but not when dissolved in ether). This observation was used as evidence in favor of the phenomenon of "asymmetric induction" by McKenzie[61] but has been attributed more logically by Jamison and Turner[62] to the formation of the diastereomeric hemiacetals **46A** and **46C** in unknown concentrations in solution.

O=C(OR*)–C(OH)(OEt)Ph ⇄ O=C(OR*)–C(=O)Ph ⇄ O=C(OR*)–C(OEt)(OH)Ph

46A **46B** **46C**

R* = (−)-menthyl

2-Octyl phenyl sulfoxide is an unusual and highly informative example. It has two centers of asymmetry, one on carbon and the other on sulfur. By the action of hydrochloric acid in dioxane, Mislow and co-workers[63] have shown that the sulfur center of asymmetry can be epimerized (*R,S*-**47A** ⇆ *R,R*-**47C**) via the achiral sulfur dichloride center in **47B** without affecting

C_6H_{13}–C(H)(Me)–S(O)–Ph ⇄ (HCl / H_2O) [C_6H_{13}–C(H)(Me)–S(Cl)(Cl)–Ph] ⇄ (H_2O) C_6H_{13}–C(H)(Me)–S(O)–Ph

R,S-(−)-**47A** (54%) **47B** *R,R*-(+)-**47C** (46%)

C_6H_{13}–C(Me)(H)–S(O)–Ph ⇄ (*t*-BuO⁻) [C_6H_{13}–C⁻(Me)–S(O)–Ph] ⇄ (*t*-BuO⁻) C_6H_{13}–C(H)(Me)–S(O)–Ph

S,R-(+)-**47A** (62%) **47D** *R,R*-(+)-**47C** (38%)

[61] A. McKenzie, *Ergebn. Enzymforsch.*, **5**, 49 (1936).

[62] M. M. Jamison and E. E. Turner, *J. Chem. Soc.*, 538 (1941).

[63] K. Mislow, M. M. Green, P. Lauer, J. T. Melillo, T. Simmons, and A. L. Ternay, Jr., *J. Am. Chem. Soc.*, **87**, 1958 (1962); cf. footnote 74.

the asymmetric carbon center to give a 54:46 ratio of the *R,S*-(−)-**47A** to *R,R*-(+)-**47C**. By the use of potassium *t*-butoxide in dimethyl sulfoxide, Cram and Pine[64] demonstrated that the carbon center could be epimerized (*S,R*-**47A** ⇆ *R,R*-**47C**) via the carbanion **47D**, without affecting the asymmetric sulfur center, to give a 62:38 ratio of *S,R*-(+)-**47A** to *R,R*-(+)-**47C**. In either case, the *S,R* or *R,S* diastereomer was thermodynamically more stable than the *R,R* or *S,S* diastereomer. Ideally the isomer ratio in the two experiments would have been the same, but the different solvent conditions could easily account for the observed difference.

1-4.3d. *Reversible Isomerism about a Double Bond.* Isomerization about a C=N or N=N double bond may lead, under proper circumstances, to racemization or mutarotation. The same can apply to the C=C bond, usually under photochemical activation. The oxime of cyclohexanone-4-carboxylic acid (**48**) exists in enantiomeric forms **48A** and **48B** by virtue of the restricted rotation around the carbon-nitrogen double bond of the oximino group.[65]

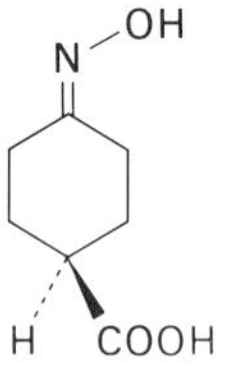

48A 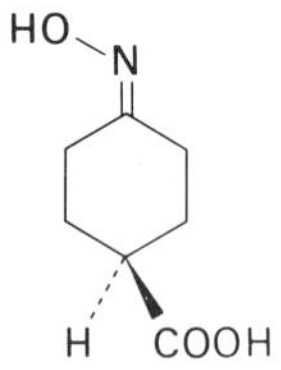 **48B**

When an aqueous solution of this racemic acid was treated with quinidine, a single salt crystallized in better than 80% yield. The original acid was regenerated in solution and had a maximum rotation as the sodium salt of $[M]_D$ −91°. This sodium salt in water had a half-life of 24 min at 20° and in 0.1*N* sodium hydroxide of 21 hr at 20°. Treatment of the original solution of **48** with brucine gave a salt from which the enantiomeric acid could be obtained as its sodium salt in solution.

1-4.4. Kinetically Controlled Asymmetric Transformations. The fourth general method for obtaining chiral compounds, namely by kinetically controlled asymmetric processes, is also the most interesting from a mechanistic viewpoint since diastereomeric transition states or intermediates (including complexes and solvates) are the controlling factors in the stereochemical course of the reaction. In these processes the ground state free energies of the reactants for the competing pathways must be identical ($\Delta G^\circ = 0$). Only the free energies of activation ($\Delta G^\ddagger$) of the two pathways

[64] D. J. Cram and S. H. Pine, *J. Am. Chem. Soc.*, **85**, 1096 (1963).
[65] W. H. Mills and A. M. Bain, *J. Chem. Soc.*, **97**, 1866 (1910); **105**, 64 (1914).

differ; the extent of asymmetric transformation depends only upon the differences in the free energies ($\Delta\Delta G^{\ddagger}$) of the competing pathways. An idealized energy profile diagram representing such kinetically controlled asymmetric processes is given in Fig. **1-3** in which the reactants are separated

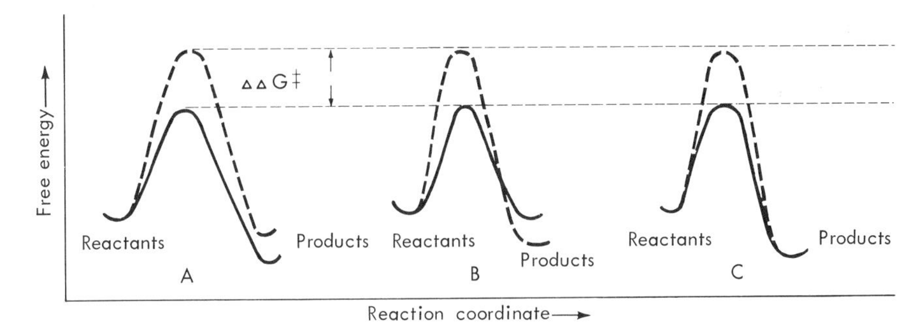

Fig. 1-3. Energy-reaction coordinate diagram for kinetically controlled asymmetric transformations.

from the products by two diastereomeric pathways. For all kinetically controlled asymmetric transformations, the reactants have the same ground state free energy (i.e., $\Delta G^{\circ} = 0$). This condition is fulfilled in one of two ways. Either a racemic substrate, (−A) (+A), reacts with a chiral reagent (R*) via two competing diastereomeric transition states $TS_{(-A)R^*}$ and $TS_{(+A)R^*}$ [the free energies of enantiomers (−A) and (+A) are always identical in an achiral environment] or a single achiral substrate (B) (or achiral group within a

$$A_{(-)} + R^* \longrightarrow [TS_{(-A)R^*}] \longrightarrow \text{products}$$
$$A_{(+)} + R^* \longrightarrow [TS_{(+A)R^*}] \longrightarrow \text{products}$$

chiral molecule) reacts with a chiral reagent (R*) via diastereomeric transition states. The ground state free energies of the products are not involved. In

$$B + R^* \begin{cases} \longrightarrow [TS_{(-B)R^*}] \longrightarrow \text{products} \\ \longrightarrow [TS_{(+B)R^*}] \longrightarrow \text{products} \end{cases}$$

Fig. **1-3A** the ground state free energy of that product formed via the higher transition state is represented as having a higher free energy than that formed via the lower transition state. This is not a requirement and in fact it is very often otherwise (Fig. **1-3B**). One well-documented exception is the self-immolative type asymmetric synthesis (see Sec. **1-1**., formulas **1** + **2** → **3**), in which the ground state free energies of the products from both transition states are identical (Fig. **1-3C**).

These two ways of fulfilling the condition that the ground state free energies of the reactants be identical, plus the fact that R* may be either a chiral reagent or a chiral physical force, permit division of kinetically controlled asymmetric processes into four categories: (a) Those processes in which there is preferential reaction of one enantiomer of a racemic substrate with a chiral reagent are commonly known as *kinetic resolutions* (Sec. **1-4.4a**) or in some cases, asymmetric decompositions. (b) The reactions of an achiral (but prochiral) substrate with a chiral reagent or of an achiral (but prochiral) group within a chiral substrate with an achiral reagent to give a chiral product are generally known as *asymmetric syntheses* (Sec. **1-4.4b**). (c) The reactions in which there is preferential destruction of one member of a racemic pair by the action of a chiral physical force are the processes commonly known as *absolute asymmetric decompositions.* (d) The reactions of an achiral substrate and achiral reagent in the presence of a chiral physical force to give a chiral product are known as *absolute asymmetric syntheses.* Absolute asymmetric decompositions and syntheses will be discussed together (Sec. **1-4.4c** and Sec. **10-4**). The same energy reaction coordinate diagrams can apply to all of these processes and thus the division into kinetic resolutions, asymmetric syntheses, absolute asymmetric decompositions, and absolute asymmetric syntheses is one of convenience based on the chemical details of the reactions, not on the kinetics or thermodynamics of the reaction. In all of these processes where $\Delta G^\circ = 0$ and $\Delta\Delta G^\ddagger \neq 0$, it follows that if stereoisomers are formed, they will be produced in unequal amounts; that is, the reaction will be stereoselective and the degree of stereoselectivity will depend upon the magnitude of $\Delta\Delta G^\ddagger$.

1-4.4a. *Kinetic Resolution.* The rates of reaction, $k_{(-)}$ and $k_{(+)}$, of the enantiomers of a racemic substrate, (−A) and (+A), with a chiral reagent must differ.[66] This difference becomes experimentally important from the standpoint of kinetic resolution when the reaction in question is incomplete with respect to the racemic substrate. Thus, if the racemic substrate reacts with insufficient chiral reagent for complete conversion to products, the remaining unconverted substrate is enriched in the less reactive isomer and no longer consists of a 50:50 racemic mixture. The same is true if there

[66] It may be, of course, that these rates are so nearly the same that they are not readily distinguishable experimentally, but more often than not the rate differences are appreciable. For instance, even the rate of esterification of racemic α-phenylbutyric anhydride with (*S*)-(+)-2-propanol-1-d_3, in which the two groups on the chiral center differ only in isotopic substitution, is faster for the utilization of one enantiomer than the other by 0.5%. It is possible that the two rates may be identical, but under unusual circumstances only. Thus, it has been observed [F. Akiyama, K. Sugino, and N. Tokura, *Bull. Chem. Soc. Japan*, **40**, 359 (1967)] that the rate of esterification of (−)-menthylacetyl chloride with (−)-menthol in acetonitrile solvent was faster than that of (+)-menthylacetyl chloride. But the reverse was true in liquid sulfur dioxide solvent. In theory there is a mixture of these two solvents in which the rates are identical.

are equivalent amounts of substrate and reagent, but the reaction is interrupted before it has reached completion. If the chiral center in the substrate is destroyed during the reaction, only one optically active isomer can be obtained; but if the chiral center is left intact by the reaction, the more reactive substrate enantiomer will be represented in the product and thus both chiral forms or their derivatives can be obtained.

The extreme case of kinetic resolution for the production of optically active compounds is encountered when $k_{(-)}$ is so much greater than $k_{(+)}$, or vice versa, that essentially complete conversion of one isomer may be achieved without appreciable loss of its enantiomer. This technique has been put to excellent use for the production of optically active amino acids by Greenstein and co-workers.[67] An N-acetyl or N-chloroacetyl derivative of a racemic amino acid (*R*-**49**, *S*-**49**) is digested with an enzyme from hog liver homogenates, Acylase I (the chiral reagent), at 37° for 12 hr and the hydrolyzed natural enantiomer (*S*-**50**) recovered from the medium by

COOH
AcNH—C—H
R
S-**49**

COOH
H—C—NHAc
R
R-**49**

Acylase I →

COOH
H_2N—C—H
R
S-**50**

COOH
H—C—NHAc
R
R-**49** (No reaction)

conventional means. The unreacted unnatural acylated enantiomer (*R*-**49**) is then isolated and hydrolyzed by chemical means to give the unnatural *R*-amino acid. The rate of hydrolysis of the natural *S*-enantiomers for naturally occurring amino acids ranged from 10,000 to 40,000 times that of the unnatural *R*-enantiomers and both isomers are obtained in purities of 99.9%. This enzyme can also be used for the resolution of some amino acids which do not occur in nature, for example, α-aminocyclohexylcarboxylic acid and isovaline, (α-methyl-α-aminobutyric acid).

Horeau has developed an extremely useful system for the correlation of configurations of secondary alcohols (or amines) based upon kinetic resolution.[68] In a typical example, (+)-α-phenylbutyric anhydride (**51**), 2.7 mmoles,

[67] J. P. Greenstein, S. M. Birnbaum, and L. Levintow, *Biochem. Prep.*, **3**, 84 (1953) and references cited therein.

[68] (a) A. Horeau, *Tetrahedron Lett.*, 506 (1962). (b) *ibid.*, 965 (1962). (c) A. Horeau and H. B. Kagan, *Tetrahedron*, 2431 (1964). (d) A. Horeau and A. Nouaille, *Tetrahedron Lett.*, 3953 (1966). (e) A. Horeau, *Bull. Soc. Chim. Fr.*, 2673 (1964). (f) R. Weidman and A. Horeau, *ibid.*, 117 (1967). Horeau has designated the method "Dedoublement Partiel," which is a kinetic resolution in the sense we use the term.

was treated with excess racemic isopropylphenylcarbinol (**52**), 15 mmoles, in cold anhydrous pyridine. The unreacted carbinol was recovered, purified, and its rotation found to be α_D^{23} $-1.655°$ (neat, 1 dm). The ester (**53**) could also be isolated, the carbinol regenerated, and its rotation determined. In practice this is not done since it can be calculated readily from the known yield of ester (determined by titration of the liberated acid and the rotation of recovered unreacted carbinol). Correcting for the enantiomeric purity

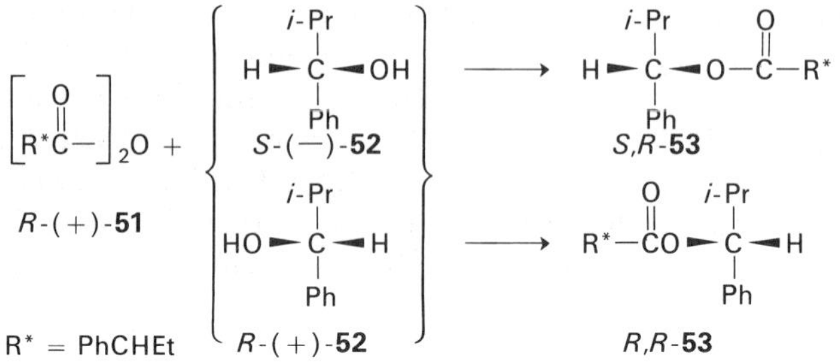

$$[R^*C(=O)-]_2O + \left\{ \begin{array}{l} H\text{—}C(i\text{-Pr})(Ph)\text{—}OH \quad S\text{-}(-)\text{-}\mathbf{52} \\ HO\text{—}C(i\text{-Pr})(Ph)\text{—}H \quad R\text{-}(+)\text{-}\mathbf{52} \end{array} \right\} \begin{array}{l} \longrightarrow H\text{—}C(i\text{-Pr})(Ph)\text{—}O\text{—}C(=O)\text{—}R^* \quad S,R\text{-}\mathbf{53} \\ \longrightarrow R^*\text{—}C(=O)O\text{—}C(i\text{-Pr})(Ph)\text{—}H \quad R,R\text{-}\mathbf{53} \end{array}$$

R-(+)-**51**

$R^* = PhCHEt$

of the chiral anhydride **51** gives the percent kinetic resolution, in this case, an amazingly large 49%. This is a measure of the stereoselectivity of the reaction and is completely analogous to the term *percent asymmetric synthesis*. By studying many such examples, Horeau found that the configuration of the recovered alcohol could be related to that of the optically active reagent. He formulated the following empirical generalization:[68e] When the anhydride made from dextrorotatory α-phenylbutyric acid is used, the unreacted recovered alcohol will have the absolute configuration represented by **54** when R_L represents the largest group and R_M, the medium sized group.

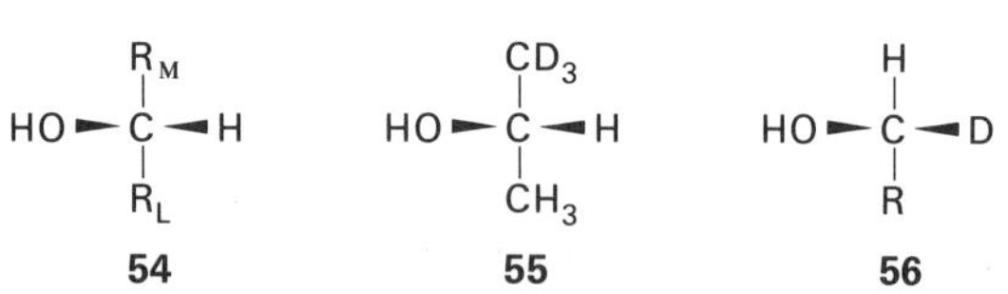

$$HO\text{—}C(R_M)(R_L)\text{—}H \quad \mathbf{54} \qquad HO\text{—}C(CD_3)(CH_3)\text{—}H \quad \mathbf{55} \qquad HO\text{—}C(H)(R)\text{—}D \quad \mathbf{56}$$

This method correctly correlated the configurations of five methylalkylcarbinols, nine phenylalkylcarbinols, and several miscellaneous secondary carbinols[68f] with stereoselectivities ranging from 8 to 79%. This method is especially valuable since the configuration of a particular secondary alcohol may be predicted without starting with an optically active sample which often is difficult to obtain.

This same procedure can be reversed so that the optically active reagent becomes a sample of the secondary carbinol and the excess racemic substrate becomes the optically inactive α-phenylbutyric anhydride.[69] In this case, however, the anhydride is used in excess and the optical rotation of the α-phenylbutyric acid recovered from hydrolysis of the unreacted anhydride is determined. It was demonstrated empirically that when the configuration of the optically active alcohol was that shown in **54**, the recovered α-phenylbutyric acid had the *S*-(+)-enantiomer in excess.[70,71] This method has proved its value by successfully predicting the configuration of secondary carbinols in a series of compounds not directly related to those from which the generalizations were developed as in the caryophyllene carbinols.[72]

This method is capable of great sensitivity as demonstrated by its application to (*S*)-(+)-2-propanol-1-d_3 (**55**) which gave a kinetic resolution of 0.4 to 0.5% in several runs. The recovered α-phenylbutyric acid (α_D + 0.02 to +0.12 ± 0.002) had the *S*-(+) configuration which required that CH_3 be larger (R_L) than CD_3 (R_M) as represented in **54** and **55**. These results, namely that deuterium exerts a smaller steric interaction than hydrogen, correspond to earlier findings on the optically active 1-deuterio alcohols[68d] (**56**) and to theoretical predictions.[71]

In certain cyclic secondary carbinols, the application of Horeau's method of kinetic resolution gave results which at first appeared anomalous. These could be rationalized, however, by a closer inspection of molecular models which showed that a "first-order" determination of R_M and R_L, based upon the extent of branching on the carbons attached to the carbinol center, did not correctly reflect the overall steric hindrance on the respective sides of the carbinol carbon atom because of greater 1,3 interactions on the "R_M" than on the "R_L" side. This finding emphasizes the necessity of using caution in applying this or related methods (see Sec. **2-2.1**). If the configuration of these latter carbinols had not been known by other methods, the

[69] In this case the anhydride from racemic α-phenylbutyric acid is a mixture of *meso* and racemic diastereomeric forms which appear to be in equilibrium under the reaction conditions. The rate of reaction of the *R,S* and *S,R* epimers with the chiral reagent must differ and also must be comparable or faster than the rate of reaction of the *meso* substrate for the method to give significant results.

[70] As carefully pointed out by A. Horeau, A. Nouaille, and K. Mislow [*J. Am. Chem. Soc.*, **87**, 4957 (1965)], this is not equivalent to saying that the alcohol **54** will always have the absolute *R* configuration. The sequential rule in the Cahn, Ingold, and Prelog notation is not based on the increasing size of groups even though it generally follows this in a closely related series, i.e., $(CH_3)_3C-$ takes precedence over $(CH_3)_2CH-$ which takes precedence over CH_3CH_2-, etc. There can be many exceptions to this; thus $CH{\equiv}C-$ would take notational precedent over $(CH_3)_2CH-$ but might very likely act as a smaller group. This has been specifically demonstrated[68f] in the cases of CD_3- versus CH_3- and D— versus H—. The deuterium substituted group takes notational precedence over the normal hydrogen isotope but *exerts a smaller* steric effect.[68f,71]

[71] K. Mislow, R. Graeve, A. J. Gordon, and G. H. Wahl Jr., *J. Am. Chem. Soc.*, **85**, 1199 (1963); **86**, 1733 (1964).

[72] A. Horeau and J. K. Sutherland, *J. Chem. Soc.* [C], 247 (1966).

original assignments obtained by the application of this kinetic resolution method might have gone unchallenged.

The application of kinetic resolution to the solution of a difficult stereochemical problem is illustrated by the proof of configuration of the ozonides from *cis* and *trans* alkenes.[73] Although it had been assumed that the ozonide from the *cis* isomer (*cis* **57**) had the *cis* structure (**58**) and from the *trans* isomer (*trans* **57**) had the *trans* structure (**59**), it was at least possible that this was reversed. When the olefin is symmetrical as is 3,4-dimethyl-3-hexene, the *cis* ozonide exists as an achiral *meso* compound (**58**) while the *trans* ozonide exists as a racemic pair (**59**). Use was made of the fact that ozonides react readily with amines (to give unspecified products). The purified ozonide from each isomer was allowed to react with less than the stoichiometric amount of brucine and the unreacted ozonide recovered, purified, and its rotation determined. Only the ozonide from the *trans* isomer showed optical

cis-**57** $\xrightarrow{O_3}$ **58** $\xrightarrow{\text{Brucine}}$?

trans-**57** $\xrightarrow{O_3}$ [**59A** + **59B**] $\xrightarrow{\text{Brucine}}$?

activity, thus unequivocally establishing its *trans* structure. It is worth noting that this method is useful even though the mechanism of the reaction is not known.

Kinetic resolution and asymmetric synthesis can take place simultaneously. When the Grignard reagent from racemic 2-phenyl-1-chlorobutane (**60**) reacts with phenyl isopropyl ketone (**61**) in the presence of the optically active ether (+)-2,3-dimethoxybutane (**62**), the ketone is reduced to the carbinol **65** and the Grignard reagent is oxidized to the olefin **64**.[74] Under

2 CH_2MgCl–CHPh–CH_2–CH_3 (**60**) + *i*-Pr–C(=O)–Ph (**61**) $\xrightarrow[\text{2) } H_2O]{\text{1) Me–C(MeO)(H)–C(H)(OMe)–Me } \textbf{62}}$ CH_3–CHPh–CH_2–CH_3 (**63**) and CH_2=C(Ph)–CH_2–CH_3 (**64**) + *i*-Pr–CHOH–Ph (**65**)

[73] R. W. Murry, R. D. Youssefyeh, and P. R. Story, *J. Am. Chem. Soc.*, **88**, 3655 (1966).

[74] J. D. Morrison and R. W. Ridgway, *Tetrahedron Lett.*, 569 (1969). The use of optically active ethers in the Grignard reduction reaction is discussed in Sec. **10-1**.

the influence of the optically active ether (**62**), one of the enantiomeric forms of the Grignard reagent reacts more rapidly with the ketone than does the other to give the optically active alcohol (**65**, 8.4% e.e.). At the same time the less reactive form of the Grignard reagent (actually the less reactive form of the diastereomeric ether-Grignard complex) builds up in solution. Upon hydrolysis this is converted to the optically active hydrocarbon (**63**, 4.9% e.e.).

1-4.4b. *Asymmetric Synthesis.* In contrast to a kinetic resolution which starts with a racemic mixture, the individual molecules of which are chiral, an asymmetric synthesis converts an achiral substrate (or an achiral group within a chiral molecule, Sec. **1-2**) into a chiral product by reaction with a chiral reagent. In certain instances the forward reaction may be an asymmetric synthesis, while the reverse is a type of kinetic resolution. The reduction of pyruvic acid (**66**) by a chiral reducing agent to give lactic acid (**67**) in which one enantiomer predominates over the other is such an asymmetric synthesis; the reaction of racemic lactic acid (*RS*-**67**) with an optically active oxidizing (or dehydrogenating) agent will be faster for one isomer than the other and is an asymmetric decomposition or a destructive kinetic resolution.

$$\underset{\mathbf{66}}{CH_3\overset{\overset{\displaystyle O}{\|}}{C}COOH} \rightleftarrows \underset{R\text{-}\mathbf{67}}{CH_3\!\blacktriangleright\!\overset{\overset{\displaystyle OH}{|}}{\underset{\underset{\displaystyle H}{|}}{C}}\!\blacktriangleleft\!COOH} + \underset{S\text{-}\mathbf{67}}{CH_3\!\blacktriangleright\!\overset{\overset{\displaystyle H}{|}}{\underset{\underset{\displaystyle OH}{|}}{C}}\!\blacktriangleleft\!COOH}$$

The enzyme-coenzyme system, lactic acid dehydrogenase-nicotinamide adenine dinucleotide (LAD-NAD), constitutes the chiral reagent which brings about the well-known enzymic process represented by these reactions, where the stereoselectivity is complete.

In Secs. **1-2** and **1-3** several characteristic examples of asymmetric syntheses already have been given; for the remaining portion of this introductory chapter, we shall outline in more detail some of the common features of all asymmetric syntheses before beginning an exhaustive treatment in the subsequent chapters.

Both qualitative and quantitative aspects are involved in asymmetric synthesis. Qualitatively, every asymmetric synthesis will give a product which has a preponderance of either the *R* or *S* configuration in the newly formed chiral center. Quantitatively, the extent to which one isomer predominates over the other can vary from experimentally nondetectable to essentially 100%, the latter being found in enzymic processes with natural substrates. Any serious study of an asymmetric synthesis will prove the variations in both qualitative (configuration of product) and quantitative (percent asymmetric synthesis) aspects caused by variation of the many possible parameters such as temperature, solvent, variations in chiral reagent, and structural variations in substrate. Far too often scanty stereochemical results have been made the basis of unwarranted generalizations concerning the topology and nature of nonbonded interactions in the transition state.

From the qualitative standpoint the binary nature of chirality dictates a priori a 50:50 chance of predicting the correct configuration of the predominant product in any one isolated asymmetric synthesis. If the observed configuration is correctly predicted in many cases, based upon certain assumptions concerning the mechanism of the reaction, this does not prove the validity of the assumptions, since almost certainly several sets of alternative assumptions could be devised which would lead to the same prediction. Nevertheless, the more examples observed over the widest possible experimental variation of parameters, the greater the possibility of developing a reliable theory. Certainly this is the first step to developing a useful empirical correlation of the results. In some respects, the most valuable observations are those which do not correlate with the initial hypothesis since they permit one to refine the hypothesis, to better define its structural variations and limitations, or to replace it, if necessary.

Since the pioneering publications of Prelog,[12] Cram,[13] Doering,[5] and Mosher,[10] attempts to correlate configuration by asymmetric synthesis have almost universally relied upon the concept of large, medium, and small groups (R_L, R_M, and R_S) which control the direction and extent of asymmetric synthesis. In connection with Horeau's kinetic resolution method (Sec. **1-4.4a**) we have commented upon the fundamental problem associated with ordering specific groups with respect to size, except when the groups fall into a simple series such as

$$CH_3 < CH_2CH_3 < CH(CH_3)_2 < C(CH_3)_3.$$

(Since one group is usually hydrogen, there is no problem with the decision concerning R_S except in the case of deuterium versus hydrogen which has been mentioned previously.[71]) The difficulty of specifying group size is especially pronounced when the groups in question differ considerably in type, i.e., *t*-butyl versus trifluoromethyl versus trimethylsilyl versus trimethylammonium, etc. One solution is to determine experimentally, in an asymmetric synthesis where the absolute configuration of the product is known, which group acts as though it were the larger in the system in question and then use this operationally determined ordering in subsequent interpretations. The ranking of phenyl and *t*-butyl groups is such an example; a priori one might conclude that the steric effect of the *t*-butyl group, situated adjacent to the site of reaction, would be greater than a phenyl group. This seems to be the case as revealed by the relative difficulties of synthesis or relative reactivities of tri-*t*-butylcarbinol versus triphenylcarbinol. This is also the conclusion deduced from comparison of the van der Waals radii of the *t*-butyl and phenyl groups when both are considered "spinning tops."[75] Furthermore, their conformational energy values[76] as determined in the cyclohexyl ring system clearly indicate that the *t*-butyl group exerts larger steric interactions than phenyl as far as equatorial versus axial orientation in the cyclohexane ring is concerned. But in the studies on asymmetric reductions of phenyl *t*-butyl ketone by optically active Grignard reagents (Sec. **5-2.3**), the reverse is clearly true, namely that in this particular system the phenyl group acts as though it were R_L and *t*-butyl as R_M (Tables **5-5** and **5-6**). The same is true for phenyl-*t*-butylcarbinol in the kinetic resolution system of Horeau.[68,77] It is important to realize that when one tries to determine a priori whether one group is "larger" or "smaller" than another, there is a 50:50 chance of correctly predicting the observed result. Thus the application of asymmetric synthesis reactions in the determinations of relative group sizes (or relative configurations of compounds) must be carried out with great caution. This is especially true when dealing with molecular environments which deviate appreciably from those for which the method was developed. The most important test of the validity of a conclusion of this type is its application to many specific examples over a suitably large variety of molecular types.

One would anticipate on stereochemical grounds that if the addition of methyl Grignard reagent to the (−)-menthyl ester of benzoylformic acid (**68**) gives (*R*)-(−)-atrolactic acid (**69**), then the addition of phenyl Grignard reagent to the (−)-menthyl ester of pyruvic acid (**70**) should give the enantiomeric (S)-(+)-atrolactic acid (**71**) which indeed it does. The percent asymmetric synthesis in the two examples above (**68** → **69** versus **70** → **71**) need not be identical[78] and, in fact, it is not even a theoretical requirement that the configurations of the products be enantiomeric, since neither the ground state free energies of the respective reactants nor the comparable transition states in the two cases are identical. Nevertheless in all reported cases in the McKenzie-Prelog system, the anticipated reversal of configuration as given by **68** → **69** and **70** → **71** is observed. But the need for exercising caution is

[75] As calculated from bond distances, bond angles, and van der Waals radii, there is a small but significant difference (0.17 A) in maximum radii of the *t*-butyl group (3.33 A) and the phenyl group (3.16 A) when they are considered as spinning around their bonding axes.

[76] E. Eliel, N. L. Allinger, S. J. Angyal, and G. A. Morrison, *Conformational Analysis* (New York: Interscience Publishers, Inc., 1965), p. 441. The value for *t*-butyl has now been determined to be 5.4 kcal/mole [N. L. Allinger, J. A. Hirsch, M. A. Miller, I. J. Tyminski, and F. A. Van-Catledge, *J. Am. Chem. Soc.*, **90**, 1199 (1968)] versus 2.6–3.1 kcal/mole for phenyl.

[77] Phenyl-*t*-butylcarbinol has not been studied in Prelog's benzoylformic ester system; however, see Sec. **8-1.2**, where the asymmetric peroxidation of sulfides is described and where the authors interpret their results on the basis that *t*-butyl is larger than phenyl.

[78] The data taken from A. McKenzie, *J. Chem. Soc.*, 365 (1906) show 21.1% asymmetric synthesis in the first case (**68** → **69**) and 25.0% asymmetric synthesis in the second, which are well beyond the polarimetric limit of experimental error.

emphasized by the observation that products of opposite configuration are obtained by the action of phenylmagnesium bromide and phenylmagnesium iodide on phenylacetoin (**72**).[79] Under otherwise identical conditions the ratio of *meso* to *dl* was 1:2.0 when phenylmagnesium iodide was used but 2.2:1 when phenylmagnesium bromide was used.

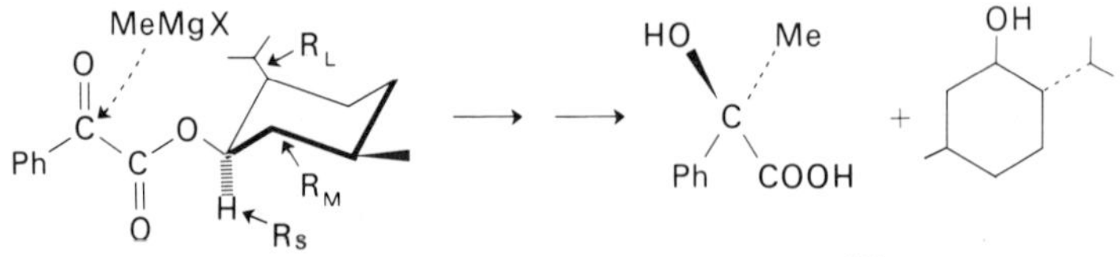

MeMgX R_L R_M R_S → → HO Me C Ph COOH + OH

68 **69**

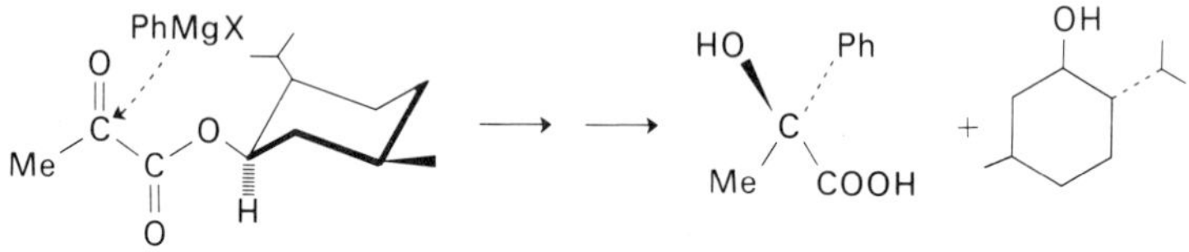

PhMgX → → HO Ph C Me COOH + OH

70 **71**

HO O Me C—C Me Ph + PhMgX → HO OH Me C—C Me Ph Ph and HO OH Me C—C Ph Ph Me

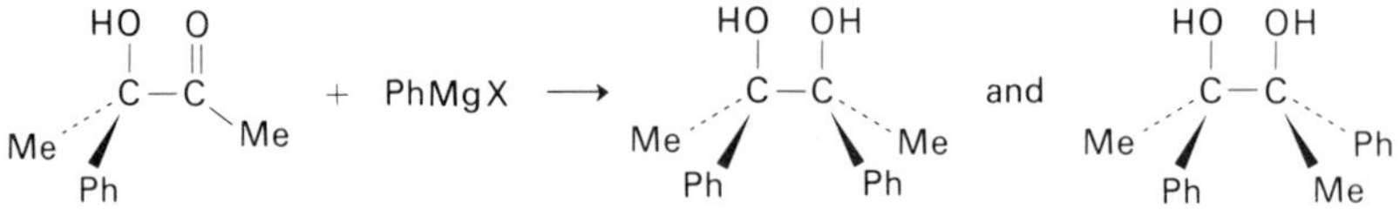

72-*dl* **73**-*meso* **74**-*dl*

Another example which indicates the necessity for exercising caution in applying the results from one system to another is given by the experiments of Mitsui and Kudo[80] on the asymmetric aldol-type condensation of propiophenone (**75**) with either (−)-menthyl or (+)-bornyl acetate (**76**) catalyzed with diethylaminomagnesium bromide. The inducing chiral reagent is

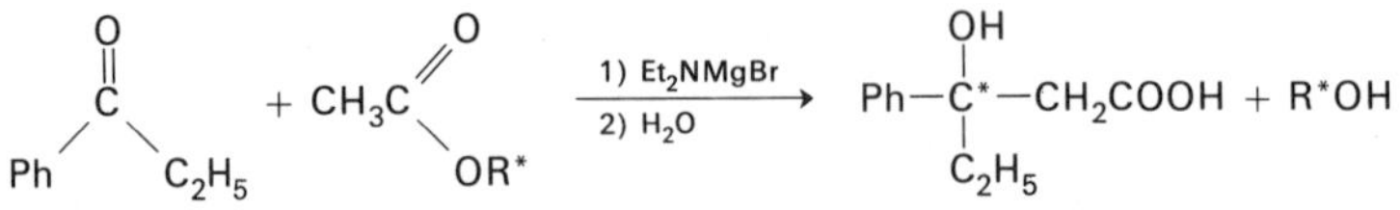

$$\underset{\mathbf{75}}{Ph-\overset{O}{\overset{\|}{C}}-C_2H_5} + \underset{\mathbf{76}}{CH_3\overset{O}{\overset{\|}{C}}-OR^*} \xrightarrow[2)\ H_2O]{1)\ Et_2NMgBr} \underset{\mathbf{77}}{Ph-\overset{OH}{\overset{|}{\underset{C_2H_5}{\underset{|}{C^*}}}}-CH_2COOH} + R^*OH$$

the optically active alcohol moiety of the acetate. According to the Prelog generalizations, (−)-menthol and (+)-borneol have opposite chiral arrangements of R_L, R_M, and R_S at the carbinol carbon and therefore, in addition

[79] J. H. Stocker, P. Sidisunthorn, B. M. Benjamin, and C. J. Collins, *J. Am. Chem. Soc.*, **82**, 3913 (1960). See J. H. Stocker, *ibid.*, **88**, 2878 (1966) for additional anomalies.

[80] S. Mitsui and Y. Kudo, *Tetrahedron*, **23**, 4271 (1967).

reactions of α-keto esters (Sec. **2-2.2a**), should and do lead to atrolactic acids of opposite configurations. However, in the aldol-type condensation, represented by **75** + **76** → **77**, (−)-menthyl acetate (**76**, R* = (−)-menthyl) leads to the *same* (*S*)-β-hydroxy-β-phenylvaleric acid (**77**) (70.2% asymmetric synthesis) as does (+)-bornyl acetate (**76**, R* = (+)-bornyl, 36.0% asymmetric synthesis) in direct contrast to the experience in the atrolactic acid system (cf. Sec. **4-2**).

Walborsky and co-workers[81] have also observed a most interesting reversal of stereoselectivity with solvent change in the case of the Michael addition of ethyl acrylate (**78**) to (−)-menthyl chloroacetate (**79**, R* = (−)-menthyl). In toluene solvent the only isomer formed in several experiments was levorotatory *trans*-cyclopropanedicarboxylate, isolated as its dimethyl

$$CH_2{=}CHCOOEt + ClCH_2\overset{O}{\overset{\|}{C}}{-}OR^* \xrightarrow[\substack{2)\ H_2O\ (OH^-)\\ 3)\ CH_2N_2}]{1)\ t\text{-}BuO^-} \text{trans-cyclopropane (COOMe, H / H, COOMe)}$$

78 **79** **80**

ester (**80**) ($[\alpha]_D^{24}$ −4.0 to −2.8°, enantiomeric purity 1.8–3.1%), but in dimethylformamide solvent it was always dextrorotatory ($[\alpha]_D^{24}$ +8.6 to +9.3°, enantiomeric purity 10.2–10.9%). This is a complex example which is discussed in Sec. **6-3.1**. Another case of reversal of stereochemistry with solvent change is found in the hydrogenation-hydrogenolysis of the imine formed from *R*-(+)-α-methylbenzylamine and α-keto-glutaric acid.[82] In ethanol solvent (*S*)-(+)-glutamic acid is formed (13% e.e.) but in dioxane the (*R*)-(−)-enantiomer is in excess (19% e.e., cf. Sec. **7-2.2**).

There also may be wide variations in the extent of asymmetric synthesis with temperature, even a reversal in the steric course of the reaction as observed by Pracejus[83] in the addition reaction of methanol to phenylmethylketene (**81**) in the presence of the alkaloid acetylquinine, to give optically active methyl α-phenylpropionate (**82**). At −110° the ratio of *R*-(−) to *S*-(+) ester was 87:13. This ratio decreased with increasing

$$O{=}C{=}C(Ph)(Me) + CH_3OH \xrightarrow{\text{acetylquinine}} Me{-}C(COOCH_3)(Ph){-}H \text{ and } H{-}C(COOCH_3)(Ph){-}Me$$

81 *S*-(+)-**82** *R*-(−)-**82**

[81] (a) Y. Inouye, S. Inamasu, M. Ohno, T. Sugita, and H. M. Walborsky, *J. Am. Chem. Soc.*, **83**, 2962 (1961). (b) H. M. Walborsky and C. G. Pitt, *ibid.*, **84**, 4831 (1962).

[82] K. Harada and K. Matsumoto, *J. Org. Chem.*, **33**, 4467 (1968).

[83] (a) H. Pracejus, *Ann.*, **634**, 9 (1960). (b) *ibid.*, **634**, 23 (1960).

temperature until at $-50°$ the percent asymmetric synthesis was essentially zero. At $-37°$ the *S*-(+) isomer predominated slightly but when the temperature was 0° and above the reaction again produced the *R*-(−) isomer in excess. For the low temperature reaction, $\Delta\Delta H^{\ddagger} \cong 2.8$ kcal/mole and $\Delta\Delta S^{\ddagger} \cong -13$ cal/deg/mole. The reversal in stereochemistry may be due solely to the entropy factor or it may be a result of a change in mechanism with temperature. Even when the entropy factor is zero, there will be significant changes in stereoselectivity with temperature. For example, if one assumes an entropy factor of zero (i.e., $\Delta\Delta G^{\ddagger} = \Delta\Delta H^{\ddagger}$), then for a particular reaction for which $\Delta\Delta G^{\ddagger} = -1.0$ kcal/mole at 100° the reaction will be 58% stereoselective (79:21 isomer ratio), at 0° it will be 72% stereoselective (86:14 isomer ratio), and at $-100°$ it will be 90% stereoselective (95:5 isomer ratio). Obviously the temperature dependence measurements of asymmetric synthesis reactions should be made whenever possible.

Although Prelog's generalization[12] of the atrolactic acid asymmetric synthesis and Cram's rule[13] for "steric control of asymmetric induction" in carbonyl addition reactions are carefully stated empirical rationalizations of experimental results, to what extent do these models represent the actual mechanisms for these reactions? Speculation concerning the nature of the transition states in asymmetric synthesis reactions is both useful and irresistible. In general, asymmetric synthesis reactions, such as those to which Prelog's generalization and Cram's rule have been applied, have energies of activation which are well above the normal rotational energy barriers encountered in flexible molecules. Thus the Curtin–Hammett[84] principle applies, namely that the product ratio depends solely upon the free energy differences, $\Delta\Delta G^{\ddagger}$, between the competing transition states and not upon the population of various ground state conformations of the reactants.

The problem is clearly one of defining quantitatively the relative energy levels of the competing transition states leading to the stereochemically different products. This is an extremely complex problem*[85], and the general approach has been, with the aid of various simplifying assumptions, to reduce the many possible transition states, with their various topologies and quantized energy levels, to a few (often two) crucial ones which seem to embody the most important interactions from a stereochemical viewpoint. These then can be dealt with intuitively or semiquantitatively.

In systems such as the asymmetric Grignard reduction (Fig. **1-4A**, see also **1** + **2** → **3**, Sec. **1-1**.) where the chiral center in the reactant is being destroyed at the same time that the chiral center in the product is being formed (designated self-immolative reactions by Mislow[24] and asymmetric

[84] D. Y. Curtin, *Record Chem. Progress*, **15**, 111 (1954).

[85] R. E. Weston, Jr., *Science*, **158**, 332 (1967).

* The complexity is better appreciated by reading Weston's detailed discussion of the three-atom transition state problem.[85]

transfer processes by Pracejus[14]), there is no difference between the ground state free energies of the reactants which give the competing transition states and there is no difference in free energies of the products formed from the competing transition states. The enantiomeric ratio of products can only be attributed to energy differences in the transition state with no possible contribution from energy differences in the starting materials and products (Fig. **1-4**), since none exists. This statement should be qualified by pointing

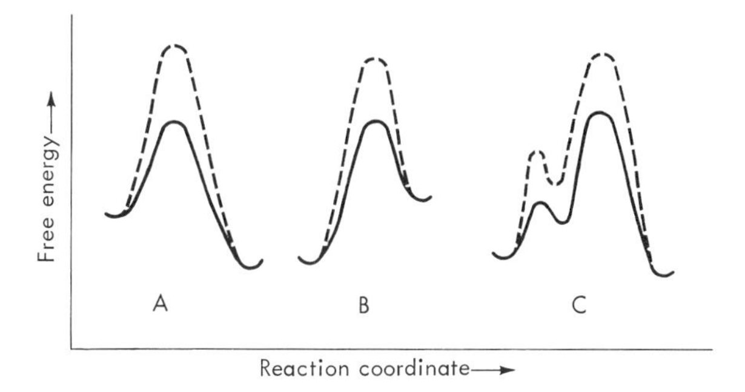

Fig. 1-4. Types of energy profiles for asymmetric syntheses.

out the possibility even in the self-immolative asymmetric synthesis for rapid, reversible pretransition-state intermediate formation (Fig. **1-4C**).

In asymmetric syntheses which produce diastereomeric products, such as the McKenzie and Prelog system (see **4** → **5** → **6**, Sec. **1-1**) or the Cram system (see **7** → **8A** + **8B**, Sec. **1-1**), one can consider the possible conformations of both reactants and diastereomeric products and their potential contributions to the transition states.

Cram's rule[13] states that in addition reactions at a carbonyl group adjacent to a chiral center, that diastereomer will predominate which would be formed by the approach of the entering groups (R) from the less hindered side of the double bond when the rotational conformation of the C—C bond is such that the double bond is flanked by the two least bulky groups attached to the adjacent chiral center as shown in **83**. This has proved to be an extremely useful generalization. Several refinements have been devised to accommodate situations in which not only the open-chain type structure

83 **84A** **84B**

83 is involved but also rigid structures (**84A**) and dipolar structures (**84B**).[86] These are discussed in detail (Sec. **3-2** and **3-3**) as well as recent proposals by Karabatsos[87a,87b] and by Felkin and co-workers[87c,87d] concerning the detailed conformation of the transition state in the Cram system. Ugi[88] has developed a quantitative treatment based on group theory, which is of great importance for certain types of asymmetric synthesis systems. With this method, one may calculate the percent asymmetric synthesis using equations based upon a model that is independent of ground state considerations. According to this mathematical model, the relative proportion of stereoisomeric products is approximated by a summation of all the point group interactions in the transition state. This model is applicable to flexible (open chain) type transition states such as those found in the benzoylformate to atrolactate conversion. It has been put on a mathematical basis by Ruch and Ugi[89a] and has been extended to other systems.[89b,89c,90] According to this mathematical model, the asymmetric synthesis reactions under consideration can be symbolized as follows:

$$\mathbf{85} \longrightarrow \begin{bmatrix} R,R' + R,S' \\ \text{Transition} \\ \text{states} \end{bmatrix} \longrightarrow R,R'\text{-}\mathbf{86} \text{ and } R,S'\text{-}\mathbf{86}$$

85   *R,R'*-**86**   *R,S'*-**86**

The theory is concerned with mathematical models for the diastereomeric transition states which we shall symbolize by $[R,R']^{\ddagger}$ and $[R,S']^{\ddagger}$. The three ligands on the original asymmetric carbon (C) of the substrate (**85**) are designated L_1, L_2, and L_3 and those on the asymmetric center (C′) formed during the reaction are designated L'_1, L'_2, and L'_3. These two centers (C and C′)

[86] D. J. Cram and D. R. Wilson, *J. Am. Chem. Soc.*, **85**, 1245 (1963).

[87] (a) G. J. Karabatsos, *J. Am. Chem. Soc.*, **89**, 1367 (1967). (b) G. J. Karabatsos and N. Hsi, *ibid.*, **87**, 2864 (1965). (c) M. Chérest, H. Felkin, and N. Prudent, *Tetrahedron Lett.*, 2201 (1968). (d) M. Chérest and H. Felkin, *Tetrahedron Lett.*, 2205 (1968).

[88] I. Ugi, *Chimia*, **19**, 89 (1965).

[89] (a) E. Ruch and I. Ugi, *Theoret. Chim. Acta*, **4**, 287 (1966). (b) E. Ruch, A. Schoenhofer, and I. Ugi, *ibid.*, **7**, 420 (1967). (c) I. Ugi and G. Kaufhold, *Ann.*, **709**, 11 (1967).

[90] The reader is referred to these articles[88,89] for the details of their mathematical treatment which are beyond the scope of this review. The short summary here does not present a true picture of the broader application envisioned by Ruch and Ugi[89a,89b] Short summaries of the Ugi treatment are also presented by Pracejus (Reference 14, pp. 541–544) and by R. Cruse in an addendum chapter in E. L. Eliel, *Stereochemie der Kohlenstoffverbindungen*, translation by A. Lüttringhaus and R. Cruse (Weinheim: Verlag Chemie, GmbH, 1966), pp. 537–544.

are not directly bonded to each other but are joined through one or more atoms symbolized by $\sim$. Formulas $[R,R']^{\ddagger}$-**86** and $[R,S']^{\ddagger}$-**86** are not in-

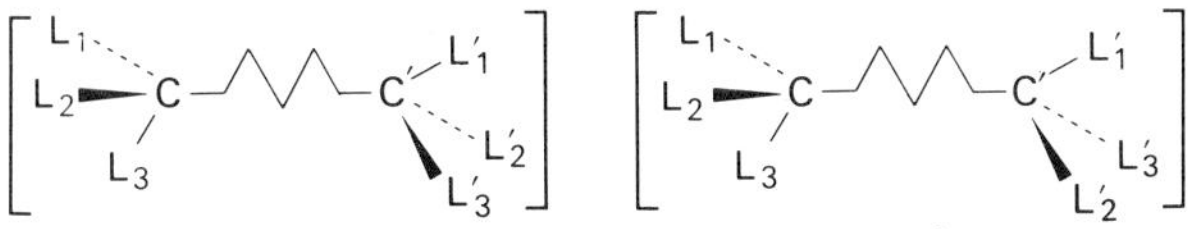

$[R,R']^{\ddagger}$-**86** $[R,S']^{\ddagger}$-**86**

tended to define specific conformations of the respective transition states but are intended to symbolize the summation of all possible interactions of the ligands in the epimeric transition states. Ugi postulates that the ratio of epimeric products, *R,R'*-**86** and *R,S'*-**86** (or the ratio of their respective enantiomers), can be calculated from a linear free energy relationship similar in concept in certain respects to the Hammett linear free energy expression for the relative rates of closely related reactions. He proposes that the difference in free energies of activation $\Delta\Delta G^{\ddagger}$ for the competing epimeric transition states is a function of the product of two chirality constants χ and χ' (Eq. 1)

$$\Delta\Delta G^{\ddagger} = -RT \ln \frac{k_{R,R'}}{k_{R,S'}} = -RT \ln Q = f[\chi \cdot \chi'] \quad (1)$$

The concept of the chirality constant is a most important innovation and is arrived at as follows. There is a ligand constant, λ, associated with each ligand L, which is assumed to be an invariant property of that specific ligand. The chirality constant χ of a specific asymmetric center, C, is a function of the associated ligands. Group theory analysis of the transition state interactions in this model indicates that the dissymmetry effect of the ligands on carbon atom C can be described by a symmetry-adapted function which is the product of differences. For the reactions of the type represented by **85** → **86** (of the $\Lambda_{3,3}$ type), this takes the following form:

$$\chi = f(\lambda_1 - \lambda_2)(\lambda_2 - \lambda_3)(\lambda_3 - \lambda_1) \quad (2)$$

The chirality constant, χ', for the emerging asymmetric center C′, is similarly defined:

$$\chi' = f(\lambda'_1 - \lambda'_2)(\lambda'_2 - \lambda'_3)(\lambda'_3 - \lambda'_1) \quad (3)$$

The sign of the chirality constant will depend upon the order in which the ligands are attached to the established chiral center. The relative magnitudes of the ligand constants (λ), not their absolute values, determine the magnitude of the chirality constant (χ). The configuration of the emerging asymmetric center, i.e., whether the *R,R'* or *R,S'* form is produced in excess, is determined by the configuration of the original asymmetric center in the substrate and thus an antisymmetric factor, δ, which must be either $+1$ or -1, is introduced.*

In the mathematical treatment, the ligand constants are considered as pure numbers although there is a strong predilection that these numbers are associated with the steric requirements of the ligands and thus their interactions represent energy terms. Ugi has arbitrarily chosen to establish his system of ligand constants on the basis of λ for hydrogen equaling 0.00 and for methyl, 1.00. For all other groups, the ligand constants are evaluated experimentally on the basis of this standard relative difference between hydrogen and methyl. This experimental evaluation is made possible by collecting all unknowns into a reaction parameter constant τ and comparing very closely related reactions under identical conditions. The reaction parameter constant τ is presumed to be a characteristic property for a specific reaction type under the same conditions of solvent, temperature, etc., but it will have a different value for each new reaction type or new set of reaction conditions. The χ' term (the chirality constant for the newly forming chiral center), though experimentally inaccessible, should be the same for all reactions of a specific type under the same conditions and thus can be incorporated into the reaction parameter constant τ. With the introduction of this constant, Eq. 1 takes the form of Eq. 4, which can now be used for the evaluation of epimeric product ratios (Q) of closely related reactions.

$$\log Q = \chi\delta\tau \tag{4}$$

This equation resembles in certain respects the Hammett equation to which has been added the antisymmetry constant δ. These parameters are evaluated from experimental results in one particular system and a set of λ values

$$Ph{-}\overset{O}{\overset{\|}{C}}{-}\overset{O}{\overset{\|}{C}}{-}O{-}CR^1R^2R^3 \xrightarrow[\text{2) } H_2O]{\text{1) MeMgI}} Ph{-}\underset{Me}{\overset{OH}{C}}{-}\overset{O}{\overset{\|}{C}}{-}O{-}CR^1R^2R^3$$

87

$$Ph{-}\overset{CH_3}{\overset{|}{C}}{=}CHCOOCR^1R^2R^3 \xrightarrow[\text{HOAc}]{H_2,\ Pt} Ph{-}\underset{H}{\overset{CH_3}{C}}{-}CH_2COOCR^1R^2R^3$$

88

* The sign of the antisymmetry factor is arbitrary and is based on the Cahn–Ingold–Prelog nomenclature sequence rules. L_1' is arbitrarily assigned to that group which has the highest priority according to the sequence rule, L_2' is intermediate, and L_3' is lowest. When the clockwise order of the groups L_1', L_2', and L_3' attached to the emergent asymmetric center C′, viewed from the side opposite the connection to C, follows the sequence rule, then $\delta = +1$; but when contrary to the sequence rule, $\delta = -1$.

developed for the ligands. These same λ values can then be used for interpreting or predicting results in a different system where, however, a new τ value will need to be determined experimentally. These λ values are based primarily upon the experimentally determined asymmetric synthesis of atrolactic esters (**87**) and β-phenylbutyric esters (**88**). The percents of asymmetric synthesis as experimentally determined and as calculated using Eq. 4 are given in Table **1-1**.

TABLE 1-1

Comparison of Stereoselectivities, Observed and Calculated by Ugi-Ruch Method[a]

$$\text{Ph}-\overset{\displaystyle O}{\overset{\|}{C}}-\overset{\displaystyle O}{\overset{\|}{C}}-O-C(R^1)(R^2)(R^3) \quad (R\text{-I}) \xrightarrow{\text{MeMgI}} \text{Ph}-\overset{\displaystyle OMgI}{\overset{|}{\underset{\displaystyle Me}{\underset{|}{C}}}}-\overset{\displaystyle O}{\overset{\|}{C}}-O-C(R^1)(R^2)(R^3) \quad (R,R\text{-II})$$

Ester groups in **87** and **88**			Ligand constants			Chirality constant	Mole % *R,R*-II	
R^1	R^2	R^3	λ_1	λ_2	λ_3	χ	Calc[b]	Found[c]
H	CH_3	$(CH_3)_3C$	0	1.00	1.49	0.730	62.4	62.0
H	CH_3	C_6H_5	0	1.00	1.23	0.283	55.0	51.5
H	CH_3	Mesityl	0	1.00	1.58	0.918	65.4	65.0
H	CH_3	1-Naphthyl	0	1.00	1.29	0.375	56.3	56.0
H	CH_3	(+)-Bornyl	0	1.00	1.86	1.599	75.2	77.0
H	CH_3	$(C_6H_5)_3C$	0	1.00	1.75	1.313	71.3	74.5
H	CH_3	—[d]	0	1.00	2.10	2.303	83.2	83.0
H	C_6H_5	(+)-Bornyl	0	1.23	1.86	1.442	73.1	70.5
H	C_6H_5	$(C_6H_5)_3C$	0	1.23	1.75	1.118	68.4	63.5

[a] Adapted from I. Ugi, *Chimia*, **19**, 91 (1965).

[b] Calculated for τ value = 0.300. The excess of mole percent over 50 equals percent asymmetric synthesis.

[c] Experimental results taken from V. Prelog, E. Philbin, E. Watanabe, and M. Wilhelm, *Helv. Chim. Acta*, **39**, 1086 (1956) and V. Prelog, O. Ceder, and M. Wilhelm, *ibid.*, **38**, 303 (1955).

[d] 2,4,6-Tricyclohexylphenyl.

The excellent correspondence between *calculated* and *found* percents asymmetric synthesis for the first seven values for the atrolactic (**87**) synthesis, if considered by themselves, would not be surprising since in each case the λ value for the R^3 group is the parameter which can be adjusted to obtain correspondence. But in the last two examples, the λ values of two groups, established in prior reactions, are used to determine the degree of asymmetric synthesis for a third and fourth reaction. Also the excellent correspondence between the *calculated* and *found* results for a second very different reaction (**88**), using a new reaction parameter constant τ but the same λ values, makes it appear highly likely that this approach is fundamentally sound. The calculated mole percent values in Table **1-1** have been reported to 0.1%, a precision which considerably exceeds the accuracy of the experimental results upon which they are based. This does not detract from the theory but emphasizes the necessity for additional quantitative experiments under controlled conditions specifically designed to test the theory. One such application is given in Sec. **7-4.2**.

The introduction and application of this quantitative treatment is a valuable development in the area of asymmetric synthesis. There are inherent limitations to the method since at present it appears to be applicable only in cases where the chiral centers are not adjacent to each other.* This excludes a large number of important reactions, such as the types covered by Cram's rule; nor can this treatment in its present form be applied to systems which proceed via rigid cyclic transition states such as asymmetric Grignard reductions. Also, since the utility of this method increases with the number of examples in any one system, it is of very limited predictive value in any new situation. Regardless of these limitations, this method is the most basic approach yet attempted and its further exploration, testing, and development is of prime importance. The successful extension of this quantitative method to a wider group of reactions, with the possibility of accommodating close range interactions of an electronic as well as steric nature, constitutes a real challenge to the future development of the Ugi–Ruch treatment.

1-4.4c. *Absolute Asymmetric Reactions.* The fascination of the subject of absolute asymmetric synthesis is found in its relationship to the origin of the first chiral organic substances of living matter, a subject concerning which there has been much intriguing speculation.[2,21,91] Along with recent developments concerning the origin of organic molecules under primordial condition,[92] there has come a renewed interest in this subject. Absolute asymmetric synthesis includes only those processes which result in the formation of a chiral product without the intervention of any other chiral *chemical*

* This is similar in principle, but for very different reasons, to the application of the Hammett equation to *para-* and *meta-*substituted benzene derivatives but not to *ortho-*substituted (without modification of the σ term).

[91] G. W. Wheland, *Advanced Organic Chemistry*, 3rd Edition, (New York: John Wiley & Sons, Inc., 1960), pp. 329–333.

[92] (a) A. I. Oparin, *Origin of Life*, 3rd Edition (New York: Academic Press, 1957). (b) J. Keosian, *The Origin of Life*, 2nd Edition (New York: Reinhold Book Corp., 1968). (c) G. Wald, *The Origin of Life. Proc. Nat. Sci. Acad. U.S.*, **52**, 595 (1964).

reagent,* and thus it is limited to those chiral processes brought about by a chiral physical force, i.e., elliptically or circularly polarized light. By broadly interpreting the term "chiral reagent" to include a chiral physical force, the original definition of asymmetric synthesis (Sec. **1-2**) encompasses those processes known by the term *absolute asymmetric synthesis.* It is required that the chiral physical force be necessary for the reaction in question, i.e., that without the physical force the reaction will not proceed. Without this condition there is no rational basis for the reaction to take a chiral course.

Two processes can be distinguished: In the first, which can be designated *absolute asymmetric decomposition,* one enantiomer of a racemic pair undergoes preferential reaction under the influence of a chiral physical force; in the second, an achiral substrate is converted into a chiral product by the action of a chiral physical force.

In the literature prior to 1930 there appear scattered reports of unsuccessful or inconclusive attempts at absolute asymmetric decomposition and absolute asymmetric syntheses. These have been reviewed.[15,18,20,93] Werner Kuhn[94a] was the first to set down clearly the conditions necessary for a successful absolute asymmetric decomposition or synthesis; Kuhn and Knopf[93b] were the first to demonstrate a practical example of absolute asymmetric destruction. Racemic N,N-dimethyl-α-azidopropionamide (**89**) was irradiated with levocircularly polarized light (2800–3100 A). After the

$$\underset{\mathbf{89}}{CH_3CH(N_3)\overset{O}{\overset{\|}{C}}NMe_2} \xrightarrow[280\text{–}300\ m\mu]{h\nu} N_2 + \underset{\mathbf{90}}{H_2N{-}\overset{O}{\overset{\|}{C}}{-}NH_2} + ?$$

decomposition was one-third complete, the residual undecomposed amide was isolated and found to have an observed rotation, α_{5791} −1.04°, which corresponds to an excess of 0.7% of the levo isomer. Dextrocircularly

* The term *absolute asymmetric synthesis* has also been used to mean the generation of optically active products without the use of any chiral agents produced by living systems. The asymmetric catalytic decomposition of methyl ethyl malonate on chiral quartz surface would qualify as an absolute asymmetric synthesis according to this viewpoint. We have not considered as absolute asymmetric synthesis those processes which involve the combination of an achiral substrate and an achiral reagent in the presence of either a homogeneous or heterogeneous chiral catalyst. The reasons for the exclusion of such homogeneous catalytic processes is clear, since the catalyst is directly involved on a molecular basis in the reaction. The exclusion of those heterogeneous reactions catalyzed by a chiral crystalline surface of a substance which is chiral only by virtue of its crystalline dissymmetry and not its molecular structure, such as quartz, is a point open to debate. Klabunovskii[15] has considered both of these catalytic processes to be "total asymmetric synthesis" but we have considered such catalysts to be chiral reagents which is in agreement with Pracejus.[14b]

[93] A. Amariglio, H. Amariglio, and X. Duval, *Ann. Chim.* (France), **3**, [14], 5 (1968).

[94] (a) W. Kuhn, in *Stereochemie*, ed. K. Freudenberg (Leipzig and Vienna: Franz Deuticke, 1933), pp. 382ff. (b) W. Kuhn and E. Knopf, *Z. physikal. Chem.*, **7B**, 292 (1930).

polarized light caused preferential destruction of the (−) enantiomer leaving the (+) enantiomer in excess.

The α-azido group shows an absorption maximum in the region 3100–2800 A. By using pure (+) and (−) isomers of **89**, it was shown that the (+) form had a 2.4% greater absorption for levocircularly polarized light than the (−) form in this region. In accord with theoretical considerations it was the (+) isomer which was preferentially destroyed by the levocircularly polarized light and vice versa. A calculation which takes into consideration the difference in absorption of the enantiomers and the amount of α-azido compound decomposed predicts 0.84% e.e. in reasonable agreement with that observed.

Attempts to demonstrate true absolute asymmetric synthesis, as contrasted to this example of absolute asymmetric decomposition, have been either negative, inconclusive, or controversial. This is discussed in detail in Chapter **10**.

1969 Addenda

Sec. 1-1. Enzymatic aspects of asymmetric synthesis have now been thoroughly reviewed in a two-volume monograph by Ronald Bentley.[95]

Sec. 1-3. The necessity for distinction between "optical purity," determined by the ratio of observed and maximum rotation, and enantiomeric composition, determined on an absolute basis, has been demonstrated by the finding[96] that there are large deviations from linearity between optical rotation and enantiomeric composition with α-methyl-α-ethylsuccinic acid.

Sec. 1-3. An experiment from the steroid field which approximates the hypothetical example used in **12** and **14**, p. 18, is the reaction of achiral crotyl bromide with estradiol to form *trans*-19-propenylandrosta-1,4-diene-17-β-ol-3-one.[97] Thermal rearrangement gives 2-(α-methylalyl)-estradiol (and the 4-substituted position isomer) which upon hydrogenation and ozonolysis gives (*S*)-(+)-α-methylbutyric acid, 75–80% e.e.

Sec. 1-4.2. The analytical resolution of racemic 2-trifluoro-acetylamino-octane by glc using a chiral stationary phase has been described.[98]

[95] R. Bentley, ed., *Molecular Asymmetry in Biology* (New York-London: Academic Press), Vol. 1 (1969), Vol. 2 in press (1970).

[96] A. Horeau, *Tetrahedron Lett.*, 3121 (1969).

[97] H. J. Hansen and H. Schmid, *Chem. in Brit.*, **5**, 111 (1969).

[98] D. A. Mitchard, *J. Org. Chem.*, **34**, 2787 (1969).

Sec. 1-4.3. Pertinent to the discussion of isomerization due to pyramidal inversion at nitrogen is the finding[99] that the nitrogen invertomers of *N*-methoxy-3,3-dimethoxycarbonyl-5-cyano-1,2-oxazoline are stable at room temperature ($E_{act} = 29$ kcal/mole). *N*-Methyl- and *N*-*t*-butyldiphenyloxaziridine have been obtained optically active from the corresponding imines by oxidation with (+)-monopercamphoric acid.[100] The rate of racemization of the *t*-butyl homolog is 8000 times that of the methyl compound at 100°. The interconversion of *syn*- and *anti*-7-chloro-7-azabenzonorbornadienes is also slow[101] ($\Delta G^* = 23.5$ kcal/mole).

Sec. 1-4.3b. Additional studies on the inversion barrier to "swiveling" in *meta* cyclophanes have been reported.[102]

Sec. 1-4.4a. A partial kinetic resolution and production of a chiral olefin of low enantiomeric purity has been accomplished by dehydration of a racemic alcohol using (+)-camphorsulfonic acid as a chiral catalyst.[103]

Sec. 1-4.4b. The Ruch-Ugi stereochemical analogy model has now been reviewed.[104]

[99] K. Müller and A. Eschenmoser, *Helv. Chim. Acta*, **52**, 1823 (1969).
[100] F. Montanari, I. Moretti, and G. Torre, *Chem. Commun.*, 1086 (1969).
[101] V. Rautenstrauch, *Chem. Commun.*, 1122 (1969).
[102] J. R. Fletcher and I. O. Sutherland, *Chem. Commun.*, 1504 (1969).
[103] S. I. Goldberg and N. C. Miller, *Chem. Commun.*, 1409 (1969).
[104] E. Ruch and I. Ugi, *Topics in Stereochemistry*, ed. E. Eliel and N. Allinger (New York: Interscience Pub. Inc., 1969), Vol. 4, pp. 99–125.

2

Reactions of Achiral Reagents with Chiral Keto Esters

2-1. Introduction

Kipping[1] and Cohen and Whiteley[2] in 1900 independently described unsuccessful attempts to obtain an asymmetric synthesis by the general process of creating a new chiral center in the acid moiety of an ester of an optically active alcohol. The original optically active alcohol was removed subsequently by hydrolysis and recovered; the acid moiety of the ester molecule was then inspected for optical activity. Among other experiments,

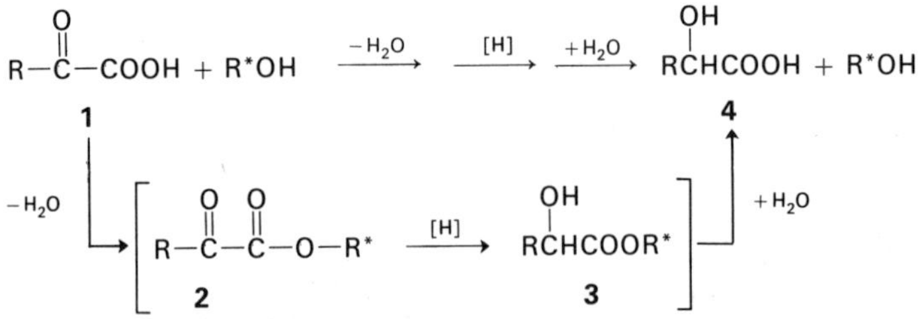

$$\underset{\mathbf{1}}{R-\overset{O}{\overset{\|}{C}}-COOH} + R^{*}OH \xrightarrow{-H_2O} \xrightarrow{[H]} \xrightarrow{+H_2O} \underset{\mathbf{4}}{R\overset{OH}{\overset{|}{C}}HCOOH} + R^{*}OH$$

$$\mathbf{1} \xrightarrow{-H_2O} \left[\underset{\mathbf{2}}{R-\overset{O}{\overset{\|}{C}}-\overset{O}{\overset{\|}{C}}-O-R^{*}} \xrightarrow{[H]} \underset{\mathbf{3}}{R\overset{OH}{\overset{|}{C}}HCOOR^{*}}\right] \xrightarrow{+H_2O} \mathbf{4}$$

Fig. 2-1. Asymmetric reductions of α-keto esters.

[1] F. S. Kipping, *Proc. Chem. Soc.*, **16**, 226 (1900).

[2] J. B. Cohen and C. E. Whiteley, *Proc. Chem. Soc.*, **16**, 212 (1900); *J. Chem. Soc.*, **79**, 1305 (1901).

Kipping described the reduction of the (−)-bornyl esters of the α-keto acids, benzoylformic acid (**1**, R = C_6H_5), and pyruvic acid (**1**, R = CH_3), while Cohen and Whiteley described the reduction of (−)-menthylpyruvate (**2**, R = CH_3, R* = (−)-menthyl). Their theoretical approach was sound, but their experiments failed for practical reasons.*

Neither Kipping nor Cohen pursued these studies further. Four years later McKenzie[3] not only repeated these same reactions but also successfully applied the then recently discovered Grignard reactions to the same system and thereby circumvented the racemization problem. This development transformed the α-keto ester asymmetric synthesis into a practical system for extensive study; McKenzie chose this as his major field of scientific investigation during the next 31 years. This has become a classical model for many asymmetric syntheses and has been developed by Prelog[4] into a valuable procedure for the configurational correlation of secondary alcohols.

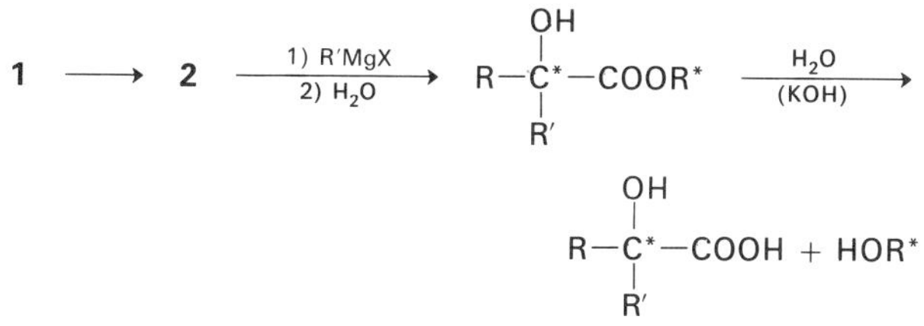

$$\mathbf{1} \longrightarrow \mathbf{2} \xrightarrow[2)\ H_2O]{1)\ R'MgX} R-\overset{\displaystyle OH}{\underset{\displaystyle R'}{C^*}}-COOR^* \xrightarrow[(KOH)]{H_2O} R-\overset{\displaystyle OH}{\underset{\displaystyle R'}{C^*}}-COOH + HOR^*$$

Fig. 2-2. Asymmetric Grignard additions to α-keto esters.

McKenzie's experiments were intimately concerned with the subject of asymmetric induction (Sec. **1-1**) which, in turn, was closely associated with the phenomenon of mutarotation, since the same optically active α-keto esters which were used in the asymmetric synthesis experiments also showed mutarotation in alcoholic solvents.[5] These two phenomena were correlated by showing that when the mutarotation was positive the asymmetric synthesis gave levorotatory lactic or atrolactic acids and vice versa; only in five out of thirty-four cases did an ester with one sign of mutarotation give a lactic or atrolactic acid of the same sign. As late as 1935, these muta-

* It has now been established (see Table **2-4**) that asymmetric syntheses did take place in these reactions. However, during the basic hydrolysis of (−)-bornyl mandelate, the mandelate ester portion was racemized by virtue of the enolizable α-hydrogen. The α-hydrogen of lactate esters is less acidic and it has been shown that lactate esters can survive hydrolysis under mild conditions without racemization. However, because these workers isolated the lactic acid (**4**, R = CH_3) from the pyruvate asymmetric reduction as its crystalline zinc salt, it is very possible that the racemic salt separated, leaving any optically active lactic acid in the mother liquors.

[3] (a) A. McKenzie, *J. Chem. Soc.*, **85**, 1249 (1904). (b) H. Wren, *J. Chem. Soc.*, 270 (1952).

[4] V. Prelog, *Helv. Chim. Acta*, **36**, 308 (1953).

[5] A. McKenzie and E. W. Christie, *Biochem. Z.*, **277**, 426 (1935).

rotations and asymmetric syntheses were interpreted in terms of an asymmetry which was "induced" in the α-keto group by the optically active alcohol moiety (Fig. **2-3**). It is clear that formulas **5** and **6**, which are as represented

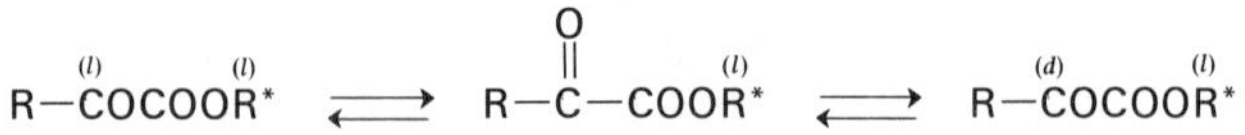

$$\underset{\mathbf{5}}{R\!-\!\overset{(l)}{C}O\overset{(l)}{COOR^*}} \rightleftarrows \underset{\mathbf{2}}{R\!-\!\overset{O}{\overset{\|}{C}}\!-\!\overset{(l)}{COOR^*}} \rightleftarrows \underset{\mathbf{6}}{R\!-\!\overset{(d)}{C}O\overset{(l)}{COOR^*}}$$

Fig. 2-3. Asymmetric induction according to McKenzie.

by McKenzie and Mitchell, were meant to show a keto group which was chiral in its ground state. The source of the mutarotation in alcoholic solvents now has been satisfactorily explained on the basis of hemiacetal formation[6,7]

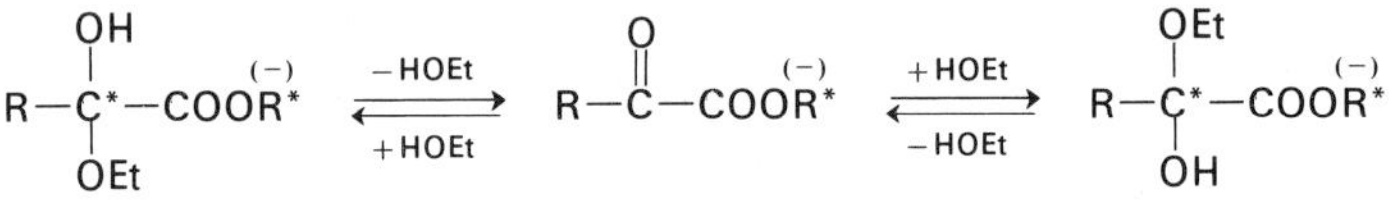

$$\underset{\mathbf{7}}{R\!-\!\overset{OH}{\underset{OEt}{C^*}}\!-\!\overset{(-)}{COOR^*}} \underset{+HOEt}{\overset{-HOEt}{\rightleftarrows}} \underset{\mathbf{2}}{R\!-\!\overset{O}{\overset{\|}{C}}\!-\!\overset{(-)}{COOR^*}} \underset{-HOEt}{\overset{+HOEt}{\rightleftarrows}} \underset{\mathbf{8}}{R\!-\!\overset{OEt}{\underset{OH}{C^*}}\!-\!\overset{(-)}{COOR^*}}$$

Fig. 2-4. Mutarotation of α-keto esters in alcoholic solution.

(Fig. **2-4**). Since **7** and **8** are diastereomers, they will be produced in unequal amounts with a resulting change in the optical rotation of the alcoholic solution of the α-keto ester **2** until a new equilibrium value is established. The idea of "asymmetric induction" as a dissymmetric physical effect which can be transmitted from one part of a molecule to another through bonds was tested by Kubitscheck and Bonner[8] who were unable to detect optical activity in the product (**10**) from either the addition of methylmagnesium iodide to (−)-menthyl *p*-benzoylbenzoate (**9**) or from its reduction. They cautiously concluded that if "asymmetric induction" in the McKenzie sense exists, it is incapable of transmission through an aromatic nucleus.

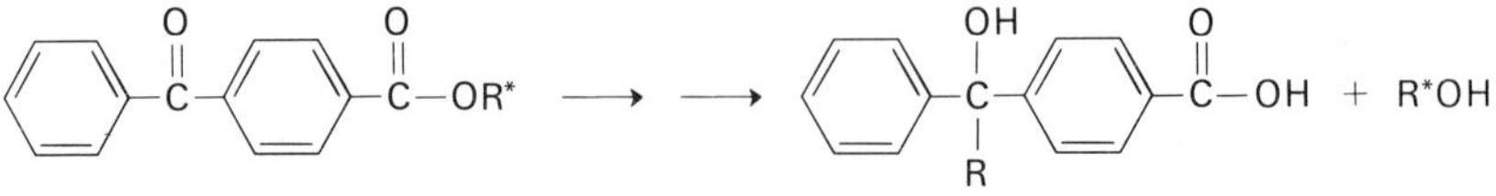

$$\underset{\mathbf{9}}{C_6H_5\!-\!\overset{O}{\overset{\|}{C}}\!-\!C_6H_4\!-\!\overset{O}{\overset{\|}{C}}\!-\!OR^*} \longrightarrow \longrightarrow \underset{\mathbf{10}}{C_6H_5\!-\!\overset{OH}{\underset{R}{C}}\!-\!C_6H_4\!-\!\overset{O}{\overset{\|}{C}}\!-\!OH} + R^*OH$$

$R^* = (-)$-Menthyl
$R = H$ or Me

[6] M. M. Jamison and E. E. Turner, *J. Chem. Soc.*, 538 (1941).
[7] E. E. Turner and M. M. Harris, *Quart. Rev.*, (London) **1**, 299 (1947).
[8] M. J. Kubitscheck and W. A. Bonner, *J. Org. Chem.*, **26**, 2194 (1961).

As shown by Prelog,[4,9] asymmetric syntheses can be rationalized on the basis of normal nonbonded interactions in the transition state. The same factors which influence the course of the kinetically controlled asymmetric addition reactions (Figs. **1-1** and **1-2**) may also be operative in the thermodynamically controlled mutarotations (Fig. **2-4**), and the respective results might therefore parallel each other.

Before a detailed consideration of the many asymmetric syntheses starting with α-keto esters, it is necessary to appreciate certain experimental details of the reaction sequence represented by Figs. **2-1** and **2-2**. There is no special problem associated with the synthesis of the optically active ester of the α-keto acid, which should of course be made from enantiomerically pure carbinol and purified before it is subjected to the asymmetric addition step. But great care must be taken with the isolation of the atrolactic acid (or other substituted lactic acid) to insure that it correctly reflects the ratio of diastereomers produced in the asymmetric synthesis step. Two factors are important: (a) the total yield from the α-keto ester stage to the recovered α-hydroxy acid and (b) the isolation procedure. If the yield is not high, there is a distinct possibility that the rotation of the isolated α-hydroxy acid does not reflect the true ratio of the epimeric ester precursors. If the initially formed epimeric α-hydroxy esters are involved in a side reaction, the remaining mixture of epimers will have a different composition than that initially obtained from the asymmetric synthesis reaction, because the epimeric esters react at different rates with both chiral and achiral reagents. For example, the initially formed halomagnesium salt of the α-hydroxy ester might react with excess Grignard reagent at the ester group thereby destroying one diastereomer faster than the other. The same considerations apply to the hydrolysis step. Thus, if the hydrolysis is interrupted before it is complete, the enantiomeric purity of the isolated α-hydroxy acid will not be an accurate measure of the ratio of diastereomers initially produced. Not only may the quantitative ratio differ but there even may be a change in the predominant isomer, as shown by Prelog.[4] Phenylmagnesium bromide was added to $(-)$-menthyl α-naphthylglyoxylate and the resulting menthyl phenyl-α-naphthylglycolate completely saponified with alcoholic potassium hydroxide (1 hr reflux) to give a 79% yield of phenyl-α-naphthylglycolic acid, $[\alpha]_D$ $+2.14°$ (c, 8.88; chloroform). However, in a duplicate run, when the saponification was carried out for 10 min, there was obtained a 23.5% yield of the glycolic acid with $[\alpha]_D$ $-2.3°$ (c, 8.68; chloroform). From a theoretical standpoint, the yield of atrolactic acid should be nearly 100%, but from a practical standpoint if the yield drops much below 80%, the results are suspcct for quantitative interpretation. If the yield is much below 60% the results cannot be relied upon, even to determine the $(+)$ or $(-)$ sense of the

[9] V. Prelog, *Bull. Soc. Chim. Fr.*, 987 (1956).

asymmetric synthesis, at least not without a study of the reaction for the purpose of determining the reason for the reduced yield.

The α-hydroxy acid from the saponification must be purified, but in a manner which will not change its isomer composition. This means it must not be crystallized! The basic saponification mixture can be extracted and/or steam-distilled to remove neutral impurities and treated with decolorizing carbon to remove colored impurities. When the basic solution is acidified, the precipitated solid acid should not be collected since preferential crystallization of one isomer or the racemate may have occurred. Instead, the mixture must be extracted exhaustively and the dried extracts evaporated to complete dryness in a vacuum below 70°* to isolate the total yield of α-hydroxy acid. The rotation should be taken on this material. If it is too deeply colored, the rotation can be taken on a sublimed or chromatographed sample; in the latter case, however, care must be taken to insure that only impurities are removed and that one enantiomer is not inadvertently separated from the racemate. Finally the temperature, solvent, concentration, and wavelength used in the determination of the rotation should be carefully noted so that valid comparison of polarimetric data can be made. This is readily apparent from the rotation of atrolactic acid[10] taken in various solvents at different concentrations and with different wavelengths of light. Furthermore, one must appreciate the limits of error in the calculated percent asymmetric synthesis. If the observed rotation of a sample of atrolactic acid is $\alpha_D^{13.8}$ $-1.00 \pm 0.02°$ (c, 5; ethanol; $l = 1$), then the specific rotation is $[\alpha]_D^{13.8}$ $-20.0 \pm 0.4°$ (c, 5; ethanol) and the calculated asymmetric synthesis is $53 \pm 1\%$. If, however, the observed rotation is $\alpha_D^{13.8}$ $-0.10 \pm 0.02°$ (c, 2; ethanol, $l = 1$), the calculated asymmetric synthesis is $13 \pm 2\%$. Polarimetric measurements can be much more precise than indicated here and modern instruments can give values within 0.001°; but unless the limits of error are specified, percent asymmetric syntheses should not be considered more precise than ± 1 to 2%. The percent asymmetric synthesis data in the following tables have been rounded to the nearest percent. These uncertainties, plus the additional ones arising from variations in experimental conditions, must be kept in mind when evaluating the following data, especially that from the older literature. As discussed in Sec. **1-3**, it is not necessary to use polarimetry to determine the enantiomeric purity of a product such as atrolactic acid, although this has been used almost universally in the past.

If the asymmetric synthesis is brought about by reduction of the optically active benzoylformate ester (**11**), instead of by treatment with a Grignard

* Above this temperature, atrolactic acid undergoes slow dehydration to atropic acid.

[10] A. McKenzie and G. W. Clough, *J. Chem. Soc.*, **97**, 1016 (1910); (−)-atrolactic acid, $[\alpha]_D^{13.8}$ −37.7° (c, 3.35; ethanol), $[\alpha]_{5461}^{20}$ −45.7 (c, 3.0; ethanol), $[\alpha]_D^{14}$ −36.5 (c, 4.14; acetone), $[\alpha]_D^{14.5}$ −51.1 (c, 2.16; water), $[\alpha]_D^{14.5}$ −53.8° (c, 0.685; water); sodium salt, $[\alpha]_D^{13.5}$ −56.5° (c, 1.9; water).

reagent (Fig. **2-2**), the resulting mandelate ester (**12**) is quite susceptible to base catalyzed racemization by virtue of the hydrogen being *alpha* to both the ester function and the aromatic ring. The epimeric purity of the mixture of diastereomeric mandelic esters (**12**) can be determined indirectly with only slight danger of racemization by lithium aluminum hydride reduction to the corresponding glycols (**13**) (cf. Table **2-4**, footnote d).

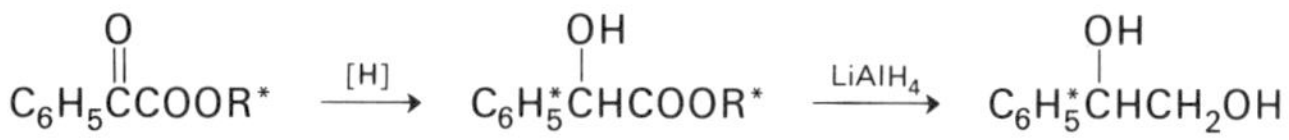

$$\underset{\mathbf{11}}{C_6H_5\overset{O}{\overset{\|}{C}}COOR^*} \xrightarrow{[H]} \underset{\mathbf{12}}{C_6H_5\overset{OH}{\overset{|}{\overset{*}{C}}}HCOOR^*} \xrightarrow{LiAlH_4} \underset{\mathbf{13}}{C_6H_5\overset{OH}{\overset{|}{\overset{*}{C}}}HCH_2OH}$$

2-2. Prelog's Generalization

Prelog's generalization[4,9] covering the relationship between the configuration of the optically active alcohol of an α-keto ester and the steric course of the asymmetric reaction has been introduced (Sec. **1-4.4b**) and can best be summarized as follows.[4,9] The α-keto ester (**14**) is considered with its two carbonyl groups in the anticoplanar conformation and with the three groups attached to the chiral carbinol carbon atom arranged so that the two smaller groups R_S and R_M are in staggered positions on either side of the ester carbonyl as shown in **14**. According to this model, the α-hydroxy acid (**15**) resulting from addition to the keto carbonyl group will have that configuration represented by approach of the reagent from the side of the smaller R_S group* as indicated in Fig. **2-5**. The reagent is represented here

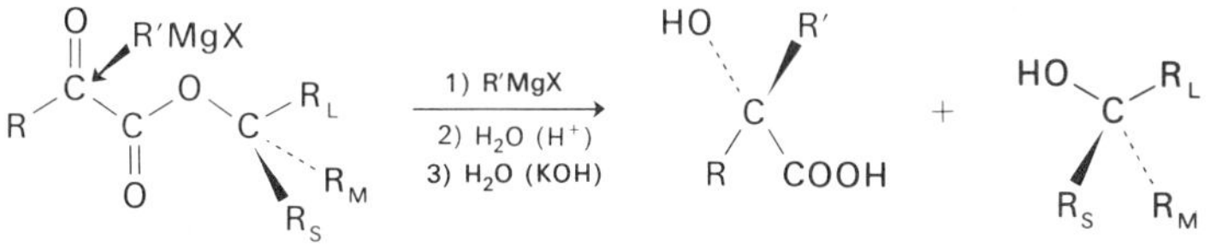

R′MgX
1) R′MgX
2) H_2O (H^+)
3) H_2O (KOH)
HO R′ C R COOH + HO C R_L R_S R_M
14 **15**

Fig. 2-5. Prelog's generalization for steric control of α-keto ester asymmetric synthesis.*

as R′MgX, but it may be sodium borohydride, lithium alkyl, etc. Subsequent hydrolysis will thus give the α-hydroxy acid with the indicated enantiomer (**15**) in excess. If the enantiomeric carbinol is used, the enantiomeric α-hydroxy acid will be produced in excess. By determining the configuration of the acid, the configuration of the alcohols used for making the keto esters can be deduced (or vice versa).

* We shall use R_S, R_M, and R_L to denote small, medium and large groups where Cram and Prelog have used S, M, and L. R_S is usually hydrogen.

This generalization for predicting the stereochemistry of the products from α-keto ester asymmetric syntheses was put forward as an empirical model and yet its remarkable success implies some correlation of the model with the conformation of the transition state. It has been suggested by Cornforth and co-workers, in connection with the dipolar model for the Cram-type asymmetric syntheses (Sec. **3-4**), that the *trans*-coplanar conformation (Fig. **2-5**, formula **14**), although it may not represent the most stable ground state, it may represent the most stable transition state because in this form the greatest separation of charge between negative centers is possible (**16A**).

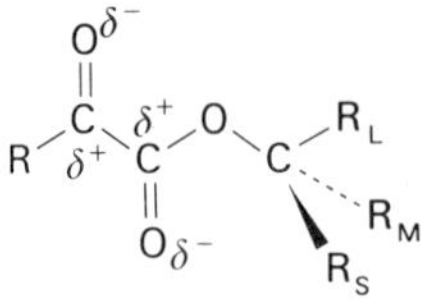

16A 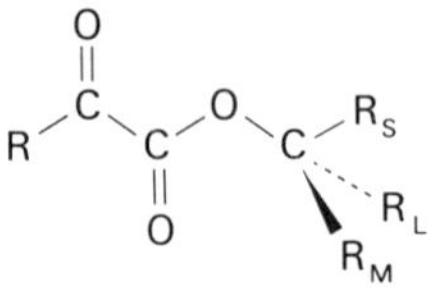 **16B** 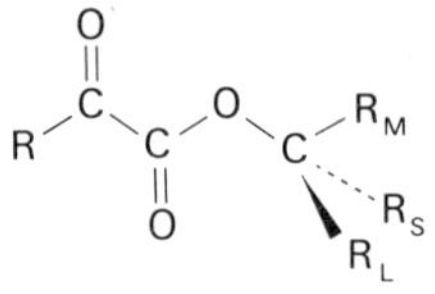 **16C**

The rationale for choosing the conformation having the small and medium groups staggered with respect to the ester carbonyl is largely intuitive. Once the staggered conformation is chosen, it follows that **16A** and **16B**, of the three conformations (i.e., **16A**, **16B**, and **16C**) corresponding to rotation around the carbon-oxygen bond, present lesser steric hindrance to the front face of the α-keto carbonyl group as compared to the back side of this carbonyl group, while only **16C** offers less steric hindrance to attack at the back face of the carbonyl group. Either model **16A** or **16B** can be used to predict the favored product. Prelog has chosen* **16A** but several authors have used **16B**. There is extensive discussion of the analogous point in connection with the consideration of Cram's rule (Sec. **3-4**).

Prelog was careful to point out that R_S, R_M, and R_L should be hydrocarbon residues (or hydrogen in the case of R_S) which are clearly different in their steric requirements. He also emphasized the necessity of having a substantial extent of asymmetric synthesis before reliable configuration conclusions could be drawn. From a practical standpoint an atrolactic acid asymmetric synthesis probably should be above 5% before it is used for a configurational correlation. A 5% asymmetric synthesis at room temperature, i.e., a reaction which produces a 52.5:47.5 ratio of isomers, corresponds to competing reactions which have a difference in their energies of activations of about fifty small calories per mole. It is unreasonable to expect

* In Prelog's original publication[4] a representation equivalent to **16B** was taken as the conformational model for this reaction; however, in his subsequent review[9] **16A** was chosen. We shall, therefore, refer to **14**, which incorporates conformation **16A**, as the Prelog model.

that, in diverse and complex situations such as this, we can accurately forecast effects which differ by less than fifty calories (0.05 kcal). On the other hand, it is amazing that we are able to predict with reasonable confidence, effects which are the result of energy differences in this range. This is possible in any one reaction because the competing processes are compared under identical conditions in the same reaction mixture and the energy of the reactants is identical (Fig. **1-3**). Success in comparing asymmetric syntheses from different α-keto esters lies in the choice of examples which do not differ too widely. This latter choice becomes a matter of judgment.

The problem of ordering groups R_S, R_M, and R_L as to relative size has been considered (Sec. **1-4.4a**). It is simplified in the present application by two considerations; since R_S has been hydrogen in almost every case studied in the α-keto ester system, a decision must be made only between R_M and R_L. If a clear distinction cannot be made between R_M and R_L, the extent of asymmetric reaction probably will be small and the method, strictly speaking, will not be applicable for configuration determination in such a case.

In the application of this method to the determination of configuration of secondary alcohols, the action of methyl magnesium iodide on optically active esters of benzoylformic acid has been used universally. When R_L

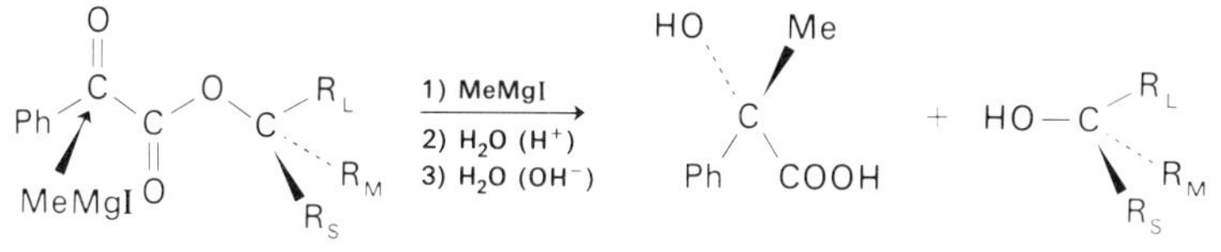

1) MeMgI
2) H_2O (H^+)
3) H_2O (OH^-)

17A (preferred) **18A** (*S*)-(+)-Atrolactic acid

Fig. 2-6. Prelog's atrolactic acid asymmetric synthesis model.

and R_M are saturated hydrocarbon residues which differ only in chain branching, then the ordering of groups for R_M and R_L will parallel the ordering of groups for the Cahn–Ingold–Prelog nomenclature scheme. As long as one chooses examples which fit this simplified situation and uses the reaction of methyl Grignard reagent on the benzoylformic ester, alcohols of *R* configuration at the carbinol carbon atom will lead to (*R*)-(−)-atrolactic acid while alcohols with the *S* configuration will lead to (*S*)-(+)-atrolactic acid. But this parallelism of group size and Cahn–Ingold–Prelog nomenclature ordering disappears with the introduction of either unsaturation or hetero atoms and thus each case must be individually analyzed. Furthermore, there may be complications due to overall molecular topology (e.g., other asymmetric centers in R_M or R_L) which will invalidate a simple analysis based upon the extent of branching at the carbon atoms adjacent to the carbinol center.

Of the 120 different examples taken from the literature and reported in Tables **2-1** through **2-6**, there are perhaps four cases which at first glance might be considered exceptions to the Prelog generalization. These are seen, on closer examination, to be examples to which the Prelog generalization does not really extend. It will be profitable to consider some of these special cases by discussing specific examples. However, we shall first consider a normal example for the purpose of comparison.

When (+)-3,3-dimethyl-2-butanol [(+)-methyl-*t*-butylcarbinol] was converted into its benzoylformate ester, which in turn was treated with the Grignard reagent from methyl iodide and the diastereomeric atrolactic esters hydrolyzed, atrolactic acid was obtained in 82% yield with $[\alpha]_D$ +8.7° (c, 7.6; ethanol).[11] This means that the (*S*)-(+)-atrolactic acid was produced in 24% excess over the racemic atrolactic acid,* i.e., a 24% asymmetric synthesis. Since (*S*)-(+)-atrolactic acid was formed in good yield with substantial stereoselectivity, this clearly belongs to the class of reactions represented by **17A** and **18A** in Fig. **2-6**. The *tert*-butyl group is obviously larger than the methyl group (as well as taking notational preference in the Cahn–Ingold–Prelog scheme). Consequently, according to the Prelog generalization, (+)-3,3-dimethyl-2-butanol must have the absolute *S* configuration† represented by **19**.

HO–C(R_L)(R_M)(R_S)

19A

H–C(R_M)(R_L)–OH

19B

H–C(Me)(C(CH_3)$_3$)–OH

19C

When 3-*β*-cholestanol (**20**) was subjected to the atrolactic acid asymmetric sequence, there was obtained an 82% yield of atrolactic acid, $[\alpha]_D$ −0.65°, corresponding to a 1.7% excess of the *R* enantiomer.[9] This extent of stereoselectivity is too low to permit any reliable assignment of configuration based upon the Prelog model. The chiral center at C-3 is known to have the *S* configuration. It is flanked on both sides by methylene groups (at C-2 and C-4) and thus the determining feature for designation of R_L

[11] V. Prelog, E. Philbin, E. Watanabe, and M. Wilhelm, *Helv. Chim. Acta*, **39**, 1086 (1956).

* The maximum rotation reported for atrolactic acid in ethanol is $[\alpha]_D^{13.8}$ −37.7° (c, 3.3; ethanol).[10] Although there may be small variations of [α] with concentration changes in ethanol, the concentration has not been reported in many cases and we shall uniformly use this value as the basis for calculating the percent asymmetric synthesis of atrolactic acid when the rotation is reported in ethanol.

† This absolute configuration has now been rigorously proved; J. Jacobus, Z. Majerski, K. Mislow, and P. von R. Schleyer. *J. Am. Chem. Soc.*, **91**, 1998 (1969).

versus R_M, as well as for the notational preference, requires that one proceed beyond the *alpha* methylenes to the *beta* carbons at C-1 and C-5. The greater branching at C-5 is definitive in designating the chiral center at

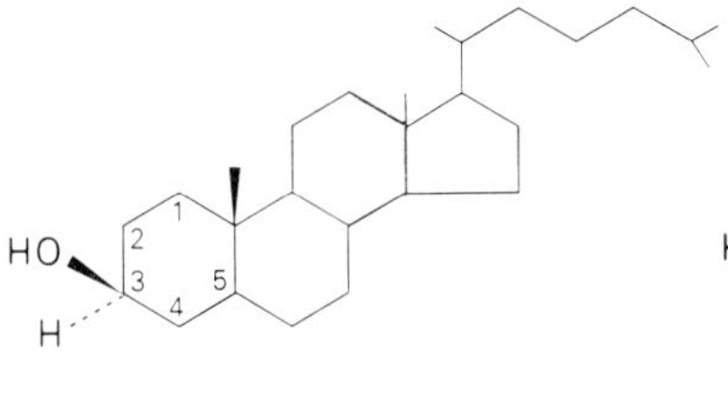

20

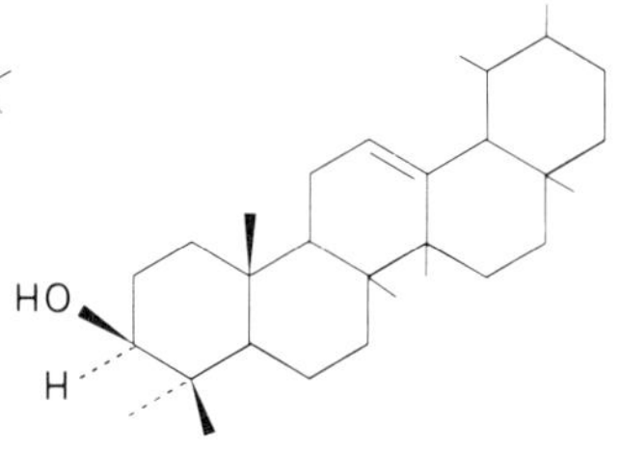

21

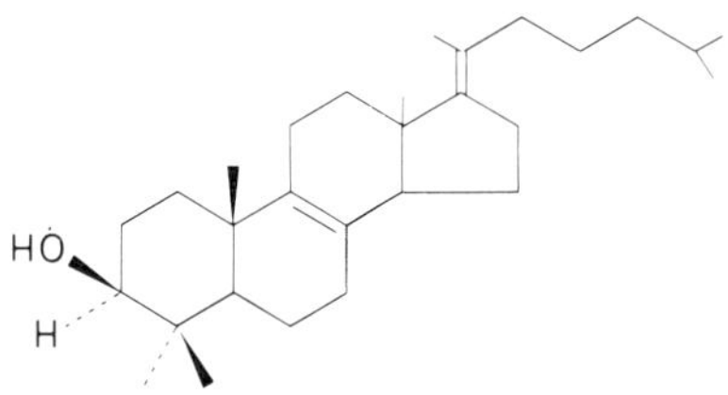

22

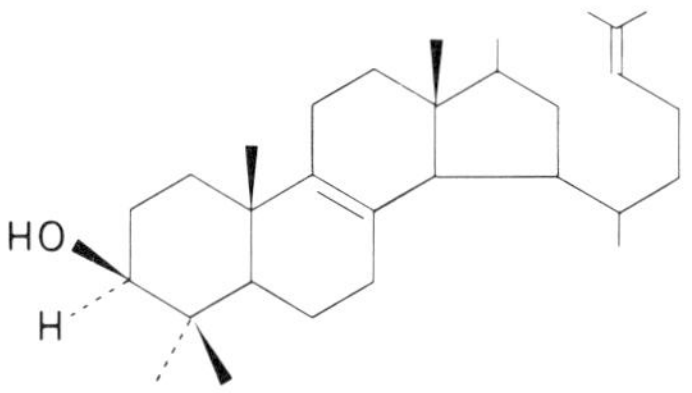

23

C-3 as *S* and in a formal sense would also be determining in designating the C-4, C-5 branch as R_L over the C-2, C-1 branch as R_M. This latter conclusion is at best presumptuous and is not obviously verified by inspecting molecular models. Consequently from the standpoint of a stereochemical analysis as well as the marginal asymmetric synthesis observed, this is an example to which the Prelog generalization should not be applied. But when the triterpenes α-amyrin (**21**), dihydrolanosterol (**22**), and euphol (**23**) were similarly treated, the asymmetric syntheses of (*S*)-(+)-atrolactic acid were substantial in each case—10, 34, and 24%, respectively.[10] The *gem*-dimethyl group at C-4 entirely changes the picture from that encountered with cholestanol (**20**) to that found with (*S*)-(+)-3,3-dimethyl-2-butanol (**19C**). Thus these results are in complete accord with the *S* configuration which has been assigned independently to **21**, **22**, and **23**.

A consideration of the configuration of the alcohol group at C-6 in the morphine series is instructive. When the benzoylformate ester of dihydrocodeine (**24A**) is treated with the Grignard reagent from methyl iodide, and the resulting atrolactic ester hydrolyzed, a 15% asymmetric synthesis of *R*-(−)-atrolactic ester is observed (10% chemical yield).[12] The configuration at C-6 is known to be *S*. If one were to designate C-5 as R_L and C-7 as R_M

[12] K. W. Bentley and H. M. E. Cardwell, *J. Chem. Soc.*, 3252 (1955).

and *formally* apply the Prelog generalization, one would predict that *S*-(+)-atrolactic acid would predominate, contrary to observation. Cardwell

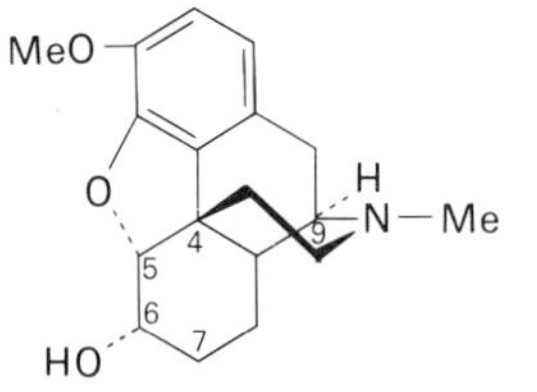

24A 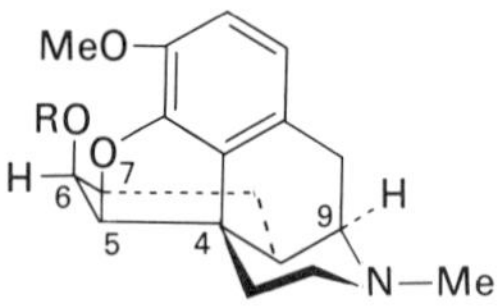 24B (R = PhCOCO) 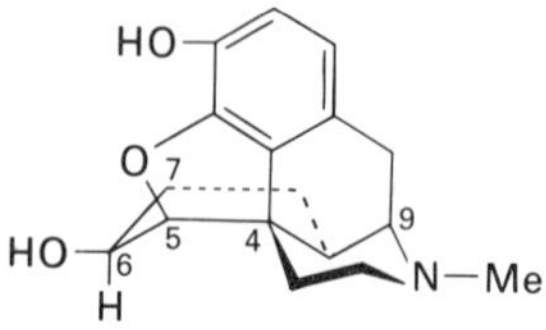 24C*

and Bentley, by inspecting molecular models, concluded that in the Prelog model the methoxy group of the aromatic ring is so situated (**24B**) that it protects one face of the α-keto group with the result that it is the (*R*)-(−)-form of atrolactic acid which should predominate, as is observed. From this they concluded that the ring juncture at C-9 must be as shown in **24B**. There are many assumptions in such an analysis but fundamentally the Prelog generalization specifies that R_S, R_M, and R_L should be hydrogen or hydrocarbon residues and this injunction specifically excludes the application of the generalization to such cases as dihydrocodeine (**24A**) which has an oxygen attached to R_L. This same situation is encountered to an even greater extent in some examples taken from the sugar series.

Kawana and Emoto[13] have carried out the atrolactic acid asymmetric synthesis sequence on a series of sugar derivatives which are typified by the benzoylformate of 1,2:5,6-di-O-isopropylidene-α-D-glucofuranose (**25A**). A 33% asymmetric synthesis of (*R*)-(−)-atrolactic acid in 78% chemical yield

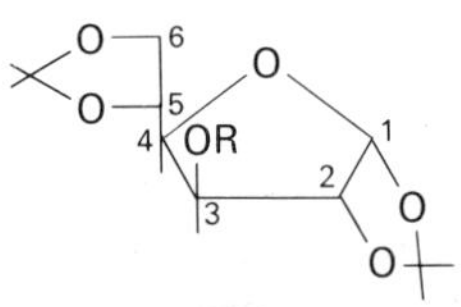

25A
R = PhCOCO⁻

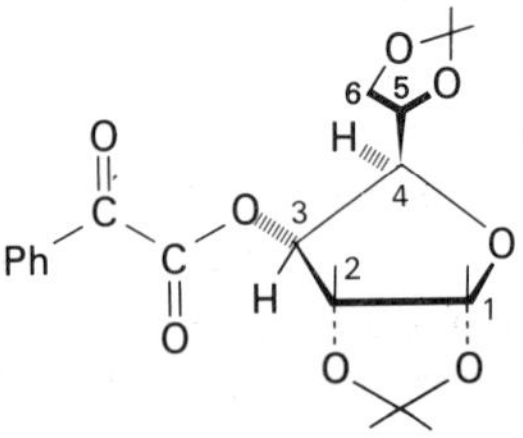

25B

was realized (an average of two closely corresponding experiments). From a steric requirement standpoint the substituent at C-4 is *cis* to the hydroxyl

* The X-ray crystallographic determined conformation of morphine [M. MacKay and D. C. Hodgkin, *J. Chem. Soc.*, 3261 (1955)] shows the crucial ring in a pseudoboat form (**24C**) in contrast to the chair form **24B** assumed for dihydrocodeine by Bentley and Caldwell.[12]

[13] (a) M. Kawana and S. Emoto, *Bull. Chem. Soc. Jap.*, **40**, 2168 (1967). (b) M. Kawana and S. Emoto, *ibid.*, **41**, 259 (1968).

group at C-3 and C-4 would thus apparently qualify as R_L as compared with the C-2 carbon atom, R_M, whose substituent is *trans* to the hydroxyl group at C-3. These assumptions would in fact bring the example into formal agreement with the Prelog generalization. The authors of these studies, however, state that inspection of molecular models indicated that in the Prelog model the isopropylidene group at C-5, C-6 covers the front face of the α-keto group. This would hinder front-side attack and would favor attack from the back which would lead to (*R*)-(−)-atrolactic acid as observed.

Kawana and Emoto therefore proposed a slightly different preferred reacting conformation, namely **25B**, in which the alcohol-oxygen bond is extended behind the plane of the page. This analysis loses sight of the point that the groups neighboring the chiral carbinol center are not hydrocarbon residues and as such are not properly included for consideration according to the Prelog generalization. The extent of complexing of the Grignard reagent with the various oxygens in such a molecule is not known and could easily have a profound effect on the steric course of the asymmetric synthesis, either by increasing the effective size of one group or another or by bringing the Grignard reagent into a more favorable position for faster reaction at one face of the α-keto group than the other. In fact, a compound such as **25** is a perfect example of the type of situation to which the Prelog generalization is completely inapplicable. Of course, examples of this type should be studied so that the multiple variables in such complex systems can be evaluated and better understood, but the Prelog model per se should not be applied to them.

When (−)-epi-catechin tetramethyl ether (**26A**)[14] was subjected to the atrolactic sequence, a 50% crude yield of atrolactic acid was obtained which was too colored for rotational studies. It was recrystallized to give an unspecified yield of (−)-atrolactic acid, $[\alpha]_D$ −16.4°, corresponding to a 43% excess of the *R* isomer. The low yield and recrystallization of product introduce serious experimental uncertainties and the several oxygens in the molecule certainly complicate the model still further. Nevertheless, the generalization was applied with R_L being assigned to the C-2 carbon atom and R_M to the C-4 methylene group. An experiment on the trimethyl ether of (−)-epi-afzelechin, the 3′-demethoxy analog of **26A**, gave very similar results. But when the same sequence was conducted on the epimeric (+)-catechin tetramethyl ether (**26B**), a 60% yield of crude material was obtained. This was colored, and after two crystallizations (unspecified yield) had $[\alpha]_D^{21}$ −1.1° corresponding to 3% excess of the *R*-(−)-isomer. Since (−)-epi-catechin (**26A**) is epimeric at C-3 with (+)-catechin, the two should have given enantiomeric atrolactic acids in excess. In a formal sense this is a failure of the Prelog generalization; but in a practical sense, the low stereoselectivity

[14] A. J. Birch, J. W. Clark-Lewis, and A. V. Robertson, *J. Chem. Soc.*, 3586 (1957).

and especially an oxygen atom near the chiral center clearly exclude this example from serious consideration. Yet an attempt was made to rationalize these results based upon studies of the conformation of catechin. Furthermore the proposed axial 3-hydroxy conformation for catechin most likely would not be valid for the benzoylformate ester.

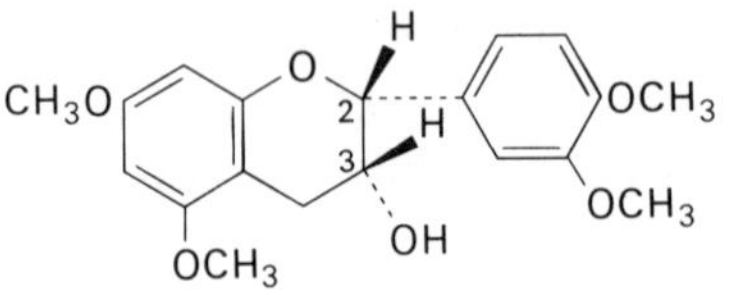

26A 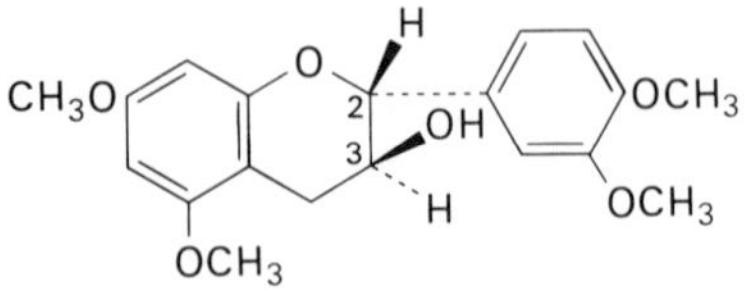 **26B**

The atrolactic asymmetric synthesis has been applied successfully to several cases in which the chirality of the molecule is due to atropisomerism rather than to an asymmetrically-substituted carbon atom at the alcohol function. The (−)-dinaphthylcarbinol, **27**, exists in enantiomeric forms by virtue of the chirality of the molecule as a whole and not because of chirality centered at the carbinol carbon atom (as can be seen by interchanging the hydroxyl and hydrogen, a process which gives no net change in the geometry of the system).[15] One cannot designate R_M and R_L in the usual way in this

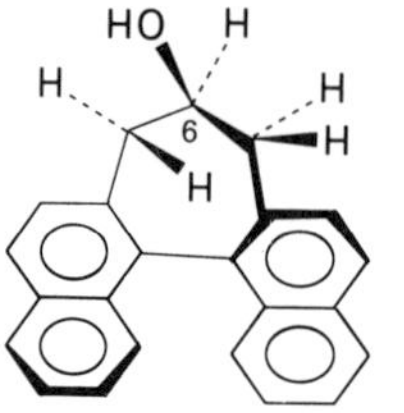

27 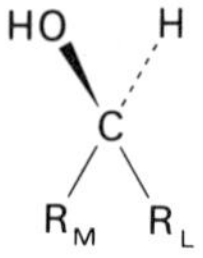 **28**

case and so this involves an extension of the original model. It is quite clear, however, in viewing the naphthyl rings from the vantage point of the hydroxyl group at C-6, that the ring which protrudes to the right will exercise much more steric influence than that which recedes to the left. Thus, for the purpose of the atrolactic asymmetric synthesis model, formula **27** is equivalent to **28**. This latter configuration in the Prelog model leads to (*R*)-(−)-atrolactic acid. When (−)-dinaphytho-2′,1′:1,2;1″,2″:3,4-cycloheptadiene-6-ol (**27**) was subjected to the atrolactic acid asymmetric synthesis sequence, atrolactic acid (69% yield, 20% e.e.) with the *R* configuration as predicted was formed. This is the same configuration as that arrived at by asymmetric reduction and optical rotatory dispersion (ORD) studies.

[15] K. Mislow, V. Prelog, and H. Scherrer, *Helv. Chim. Acta*, **41**, 1410 (1958).

The highest stereoselectivities of this type have been observed by Berson and Greenbaum[16] who subjected the (+)-biphenyl **29** and the (+)-binaphthyl **30** to the atrolactic acid sequence and obtained (−)-atrolactic acid from both with 93 and 85% asymmetric syntheses respectively (61 and 41% yields).

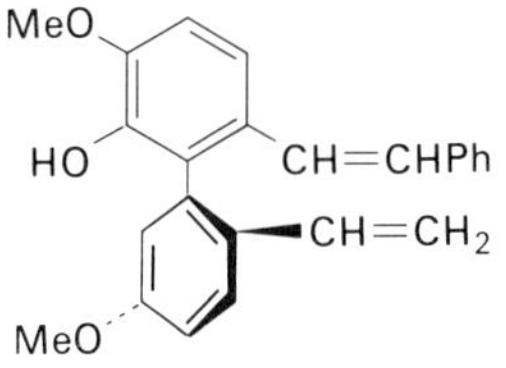

R-(+) **29** 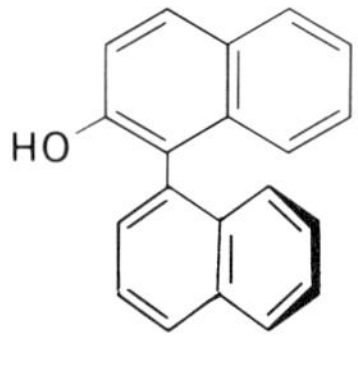 *R*-(+) **30**

In accord with the extended Prelog rule, these biphenyls were assigned the *R*-configurations which have been confirmed by independent evidence (Sec. **9-2.1**). Obviously this method can be of great value for configurational correlations of the atropisomer types.

When (−)-menthyl β-phthalonate (**31**) was treated with the Grignard reagent from methyl iodide, the stereochemical results were reversed from those expected from the Prelog model.[17] Thus there was obtained a 62%

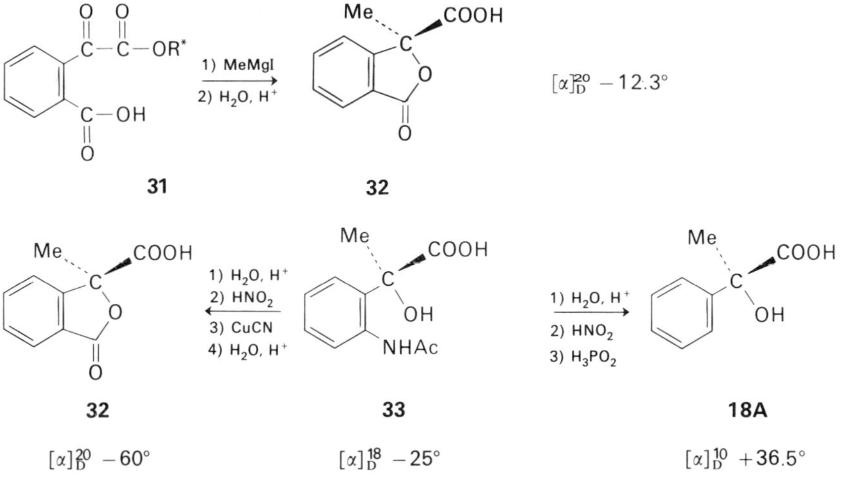

31 → 1) MeMgI; 2) H_2O, H^+ → **32** $[\alpha]_D^{20}$ −12.3°

32 $[\alpha]_D^{20}$ −60° ← 1) H_2O, H^+; 2) HNO_2; 3) CuCN; 4) H_2O, H^+ ← **33** $[\alpha]_D^{18}$ −25° → 1) H_2O, H^+; 2) HNO_2; 3) H_3PO_2 → **18A** $[\alpha]_D^{10}$ +36.5°

Fig. 2-7. Asymmetric synthesis of (−)-3-methyl-3-carboxyphthalide and its configurational correlation with (+)-atrolactic acid.

[16] J. A. Berson and M. A. Greenbaum, *J. Am. Chem. Soc.*, **79**, 2340 (1957); (b) *ibid.*, **80**, 445 (1958); (c) *ibid.*, **80**, 653 (1958); (d) *ibid.*, **81**, 6456 (1959).

[17] M. N. Kolosov, A. I. Gurevich, and Yu B. Shvetsov, *Izv. Akad. Nauk SSSR, Otd. Khim. Nauk*, 701 (1963); *Chem. Abst.*, **59**, 8675 (1963).

yield of (−)-3-methyl-3-carboxyphthalide (**32**), $[\alpha]_D^{20}$ −12.3°, corresponding to a 20% e.e. The configuration and enantiomeric purity were established as indicated in Fig. **2-7** by the conversion of the common intermediate, *o*-acetylaminoatrolactic acid (**33**), into both (*S*)-(+)-atrolactic acid (**18A**) and (*S*)-(−)-3-methyl-3-carboxyphthalide (**32**). It is not surprising that an *ortho*-carboxyl group should have a considerable although unpredictable influence on the steric course of this reaction. However, as indicated in Table **2-2**, *p*-methoxy and *p*-methyl benzoylformates followed the Prelog model as expected.

2-3. Quantitative Aspects of α-Keto Ester Asymmetric Syntheses

We shall now survey all the recorded examples of the α-keto ester asymmetric synthesis with emphasis, whenever possible and justified, on the quantitative results. We shall systematically consider the examples which bear on (a) the effect of variations in the chiral alcohol moiety on asymmetric synthesis with α-keto esters; (b) the effect of variations in the R group of the α-keto acid; (c) the effects upon stereoselectivity resulting from changes in the addition reagent; (d) asymmetric reductions of chiral esters of α-keto acids; (e) the determination of configuration by the atrolactic asymmetric synthesis method; and (f) the variety of molecular types to which this reaction has been applied.

A survey such as this must compare data on experiments reported from different laboratories, carried out under various conditions, over a considerable time span. The solvent, concentration, and temperature of rotations have not always been properly recorded; thus due allowance should be made for these variables in evaluating the data.

2-3.1. Effect of the Chiral Alcohol Moiety on Asymmetric Synthesis with α-Keto Esters. In spite of the large number of examples of this type of asymmetric synthesis which have been reported, only one study[11] has been designed to specifically and systematically study the effect of changes in the size of the ligands R_L and R_M on the extent of asymmetric synthesis. A comparison of the esters from some methylcarbinols (Table **2-1**, Nos. 1–8) shows that the stereoselectivity increases from 3 to 60% in progressing from the methylphenylcarbinyl to methyl-2,4,6-tricyclohexylphenylcarbinyl ester. It is surprising that the extent of asymmetric synthesis is greater in the case of the methyl-*n*-hexyl ester than the phenyl and α-naphthyl esters. With this exception, the increase in asymmetric synthesis follows the increase in steric requirement of the R_L ligand as would be predicted for competitive reactions whose relative rates are dependent upon steric factors.

TABLE 2-1

The Effect of the Size of the Ligands, Attached to the Chiral Alcohol Center, on the Extent of Asymmetric Synthesis

Ph–CO–CO–O–C(H)(R_L)(R_M) → 1) MeMgI 2) H_2O (H^+) 3) H_2O (KOH) 4) H_2O (H^+) → Ph–C(Me)(OH)–COOH (*S*)-(+) + HO–C(H)(R_L)(R_M)

No.	R_M	R_L	Carbinol config.	(*S*)-(+)-Atrolactic acid: Asym. syn. % e.e.	(*S*)-(+)-Atrolactic acid: Yield %	Ref.
1	Me	Ph	*S*	3	78	18, 19a
2	Me	α-$C_{10}H_7$	*S*	12	90	11
3	Me	*n*-C_6H_{13}	*S*[a]	18	43	11
4	Me	$C(CH_3)_3$	*S*	24	82	19b
5	Me	Mesityl	*S*[b]	30	80	11
6	Me	$C(C_6H_5)_3$	*S*[a,b]	49	—	11
7	Me	(−)-Bornyl	*S*[b,d]	54	90	22
8	Me	$C_{24}H_{35}$[c]	*S*[a,b]	60	75	11
9	Steroid (**34**)		*S*[a,b,d]	52	91	20
10	Steroid (**35**)		*S*[b,d]	18	85	21
11	Ph	(−)-Bornyl	*R*[d]	41	75	22
12	Ph	$C(C_6H_5)_3$	*S*[a,b]	27	85	11

[a] The actual experiment was carried out on the enantiomer.
[b] The configuration of these alcohols has been assigned primarily upon the basis of this asymmetric synthesis.
[c] $C_{24}H_{35}$ stands for the 2,4,6-tricyclohexylphenyl group.
[d] Configuration at the carbinol carbon atom.

The decrease from 49 to 27% stereoselectivity in progressing from the methyltriphenylmethylcarbinyl to the phenyltriphenylmethylcarbinyl ester is as anticipated as a result of the reduction of the *difference* in the steric

[18] S. Mitsui and A. Kanai, *Nippon Kagaku Zasshi*, **86**, 627 (1965); *Chem. Abst.*, **65**, 2101 (1966).
[19] (a) A. McKenzie and H. A. Müller, *J. Chem. Soc.*, **95**, 544 (1909). (b) A. McKenzie and P. D. Ritchie, *Biochem. Z.*, **237**, 1 (1931).
[20] V. Prelog and G. Tsatsas, *Helv. Chim. Acta*, **36**, 1178 (1953).
[21] J. C. Danilewicz, D. C. F. Garbutt, A. Horeau, and W. Klyne, *J. Chem. Soc.*, 2254 (1964).
[22] V. Prelog, O. Ceder, and M. Wilhelm, *Helv. Chim. Acta*, **38**, 303 (1955).

requirements of R_M versus R_L and is also in line with the reduction from 54 to 41% stereoselectivity noted in going from the methyl-(−)-bornylcarbinyl (No. 7) to the phenyl-(−)-bornylcarbinyl ester (No. 11). It is also of interest to note that the extreme difference in steric requirements of methyl and 2,4,6-tricyclohexylphenyl has resulted in a high stereoselectivity of 60% corresponding to an energy difference, $\Delta\Delta G^\ddagger$, of about 0.8 kcal/mole for the competing transition states. One can speculate further that a still greater difference in steric hindrance, as would be encountered with the benzoylformate of methyl-2,4,6-tri-*t*-butylphenylcarbinol, for instance, might result in enhanced stereoselectivity.

It seems certain that it is not the total bulk of the ligand R_L which is most important in controlling the extent of asymmetric synthesis but the steric requirements proximate to the chiral carbinol center. Numbers 9 and 10 in Table **2-1** have large steroidal R_L ligands as seen in structures **34** and **35**. In **34**, where the group is attached *cis* to the C-17 angular methyl, the asymmetric synthesis was 52%; but in **35** where these groups are *trans*, this fell to 18% which is even below that found for the *t*-butyl case (Table **2-1**, No. 4).

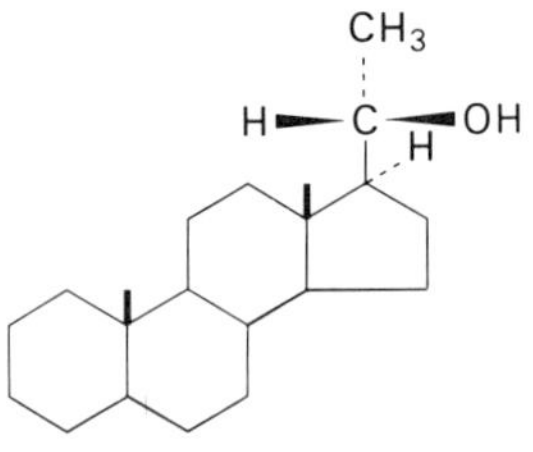

CH_3 H—C—OH H

34

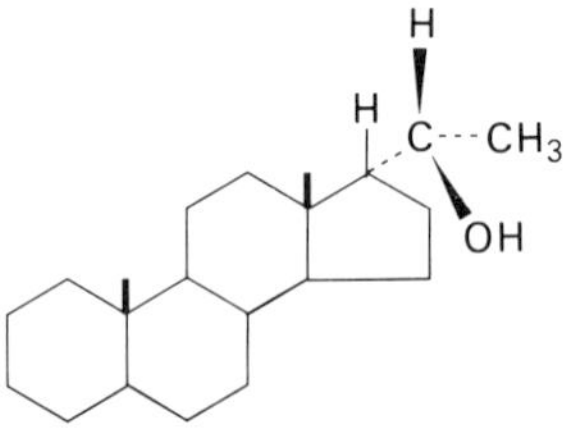

H H C---CH_3 OH

35

In examples 9 and 10 there is still a hydrogen on the α-carbon of the R_L ligand. These results demonstrate that it is not the total size of the group which is important in determining the extent of asymmetric synthesis but the relative steric requirement immediately on each side of the chiral center. The large difference observed in these two cases also demonstrates that it is not sufficient to look at the difference in branching at the α-carbon (i.e., primary, secondary, or tertiary group) but that the stereochemistry of the molecule as a whole must be considered. This is of course emphasized by the high stereoselectivity in the atropisomer examples, **29** and **30**.

2-3.2. Effect of Variations in the α-Keto Acid Moiety upon the Extent of Asymmetric Synthesis. The experiments from which one can draw conclusions concerning the influence of the R group of the α-keto ester on the extent of asymmetric synthesis are collected in Table **2-2**. The examples are limited in scope and it is difficult to draw any broad conclusions. The substitution of *p*-tolyl or *p*-anisyl for phenyl, as anticipated, does not cause a significant change in the stereoselectivity of the reaction.

Some appreciable difference might have been expected when α-naphthyl (No. 4) was substituted for phenyl but the effect is at least minimal if real.*

TABLE 2-2

The Effect of the R Group of the α-Keto Acid Moiety on the Extent of Asymmetric Synthesis

$$R-\overset{O}{\overset{\|}{C}}-COOR^* \xrightarrow[2)\ H_2O\ [H^+]]{1)\ R'MgX^a} \xrightarrow[H_2O\ [H^+]]{H_2O\ [KOH]} R-\underset{R'}{\underset{|}{\overset{OH}{\overset{|}{C}}}}-COOH + HOR^*$$

No.	R	R′	R*	α-Hydroxy acid			
				Config.	Asym. syn.[b] % e.e.	Yield %	Ref.
1	Phenyl	Methyl	(−)-Menthyl	*R*	22–30	77–97	—[d]
2	*p*-Tolyl	Methyl	(−)-Menthyl	*R*	25	91	5
3	*p*-MeO-Ph	Methyl	(−)-Menthyl	*R*	26	65	23
4	α-Naphthyl	Methyl	(−)-Menthyl	*R*	29[c]	68	4
5	*o*-C_6H_4COOH	Methyl	(−)-Menthyl	*S*	10	62	21
6a	Methyl	Phenyl	(−)-Menthyl	*S*	14, 18	79, 50	24, 25
6b	Phenyl	Methyl	(−)-Menthyl	*R*	22–30	77–97	—[d]
7a	Methyl	*p*-Tolyl	(−)-Menthyl	*S*	13	—[e]	5
7b	*p*-Tolyl	Methyl	(−)-Menthyl	*R*	25	91	5
8a	Methyl	*p*-Anisyl	(−)-Menthyl	*S*	15	30	23
8b	*p*-MeO-Ph	Methyl	(−)-Menthyl	*R*	26	60	23
9a	Phenyl	*p*-Tolyl	(−)-Menthyl	*S*	15	63	5
9b	*p*-Tolyl	Phenyl	(−)-Menthyl	*R*	25	28	5
10a	Phenyl	*p*-Tolyl	(−)-Bornyl	*R*	7	—[e]	5
10b	*p*-Tolyl	Phenyl	(−)-Bornyl	*S*	4	—[e]	5
11a	Methyl	*p*-Tolyl	(−)-Bornyl	*R*	—[f]	—[e]	5
11b	*p*-Tolyl	Methyl	(−)-Bornyl	*S*	10	36	5
12a	Methyl	Phenyl	(+)-2-Octyl	*S*	9, 13	38, 20	19
12b	Phenyl	Methyl	(+)-2-Octyl	*R*[g]	18	43	19

[a] All Grignard reagents are the bromides except methyl which is the iodide.
[b] See text for discussion of limits of error of these figures.
[c] An estimate of a maximum value of 29 % e.e. can be made but the complete resolution of α-(1-naphthyl)-mandelic acid is in doubt.[19]
[d] See Table **2-5** for the individual experiments and references.
[e] No yield was reported in the literature.
[f] No significant observable rotation.
[g] The actual experiment reported in the literature was carried out on the enantiomer.

* See Table **2-2**, Footnote c.
[23] A. McKenzie and P. D. Ritchie, *Biochem. Z.*, **250**, 376 (1932).
[24] A. McKenzie, *J. Chem. Soc.*, **89**, 365 (1906).
[25] J. A. Reid and E. E. Turner, *J. Chem. Soc.*, 3219 (1951).

Replacement of the phenyl group with the *o*-carboxyphenyl group (No. 5) did not greatly alter the extent of asymmetric synthesis although it did reverse the direction of asymmetric synthesis as already discussed (Sec. **2-2.1**, formulas **31–33**). Perhaps the surprising observation is that the stereoselectivities of the first five examples all lie between 20 and 30%. This is about the same range as the values found in the literature for the very first example which has been reported independently in six different publications.

When the R group of the α-keto ester and the R′ group of the Grignard reagent are interchanged, there is a reversal in the direction of the asymmetric synthesis as is expected but not absolutely required (Sec. **1-4.4b**). The paired reactions in Table **2-2** (Nos. 6a versus 6b, 7a versus 7b, 8a versus 8b, 11a versus 11b, and 12a versus 12b), which lead to a mixture of the same diastereomers, uniformly show that the extent of asymmetric synthesis is lower for the pyruvate esters than for the benzoylformate esters. The fact that the ratio of diastereomeric products changes with the interchange of R and R′ indicates that "product development control" (Sec. **3-3.1**) does not exert an appreciable influence in this reaction.

The low or unspecified yields in the last three pairs of examples in Table **2-2** make it unwise to draw any conclusions beyond the observation that the menthyl esters seem to give higher asymmetric syntheses and better yields than do the 2-octyl and bornyl esters. This, in addition to its ready availability, is undoubtedly the reason that (−)-menthol has been used in this type of asymmetric synthesis much more than any other optically active carbinol. Again it is seen that the existing examples do not really encompass the extreme steric effects which might be useful in gaining a better insight into the effect of these variables. For instance, examples involving the mesityl and *t*-butyl groups are missing from this table.

2-3.3. The Effect upon Stereoselectivity Resulting from Changes in the Addition Reagent. In Table **2-3** are listed results which allow evaluation of the effects of various reagents upon the extent of asymmetric addition to (−)-menthyl benzoylformate. It is apparent that the Grignard reagent from methyl iodide shows higher selectivity than the other reagents reported with the exception of the acetylene Grignard (but not lithium acetylide) which appears to be slightly better. This includes reductions with lithium aluminum hydride, sodium borohydride, and heterogeneous reductions with aluminum amalgam, data for which are collected in Table **2-4**. There appears to be little difference between the Grignard reagent from methyl iodide and bromide. Neither the Grignard reagent from methyl chloride nor dimethylmagnesium have been reported in this reaction. Deductions concerning the effect of the addition reagent are hampered by the fact that α-*n*-propyl-, α-*i*-butyl-, α-*t*-butyl-, and α-(α-naphthyl)-mandelic acids have not been resolved and thus the extent of asymmetric synthesis in these cases cannot be determined. Scattered experiments reported in the

TABLE 2-3

The Effect of the Addition Reagent on the Extent of Asymmetric Synthesis[a]

$$Ph-\overset{O}{\overset{\|}{C}}-\overset{O}{\overset{\|}{C}}-O\text{-}(-)\text{-Menthyl} \xrightarrow[\substack{3)\ H_2O\ (KOH)\\ 4)\ H_2O\ (H^+)}]{\substack{1)\ \text{Reagent}\\ 2)\ H_2O\ (H^+)}} Ph-\overset{HO\quad R}{C}-COOH + (-)\text{-Menthol}$$

No.	Reagent[a]	α-Substituted mandelic acid: Substituent R	Asym. syn. % e.e.	Yield %	Ref.
1	CH_3MgI	CH_3	22–30	77–97	—[b]
2	CH_3MgBr	CH_3	22	97	25, 26
3	C_2H_5MgBr	C_2H_5	10[c]	—	3a
4	n-C_3H_7MgI	n-C_3H_7	6[d]	59	24
5	i-C_4H_9MgI	i-C_4H_9	—[e]	55, 58[e]	24
6	t-C_4H_9MgI	t-C_4H_9	—[f]	29	24
7	p-TolylMgBr	p-Tolyl	15	—	5
8	CH≡CMgBr	CH≡C—	32	—	27
9	CH≡CLi	CH≡C—	0.4	—	27
10	α-NaphthylMgBr	α-Naphthyl	—[g]	—	24

[a] For reductions, see Table **2-4**.
[b] See Table **2-5** for data and references.
[c] A minimum value of 10 % asymmetric synthesis can be estimated (from the incomplete data of McKenzie,[3a] α_D^{18} −0.41°), based on the now known rotation of enantiomerically pure α-ethylmandelic acid [S. Mitsui, S. Imaizumi, Y. Senda, and K. Konno, *Chem. Ind.* (London), 233 (1964) and T. R. Emerson, et al., *J. Chem. Soc.*, 4007 (1965)].
[d] If the optical rotation of α-propylmandelic acid approximates that of α-ethylmandelic acid (Footnote c), a value of 6 % e.e. can be estimated.
[e] In three experiments, McKenzie[24] reported α-isobutylmandelic acid with $[\alpha]_D$ −16.9°, −5.0°, and −1.2° in unspecified, 55, and 58 % yields, respectively.
[f] McKenzie[24] obtained a 29 % chemical yield of α-*t*-butylmandelic acid, α_D^{16} −0.09° (EtOH, c ?). According to Prelog's generalization this should have the *R* configuration. K. Shingu, S. Hogishita, and M. Nakagawa [*Tetrahedron Lett.*, 4371 (1967)] have assigned the *S* configuration to (−)-α-*t*-butyl-α-hydroxyphenylacetic acid, $[\alpha]_D$ −50° (methanol) based on a stereospecific synthesis and circular dichroism (CD) studies. This discrepancy should be investigated.
[g] A sample of the barium salt of α-(1-naphthyl)mandelic acid, $[\alpha]_D$ −11.0°, was obtained after crystallization but this acid has not been critically resolved.

literature (see Tables **2-5** and **2-6**) on bornyl benzoylformate, 2-octyl pyruvate, and menthyl pyruvate using some of these same Grignard reagents are in agreement with these deductions made from the data in Table **2-3**. But it is apparent that further information on this point should be collected.

[26] D. M. Bovey and E. E. Turner, *J. Chem. Soc.*, 3223 (1951).
[27] I. Iwai and Y. Yura, *Yakugaku Zasshi*, **80**, 1193, 1199 (1960); *Chem. Abst.*, **55**, 3646, 3647 (1961).

TABLE 2-4

Asymmetric Reductions of Chiral α-Keto Esters

$$R-\overset{O}{\overset{\|}{C}}-COOR^* \xrightarrow{[H]} \xrightarrow{H_2O} R-\overset{OH}{\overset{|}{C}}{}^*HCOOH \text{ or } \left[RC^*H(OH)CH_2OH\right] + HOR^*$$

No.	R	R*	Reagent	α-Hydroxy acid Config.	Asym. syn. % e.e.	Yield %	Ref.
1	Me	(−)-Menthyl	Al-Hg	*R*	—[a]	—	28
2			Al-Hg	*R*	8	—[b]	29a
3			Na-Hg	*R*	—[a]	—	28
4			Zn, HOAc	*R*	—[a]	—	3a
5			H_2, (Pt; Ni; Pd)	*R*	21–33	—[b]	29a
6			$NaBH_4$	*R*	3	—[b]	29a
7	Me	(−)-2-Me-butyl	Al-Hg	*S*	11[c]	—	19a
8	*t*-Bu	(−)-Menthyl	H_2 (Pt, Ni, Pd)	*R*	17–41	—[b]	29a
9			H_2 (Pd·C·HCl)	*R*	3	—[b]	29a
10			Al-Hg	*R*	7	—[b]	29a
11			$NaBH_4$	—	0	—	29a
12	Ph	(−)-Menthyl	$LiAlH_4$	*R*[d]	6[d]	85	30
13			$LiAlH_4$	*R*[d]	8[d]	89	31
14			$LiAlH_4$	*R*[d]	7[d]	73	8
15	Ph	(+)-Bornyl	$LiAlH_4$	*S*	3[d]	90	30
16	Ph	(−)-Menthyl	$NaBH_4$	*R*[d]	29[d]	51	31
17			$NaBH_4$	*R*[d]	5[d]	65	8
18			$NaBH_4$	*R*[d]	7[d]	81	8
19			Al-Hg	*R*	3–10[e]	86	32
20			Al-Hg	*R*[d]	7[d]	70	31
21			Al-Hg	*R*	7[f]	—	3a
22			Al-Hg	*R*	6[f]	—	3a
23			Na-Hg	S[g]	15[g]	—	3a
24			H_2 (Ni)	*S*	14	—	13b
25			H_2 (Pd·base)	*R*	13	—	13b
26			H_2 (Pd·acid)	*S*	13	—	13b
27	Ph	(*S*)-(−)-1-Ph-ethanol	$NaBH_4$	*S*	7.5[b]	—	18, 29a
28			H_2 (Ni, PtO_2)	*S*	2–4	—	18
29			H_2 (Pd·base)	*S*	1	—	18
30			H_2 (Pd·acid)	*R*	1	—	18
31	Ph	1,2:5,6-di-O-Cyclohexylidene-D-glucofuranose	H_2 (Ni)	*S*	45	—	13b
32			H_2 (Pd·base)	*R*	23	—	13b
33			H_2 (Pd·acid)	*S*	24	—	13b

[28] A. McKenzie, *J. Chem. Soc.*, **87**, 1373 (1905).

[29] (a) S. Mitsui and A. Kanai, *Nippon Kagaku Zasshi*, **87**, 179 (1966); *Chem. Abst.*, **65**, 17006 (1966). (b) S. Mitsui and Y. Imai, *Nippon Kagaku Zasshi*, **88**, 86 (1967); *Chem. Abst.*, **67**, 43934 (1967).

TABLE 2-4—cont.

No.	R	R*	Reagent	α-Hydroxy acid Config.	Asym. syn. % e.e.	Yield %	Ref.
34	Ph	1,2:5,6-di-O-Isopropylidene-D-glucofuranose	H_2 (Ni)	*S*	37	—	13b
35			H_2 (Pd·base)	*R*	15	—	13b
36			H_2 (Pd·acid)	*S*	20	—	13b
37	Ph	5-O-ethyl-1,2-Isopropylidene-D-glucofuranose	H_2 (Ni)	*S*	20	—	13b
38			H_2 (Pd·base)	*S*	1	—	13b
39			H_2 (Pd·acid)	*S*	9	—	13b
40	Ph	α-Amyrinyl	$LiAlH_4$	*S*[d]	6	84	30
41	Ph	Cholesteryl	$LiAlH_4$	*S*[d]	3	49	31
42			$NaBH_4$	*S*[d]	3	45	31
43			Al-Hg	*S*[d]	4	43	31
44			Al(O-*i*-Pr)$_3$	*S*[d]	2	30	31
45	Ph	(+)-Phenyldi-hydrothebainyl	$LiAlH_4$	*S*[d]	69[h]	45	16d
46			$NaBH_4$	*S*[d]	14[h]	71	16d
47	α-Naph	(−)-Menthyl	Al-Hg	*R*	8	—	29b
48			$NaBH_4$	*R*	3	—	29b
49			Al(O-*i*Pr)$_3$	—	0	—	29b
50			H_2 (PtO_2)	*R*	20–55	—	29b
51			H_2 (Co, Cu, Ni)	*S*	5–22	—	29b
52			H_2 (Pd)	*R*	0–14	—	29b

[a] Only the sign of rotation of atrolactic acid reported.[28]

[b] Results from several experiments under different conditions; yield not given but presumably high; % e.e. determined by GLC.

[c] This is a minimum value estimated from incomplete data.[19a]

[d] In these cases the mandelic ester was not hydrolyzed but was reduced to the glycol; $[\alpha]_D^2$ 40.6 (EtOH) taken as rotation of pure glycol.[30] The warning that racemization occurs during the reduction[16d] is unfounded according to our experience.

[e] Obtained by hydrolysis of the O-acetylmandelate which is more resistant to racemization than the mandelic ester itself.

[f] Calculated from the rotation of the mixture of diastereomeric mandelic esters and the known rotations of the pure diastereomers.

[g] This would be a reversal of the Prelog generalization but McKenzie[3a] expressed some reservation concerning the experiment.

[h] Stereoselectivities of 48 and 10% were reported. These values are recalculated on the basis of Footnote d.

[30] V. Prelog, M. Wilhelm, and D. B. Bright, *Helv. Chim. Acta*, **37**, 221 (1954).

[31] S. P. Bakshi and E. E. Turner, *J. Chem. Soc.*, 168 (1961); for catalytic reductions, see T. Kamaishi and S. Mitsui, *Nippon Kagaku Zasshi*, **86**, 623 (1965); *Chem. Abst.*, **65**, 2326 (1966).

[32] A. McKenzie and H. B. P. Humphries, *J. Chem. Soc.*, **95**, 1105 (1909).

2-3.4. The Asymmetric Reduction of Chiral Esters of α-Keto Acids. In Table **2-4** are listed the known reductions of chiral esters of α-keto acids by various achiral reducing agents. With the exception of certain heterogeneous catalytic reductions (Nos. 5, 8, 9, 24–26, 28–39, 50–52) the percent stereoselectivities are substantially lower than those encountered with the comparable addition of the methyl Grignard reagent (Table **2-1**). From the one case available, No. 14 versus No. 15, it seems that the stereoselectivity in the reduction of the (+)-bornyl ester with lithium aluminum hydride is lower than that found for the (−)-menthyl ester, and the stereochemical course of the reactions are reversed, in a manner parallel to that observed for the Grignard addition reaction (Table **2-5**).

The asymmetric reductions are less satisfactory from an operation standpoint because of the racemization which usually occurs upon basic hydrolysis of the mandelate esters and which can occur upon hydrolysis of the pyruvate esters. This difficulty is partially circumvented by converting to the O-acetyl derivatives before saponification.[32] Thus the hydrolysis of mixed O-acetyl mandelate esters from the aluminum amalgam reduction of (−)-menthyl benzoylformate (Table **2-4**, No. 19) gave (−)-mandelic acid but the correspondence of results from duplicate runs was unsatisfactory. Also the reduction of the mandelate esters with lithium aluminum hydride to phenylethylene glycol (**12** → **13**, Sec. **2-2**, Table **2-4**; cf. footnote d)[31] circumvents this difficulty. However it gives a neutral product which must be purified with care. A more serious consideration stems from the report[16d] that both lithium aluminum hydride and sodium borohydride reductions of the benzoylformate ester of phenyldihydrothebaine* (Fig. **2-8**) give (*S*)-(+)-phenylethylene glycol (**13**) while treatment with methylmagnesium iodide gives (*R*)-(−)-atrolactic acid.[16b] Since *S*-(+)-phenylethylene glycol and *R*-(−)-atrolactic acid have opposite relative configurations, the asymmetric syntheses have taken place in the opposite sense! The reversal of stereoselectivity has not been observed with other benzoylformates (compare Tables **2-1** and **2-4**).

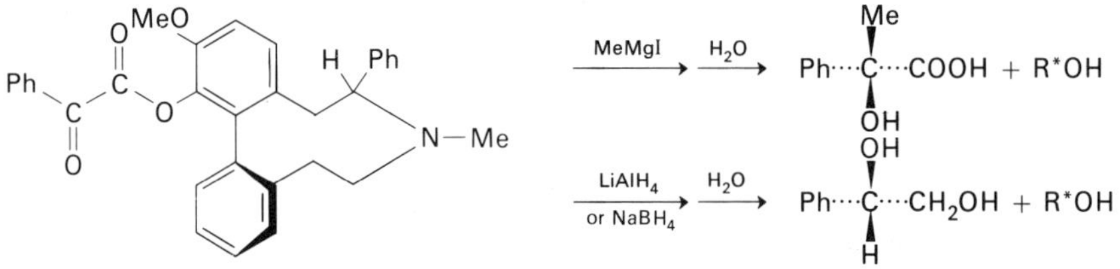

Fig. 2-8. Conflicting asymmetric syntheses; Grignard addition (70% e.e.) and metal hydride reduction ($LiAlH_4$, 48% e.e.; $NaBH_4$, 10% e.e.).

* See Sec. **9-2.1** for an interesting asymmetric synthesis of this compound.

Further investigations[16d] suggested that the ester carbonyl group is reduced faster than the α-keto group. If this is indeed so, altogether different factors are involved in the asymmetric synthesis. The hemiacetal which would be formed initially from the ester carbonyl would be chiral and an asymmetric synthesis altogether different from that of a regular α-keto ester asymmetric synthesis would then occur on reduction of the α-keto hemiacetal (Fig. **2-9**). It is impossible to predict with any certainty the course of such a reaction.*

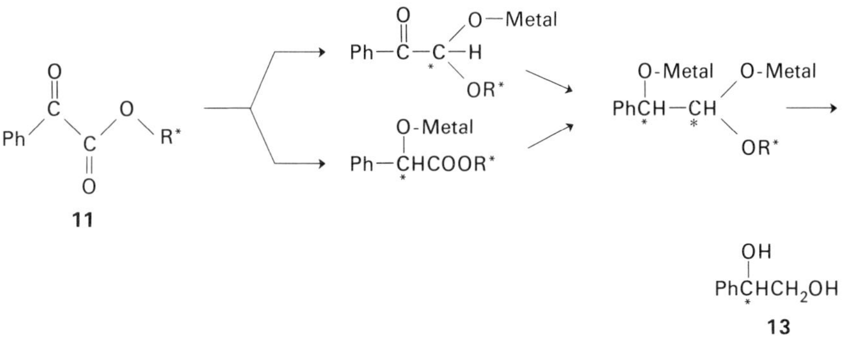

Fig. 2-9. Competing pathways for the metal hydride asymmetric reduction of α-keto esters.

"Double asymmetric induction" arises in the reduction of a chiral benzoyl formate ester with a chiral reducing agent; this is discussed in Sec. **5-1.5**, Table **5-3** and Sec. **5-3.1**, Table **5-11**.

2-3.5. Determination of Configuration by the Atrolactic Asymmetric Synthesis. In Sec. **2-3.1** the effects upon extent of asymmetric synthesis resulting from systematic changes in R_L and R_M of the inducing alcohol were discussed in connection with the examples in Table **2-1**. In Table **2-5** are listed those remaining published examples in which the action of the Grignard reagent from methyl iodide on the benzoylformate ester of a chiral alcohol has been studied. A survey of Tables **2-1** and **2-5** shows that there are no exceptions to the Prelog generalization when the boundary conditions for the model are observed.† This does not mean that no exceptions will be found but it constitutes an impressive record and adds greatly to the value of the model as a predictive method.

* We have observed that the asymmetric reduction of other chiral benzoylformate esters with LAH and $NaBH_4$ proceeds normally; cf. also Table **2-4**. Sodium borohydride reductions of ethyl benzoylformate have been repeated (J. A. Dale and H. S. Mosher, unpublished results) under conditions reported;[16d] ethyl mandelate was produced as expected but no phenylethylene glycol was found.

† Examples such as Nos. 18, 19, and 26 in Table **2-5** and Nos. 31–39 in Table **2-4** which might be considered exceptions have been discussed in Sec. **2-2**.

TABLE 2-5

Determination of Carbinol Configuration by the Atrolactic Asymmetric Synthesis[a]

$$\mathrm{Ph{-}\overset{\overset{\displaystyle O}{\|}}{C}{-}COOR^*} \xrightarrow[\text{2) } H_2O\ (H^+)]{\text{1) MeMgI}} \xrightarrow[(KOH)]{H_2O} \mathrm{Ph{-}{}^*\overset{\overset{\displaystyle OH}{|}}{C}(CH_3)COOH + HOR^*}$$

No.	Alcohol, R*OH: Name	Formula	Config.	Atrolactic acid: Config.	Asym. syn.[b] % e.e.	Yield %	Ref.
1	(−)-2-Methyl-1-butanol	$C_5H_{12}O$	*S*[c]	*S*	(+)[c]	69	18
2	(−)-2-Octanol	$C_8H_{18}O$	*R*	*R*	18	43	19
3	(−)-Menthol	$C_{10}H_{20}O$	*R*	*R*	22	—[d]	3a
			R	*R*	25	—[d]	3a
			R	*R*	23	79	24
			R	*R*	25	97	34
			R	*R*	29	30	8
			R	*R*	29	77	13
			R	*R*	30	78	13
4	(+)-Neomenthol	$C_{10}H_{20}O$	*S*	*S*	12	94	34
5	(−)-Borneol	$C_{10}H_{18}O$	*R*	*R*	5	77	24
			R	*R*	4	90	24
6	(+)-Borneol	$C_{10}H_{18}O$	*S*	*S*	11	90	34
7	(−)-Isoborneol	$C_{10}H_{18}O$	*R*	*R*	8	98	34
8	1(*S*),8(*S*)-*cis*-8-Methylhydrindanol	$C_{10}H_{18}O$	*S*	*S*	16	75	35
			S	*S*	16	90	35
9	1(*S*),8(*R*)-*cis*-8-Methylhydrindanol	$C_{10}H_{18}O$	*S*	*S*	12	94	35
10	5(*S*),8(*S*)-*cis*-8-Methylhydrindanol	$C_{10}H_{18}O$	*S*	*S*	6	62	35
			S	*S*	6	—[d]	35
11	1(*S*),9(*S*)-*trans*-1-Decalol	$C_{10}H_{18}O$	*S*	*S*	15	82	33
12	1(*S*),9(*R*)-*trans*-1-Decalol	$C_{10}H_{18}O$	*S*	*S*	2	100	33
13	Dimethyl jaconecate	$C_{11}H_{18}O$	*S*	*S*	11	60	41b
14	(−)-3(3′,5′-Dimethoxy-2′-methylphenyl)-2-butanol	$C_{13}H_{20}O_3$	*R*	*R*	37	80	36
15	11-Noreusantan-11-ol	$C_{14}H_{26}O$	*S*	*S*	15	—[d]	37
16	Dihydrocodeine (**24**)	$C_{18}H_{22}O_3N$	*S*	*R*	15	10	12
17	(−)-Epiafzelchin-5:7:4′-trimethyl ether	$C_{18}H_{20}O_5$	*R*	*R*	(−)[e]	—[d]	14
18	(−)-Epicatechin-5;7:3′;4′-tetramethyl ether (**26a**)	$C_{19}H_{22}O_6$	*R*	*R*	(−)[e]	—[d]	14

TABLE 2-5—cont.

No.	Alcohol, R*OH Name	Formula	Config.	Atrolactic acid Config.	Asym. syn.[b] % e.e.	Yield %	Ref.
19	(+)-Catechin-5:7:3′:4′-tetramethyl ether (**26b**)	$C_{19}H_{22}O_6$	*R*	*R*	(−)[f]	60	14
20	17-β-Androstanol	$C_{19}H_{32}O$	*S*	*S*	16	45	38
21	Methyl tetrahydrogibberellate	$C_{20}H_{30}O_6$	*S*	*S*	10	71	41a
22	Methyl 3-epitetrahydrogibberellate	$C_{20}H_{30}O_6$	*R*	*R*	6	68	41a
23	20-β-Oxy-5-α-pregnane	$C_{21}H_{34}O$	*R*	*R*	52	91	20
24	Yohimbine	$C_{21}H_{26}ON_2$	*S*	*S*	4	—[d]	39
25	Methyl reserpinate	$C_{22}H_{28}O_4N_2$	*R*	*R*	10	—[d]	39
26	3β-Cholestanol (**20**)	$C_{27}H_{48}O$	*S*	*R*	2	82	38
27	7α-Cholestanol	$C_{27}H_{48}O$	*R*	*R*	13	93	38
28	7β-Cholestanol	$C_{27}H_{48}O$	*S*	*S*	69	86	38
29	22α-Oxy-3-methoxy-5-cholestene	$C_{28}H_{48}O_2$	*R*	*S*	18	70	40
30	22β-Oxy-3-methoxy-5-cholestene	$C_{28}H_{48}O_2$	*R*	*R*	33	52	40
31	Euphol	$C_{30}H_{50}O$	*S*	*S*	24	69	38
32	α-Amyrin	$C_{30}H_{50}O$	*S*	*S*	10	71	38
33	Dihydrolanosterol	$C_{30}H_{52}O$	*S*	*S*	34	58	38
34	5-Deoxy-1,2-O-isopropylidine-D-xylose	$C_8H_4O_4$	*S*	*R*	22	77	13
			S	*R*	22	53	13a
35	5-Deoxy-5-*S*-ethyl-1,2-O-isopropylidine-D-xylose	$C_{10}H_8O_4S$	*S*	*R*	25	81	13a
			S	*R*	26	73	13a
36	5-O-Ethyl-1,2-O-isopropylidine-D-xylose	$C_{10}H_{18}O_5$	*S*	*R*	28	77	13a
			S	*R*	28	70	13a
37	1,2:5,6-Di-O-isopropylidine-D-glucose	$C_{12}H_{20}O_6$	*S*	*R*	34	79	13a
			S	*R*	33	77	13a
38	1,2:5,6-Di-O-cyclohexylidine-D-glucose	$C_{18}H_{28}O_6$	*S*	*R*	38	72	13a
			S	*R*	37	68	13a
39	(+)-1-(α-Naphthyl)-2-naphthol	$C_{20}H_{14}O$	*R*	*R*	85	41	16c
40	(−)-Dinaphtho-2′,1:1,2:1″,2″:3,4-cycloheptadien-6-ol	$C_{23}H_{18}O$	*R*	*R*	20	69	15

[33] W. R. Feldman and V. Prelog, *Helv. Chim. Acta*, **41**, 2396 (1958); *ibid.*, **42**, 397 (1959).
[34] V. Prelog and H. L. Meier, *Helv. Chim. Acta*, **36**, 320 (1953).
[35] W. Acklin and V. Prelog, *Helv. Chim. Acta*, **42**, 1239 (1959).
[36] P. P. Mehta and W. B. Whalley, *J. Chem. Soc.*, 3777 (1963).

TABLE 2-5—cont.

No.	Alcohol, R*OH			Atrolactic acid			Ref.
	Name	Formula	Config.	Config.	Asym. syn.[b] % e.e.	Yield %	
41	(+)-α-Phenyldihydro-thebaine bismethine	$C_{24}H_{22}O_3$	*R*	*R*	93	61	16b
42	(+)-α-Phenyldihydro-thebaine dihydromethine	$C_{26}H_{31}O_3N$	*R*	*R*	89	63	16b
43	(+)-α-Phenyldihydro-thebaine isomethine	$C_{26}H_{29}O_3N$	*R*	*R*	91	70	16b
44	(+)-α-Phenyldihydro-thebaine	$C_{25}H_{27}O_3N$	*R*	*R*	71	78	16b
			R	*R*	70	67	16b

[a] This table is an extension of Table **2-1**.

[b] Based upon $[\alpha]_D^{13.8}$ 37.7° (c, 3.3, ethanol) for rotation of enantiomerically pure atrolactic acid.[9] See text for discussion of limits of errors of these values.

[c] This is a special case, achiral at the carbinol center but *S* at the carbon proximate to the carbinol carbon. See text for discussion.

[d] No yield could be obtained from the data reported.

[e] Atrolactic acid of $[\alpha]_D$ −16.4° from epi-afzelchin and $[\alpha]_D$ −30.4° from (−) epicatechin was isolated after crystallization but amounts were not given and no asymmetric synthesis can be calculated although it is probably substantial.

[f] Atrolactic acid with $[\alpha]_D$ −1.1° (3 % e.e.) was isolated after two crystallizations.

[37] Y. Abe, T. Miki, M. Sumi, and T. Toga, *Chem. Ind.* (London), 953 (1956).

[38] W. G. Dauben, D. F. Dickel, O. Jeger, and V. Prelog, *Helv. Chim. Acta*, **36**, 325 (1953).

[39] Y. Ban and O. Yonemitsu, *Chem. Ind.* (London), 948 (1961).

[40] K. Tsuda and R. Hayatsu, *Chem. Pharm. Bull.* (Tokyo), **6**, 580 (1958).

[41] (a) S. Masamune, *J. Am. Chem. Soc.*, **83**, 1515 (1961); (b) *ibid.*, **82**, 5253 (1960).

Only a few additional comments need to be made. The addition of methylmagnesium iodide to (−)-2-methylbutyl benzoylformate (Table **2-5**, No. 1) is a special case since it involves the ester of a primary alcohol. Routine application of Prelog's generalization would predict a racemic product since in the model (**36**) the probability of attack from either side would be equal because R_S and R_M are the same, namely hydrogen. But this is a conclusion that is completely unwarranted and a use of the generalization which is, of course, not intended. The two faces of the α-keto group are diastereotopic when R_L is chiral as in **37** whether the asymmetric center is on the carbinol

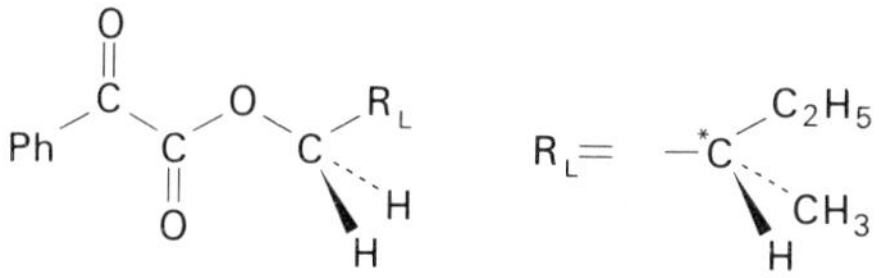

36 **37**

carbon or is still further removed. Although the extent of asymmetric synthesis may be small, it must still exist and was in fact observed.

Examples 5 and 6 (Table **2-5**) represent the same asymmetric synthesis reaction on enantiomeric bornyl benzoylformates. Symmetry principles demand that, under identical conditions, epimeric products must result from enantiomers and that the *extent* of asymmetric synthesis in each case be the same. Epimers are indeed formed but the extent of asymmetric synthesis reported in one case was 4–5% and the other, 11%. This difference remains unexplained but probably represents experimental variation in the earlier results when interest was centered on establishing the fact of an asymmetric synthesis, not its extent. The correspondence should be much better than this and with modern laboratory techniques undoubtedly would be.

In several cases, isomer pairs which are epimeric at the carbinol carbon have been studied: (−)-menthol and (+)-neomenthol (Table **2-5**, Nos. 3 and 4; (+)-borneol and (−)-isoborneol (Nos. 6 and 7); (+)-catechin tetramethyl ether and (−)-epicatechin tetramethyl ether (Table **2-5**, Nos. 18 and 19); methyl tetrahydrogibberellate and methyl 3-epitetrahydrogibberellate (Nos. 21 and 22); 7α- and 7β-cholestanol (Nos. 27 and 28); and 22-α-oxy- and 22-β-oxy-3-methoxy-5-cholestene (Nos. 29 and 30). Prelog and co-workers[32] have commented on the fact that axially substituted benzoylformate esters generally yield lower stereoselectivities than the equatorial epimers. This seems to be a valid generalization, based on the relatively few examples available; but when dealing with compounds with multiple asymmetric centers, the complete molecular topology in the neighborhood of the carbinol center should be analyzed since the effect is not solely a property of the equatorial or axial nature of the attachment but the total steric environment in the vicinity of the attachment.

2-3.6. Miscellaneous Asymmetric Syntheses with Chiral Esters of α-Keto Acids. There is a group of asymmetric syntheses which for various reasons do not fit conveniently into the previous tables; they are now collected in Table **2-6**. In a few cases, the same example has been included in two different tables for convenient comparison. Many of these examples involve an α-hydroxy acid whose enantiomeric purity is not known and the percent asymmetric synthesis in these cases cannot be calculated. The configuration reported for the α-hydroxy acids in many cases depends upon the application of the Prelog generalization to the reactions under consideration. This reaction has proved to have utility for configurational correlations and its extensive use should continue.

2-3.7. Asymmetric Synthesis with Chiral Benzoylformamides. A comparison of asymmetric syntheses using chiral α-keto-amides corresponding to the chiral benzoylformate esters is hampered by the limited data. Catalytic reduction of (*S*)-(−)-N-(α-methylbenzyl)benzoylformamide[18] (Fig. **2-10**) using Raney nickel, palladium, or palladium on carbon in acidic,

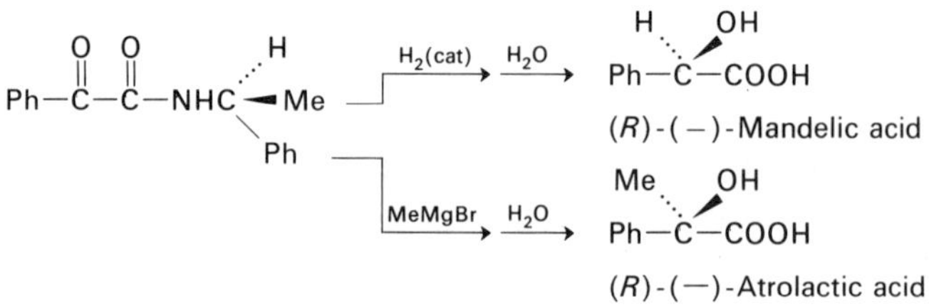

Fig. 2-10. Asymmetric reaction of (*S*)-(−)-*N*-(α-methylbenzyl)benzoylformamide.

basic, or neutral solution gives an excess of (*R*)-(−)-mandelic acid (5–25% e.e.). This is contrary to what might have been expected on the basis of a naïve extension of Prelog's model from that for the corresponding ester (Fig. **2-5**). Under the basic and neutral conditions, catalytic reduction of the corresponding ester (Table **2-4**, Nos. 29 and 28) follows Prelog's generalization (1–4% e.e.); however, the generalization is not followed under acid conditions (Table **2-4**, No. 30). Sodium borohydride reduction of this amide gave, after hydrolysis, (*R*)-(−)-mandelic acid in excess and the action of methylmagnesium iodide gave an excess of the (*R*)-(−)-atrolactic acid. These are the opposite configurations to those obtained by the comparable reactions on the ester. It is apparent that the Prelog generalization cannot be extended directly to the amide system but much more data must be collected before any useful generalizations concerning this system can be made.

TABLE 2-6

Miscellaneous Asymmetric Synthesis with α-Keto Esters

$$R-\overset{O}{\overset{\|}{C}}-COOR^* \xrightarrow[H_2O\,(H^+)]{R'\text{ reagent}} \xrightarrow[(KOH)]{H_2O} R-\overset{OH}{\overset{|}{C}}R'COOH + HOR^*$$

No.	Keto ester R	Reagent R′	Alcohol R*	Alcohol Config.	α-Hydroxy acid Config.	α-Hydroxy acid Asym. Synthesis [%e.e. or α_D°]	α-Hydroxy acid Yield %	Ref.
1	CH_3	*t*-BuMgBr	(−)-Menthyl	*R*	*S*[a]	+0.91	—	42
2	CH_3	C_2H_5MgBr	(−)-2-Methyl butyl	—[b]	*S*	(−)	—	18
3	CH_3	C_6H_5MgBr	(−)-2-Methyl butyl	—[b]	—	0[c]	—	18
4	CH_3	α-$C_{10}H_7MgBr$	(−)-2-Methyl butyl	—[b]	—	0[c]	—	18
5	CH_3	C_6H_5MgBr	(+)-2-Octyl	*R*	*R*	9	38	19
6	CH_3	C_6H_5MgBr	(+)-2-Octyl	*R*	*R*	13	20	19
7	CH_3	*p*-TolylMgBr	(−)-Bornyl	*R*		0[c]	—	5
8	CH_3	HC≡CMgBr	(−)-Menthyl	*R*	*S*[a]	−6.5[e]	33	27
8a	CH_3	HC≡CLi	(−)-Menthyl	*R*		0.0[d,e]	11	27
9	CH_3	HC≡CMgBr	(+)-Bornyl	*S*	*R*[a]	+6.6[e]	24	27
10	C_2H_5	*i*-$C_6H_{13}MgBr$	(−)-Menthyl	*R*	*S*[a,f]	+1.1	63	43,44
11	*i*-C_6H_{13}	C_2H_5MgBr	(−)-Menthyl	*R*	*R*[a,f]	−0.6	87	43,44
12	C_2H_5	C_6H_5MgBr	(−)-Menthyl	*R*	*S*	28[g]	30	45
13	C_6H_5	C_2H_5MgBr	(−)-Menthyl	*R*	*R*	35[g]	38	45
14	C_6H_5	CH_3MgI	(−)-2-Methyl butyl	—[b]		0[c]	—	18
15	C_6H_5	C_2H_5MgI	(−)-2-Methyl butyl	—[b]	*S*	(+)	—	18
16	C_6H_5	α-$C_{10}H_7MgBr$	(−)-2-Methyl butyl	—[b]	*R*[a]	(+)	—	18
17	C_6H_5	HC≡CMgBr	(−)-Menthyl	*R*	*R*	32 −14	31	27
18	C_6H_5	HC≡CMgBr	(+)-Bornyl	*S*	*S*	15 +6.6	15	27
18a	C_6H_5	HC≡CLi	(−)-Menthyl	*R*		0.4[d] (−)	43	27
18b	C_6H_5	HC≡CLi	(+)-Bornyl	*S*		0.2[d] (+)	56	27
19	C_6H_5	HC≡CLi	Cyclolaudenyl	*R*[a]		0.2 (+)[d]	18	27
20	C_6H_5	PhC≡CLi	(−)-Menthyl	*R*		−0.1[d]	36	27
21	C_6H_5	PhC≡CLi	(+)-Bornyl	*S*		+0.4[d]	19	27
22	C_6H_5	$PhCH_2CH_2Li$	(−)-Menthyl	*R*		−0.1[d]	50	27
23	C_6H_5	$PhCH_2CH_2Li$	(+)-Bornyl	*S*		0.1[d]	52	27
24	C_6H_5	C_2H_5MgI	(−)-Bornyl	*R*	*R*	12	57	24
25	C_6H_5	*i*-BuMgI	(+)-Bornyl	*S*	*S*[a]	+2.5	67	4
26	C_6H_5	*i*-BuMgI	(−)-Bornyl	*R*	—[h]	+1.0	45	24
27	C_6H_5	α-$C_{10}H_7MgBr$	(+)-Bornyl	*S*	*R*[a]	+1.1	66	4
28	C_6H_5	α-$C_{10}H_7MgBr$	(−)-Bornyl	*R*	—[h]	(+)	—	24
29	C_6H_5	α-$C_{10}H_7MgBr$	(+)-2-Octyl	*R*	*R*[a]	−0.5	80	19
30	C_6H_5	*p*-TolylMgBr	(−)-Bornyl	*R*	*S*[a]	7	—	5

TABLE 2-6—cont.

No.	Keto ester R	Reagent R′	Alcohol R*	Alcohol Config.	α-Hydroxy acid Config.	Asym. Synthesis [%e.e. or α_D°]	Yield %	Ref.
31	α-$C_{10}H_7$	CH_3MgI	(−)-Bornyl	*R*	*R*[a]	−3.5	95	46
32	α-$C_{10}H_7$	CH_3MgI	(−)-Menthyl	*R*	*R*[a]	10.0	68	46
33	α-$C_{10}H_7$	C_2H_5MgBr	(−)-Menthyl	*R*	*R*[a]	6.8	70	46
34	α-$C_{10}H_7$	C_6H_5MgBr	(−)-Menthyl	*R*	*R*[a]	−12.7	20	46
35	CH_3	C_6H_5MgI	(−)-Bornyl-methylcarbinyl	*S*[a]	*R*	23	72	22
36	C_6H_5	CH_3MgI	(−)-Phenyl-triphenyl-silylcarbinyl	*R*[a]	*S*	28	93	47

[a] The configuration of this alcohol or α-hydroxy acid has been assigned primarily upon the basis of the Prelog generalization.

[b] Primary alcohol, see Sec. **2-3.5**.

[c] No significant rotation observed.

[d] The reactions with the lithium alkynes were done at −40 to 0° in THF solvent.

[e] The resulting alkyne was reduced to α-ethyllactic acid whose configuration is known but the data is not sufficient for calculating the percent asymmetric synthesis.

[f] These configurations were mistakenly reversed in the original publication[43] but subsequently corrected.[44]

[g] These two experiments involve the sequential addition of first one Grignard reagent and then the other to ethyl (−)-menthyl oxalate; cf. footnote c, Table **2-3**.

[h] This is one of the experiments that did not check the Prelog generalization and was repeated with opposite results.[4]

2-4. Asymmetric Synthesis Studies with β-, γ-, δ-, and Higher Keto Esters.

Table **2-7** contains the available data concerning asymmetric synthesis studies using β, γ, δ, and other keto esters. β-Keto esters with α-hydrogens (Nos. 1 and 2) give no addition products with Grignard reagents because of the much faster enolization reaction; therefore no asymmetric synthesis information is available for this type. However, McKenzie[24] was able to show (Nos. 3, 4, and 5) that asymmetric synthesis did occur by using (−)-menthyl α,α-diethylacetoacetate which contains no α-hydrogens and therefore cannot enolize. No quantitative information was given and the implication was that the observed optical rotation was low.

(−)-Menthyl 4-oxo-pentanoate (Table **2-7**, No. 6) and (−)-menthyl 4-oxo-4-phenylbutanoate (No. 10) both gave substantial asymmetric

[42] R. J. D. Evans and S. R. Landor, *Proc. Chem. Soc.*, 182 (1962).

[43] V. Prelog and E. Watanabe, *Ann. Chem.*, **603**, 1 (1957).

[44] R. H. Cornforth, J. W. Cornforth, and V. Prelog, *Ann. Chem.*, **634**, 197 (1960).

[45] G. Vavon, G. Quesnel, and Y. Runavot, *Compt. rend.*, **237**, 617 (1953).

[46] A. McKenzie and P. D. Ritchie, *Biochem. Z.*, **231**, 412 (1931).

[47] M. Biernbaum and H. S. Mosher, *Tetrahedron Lett.*, 5789 (1968).

synthesis upon treatment, respectively, with phenyl and methyl Grignard reagents. These results are lower (5–13% and 10–17%) than thosc for the analogous α-keto ester reactions (14–18% and 22–30%). Even so, the extent of asymmetric synthesis is surprisingly large, considering the fact that there are now four atoms separating the developing chiral center from the established chiral center in the γ-keto esters, while there are only two atoms separating the same centers in the α-keto esters. It is also of interest that the extent of asymmetric synthesis is higher with the attack of the methyl Grignard reagent on the phenyl keto ester than with the attack of the phenyl Grignard reagent on the methyl keto ester in both the α- and γ-systems. As would be expected, interchanging the R groups of the Grignard reagent and the keto ester resulted in formation of enantiomeric products.

The resolution of 4-hydroxy-4-phenylpentanoic acid has been confirmed;[48] ORD studies[49] suggest that the (−) acid (and the related (+) lactone) have the absolute *R* configuration as shown (Fig. **2-11**). By analogy, the same may also hold for (−)-5-hydroxy-5-phenylhexanoic acid and its

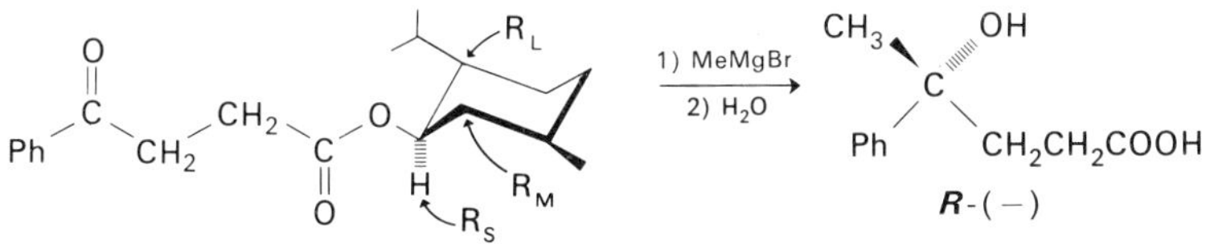

Fig. 2-11. Asymmetric synthesis with (−)-menthyl γ-keto-γ-phenylbutyrate.

related (+) lactone. If this is so, then the comparable faces of the γ and δ keto groups in the respective (−)-menthyl esters undergo preferential attack. Because of the limited data available, however, a configurational correlation model for these reactions is not warranted.

Comparison of extents of asymmetric synthesis in the γ- and δ-keto esters may not be valid. The values listed for Nos. 7 and 12 represent maximum values based upon a resolution[25] of 5-hydroxy-5-phenylhexanoic acid (lactone $[\alpha]_{5461}^{25}$ +15.67°) which may be incomplete as suggested by the value of $[\alpha]_{5461}^{25}$ +61.9° found for the lactone of 4-hydroxy-4-phenyl-pentanoic acid.

With the (−)-menthyl esters of ε-keto acids, the optical rotation of the resulting ε-hydroxy acid was marginal; ι-keto esters in which eight methylenes separate the keto and ester groups gave racemic products.

Bovey, Reid, and Turner[50] reasoned that an electronic inductive force

[48] J. Kenyon and M. C. R. Symons, *J. Chem. Soc.*, 3580 (1953).

[49] L. Verbit, S. Mitsui, and Y. Senda, *Tetrahedron*, **22**, 753 (1966); L. Verbit, private communication.

[50] D. M. Bovey, J. A. Reid, and E. E. Turner, *J. Chem. Soc.*, 3227 (1951).

TABLE 2-7

Asymmetric Synthesis with (−)-Menthyl Esters of β-, γ-, δ-, and Higher Keto Acids

$$R-\overset{O}{\overset{\|}{C}}-[Q]-COO\text{-}(-)\text{-Menthyl} \xrightarrow[H_2O\,(H^+)]{R'MgBr} \xrightarrow[(KOH)]{H_2O} R-\overset{OH}{\overset{|}{C}}R'-[Q]-COOH + \text{Menthol}$$

No.	Keto ester R	Keto ester Q	Reagent R′	Hydroxy acid or lactone: Rotation	Asym. syn. %	Yield %	Ref.
1	CH_3	CH_2	Ph	—	—[a]	—	24
2	CH_3	CHEt	Et	—[a]	—[a]	—	24
3	CH_3	CEt_2	Et	—[b]	—[b]	—	24
4	CH_3	CEt_2	Ph	(−)	—[c]	—	24
5	CH_3	CEt_2	α-$C_{10}H_7$	(−)	—[c]	—	24
6	CH_3	$(CH_2)_2$	Ph	(−)[e]	5–13[d,e]	40–60	25
7	CH_3	$(CH_2)_3$	Ph	(−)[e]	1.6–16[d,e,f]	70	25
8	CH_3	$(CH_2)_4$	Ph	α_{5780}^{25} +0.09°	—	84	25
9	CH_3	$(CH_2)_8$	Ph	—[b]	—[b]	100+	25
10	Ph	$(CH_2)_2$	Me	(+)[e]	10–17[g,e]	40	26
11	Ph	$(CH_2)_2$	Et	α_{5461}^{25} +8.5°	—	—	26
12	Ph	$(CH_2)_3$	Me	—[f,h]	13–19[f,h]	86–96	26
13	Ph	$(CH_2)_3$	Et	—[b]	—[b]	—	26
14	Ph	$(CH_2)_4$	Me	—[i]	—[b]	43–98	26
15	Ph	$(CH_2)_4$	Et	(−)[j]	—[j]	—	26
16	Ph	$(CH_2)_8$	Me	—[b]	—[b]	—	26
17	Ph	$(CH_2)_8$	Et	—[b]	—[b]	—	26
18	Ph	$[-C_6H_4-]$	$LiAlH_4$	—[b]	—[b,k]	—	8
19	Ph	$[-C_6H_4-]$	$NaBH_4$	—[b]	—[b,k]	84	8
20	Ph	$[C_6H_4]$	$LiAlH_4$	—[b]	—[b,l]	86	51
21	Ph	$[C_6H_4]$	$NaBH_4$	—[b]	—[b,m]	68	51

[a] No addition product, only enolization, i.e., recovered *β*-keto ester.
[b] No observable rotation.
[c] The solution was colored and the extent of rotation in doubt, but the sign of rotation was clearly (−).
[d] Represents five runs under varying experimental conditions.

[51] W. A. Bonner, *J. Am. Chem. Soc.*, **85**, 439 (1963).

should be damped out by two intervening methylene groups and thus these results constitute evidence against "induced asymmetry" as propounded by McKenzie.

The last examples in Table **2-6** have already been discussed (Sec. **2-1**). Bonner reasoned that an electronic inductive force should be transmitted through a conjugated system. The lack of asymmetric synthesis in the case of (−)-menthyl *p*-benzoylbenzoate was therefore taken as evidence that "asymmetric induction" in the McKenzie sense, if it exists, is incapable of transmission through an aromatic nucleus.

1969 Addenda

Sec. 2-3. Weill-Raynal and Mathieu have given a semi-quantitative interpretation to the observed extents of asymmetric syntheses reported by Prelog et al. in the benzoylformate to atrolactate system.[52] Their method consists of making an intuitive estimate of the relative population of three transition state conformations according to the original Prelog interpretation of this reaction.

---

[e] Rotation and percent asymmetric synthesis based on the lactone instead of acid; (+)-acid gave (−)-lactone.

[f] Maximum value based upon $[\alpha]^{25}_{5461}$ +15.67° for the lactone from a resolution of 5-hydroxy-5-phenylhexanoic acid which probably is only partially complete.

[g] Represents eighteen experiments made under varying reaction conditions.

[h] Represents seven experiments under varying reaction conditions.

[i] Three experiments gave (±), $[\alpha]^{25}_{5461}$ +0.08° (c, 11) and (±) acid when the Grignard was present in 1.25, 2.0, and 4.0 molar excess respectively.

[j] The rotation was just discernible, α^{25}_{D} −0.03° (c, 12.85, ethanol).

[k] The product isolated was optically inactive diol, *p*-hydroxymethylbenzyhydrol.

[l] The isolated product was optically inactive *o*-hydroxymethylbenzhydrol.

[m] The product isolated was 3-phenylphthalide.

[52] J. Weill-Raynal and J. Mathieu, *Bull. Soc. Chim. Fr.*, 115 (1969).

3

Reactions of Achiral Reagents with Chiral Aldehydes and Ketones

3-1. Introduction

The asymmetric syntheses which result from the reactions of achiral reagents (**1**) with chiral aldehydes and ketones (**2**) under kinetic control to give mixtures of diastereomeric products (**3**) are of the same fundamental

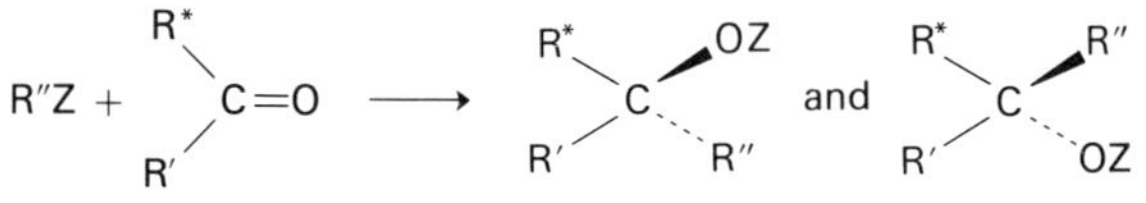

$$R''Z + \begin{matrix} R^* \\ \quad \diagdown \\ \quad C{=}O \\ \quad \diagup \\ R' \end{matrix} \longrightarrow \begin{matrix} R^* \quad OZ \\ C \\ R' \quad R'' \end{matrix} \text{ and } \begin{matrix} R^* \quad R'' \\ C \\ R' \quad OZ \end{matrix}$$

1 **2** **3A** **3B**

Fig. 3-1. Asymmetric synthesis with chiral aldehydes and ketones.*

type as the numerous asymmetric syntheses of atrolactic esters which were treated in Chapter 2. However there are some special aspects of the reactions of these aldehydes and ketones as contrasted with those of keto esters which

* In this generalized equation (Fig. **3-1**), Z most often symbolizes the metallo group of a Grignard reagent, metal alkyl, complex metal hydride, or aluminum alkoxide, etc., and R″ is a group transferred from the R″Z reagent; but R″Z may also represent a reagent such as HCN in the cyanohydrin synthesis or an aldehyde molecule as its enolate salt in the aldol condensation or as its sodium hydroxide adduct in the Cannizzaro reaction. R* represents a chiral group and R, a hydrogen, alkyl, or aryl group.

set them apart and which require that they be treated separately. For example, the chiral center in **2** is usually adjacent to the carbonyl group, resulting in what has been termed "1,2-asymmetric induction;" while in the benzoylformate esters, the chiral center and carbonyl group are separated by the carboxyl function giving rise to a "1,4-asymmetric induction." Also, no general chemical method analogous to hydrolysis in the case of the atrolactic esters has been developed for separating the inducing chiral group (R*) from the product molecule. Finally, the interpretation of the atrolactate asymmetric synthesis has been dominated by the Prelog generalization, while interpretations of nucleophilic addition reactions of chiral aldehydes and ketones have been dominated in the past by Cram's rule.

According to the definition we have adopted (Sec. **1-2**, pages 4–5), reactions starting with racemic materials do not qualify as asymmetric syntheses. Nevertheless, when suitable methods of analysis are available, such reactions can under certain circumstances render much the same information as those employing a chiral reagent or substrate. We shall find it convenient in the following section to refer to many reactions using racemic substrates in lieu of available information on true asymmetric synthesis using chiral substrates. But the vast number of reactions of this type employing racemic substrates precludes their exhaustive treatment here. A chiral substrate has several advantages: It can be used to demonstrate unequivocally that no racemization has taken place at the inducing center during the reaction;* it can be of special value in establishing the stereochemistry of the diastereomeric products as, for instance, in those cases in which one product is *meso*; and it can be of value as the basis of the analytical method for the determination of the percent stereoselectivity of the reaction. In spite of the advantages of using a chiral substrate, the added difficulties of obtaining such chiral materials has dictated the use of racemic (and therefore achiral) materials in many studies.

Consider the lithium aluminum hydride reduction of racemic 2-methylcyclopentanone (**4**) to give a mixture of *trans*-**5** and *cis*-**5** racemic products. The product mixture has been shown to contain approximately 75% *trans* racemate and 25% *cis* racemate[1] (Table **3-9**). The stereoselectivity of the reaction is therefore 50%, i.e., 50% excess of the *trans* diastereomer over the *cis*. If, instead of starting with the racemate *RS*-**4**, pure *S*-**4** enantiomer had been reduced with lithium aluminum hydride, the epimeric *S,S*-**5** and *S,R*-**5** carbinols would have been formed in the same 75:25 ratio (i.e., 50% e.e.). Thus it is seen that for determination of percent stereoselectivity it is

* Such a racemization process may go undetected when using a racemic substrate, especially in those cases in which there is an asymmetric center bearing a hydrogen on the carbon atom *alpha* to the carbonyl function.

[1] (a) J. B. Umland and B. W. Williams, *J. Org. Chem.*, **21**, 1302 (1956). (b) J. B. Umland and M. I. Jefraim, *J. Am. Chem. Soc.*, **78**, 2788 (1956).

unnecessary to work with optically active material provided suitable methods for separation (or analysis) and identification of the components of the racemic products are available.

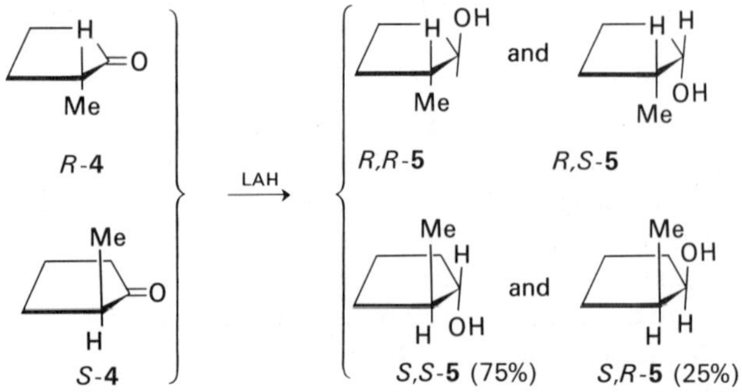

R-4 ; S-4 —LAH→ R,R-5 and R,S-5 ; S,S-5 (75%) and S,R-5 (25%)

Fig. 3-2. Reduction of a racemate with the formation of a new chiral center.

As illustrated by the reduction of cholestanone (Sec. **1-3**), the original inducing center in the carbonyl substrate (R* in Fig. **3-1**) need not be removed subsequently, as required by Marckwald's classical definition, for the reaction to be classified as an asymmetric synthesis. It is instructive, however, to begin by considering an example that does fit the Marckwald operational scheme. When (*R*)-(−)-benzoin (**6**) is treated with ethylmagnesium bromide, either diastereotopic face* of the carbonyl group may be attacked to produce a mixture of epimeric ethylhydrobenzoins (*R*,*S*-**7** and *R*,*R*-**7**).[2] In practice, only one isomer, (+)-*erythro* (*R*,*S*-**7**), was isolated in pure form† from this reaction.

In the reaction going from **6** to **7**, a prochiral carbonyl center has been converted into a chiral center under the influence of the existing asymmetric center in **6**. In the subsequent reaction going from **7** to **8**, the original inducing asymmetric center has been destroyed by oxidation to the achiral keto

* Either the *re* or *si* face, according to the prochiral nomenclature scheme of K. R. Hanson, *J. Am. Chem. Soc.*, **88**, 2731 (1966).

† In several studies of this reaction, only the *R*,*S*-**7** isomer has been isolated in pure form. This has led to the statement that this isomer is formed to the exclusion of its epimer. Since purification was accomplished by fractional crystallization and the yield of the isolated *R*,*S*-**7** form was reported to be about 50%, the epimer almost certainly was present in the mother liquors. In the comparable reaction using methylmagnesium iodide in which analysis of the diastereomeric product was carried out by a carbon-14 dilution technique,[3] the corresponding ratio of isomers (*R*,*S*-**7** to *R*,*R*-**7**, ethyl replaced with methyl) was 28 : 1.

[2] (a) A. McKenzie and H. Wren, *J. Chem. Soc.*, **97**, 473 (1910). (b) M. Tiffineau and J. Lévy, *Bull. Soc. Chim. Fr.*, **41**, 1351 (1927). (c) R. Roger, *J. Chem. Soc.*, 1048 (1937). (d) R. Roger, *ibid.*, 108 (1939). (e) S. M. Partridge, *ibid.*, 1201 (1939).

[3] J. H. Stocker, P. Sidisunthorn, B. M. Benjamin, and C. J. Collins, *J. Am. Chem. Soc.*, **82**, 3913 (1960).

group, thereby unequivocally establishing that asymmetric synthesis has been achieved.

Ph–C(=O)–C(H)(OH)–Ph $\xrightarrow[\text{2) } H_2O]{\text{1) EtMgBr}}$ HO–C(Ph)(Et)–C(H)(OH)–Ph + Et–C(Ph)(OH)–C(H)(OH)–Ph $\xrightarrow{[O]}$ HO–C(Ph)(Et)–C(=O)–Ph + Et–C(Ph)(OH)–C(=O)–Ph

R-(−)-**6** *R,R*-**7** *R,S*-**7** *R*-(+)-**8** *S*-(−)-**8**

If the order of introducing the groups is reversed, i.e., the ethyl ketone analogous to **6** is treated with phenyl Grignard reagent, then the predominant product is the *R,R* instead of the *R,S* diastereomer; upon oxidation, (*R*)-(+)-**8** is formed instead of (*S*)-(−)-**8**. It is apparent that this type of asymmetric synthesis is very similar to the atrolactic ester type, but in this case the inducing center is directly attached to the center which, in the transition state, is transformed into the new chiral center. Steric and dipolar interactions between the groups under these circumstances must be much more direct than in the atrolactic ester asymmetric synthesis system in which the two centers are separated by the carboxy moiety; one would therefore anticipate higher stereoselectivities.

3-2. Models for Stereochemical Control of Asymmetric Additions to Chiral Aldehydes and Ketones

The Cram rule[4] which has been used to correlate the stereochemical course of a large number of 1,2-asymmetric addition reactions of aldehydes and ketones, primarily with organo-metallic and complex metal hydride reagents, states that "in noncatalytic (kinetically controlled) reactions of the type shown (**9** → **10**) that diastereomer will predominate which would be formed by the approach of the entering group (R″ of the reagent R″Z) from the less hindered side of the double bond when the rotational conformation of the C—C bond is such that the double bond is flanked by the two least

[4] D. J. Cram and F. A. Abd Elhafez, *J. Am. Chem. Soc.*, **74**, 5828 (1952). The widespread use of this correlation is indicated by the fifty references to this paper appearing during the 3-year period 1964, 1965, and 1966, some 12–15 years after its original publication.

hindered bulky groups (R_S and R_M) attached to the asymmetric center." A Newman projection for this model, as published in a subsequent paper,[5] is shown in **11**.

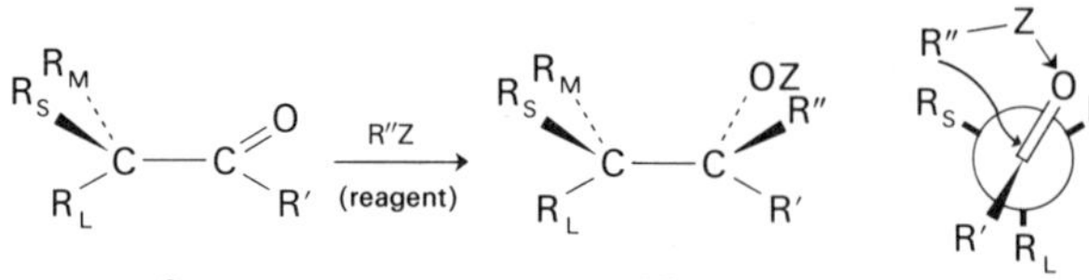

9 10 11
Predominant isomer

Fig. 3-3. Open-chain model for Cram's rule of steric control of asymmetric induction.

A logical consequence of this model is that the extent of stereoselectivity should depend upon the differences in magnitudes of the interactions of $R'' \leftrightarrow R_S$ and $R'' \leftrightarrow R_M$. It was recognized from the beginning that this model might not apply in circumstances in which one of the groups on the adjacent chiral center, for instance hydroxy or amino, could strongly complex with the reagent. For these cases a "cyclic model" (which can be represented as either **12A** or **12B**) was proposed. Still another model was introduced by Cornforth, et al.[6] to accommodate those examples in which

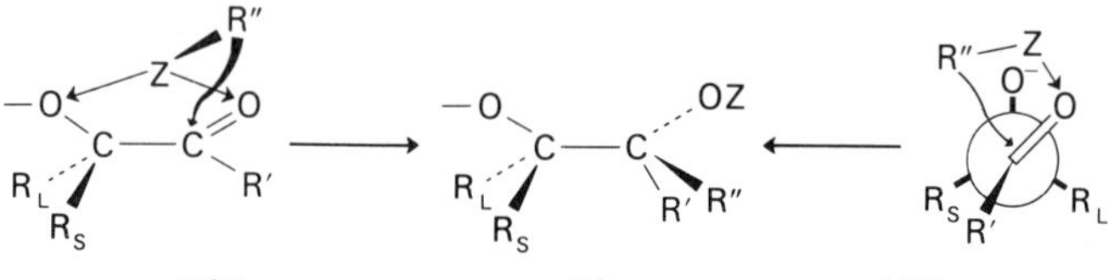

12A 13 12B
Predominant isomer

Fig. 3-4. Cyclic model for steric control of asymmetric induction where the reagent can complex with an OR (or NH_2) group.

a highly polarizable group, such as a halogen, is attached to the chiral center. This has been termed the "dipolar model" (**14A** or **14B**) and is characterized by a *trans* coplanar arrangement of the adjacent dipoles.

The success (and limitations) of these models will become apparent from the following discussion and examples, but to what extent these models represent the real stereochemistry of the transition state is another question

[5] D. J. Cram and K. R. Kopecky, *J. Am. Chem. Soc.*, **81**, 2748 (1959).

[6] J. W. Cornforth, R. H. Cornforth, and K. K. Mathew, *J. Chem. Soc.*, 112 (1959); cf. Sec. **2-2** for another application.

which will be considered later (Sec. **3-4**) in connection with the modifications of the Cram rule proposed by Karabatsos[7] and by Felkin and co-workers.[8]

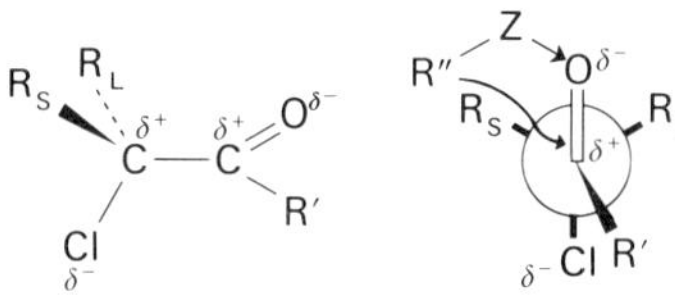

14A **14B**

Fig. 3-5. Dipolar model for steric control of asymmetric induction.

In the application of these models, one is immediately confronted with two decisions: How are the groups assigned with respect to R_L, R_M, and R_S? Which model, the open-chain (**11**), cyclic (**12**), or dipolar (**14**) should be used? The problem of ordering various groups with respect to size has already been considered (Sec. **1-4.4a**). Here, as in the application of the Prelog generalization (Sec. **2-2.1**), one of the groups is most often hydrogen and therefore its designation as R_S presents no problem. The phenyl group has almost always been considered larger than all alkyl groups; thus it is considered larger than a long aliphatic chain, in spite of the fact that the latter occupies a larger total volume, because the effective bulk of the phenyl group is concentrated in the vicinity of the site of reaction, while the long aliphatic chain can spread out away from the carbonyl group and has only a $-CH_2$ adjacent to the reactive site. For the same reason, phenyl is considered larger than benzyl. Examples testing the effective bulk of phenyl versus isopropyl have shown phenyl to be larger (Table **3-1**, Nos. 9–12). The relative effective bulks of phenyl versus *t*-butyl have not been directly tested in systems where either the Cram or Prelog models apply, but in several other systems phenyl acts as though it is larger than *t*-butyl (Sec. **1-4.4b**).

When no halogen or hetero atom is involved in R_S, R_M, or R_L of formula **9**, the open-chain model is clearly indicated; when one of these groups is chlorine, the dipolar model is indicated; and when one of these groups carries nitrogen or oxygen (or sulfur, presumably) directly bound to the chiral center, the first choice is the cyclic model. This cyclic model must be applied with considerable care because, as seen in Secs. **3-2.2** and **3-2.4**, such examples are sensitive to variables of solvent, reagent, and substituents on the hetero atom. Since there are a number of exceptions, it cannot be assumed that these types, with an oxygen-containing or nitrogen-containing moiety substituted on the chiral center adjacent to the carbonyl group, will always follow the expectations of the cyclic model.

[7] G. J. Karabatsos, *J. Am. Chem. Soc.*, **89**, 1367 (1967).
[8] M. Chérest, H. Felkin, and N. Prudent, *Tetrahedron Lett.*, 2201 (1968).

Superimposed upon the uncertainties of designation of R_L, R_M, and R_S in the substrate and the choice of the various models are the considerations of effect of solvent, reagent, and temperature upon the stereochemical course of the asymmetric synthesis. Since in isolated cases it has been clearly demonstrated that a change of reagent, temperature, or solvent can bring about a reversal of the stereochemistry, this obviously represents a complicated situation from which generalizations can be drawn only with great caution after considering many examples under various conditions.

3-2.1. The Open-Chain Model. The data in Table **3-1** illustrate the variations in percent asymmetric synthesis for examples which fit the Cram open-chain model (**11**). All except two (Nos. 24 and 31) give as the predominant isomer that one predicted by the Cram rule. In the first of the two exceptions, the lithium aluminum hydride reduction of 3-methyl-2-pentanone, a 2% excess of the predicted favored diastereomer (on the basis that ethyl is bigger than methyl) was obtained at −70°, but at 0° and +35° a 3% excess of the epimer, which is predicted to be the minor product, was found.[9] These small differences are probably real but marginal since it was estimated that the accuracy of the gas chromatographic analysis was ±2%. A 2% asymmetric synthesis at −70° represents a difference in the energies of activation for the competing transition states of only 15 cal/mole. If one adds the requirement, as done in the case of the Prelog generalization, that the stereoselectivity must be substantial* (see Sec. **2-2.1**) before this model can be used for interpreting the stereochemistry of a particular experiment, then this case (No. 24) would be excluded. However, the second exception, namely the aluminum isopropoxide reduction of 3-cyclohexyl-2-butanone (No. 31), remains as a clear violation of the Cram rule for the open-chain model. Since it was demonstrated that this particular reaction was under kinetic control,[10] it was concluded that the rule was not applicable to the Meerwein–Ponndorf–Verley reduction. Nevertheless, two other examples follow the rule (Table **3-1**, Nos. 6 and 22) and this point should be explored further.

As formulated, this rule was devised for predicting the configuration of the predominant product and not for quantitative assessment of the extent of asymmetric synthesis. The very basis of the model, however, implies that the greater the differences in steric bulk between the groups on the chiral center, the greater should be the stereoselectivity of the reaction. It can be

[9] Y. Gault and H. Felkin, *Bull. Soc. Chim. Fr.*, 1342 (1960).

[10] D. J. Cram and F. D. Greene, *J. Am. Chem. Soc.*, **75**, 6005 (1953).

* In the case of the benzoylformate to atrolactate asymmetric synthesis, a stereoselectivity greater than 5% was deemed necessary for the Prelog generalization to be applied safely to the stereochemical interpretation of the results. We believe that at least the same minimum stereoselectivity should be used in applications of the Cram rule.

seen from the first four examples in Table **3-1** that the extent of stereoselectivity is definitely dependent upon the reagent as well as the R_L, R_M, and R_S groups attached to the chiral center. Thus the observed stereoselectivity in the addition of Grignard reagents to α-phenylpropionaldehyde ($R' = H$) increased from 33 to 50 to 60% in progressing from methylmagnesium iodide to ethylmagnesium bromide to phenylmagnesium bromide. Other examples in Table **3-1** support this observation but it would be desirable to have further closely comparable data to document this point.

The effect of systematic variations in the bulk of the achiral group attached to the carbonyl center has been explored.[8] A dramatic increase in stereoselectivity, from 23% where R′ is methyl to 96% where R′ is *t*-butyl, is observed in lithium aluminum hydride reductions at 35° of the α-phenyl ketones represented in Fig. **3-6**. When the reaction in which R′ is *t*-butyl

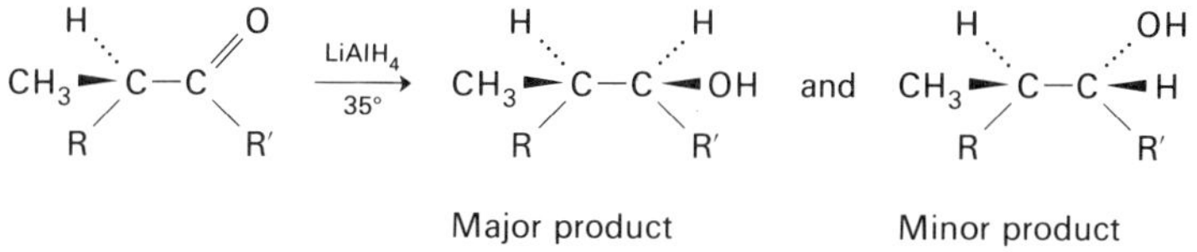

$LiAlH_4$, 35°

Major product and Minor product

R′	% Stereoselectivity, 35°	
	R = Ph	R = Cyclohexyl
Me	48	26
Et	52	33
i-Pr	70	60
t-Bu	96	23

Fig. 3-6. Increase in stereoselectivity with increased steric bulk of achiral group attached to the carbonyl center.[8]

is conducted at −70°, the major diastereomer is obtained in 99.6% excess (Table **3-1**, No. 37). In the analogous cyclohexyl ketones a similar trend begins to develop but breaks down with the *t*-butyl example.

There is essentially no stereoselectivity observed when R_M and R_L are phenyl and *p*-tolyl, respectively (Nos. 38 and 39), as anticipated from the fact that the methyl group of *p*-tolyl is remote from the reaction site.

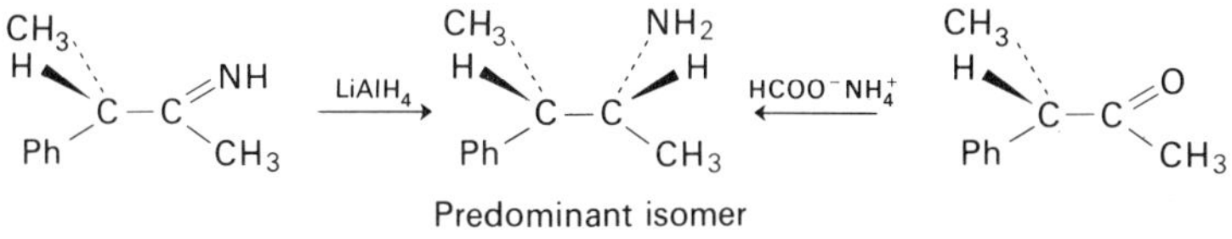

$LiAlH_4$ → Predominant isomer ← $HCOO^-NH_4^+$

Fig. 3-7. Stereoselective imine reduction and Leuckart reaction.

Numbers 25 and 26 warrant special attention. In the first of these it was shown that an imine underwent lithium aluminum hydride reduction to give the amine in a manner completely comparable to the reduction of the corresponding ketone. In the second it was shown in this single example that the Leuckart reaction for the conversion of a keto group to an amine followed the stereochemistry predicted by the Cram rule (Fig. **3-7**).

TABLE 3-1

The Effect of Variations of Substrate and Reagent on Stereoselectivities in the Cram Open-Chain System[a]

$R_S R_M R_L C{-}C(=O)R'$ $\xrightarrow[2)\ H_2O]{1)\ R''Z}$ $R_S R_M R_L C{-}C(OH)(R'')R'$ (**A**) and $R_S R_M R_L C{-}C(R'')(OH)R'$ (**B**)

No.	R_S	R_M	R_L	R'	$R''Z$[b]	A:B Ratio[a]	% Stereo-selectivity[c]	Ref.
1	H	Me	Ph	H	MeMgBr (−50°)	2.4	41	9
					(0°)	2.3	40	9
					(35°)	2.0	33	9
2	H	Me	Ph	H	MeMgI	2.0	33	4
3	H	Me	Ph	H	EtMgBr	3.0	50	4
4	H	Me	Ph	H	PhMgBr	>4	60	4
5	H	Me	Ph	Me	$LiAlH_4$ (−70°)	5.6	70	9
					(0°)	3.0	50	9
					(35°)	2.6	45	9
					(35°)	2.5	43	4
					(35°)	2.8	48	8
6	H	Me	Ph	Me	$Al(O\text{-}i\text{Pr})_3$	1.6	22	9
7	H	Me	Ph	Et	$LiAlH_4$	2.0	33	4
						3.2	52	8
8	H	Me	Ph	Ph	$LiAlH_4$	>4	60	4
9	H	*i*Pr	Ph	H	*i*PrMgBr + $MgBr_2$	1.3	13	11a
10	H	*i*Pr	Ph	H	*i*PrMgBr	1.9	32	11a
11	H	*i*Pr	Ph	H	*i*PrLi(pentane)	1	0	11a
12	H	*i*Pr	Ph	*i*Pr	$LiAlH_4$	10	82	11a
13	Me	Et	Ph	Ph	$LiAlH_4$	2.2	38	11b
14	Me	Et	Ph	H	PhMgBr	2.9	32	11b
						2.1	36	11b
15	Me	Et	Ph	Me	EtLi	2.3	40	12
						2.4	42	12
16	Me	Et	Ph	Et	MeLi	10	82	12
						8	78	12
17	H	Et	Ph	H	MeMgI	2.5	42	4
18	H	Et	Ph	H	EtMgBr	3.0	50	13

TABLE 3-1—cont.

No.	R_S	R_M	R_L	R′	R″Z[b]	**A:B** Ratio	% Stereo-selectivity[c]	Ref.
19	H	Et	Ph	Me	$LiAlH_4$	3.0	50	13
20	H	Et	Ph	Et	$LiAlH_4$	3.0	50	13
21	H	Me	Et	H	MeMgBr	1.5	20	9
22	H	Me	Et	Me	Al(O-*i*Pr)$_3$	1.3	13	9
23	H	Me	Et	Me	*i*PrCH$_2$MgBr	1.8	28	9
24	H	Me	Et	Me	$LiAlH_4$ (−70°)	1.03	2	9
					(0°)	0.95	(3)[d]	9
					(35°)	0.95	(3)[d]	9
25[e]	H	Me	Ph	Me	$LiAlH_4$[e]	2.0	33	14
26[f]	H	Me	Ph	Me	HCO_2NH_4	1.9	31	14
27	H	Me	C_6H_{11}[g]	H	MeMgI	1.9	31	10
28	H	Me	C_6H_{11}[g]	H	MeLi(pentane)	1.5	20	10
					(ether)	1.2	9	10
29	H	Me	C_6H_{11}[g]	Me	$NaBH_4$	1.7	25	10
30	H	Me	C_6H_{11}[g]	Me	$LiAlH_4$	1.4	16	10
31	H	Me	C_6H_{11}[g]	Me	Al(O-*i*Pr)$_3$	0.6	(31)[d]	10
32	H	Me	C_6H_{11}[g]	Me	$LiAlH_4$	1.6	23	8
33	H	Me	C_6H_{11}[g]	Et	$LiAlH_4$	2.0	33	8
34	H	Me	C_6H_{11}[g]	*i*Pr	$LiAlH_4$	4.1	60	8
35	H	Me	C_6H_{11}[g]	*t*Bu	$LiAlH_4$	1.6	23	8
36	H	Me	Ph	*i*Pr	$LiAlH_4$	5.0	70	8
37	H	Me	Ph	*t*Bu	$LiAlH_4$ (35°)	49	96	8
					(−70°)	499	99.6	8
38[h]	H	Ph	*p*-Tolyl	Ph	$LiAlH_4$	1.00 ± 0.06	0	3
39[h]	H	Ph	*p*-Tolyl	H	PhMgBr	1.00 ± 0.05	0	3

[a] The table is so arranged that the **A**:**B** ratios are greater than 1 when Cram's rule is followed.
[b] RMgX is used to symbolize the Grignard reagent without necessarily implying that this is the active monomeric reagent in solution.
[c] Percent stereoselectivity is the percent excess of isomer **A** over isomer **B**.
[d] Product of configuration **B** predominates.
[e] Reaction of imine to give the amine instead of ketone to give alcohol.
[f] The Leuckart reaction to form the amine.
[g] C_6H_{11} represents the cyclohexyl group.
[h] Configuration unknown.

The variety of groups which have been explored in the open-chain model could be expanded profitably to include examples of groups such as *t*-butyl, suggested by Karabatsos,[7] as well as mesityl, trityl, and naphthyl.

[11] (a) D. J. Cram, F. A. Abd Elhafez, and H. L. Nyquist, *J. Am. Chem. Soc.*, **76**, 22 (1954). (b) D. J. Cram and J. Allinger, *ibid.*, **76**, 4516 (1954).
[12] D. J. Cram and J. D. Knight, *J. Am. Chem. Soc.*, **74**, 5835 (1952).
[13] D. J. Cram, F. A. Abd Elhafez, and H. Weingartner, *J. Am. Chem. Soc.*, **75**, 2293 (1953).
[14] D. J. Cram and J. E. McCarty, *J. Am. Chem. Soc.*, **76**, 5740 (1954).

3-2.2. The Cyclic Model. In the addition reactions of ketones or aldehydes which contain a hetero atom on the chiral center adjacent to the carbonyl group, the stereochemical interpretation is much more difficult than with unsubstituted substrates. With suitable changes in reagent, solvent, hetero group, and substrate, large variations in diastereomeric product ratios can be achieved, even a reversal in expected stereochemistry, so that results can be obtained which conform to expectations based upon any one of the three models, cyclic (**14**), dipolar (**13**), or open-chain (**11**). A thorough study has been made[3,15] of the effect upon diastereomer product ratios resulting from changes in reagent, solvent, and substituents (Table **3-2**).

15 $\xrightarrow{R''Z}$ **16** and **17**

A, R′ = CH_3, Q = OH
B, R′ = CH_3, Q = OCH_3
C, R′ = Ph, Q = OH
D, R′ = Ph, Q = OCH_3

Some conclusions which can be drawn from the data presented in Table **3-2** are as follows: (a) Stereoselectivities are generally highest in the reactions of the phenyl reagents with the α-methoxy methyl ketone (Table **3-2**, Column B). The diastereomer ratio of 19:1 in the reaction of phenylmagnesium bromide with the α-methoxy methyl ketone (Column B, No. 10) corresponds to formation of 95% of the *dl-threo* isomer (**16**) versus 5% of the *dl-erythro* isomer (**17**). But in the counterreaction of the methyl reagents with α-hydroxy or α-methoxy phenyl ketones (Columns C and D, No. 10) the α-hydroxy ketones show higher stereoselectivities. (b) The stereoselectivities of the lithium reagents are higher in ether than in pentane* (Nos. 2 and 1) and in those reactions represented by the data in Columns A, C, and D the lithium reagents are more stereoselective than the corresponding Grignard reagents. (c) The steric course of the reaction of the phenyl Grignard reagents with the α-hydroxy methyl ketone (Column A, Nos. 7, 8, and 9) is reversed in progressing from the reagent made from the iodide to that made from the bromide or chloride.† But the stereoselectivities shown by the Grignard reagents made from the methyl halides (Column C, Nos. 7, 8, and 9) are practically

[15] D. J. Cram and D. R. Wilson, *J. Am. Chem. Soc.*, **85**, 1245 (1963).

[16] J. H. Stocker, *J. Am. Chem. Soc.*, **88**, 2878 (1966).

* The lithium reagents in pentane are highly associated and at least the methyl lithium reaction in pentane was heterogeneous,[8] while in ether these reagents are probably dimers.

† This is not an isolated case; Stocker[16] found that the reaction of phenyllithium with 1,2-dicyclohexylethanedione gave a 57.6% yield of the racemic *threo* glycol and less than 0.25% of the *meso* glycol, while phenylmagnesium bromide gave a 14.8% yield of the *meso* glycol and less than 0.25% of the diastereomer. Analysis was by a very reliable carbon-14 dilution method.

TABLE 3-2

The Effect of Variations in Reagent, Solvent, and Substrate upon Stereoselective Additions to Racemic Ketones[a]

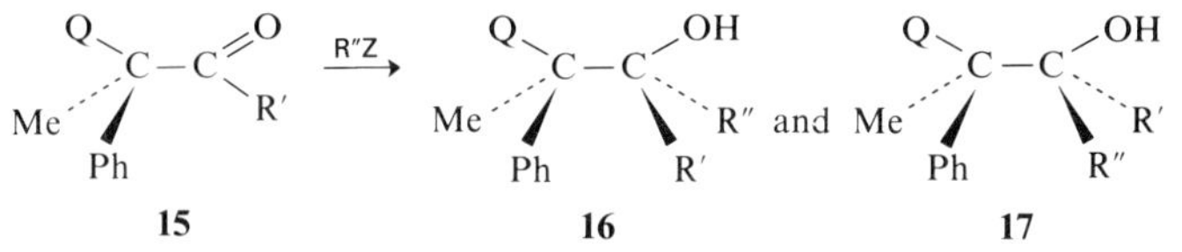

15 $\xrightarrow{R''Z}$ **16** and **17**

No.	Reagent R″Z	Solvent	Diastereomer ratio			
			16:**17**[b] R′ = Me, R″ = Ph		**16**:**17**[c] R′ = Ph, R″ = Me	
			Q = OH Column A	Q = OCH_3 Column B	Q = OH Column C	Q = OCH_3 Column D
1	R″Li	Pentane	5.0	—	3.4	—
2	R″Li	Ether	7.0	9	11	2
			10		8	
3	R″Li	Ether, TMEDA[d]	4		3.3	
4	R''_2Mg	Ether	1.8	7	3.5	1.7
5	R''_2Mg	Ether, TMEDA[d]	2.7	13	4.6	0.44
6	R''_2Mg	THF	1.6	13	6	1.1
7	R″MgI[e]	Ether	2.0		2.0	
8	R″MgBr[e]	Ether	0.56	6	2.7	2
			0.42		3.0	
9	R″MgCl[e]	Ether	0.30		2.9	
10	R″MgBr[e]	THF	1.40[f]	19	10	1.2

[a] These data are taken from References 3 and 15. The data are so presented that reactions which conform to the Cram cyclic model will give diastereomer ratios greater than 1.0 and those which are contrary to the Cram model, less than 1.0.

[b] This is the *dl-threo*:*dl-erythro* ratio.

[c] This is the *dl-erythro*:*dl-threo* ratio.

[d] TMEDA represents the addition of tetramethylethylenediamine to the reagent.

[e] RMgI, RMgBr, and RMgCl are used to symbolize the Grignard reagents made from the respective alkyl or aryl halides without specific implication concerning the reactive species in solution.

[f] This same reaction in dimethoxyethane solvent using phenylmagnesium bromide gave a diastereomer ratio (**16**: **17**) of 1.3:1.

independent of the halide used. (d) A reversal in the stereochemistry is evident in the reaction of dimethylmagnesium with the α-methoxy phenyl ketone (Column D, No. 5) in the presence of tetramethylethylenediamine (TMEDA). The addition of this coordinating reagent also reduces but does not reverse the stereoselectivity of the reactions involving the lithium reagents. On the other hand, TMEDA enhances somewhat the stereoselectivity of the reactions using diphenylmagnesium. A reversal of the *threo*: *erythro* ratio from

1.40 to 0.56 also is encountered in the reaction of phenylmagnesium bromide in going from tetrahydrofuran (THF) to ether solvent (Column A, Nos. 10 and 8).

Cram and Wilson[15] interpreted these complex results in terms of a competition of the cyclic model versus the dipolar and open-chain models. In every system the cyclic and dipolar models predict opposite stereochemistry. The stereochemistry predicted by the open-chain model depends upon whether one considers the methoxy and hydroxy groups larger or smaller than the methyl group. Based upon conformational energy values determined in monosubstituted cyclohexanes, these groups are significantly "smaller" than methyl but it is by no means certain that the same would be true of a hydroxyl or methoxy group which was complexed to a magnesium salt, or a hydroxyl group in the form of an alkoxide. Since the three models from which to choose and the variable effects of solvents, substrates, and reagents provide ample adjustable parameters to rationalize almost any results that might be obtained, it may be premature to try to do so. In spite of this, these models have been useful and it is important to determine where and how well they apply.

If we exclude those examples in Table **3-2** which give ratios less than one (Column A, Nos. 8 and 9; Column D, No. 5), we see that all the remaining examples including the α-hydroxy and α-methoxy ketones[17] in Table **3-4**, give as the predominant isomer that one predicted upon the basis of the cyclic model. The exceptions among the substituted α-amino ketones (Table **3-5**) can be rationalized. To this extent it is a valuable empirical correlation, but the exceptions should be carefully noted to indicate under what conditions the predictions based upon the rule are suspect.

It appears that the cyclic model is most reliable when applied to the reactions of the phenyl reagents with the α-methoxy methyl ketones (Table **3-2**, Column B); in these cases the dipolar model, which would predict the opposite stereochemistry, must not operate to any significant extent. The high stereoselectivities encountered in this series are also in accord with expectations for a rather constrained rigid transition state. Therefore the magnitudes of the observed stereoselectivities in this series substantiate the postulate of the cyclic model as a representation of the transition state. Under otherwise equivalent conditions one might anticipate that approach of a phenyl reagent from the "phenyl" or "non-phenyl" face of the carbonyl group would involve larger differences in nonbonded repulsive interactions in the competing diastereomeric transition states than if a methyl reagent were involved in the comparable competing reaction pathways. This

[17] D. Y. Curtin, E. E. Harris, and E. K. Meislich, *J. Am. Chem. Soc.*, **74**, 2901 (1952). This paper, which preceded and to some extent foreshadowed the publication of Cram and Elhafez,[4] presents a list of thirteen examples, reported in Table **3-5**, all of which fit the cyclic model.

expectation is clearly realized in the cases involving methoxy derivatives but is hardly evident in the corresponding hydroxy compounds.

The lowered stereoselectivities represented by the data in Columns A, C, and D are probably best explained by invoking a greater participation of the dipolar model in these reactions. It seems most likely that the open-chain model is not seriously involved in the reactions of the α-hydroxy ketones. Organometallic reagents presumably react with the active hydrogen of the hydroxyl group to form an alkoxide which has a strong dipolar interaction with the polarized carbonyl group. The transition state symbolized by the dipolar model can then effectively compete with the transition state symbolized by the cyclic model with a resulting reduction in the stereoselectivity of the reaction. In the case of the phenyl chloride and phenyl bromide Grignard reagents, the dipolar conformation presumably becomes dominant. Although it is not entirely clear why this same trend is not shown by the methyl Grignard reagents, one can postulate that the smaller methyl Grignard reagents are better accommodated by the dimensions of the cyclic transition state than are the phenyl Grignard reagents. The dipolar model would be expected to be less sensitive to variations in steric effects than the cyclic model and thus could become dominant in the case of the more bulky phenyl reagents.

It is of special interest that the stereoselectivity was altered significantly by changes in the solvent media. It is understandable that the basic reagent, tetramethylethylenediamine, could compete with the less basic α-methoxy group in coordinating with the Grignard reagent and thus interfere with the effective operation of the cyclic transition state. To the extent that this happens, the stereoselectivity should decrease. This was the effect in some cases but it did not occur uniformly. In the same way, the more basic THF (versus ether) might be expected to interfere with the cyclic model, but the results indicate the opposite.

In spite of the theoretical complexities and uncertainties, it is important to appreciate the significance of these studies from a synthetic standpoint. This is cogently illustrated by the reaction[18] of tolylmagnesium bromide with racemic 2-aminopropiophenone (**18**); the product (**19**), predicted on the basis of the cyclic model, was formed in 70% yield to the practical exclusion of the diastereomer (**20**).* An ingenious application of this reaction employed the amino ketone (**18**) which was labeled with carbon-14 in the phenyl ring and used the unlabeled phenyl Grignard reagent instead of the *p*-tolyl Grignard reagent. Under these circumstances the newly formed carbinol center was chiral by virtue of the carbon-14 isotope labeling.

[18] B. M. Benjamin, H. J. Schaeffer, and C. J. Collins, *J. Am. Chem. Soc.*, **79**, 6160 (1957).

* The analysis was performed by a carbon-14 dilution method and was therefore accurate; less than 1% of **20** was formed. These and other related results as well as the experimental method have been reviewed: C. J. Collins, *Adv. Phys. Org. Chem.*, **2**, 3–91 (1964).

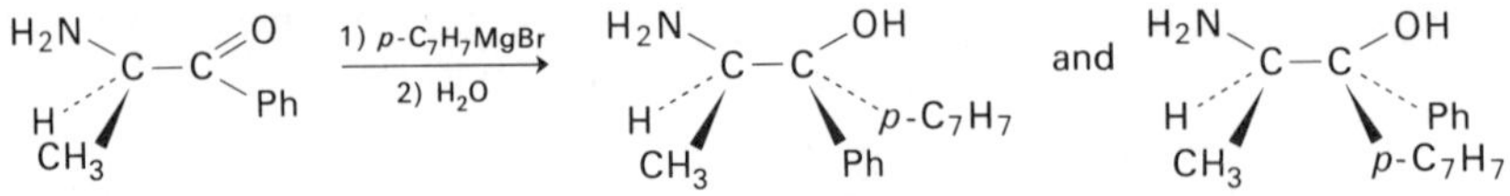

1) p-C_7H_7MgBr 2) H_2O

18 → 19 and 20

3-2.3. The Dipolar Model. When a halogen is on the chiral center adjacent to a carbonyl group, it has been proposed[6] that the controlling conformation for addition reactions is the *trans* dipolar structure **14**. It was reasoned that the carbonyl carbon should be the most electrophilic and therefore the most reactive when in the *trans* dipolar conformation with respect to the α-halogen. For the same reason, it was suggested that the rationale for the success of the Prelog generalization concerning the benzoylformate-atrolactate asymmetric synthesis lies in the fact that the *trans* dipolar conformation of the adjacent carbonyl groups is more reactive although it may not be the most stable ground state conformation (Sec. **2-4**). This idea has been put to good use in the development of a highly stereoselective epoxide and olefin synthesis[6] based upon the addition of Grignard or alkyl lithium reagents to α-chloro ketones at −70°.

Both open-chain and dipolar models predict the same stereochemistry only if the polar group is also R_L as would be the case in **21**. Opposite preferred stereochemistry is predicted for the reactions proceeding according to the dipolar and open-chain models when the polar group is R_M as the case probably is in **22**, where attack according to the dipolar model is indicated by the dashed arrow from behind the plane of the page but from in front in the open-chain model. The polar (**23**) and cyclic (**24**) models will of necessity in all circumstances predict opposite stereochemistries based on the reagent approach from the less hindered side.

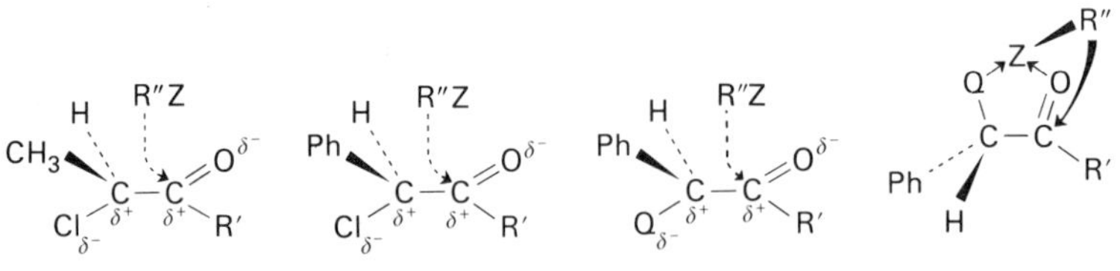

21 22 23 24

The acyclic examples involving α-chloro substituents are collected in Table **3-3**. In the reaction of 2-chlorohexanal with either ethylmagnesium bromide or ethyl lithium, a 70:30 *erythro*:*threo* mixture of diastereomers is formed (Nos. 1 and 2) in which the preponderant *erythro* isomer is the one predicted on the basis of the dipolar model and opposite to the one expected on the basis of either of the other models. In the analogous sodium borohydride reduction of 3-chloro-4-octanone (No. 3), the ratio of isomeric products formed was (80–85:20–15), with the isomer predicted by the dipolar model again predominating. The results from the other aliphatic examples (Nos. 2–7) also conform to the dipolar model. But when one of the groups on the chiral center is phenyl, the conclusions are not clear. The two accounts of the lithium aluminum hydride reduction of α-chlorodesoxybenzoin (**22**, R = Ph, Nos. 8 and 9) are conflicting, one [19a] reporting an approximately 50:50 mixture of diastereomers and the other[19b] reporting only the *threo* isomer, the one predicted on the basis of the dipolar model (chlorine acting as the polar group). But the reduction of α-chlorophenylacetone (**22**, R = Me, Table **3-3**, Nos. 10 and 11) gives as the predominant

TABLE 3-3

Asymmetric Reactions with Acyclic α-Chlorocarbonyl Compounds[a]

H, R, Cl–C–C(=O)R′ —R″Z→ H, R, Cl–C–C(R″)(OH)R′ (**A**) and H, R, Cl–C–C(OH)(R″)R′ (**B**)

No.	R	R′	R″Z	Yield	Ratio A:B	Ref.
1	*n*-Bu	H	EtMgBr		70:30	6
2	*n*-Bu	H	EtLi		70:30	6
3	Et	*n*-Bu	$NaBH_4$		(80–85):(20–15)	6
4	Et	H	*n*-BuMgBr	68	Mainly **A**	6
5	Et	H	*n*-BuLi	65	Mainly **A**	6
6	Me	Me	*n*-BuMgBr	65	Mainly **A**	6
7	Me	Me	*n*-BuLi	79	Mainly **A**	6
8	Ph	Ph	$LiAlH_4$		About 50:50	19a
9	Ph	Ph	$LiAlH_4$		Only **A**	19b
10	Ph	Me	$LiAlH_4$ (−40°)		40:60	20
11	Ph	Me	$LiAlH_4$ (+35°)		43:57	20
12	Me	Ph	$LiAlH_4$		75:25	20
13	Me	Ph	$LiBH_4$		70:30	20
14	Me	Ph	Al(O-*i*Pr)$_3$		66:34	20

[a] The reaction is so written that the preponderant product will be **A** when the dipolar model is followed.

product that isomer opposite to the one predicted on the basis of the dipolar model unless one assumes that the phenyl is more polarizable than the chloro group. No mechanistic conclusions can be deduced from Nos. 12–14

[19] (a) H. Felkin, *Compt. rend.*, **231**, 1316 (1950). (b) R. E. Lutz, R. L. Wayland, and H. G. France, *J. Am. Chem. Soc.*, **72**, 5511 (1950).

[20] H. Bodot, E. Dieuzeide, and J. Jullien, *Bull. Soc. Chim. Fr.*, **27** [5], 1086 (1960).

since either the polar or open-chain model predicts the preferred formation of **A** as observed.

The results from these limited examples seem to support the dipolar model for purely aliphatic α-chloro carbonyl compounds but the model is not applicable, at least not without modification based on further experiments, when there is a phenyl group on the chiral center. As discussed in Sec. **3-2.2**, results with α-hydroxy ketones are, for the most part, best rationalized on the basis of the cyclic model which however may be competing to a greater or less extent with the dipolar model so that the predicted stereoselectivity may be significantly reduced and under some circumstances even reversed.

3-2.4. Examples Containing α-Oxygen Substituent. Other examples involving α-oxygen and α-nitrogen substituted ketones are summarized in Tables **3-4** and **3-5**. Only a few exceptional cases will be considered specifically. Di- additions to α-diketones lead to the production of *meso* and *dl-threo* mixtures (as shown in **25A** → **25B** → **25C** and **25D**). But since the original substrate and reagent are composed of achiral molecules, there is no possibility of asymmetric synthesis (unless the reaction were conducted in a chiral environment). Stocker[21] has made a thorough study of this sequence; when the reactant is either the diketone or the hydroxy ketone (**25A** or **25B**, R = cyclohexyl, R″ = methyl), the same stereoselectivity, within experimental error, is observed. The same has been observed starting with benzoin (**25A**, R = Ph) or benzil (R″ = H); thus the final products (**25C** and **25D**) presumably are produced via the same intermediate. Since for synthetic purposes the hydroxy ketones are often not readily available, the reaction starting with the diketone has been used to make a variety of symmetrical pinacols. The stereochemistry of the conversion is therefore important in determining the *meso* versus *dl-threo* nature of the pinacols

25A $\xrightarrow{R''Z}$ **25B** $\xrightarrow{R''Z}$ **25C** (*meso*) and **25D** (*dl-threo*)

made by this process. Stocker[21] has found in the diketone reaction, where R is cyclohexyl and the reagent R″Z is phenyl lithium, that the *dl-threo* form, predicted on the basis of the cyclic model, was obtained (57.6 % yield) to the virtual exclusion of the *meso* form (less than 0.25 % by carbon-14 dilution method) but that the substitution of the Grignard reagent from phenyl bromide for phenyl lithium reversed the situation and the *meso* compound

[21] J. H. Stocker, *J. Org. Chem.*, **29**, 3593 (1964).

was formed (15% yield) to the virtual exclusion of the *dl-threo* isomer. This constitutes another well-substantiated example of the reversal of stereochemistry with change in reagent. With methyl lithium and methyl Grignard reagents (Table **3-4**, Nos. 1–4) the product predicted on the basis of the cyclic model was predominant in all cases.

The reactions of sugar aldehydes and ketones[22,23] have been interpreted in terms of the Cram cyclic model. The monomeric dialdo sugar

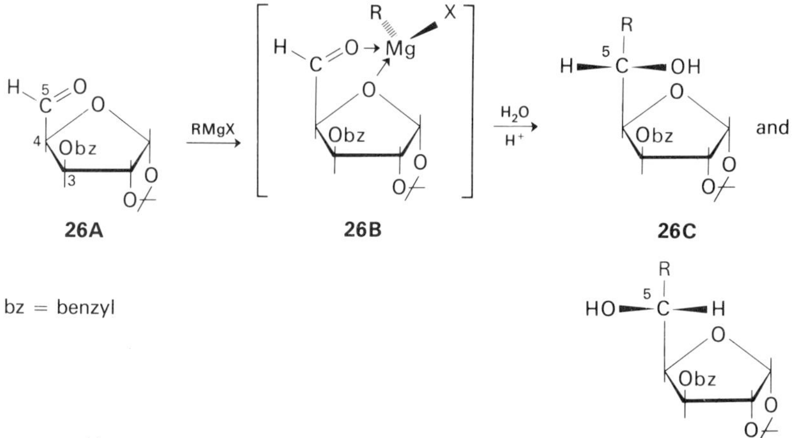

26A **26B** **26C** **26D**

bz = benzyl

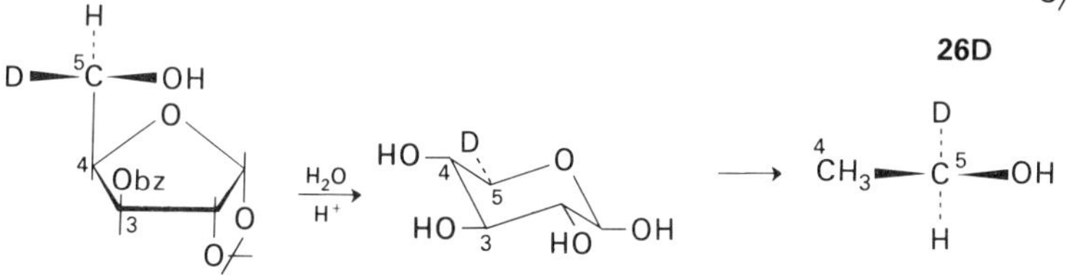

27A **27B** **27C**

derivative **26A** has been the substrate for important asymmetric syntheses. Reaction of this aldehyde with excess methylmagnesium iodide in ether is highly stereoselective. Wolfrom and Hanessian[22a] isolated only the deoxy sugar (**26C**, R = Me) of the *L*-ido series (60% yield, recrystallized product; chromatography failed to detect the C-5 epimer). Inch[23a] obtained a 69% yield of **26C** and a 9% yield of **26D** (R = Me). The comparable phenyl Grignard reagent gave a 72:5% ratio of the corresponding C-5 phenyl sugars **26C** and **26D** (R = Ph) in ether solvent but a 43:13% ratio in THF solvent. In additional studies, the methyl and phenyl keto homologs of

[22] (a) M. L. Wolfrom and S. Hanessian, *J. Org. Chem.*, **27**, 1800 (1962). (b) R. U. Lemieux and J. Howard, *Can. J. Chem.*, **41**, 308 (1963).

[23] (a) T. D. Inch, *Carbohydrate Research*, **5**, 45 (1967). (b) T. D. Inch, R. V. Ley, and P. Rich, *J. Chem. Soc.*, (C) 1683 (1968). (c) T. D. Inch, R. V. Ley, and R. Rich, *ibid.*, 1693 (1968). (d) T. D. Inch, R. V. Ley, and P. Rich, *Chem. Commun.*, 865 (1967).

TABLE 3-4

Addition Reactions of α-Hydroxy Ketone Derivatives[a]

$R_L R_S C(OR)-C(=O)R' \xrightarrow{R''Z} R_L R_S C(OR)-C(OH)R'R''$ (**A**) and (**B**)

No.	R	R_S	R_L	R′	R″Z[b]	A:B	Ratio[c,d]	Ref.
1	H	Me	C_6H_{11}	C_6H_{11}	MeLi	8	[6.9][e]	21
2	H	Me	C_6H_{11}	C_6H_{11}	MeMgCl	7	[8.9]	21
3	H	Me	C_6H_{11}	C_6H_{11}	MeMgBr	4	[4.5]	21
4	H	Me	C_6H_{11}	C_6H_{11}	MeMgI	3	[3]	21
5	H	H	Ph	*p*-Tol	$LiAlH_4$	5.5	(88:16)	3
6	H	H	*p*-Tol	Ph	$LiAlH_4$	4.9	(85:17)	3
7	H	H	Ph	*p*-Tol	PhMgBr	64	(71:1.1)	3
8	H	H	Ph	Ph	*p*-TolMgBr	55	(66:1.2)	3
9	Ac	H	Ph	Ph	*p*-TolMgBr	28	(50:1.8)	3
10	H	H	Ph	Me	PhMgBr	27	(67:2.5)	3
11	H	H	Ph	Ph	MeMgI	28	(81:2.9)	3
12	H	H	Ph	Ph	*o*-TolMgBr	17	(57:3.4)	3
13	H	H	Ph	*o*-Tol	PhMgBr	19	(18:.94)	3
14	H	Me	Ph	Ph	MeLi	8.3	(89:11)	3
							[7.8]	3
					MeLi[e]	3.4	(68:20)	3
15	H	Me	Ph	Me	PhLi	10.4	(91:9)	3
							[8.8]	3
					PhLi[e]	5.2	(63:13)	3
16	H	Me	Ph	Ph	MeMgI	1.98	(40:20)	3
							[1.93]	3
17	H	Me	Ph	Ph	MeMgCl	2.85	(41:14)	3
							[2.53]	3
18	H	Me	Ph	Me	PhMgI	2.0	(37:19)	3
							[2.0]	3
					PhMgI, MgI_2	1.69	(33:79)	3
19	H	Me	Ph	Ph	MeMgBr	2.95	(76:26)	3
							[2.36]	3
20	H	Me	Ph	Me	PhMgBr	2.36	(66:28)	3
							[2.36]	3
					PhMgBr, $MgBr_2$	2.28	(53:23)	3
21	H	Me	Ph	Me	PhMgCl	3.33	(62:19)	3
22	Me	H	Me	Ph	$NaBH_4$	2.7		24
23	Me	H	CH_2OH	Ph	$NaBH_4$	1.4		24
24	Me	H	CH_2NH_2	Ph	$NaBH_4$	0.67		24
25	H	H	Ph	Ph	*p*-MeO-PhMgBr		(70% **A**)	17
26	H	H	Ph	*p*-MeO-Ph	PhMgBr		(70% **A**)	17
27	Me	H	Ph	Ph	*p*-MeO-PhMgBr		(65% **A**)	17
28	Me	H	Ph	*p*-MeO-Ph	PhMgBr		(60% **A**)	17

26A were treated with phenyl and methyl Grignard reagents, respectively. The isomer obtained in each case was the one resulting from attack on the comparable side of the carbonyl group observed for the aldehyde **26A**. In these two cases none of the C-5 epimer could be detected.

If the Grignard reagent complexes with both the free aldehyde and ring oxygen atoms, then the methyl group has freer access to the back face of the carbonyl group, away from the benzyloxy group at C-3 which protects the front face, as depicted in **26B**. This rationalization conforms to the Cram cyclic model and ignores possible complexing with the benzyloxy group at C-3. In spite of the logic of this explanation, when this same sugar derivative (**26A**) was reduced with lithium aluminum deuteride, it was shown[22b] that the deuterium was preferentially introduced from the opposite face of the aldehyde group, giving a 30% excess of **27A** over its C-5 epimer. This derivative was converted into 5-deuterio-D-xylopyranose (**27B**) whose tetraacetate was shown by nmr studies to have the deuterium in the equatorial position (5-*R*) and this was ultimately transformed into ethanol-1-*d* (**27C**), thereby proving its absolute configuration as *R*.

3-2.5. Examples Containing α-Nitrogen Substitution. The reactions of chiral α-amino ketones have been of great importance in studies of the ephedrin-type alkaloids and of the antibiotic Chloromycetin. Simple α-aminopropiophenones (Table **3-5**, Nos. 1–11) generally give high stereoselectivities favoring the isomer predicted on the basis of the cyclic model, as found for the corresponding α-hydroxy derivatives (Table **3-2**, Column C). When the α-amino group is alkylated or acylated, the tendency is for the stereochemical control to switch from that predicted by the five-membered cyclic model to that predicted by the open-chain model in which the bulky substituted nitrogen group functions as R_L. This is best illustrated by comparing No. 1 with Nos. 18–21 in Table **3-5**; the stereochemistry during sodium borohydride reduction was changed from a ratio of 11 : 1 in favor of the *erythro* isomer using α-aminopropiophenone to a 1 : 10 ratio in favor of the *threo* isomer using the N-methyl-N-benzyl or N-benzyl-N-benzoyl analog. Similarly, Müller, et al.[25] reported that a change in stereoselectivity

[a] The reactions are so written that the **A** : **B** ratio will be greater than one when the Cram cyclic model is followed and the reagent approaches from the side of R_S rather than R_L.

[b] The solvent is diethyl ether unless otherwise indicated. Ph, Tol, and *p*-MeO-Ph represent phenyl, tolyl, and *p*-anisyl groups, respectively, and C_6H_{11} stands for the cyclohexyl group.

[c] The number in brackets, [ ], is the isomer ratio when the reactant was the corresponding diketone rather than hydroxy ketone.

[d] The ratios in parentheses, ( ), are the ratios of actual percent yields. When only one number is given in the parentheses, only one isomer was isolated in this percent yield.

[e] Pentane solvent.

[24] S. Yamada and K. Koga, *Tetrahedron Lett.*, 1711 (1967).

[25] (a) H. K. Müller, I. Jarchow, and G. Rieck, *Ann. Chem.*, **613**, 103 (1958). (b) H. K. Müller and G. Rieck, *Ann. Chem.*, **639**, 89 (1961).

TABLE 3-5

Addition Reactions of α-Amino Ketone Derivatives[a]

$R_L(H)C(Q)-C(=O)-R' \xrightarrow{R''Z}$ $R_L(H)C(Q)-C(OH)(R')(R'')$ and $R_L(H)C(Q)-C(OH)(R'')(R')$

A (*erythro*) **B** (*threo*)

No.	Q	R_L	R′[b]	R″Z[b,c]	A:B	Ratio[d]	Ref.
1	NH_2	Me	Ph	$NaBH_4$	11	—	26
2	NH_2	Me	Ph	α-NaphMgBr	—	(45% **A**)	4
3	NH_2	Me	Ph	*p*-TolMgBr	99	(70:0.8)	17,18
4	NH_2	Me	*p*-Tol	PhMgBr	49	—	17,18
5	NH_2	Ph	Ph	*p*-MeO-PhMgBr	—	(60% **A**)	17
6	NH_2	Ph	*p*-MeO-Ph	PhMgBr	—	(50% **A**)	17
7	NH_2	Ph	Ph	*p*-ClPh	—	(30% **A**)	17
8	NH_2	Ph	*p*-ClPh	PhMgBr	—	(40% **A**)	17
9	NH_2	Ph	Ph	α-NaphMgBr	—	(50% **A**)	17,4
10	NH_2	Ph	Ph	*p*-TolMgBr	—	(60% **A**)	17
11	NH_2	Ph	*p*-Tol	PhMgBr	—	(85% **A**)	17
12	NH_2	CH_2NH_2	Ph	$NaBH_4$	1.8	—	24
13	NH_2	CH_2OH	Ph	$NaBH_4$	1.3	—	24
14	$NHCH_2Ph$	Me	Ph	$NaBH_4$	7–8	—	24
15	NMe_2	Me	Ph	$NaBH_4$	0.8–0.9	—	24
16	$NMeCH_2Ph$	Me	Ph	$NaBH_4$	0.33	—	24
17	$NHCOCF_3$	Me	Ph	$NaBH_4$	3.3	—	24
18	NHAc	Me	Ph	$NaBH_4$	3.3	—	24
19	NHBz[e]	Me	Ph	$NaBH_4$	3.3	—	24

20	$NMeCH_2Ph$	Me	Ph	$NaBH_4$	0.11–0.16	—	24
21	NCH_2PhBz[e]	Me	Ph	$NaBH_4$	0.11	—	24
22	NHAc	Me	p-NO_2Ph	Al(O-iPr)$_3$	—	(72% **A**)	27
23	$NC_8H_4O_2$[e]	Me	p-NO_2Ph	Al(O-iPr)$_3$	—	(67% **B**)	27
24	$NC_8H_4O_2$[e]	Me	p-NO_2Ph	Al(O-iPr)$_3$	0.37	—	27b
25	$NC_8H_4O_2$[e]	CH_2Me	p-NO_2Ph	Al(O-iPr)$_3$	0.42	—	27b
26	$NC_8H_4O_2$[e]	OMe	p-NO_2Ph	Al(O-iPr)$_3$	0.39	—	27b
27	$NC_8H_4O_2$[e]	SMe	p-NO_2Ph	Al(O-iPr)$_3$	0.39	—	27b
28	NHAc	CH_2OH	Ph	Al(O-iPr)$_3$	—	(47% **B**)	25
29	NHAc	CH_2OH	Ph	Al(O-iPr)$_3$-$AlCl_3$	—	(56% **B**)	25
30	NHBz[e]	CH_2OH	Ph	Al(O-iPr)$_3$	—	(63% **B**)	25
31	NHBz[e]	CH_2OH	p-NO_2Ph	Al(O-iPr)$_3$	—	(28% **B**)	28
32	$NHCOCHCl_2$	CH_2OH	p-NO_2Ph	Al(O-iPr)$_3$	—	(**B**)	28
33	$NHCOOCH_2Ph$	CH_2OH	p-NO_2Ph	Al(O-iPr)$_3$	—	(34% **B**)	28
34	NHAc	CH_2OAc	Ph	Al(O-iPr)$_3$	—	(56% **A**)	25
35	NHBz[e]	CH_2OBz	Ph	Al(O-iPr)$_3$	—	(51% **A**)	25
36	$NHCOCHCl_2$	CH_2OAc	p-NO_2Ph	Al(O-iPr)$_3$	—	(68% **A**)	28
37	$NHCOCHCl_2$	CH_2SBz	p-NO_2Ph	Al(O-iPr)$_3$	—	(**A**)	28
38	$NHCOCHCl_2$	CH_2OCPh_3	p-NO_2Ph	Al(O-iPr)$_3$	—	(64% **A**)	28
39	$NHCOCHCl_2$	CH_2Cl	p-NO_2Ph	Al(O-iPr)$_3$	—	(56% **A**)	28
40	$NHCOCHCl_2$	Me	p-NO_2Ph	Al(O-iPr)$_3$	—	(90% **A**)	28
41	NHAc	Me	p-AcPh	Al(O-iPr)$_3$	—	(82% **A**)	25
42	NHAc	Me	p-OHPh	Al(O-iPr)$_3$	—	(30% **A**)	25
43	NH_2	Me	n-$C_{15}H_{31}$	$LiAlH_4$	—	(66% **A**)	27
44	Me	CH_2NH_2	Ph	$NaBH_4$	0.11	—	24
45	Me	CH_2OH	Ph	$NaBH_4$	0.41	—	24

[a] The reaction is so written that the **A**:**B** ratio will be greater than one when the Cram cyclic model is followed and the reagent approaches from the side of H.

[b] Ph, Tol, p-MeO-Ph, and Naph represent the phenyl, tolyl, p-anisyl, and naphthyl groups, respectively.

[c] All Grignard reactions were carried out in ether, sodium borohydride reductions in methanol, and Meerwein–Ponndorf reductions in isopropyl alcohol.

[d] When the ratio is not given, but a number in parentheses, then only one diastereomer was reported and in the yield given.

[e] $NC_8H_4O_2$ represents the phthalimido group; Bz, the benzoyl group; and CH_2PhBz, the p-benzoylbenzyl group.

(72% yield of *erythro* isomer to 67% yield of *threo* isomer) resulted when the acetyl group was replaced by the more bulky phthalimido group (Nos. 22 and 23).

The synthesis of the amino alcohol **29** has been realized by two different routes:[4] by the addition of α-naphthylmagnesium bromide to the amino ketone **28** and by the sodium-amalgam reduction of the oxime **30**; the erythro isomer, **29**, predominated in both cases.

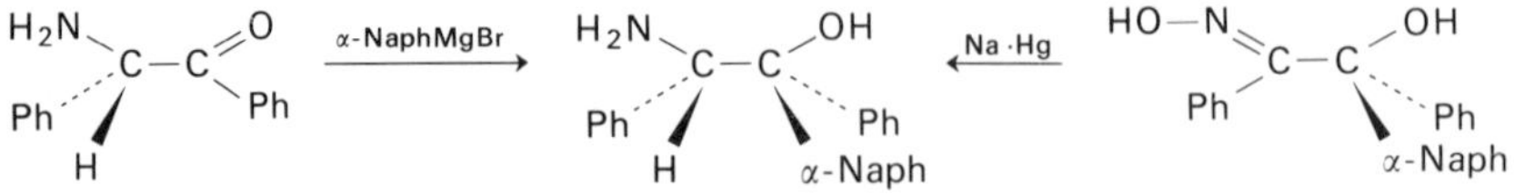

28 —α-NaphMgBr→ **29** ←Na·Hg— **30**

In an analogous manner, (+)-2-amino-3-octadecanone and its oxime are reduced[27] by lithium aluminum hydride, respectively, to the *erythro* amino carbinol (66% yield, Table **3-5**, No. 43) and the *erythro* diamine (83% yield).

Another example in which the asymmetric reaction involves addition to an imine instead of a carbonyl group is given by the reduction of certain aldimines of isopropylidene D-glyceraldehyde (**31**).[29] When R = H phenyllithium gives a 74:26 ratio of the *threo* (**32A**) over the *erythro* (**32B**) isomer; while the Grignard reagent from phenyl bromide gives a 9:91 ratio of

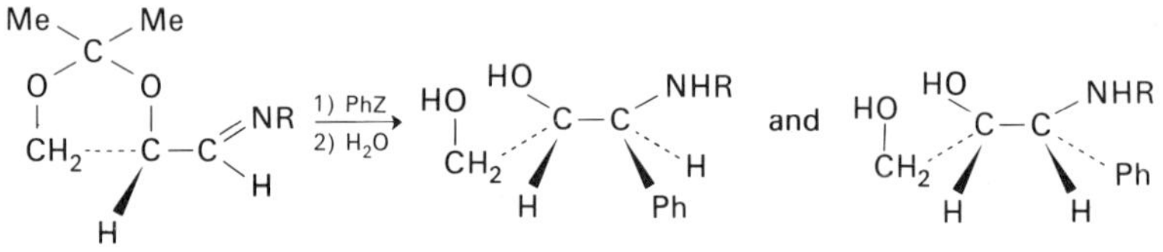

31 —1) PhZ, 2) H_2O→ **32A** (*threo*) and **32B** (*erythro*)

threo to *erythro* isomer. The same reversal of stereochemistry with change from phenyllithium to phenylmagnesium bromide is observed when R is benzyl (*threo*:*erythro* isomer ratio 63:37 changed to 22:78). Although one can say that the phenyllithium reactions are rationalized on the basis of the Cram cyclic model and the phenylmagnesium bromide reactions on the basis of the dipolar model or the six-membered cyclic model (**33**), this

[26] M. Tiffeneau, J. Lévy, and E. Ditz, *Bull. Soc. Chim. Fr.*, [5] **2**, 1855 (1935).

[27] (a) M. Proštenik and P. Alaupović, *Croat. Chem. Acta*, **29**, 393 (1957); *Chem. Abst.*, **53**, 1131 (1959). (b) D. Fleš, B. Majhofer, and M. Kovač, *Tetrahedron*, **24**, 3053 (1968).

[28] J. Sicher, M. Svoboda, M. Hrdá, J. Rudinger, and F. Šorm, *Collect. Czech. Chem. Commun.*, **18**, 487 (1953).

[29] J. Yoshimura, Y. Ohgo, and T. Sato, *J. Am. Chem. Soc.*, **86**, 3858 (1964).

amounts to little more than a restatement of the experimental observations.

When a β-amino or β-hydroxy group is added, the situation is quite different. Sicher et al.[28] found during a study on the Meerwein–Ponndorf–Verley reduction of a series of Chloromycetin precursors that regardless of the α-acylamino substituent on the chiral α-carbon, when there was a free hydroxy group in the *beta* position (Table **3-5**, Nos. 28–33), products of the *threo* series predominated: but when the *beta*-hydroxy group was protected (Nos. 34–39), products of the *erythro* series predominated as predicted on the basis of Cram's cyclic model. Sicher proposed a *six-membered cyclic* mechanism, **33A**, to account for this reversal in stereochemistry encountered with the β-hydroxy derivatives. Evidence in support of this type of mechanism is furnished by the sodium borohydride reductions[24] of β-amino-α-methylpropiophenone (**33B**, Q = NH_2) and β-hydroxy-α-methylpropiophenone (**33B**, Q = OH). In these examples, Cram's *five-membered* ring

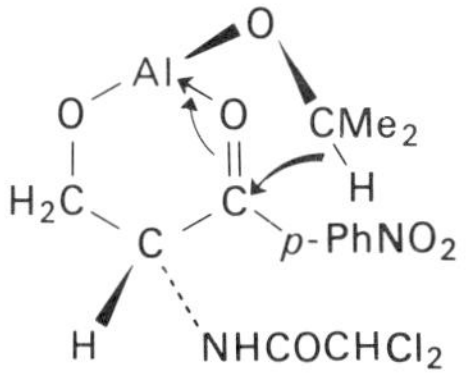

33A 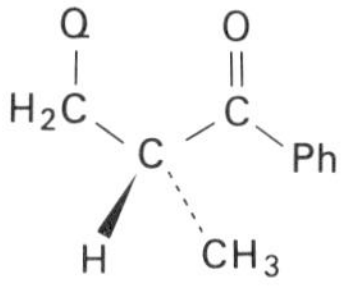 **33B**

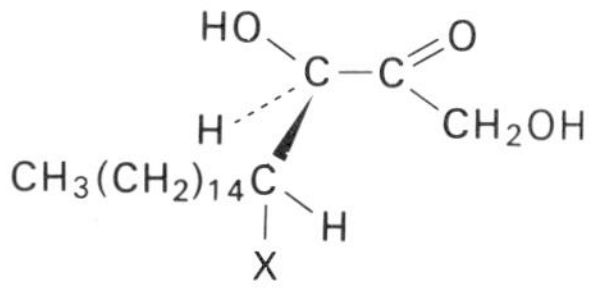

34A, X = H
34B, X = (*R*)-OH
34C, X = (*S*)-OH

cyclic model and the dipolar model are clearly inapplicable since there is no hetero substituent on the α-carbon. Thus this additional six-membered ring model must be taken into account when considering the reactions of β-amino, β-hydroxy, α,β-dihydroxy, α,β-diamino, α-amino-β-hydroxy, and α-hydroxy-β-aminoketones. Since substrates with a β-hetero substituent appear to react via model **33A** in opposition to the Cram five-membered cyclic model, the predicted effect of the β-substituent is either to reverse the stereochemistry (as seen in Table **3-4**, No. 24; Table **3-5**, Nos. 28–33, 44, and 45) or to decrease the stereoselectivity from that observed in the compound without the β-substituent (Table **3-4**, No. 23; Table **3-5**, Nos. 12 and 13). But the Meerwein–Ponndorf–Verley reductions of the phthalimido derivatives (Table **3-5**, Nos. 24–27) give a reasonably constant **A**:**B** ratio of about 0.4 whether or not there is a hetero atom in the β-position (R_L). Fleš, et al.[27b] concluded that reactions of this type are best interpreted in terms of the simple open-chain model.

In this connection it is of interest to consider the lithium aluminum hydride reduction of the compounds represented by **34A** and **34B** reported by Palameta and Zambeli.[30] The reduction of **34A** gave a *dl-threo* triol whose configuration was established from other work. On this basis these authors invoked the dipolar model to rationalize the results and also to assign the comparable *threo* configuration (for the 2,3-dihydroxy groups) in the tetrol obtained by reductions of **34B**. In this latter case, however, the situation is more complex. First there is another chiral center *beta* to the carbonyl group and, second, the additional hydroxy group in the β-position may also complex with the reagent in the same manner as shown in the six-membered cyclic model **33A** or **33B**. This model, in fact, also may be used to rationalize the formation of the *threo* form but there are so many opportunities for complexing of the reagent with the various hydroxyl groups that extrapolating the stereochemistry from **34A** to **34B** is not justified.

3-2.6. Examples of 1,3-Asymmetric Induction. These last examples, **34B** and **34C**, bring up the question of the effect upon the extent and sense of the asymmetric synthesis caused by the introduction of a group between the carbonyl group and the inducing chiral center. Three studies[31–33] specifically dealing with the situation in which the chiral center is *beta* to the carbonyl group have appeared. This process has been termed 1,3-asymmetric induction[31] and is illustrated by the addition of phenylmag-

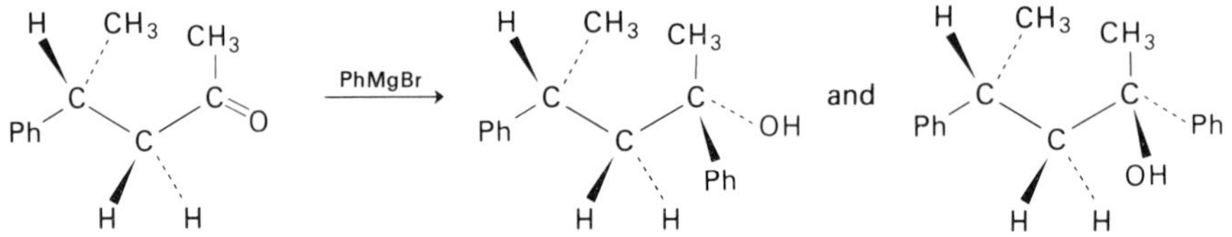

H CH$_3$ CH$_3$ Ph C C C=O H H **35** —PhMgBr→ H CH$_3$ CH$_3$ Ph C C C OH Ph H H **35A** and H CH$_3$ CH$_3$ Ph C C C Ph OH H H **35B**

nesium bromide to 4-phenyl-2-pentanone (**35**) to give a mixture of the diastereomeric carbinols **35A** and **35B** in the ratio of 83:17 (Table **3-6**, No. 5, 66% asymmetric synthesis).

Tables **3-6** and **3-7** contain selected examples in which the chiral center in the *beta* position has no hetero atom and therefore these examples are not complicated by the considerations of cyclic or dipolar transition states. The following conclusions can be drawn from the data on the ketones reported in Table **3-6**. The extent of asymmetric synthesis, although not as high as in the comparable cases of 1,2-asymmetric induction (see Tables **3-1** and **3-2** and Fig. **3-6**), are substantial in spite of the methylene group which

[30] B. Palameta and N. Zambeli, *J. Org. Chem.*, **29**, 1031 (1964).
[31] T. J. Leitereg and D. J. Cram, *J. Am. Chem. Soc.*, **90**, 4011 (1968).
[32] T. J. Leitereg and D. J. Cram, *J. Am. Chem. Soc.*, **90**, 4019 (1968).
[33] M. Brienne, C. Quannis, and J. Jacques, *Bull. Soc. Chim. Fr.*, 1036 (1968).

TABLE 3-6

1,3-Asymmetric Induction: Carbonyl Compounds with No Polar Substituents[a,b]

$$\text{Ph–CH(CH}_3\text{)(H)–CH}_2\text{–C(=O)–R}' \xrightarrow{R''Z} \text{Ph–CH(CH}_3\text{)(H)–CH}_2\text{–C(OH)(CH}_3\text{)–Ph} \;\text{and}\; \text{Ph–CH(CH}_3\text{)(H)–CH}_2\text{–C(OH)(Ph)–CH}_3$$

A and B

No.	R′	R″Z	**A**:**B** Ratio	Temp. (°C)
1	Me	PhLi	56:44	0
2	Me	PhMgBr	69:31	35
3	Me	PhMgBr	67:33	0
4	Me	PhMgBr	76:24	−78
5	Me	PhMgBr	83:17	−110
6	Ph	MeLi	50:50	0
7	Ph	MeMgCl	44:56	0
8	Ph	MeMgI	42:58	0
9	Ph	MeMgBr	42:58	35
10	Ph	MeMgBr	41:59	−78

[a] Reactions were carried out in ether solvent with excess reagent. Racemic ketone was used but only one set of isomers is represented for convenience in presentation.

[b] Examples selected from Reference 31.

"insulates" the carbonyl group from the chiral center. In the examples of Table **3-6**, it is seen that reversing the sequence of introduction of the phenyl and methyl groups at the new chiral center produces the opposite diastereomer in excess. For instance, if instead of treating the methyl ketone **35** with phenylmagnesium bromide, the phenylketone is treated with methylmagnesium bromide (No. 10), then **35B** is produced in excess instead of **35A**. This is the pattern almost uniformly encountered in the Cram open-chain 1,2-asymmetric induction system and in the Prelog benzoylformate-atrolactate (1,4-asymmetric induction) system.

These results were rationalized[31] by an open-chain model in which the two smallest groups on the adjacent carbon atom are staggered with respect to the carbonyl group as shown in **35**. According to this model, the favored approach for the entering group is toward that face of the carbonyl carbon which is least hindered by the smallest group on the chiral center: i.e., in the case of **35** this would result in the approach of the reagent from in front of the plane of the page.

TABLE 3-7

1,3-Asymmetric Induction: Lithium Aluminum Hydride Reductions[a]

R_M H O; R_L C C C R'; R R $\xrightarrow{LiAlH_4}$ R_M H OH H; R_L C C C R'; R R **A** and R_M H H OH; R_L C C C R'; R R **B**

No.	R_L	R_M	CR_2	R'	**A**:**B** Ratio
1	Ph	Me	CH_2	Me	42:58 [49:51][b]
2	*p*-MeO-Ph	Me	CH_2	Me	43:57
3	Ph	Me	CH_2	Ph	46:54 [47:53][b]
4	Ph	Et	CH_2	Me	41:59
5	Ph	Et	CH_2	Ph	48:52
6	Ph	Et	CH_2	*t*-Bu	45:55
7	C_6H_{11}[c]	Et	CH_2	Ph	43:57
8	C_6H_{11}[c]	Et	CH_2	*t*-Bu	43:57
9	Ph	Me	CMe_2	Me	26:74 [42:58][b]
10	Ph	Me	CMe_2	Ph	22:78 [25:75][b]
11	Ph	Et	CMe_2	Me	24:76
12	Ph	Me	CPh_2	Me	40:60
13	Ph	Et	CPh_2	Me	36:64

[a] All data are taken from Reference 33. Experiments were carried out at 35° in ether. Racemic ketone was used but only one set of isomers is depicted in the equation for convenience.

[b] The numbers in brackets give the isomer ratio produced when the corresponding aldehyde was treated with either methylmagnesium iodide or phenylmagnesium bromide. Note that the isomer which predominated was unchanged by the reversal in sequence of introducing these groups.

[c] C_6H_{11} represents the cyclohexyl group.

Although this model is compatible for the examples of Table **3-6**, it does not hold for the lithium aluminum hydride reductions summarized in Table **3-7**. This is clearly shown by the figures in brackets (Table **3-7**; Nos. 1, 3, 9, 10) which represent the isomer ratio when the corresponding aldehyde was treated with the appropriate Grignard reagent instead of the ketone being reduced by lithium aluminum hydride. *The same predominant diastereomer results from either sequence!* A similar dependence of the stereochemical course of a reaction upon the reagent is observed in the cyclohexane series (discussed in connection with Table **3-11**) but is rare in the open-chain compounds. The latter results have been rationalized[33] in terms of the model proposed by Felkin[8] which is discussed in Sec. **3-4**. In

TABLE 3-8

1,3-Asymmetric Induction: Carbonyl Compounds with Polar Substituents[a]

No.	Q	R	R″Z	**A**:**B** Ratio	Temp. °
1	OH	Me	PhLi	68:32	0
2	OH	Me	PhMgBr	85:15	0
3	OH	Me	PhMgBr	88:12	−75
4	OH	Me	PhMgBr	78:22	0 (heptane)
5	OH	Ph	MeLi	50:50	0
6	OH	Ph	MeMgBr	41:59	35
7	OH	Ph	MeMgBr	56:44	−78
8	OH	Ph	MeMgI	48:52	0
9	OMe	Me	PhLi	43:57	35
10	OMe	Me	PhLi	58:42	−78
11	OMe	Me	PhMgBr	44:56	35
12	OMe	Me	PhMgBr	51:49	−78
13	OMe	Ph	MeLi	70:30	0
14	OMe	Ph	MeMgCl	71:29	0
15	OMe	Ph	MeMgI	57:43	0
16	OMe	Ph	MeMgBr	67:33	35
17	OMe	Ph	MeMgBr	74:26	−78
18[b]	OMe	Ph	MeMgBr	68:32	0 (DME)[b]

[a] Data selected from Reference 32. Experiments were carried out in ether with reagent in excess. Racemic ketone was used and enantiomeric pairs of diastereomers were obtained but only one set of isomers is shown.

[b] 1,2-Dimethoxyethane (DME) solvent.

view of the complexities encountered in rationalizing the results of 1,2-asymmetric induction, it appears premature to speculate further about the situation in the conformationally more flexible and uncertain 1,3-systems.

A clear conclusion from the data in Table **3-7** is that geminal methyl substituents on the carbon atom separating the carbonyl group and chiral center increase the stereoselectivity of the reduction reaction (compare Nos. 1 and 9, 3 and 10). The effect of geminal phenyl substituents is not nearly so pronounced. Increasing the steric bulk of the substituent on the carbonyl carbon from methyl to *t*-butyl (compare Nos. 4 and 6, 7 and 8) does not cause significant changes in contrast to the findings for 1,2-asymmetric induction (see Fig. **3-6**).

In the reactions of ketones with a hydroxy substituent on the chiral β-carbon atom (Table **3-8**), the predominant isomer was determined in all but one case (No. 7) by the sequence used for introduction of the phenyl and methyl groups. The role played in these reactions by open-chain, polar, and cyclic models, similar to those proposed for 1,2-asymmetric induction (compare Figs. **3-3**, **3-4**, and **3-5**), have been discussed.[32] It might have been expected that a relatively rigid six-membered cyclic transition state leading to high stereoselectivities would be involved, but the magnitude of the results reported in Table **3-8** are not appreciably different from those in Table **3-6** where such a mechanism is clearly impossible. It was concluded[32] that a polar model similar to Fig. **3-4** best rationalized the results.

In the reactions of ketones with methoxy substituents on the chiral β-carbon atom, the predominant isomer formation seemed to depend more on other factors than on the sequence whereby the methyl and phenyl substituents were introduced onto the newly created chiral center. Thus the ratio of isomers formed by the action of phenyllithium and phenylmagnesium bromide (Nos. 9–12) ranged on both sides of 50:50, depending upon the temperature. The possible role of open-chain, polar, and cyclic models in these asymmetric reductions have been discussed.[32] The fact that a change in temperature caused a reversal of the stereochemical course of the reaction indicates that there may be two competing mechanisms which have quite different entropies of activation; possibly these are represented by the cyclic versus open-chain models. The relevance of these studies to the mechanism(s) of stereoselective polymerizations has been considered.[31,32]

A final example of 1,3-asymmetric induction is the reaction of the chiral ketone **36** with Grignard reagents.[26]

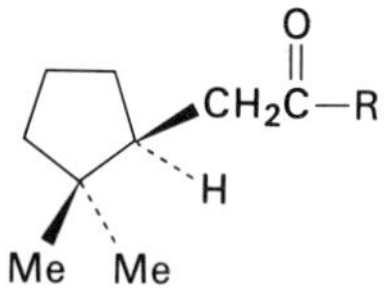

$CH_2C(=O)-R$, H, Me, Me

36A, R = Ph
36B, R = Et

When **36A** was treated with ethylmagnesium bromide, only one diastereomer was isolated (in 60% yield) and when **36B** was treated with phenylmagnesium bromide, only the other diastereomer was isolated (also in 60% yield). Thus the reaction shows high stereoselectivity but the configurations of the products are not known.

3-3. The Transition State for Addition Reactions to Chiral Ketones

To what extent do the open-chain, cyclic, and dipolar models reflect the actual transition states of the reactions to which they have been applied? As indicated in the preceding sections, the situation is complex when there is a hetero substituent in addition to the carbonyl group in the molecule. This introduces the additional factors of extent of solvation, extent of complexing, and electrostatic effects as well as the steric effects. At present, these reactions must be interpreted largely within the framework of experimental results for a closely related series. However, the reactions to which the open-chain model has been applied (Table **3-1**) offer a much better chance for successful theoretical treatment since, in these cases, hydrogen bonding, solvation, complexing, and unusual electrostatic effects are at a minimum.*

The following figures represent twelve of an infinite number of ground state conformations of one enantiomer of an open-chain compound which has a chiral function adjacent to a carbonyl group. These formulas constitute three sets; the first set (**A**–**C**) is made up of three staggered carbonyl conformations; the second set (**D**–**F**) is made up of three eclipsed carbonyl conformations; and the third set (**G**–**L**) is made up of six conformations which have no eclipsed groups. Let us assume that the transition states for the addition reactions under consideration closely resemble these ground state conformations and that the entering reagent approaches at right angles

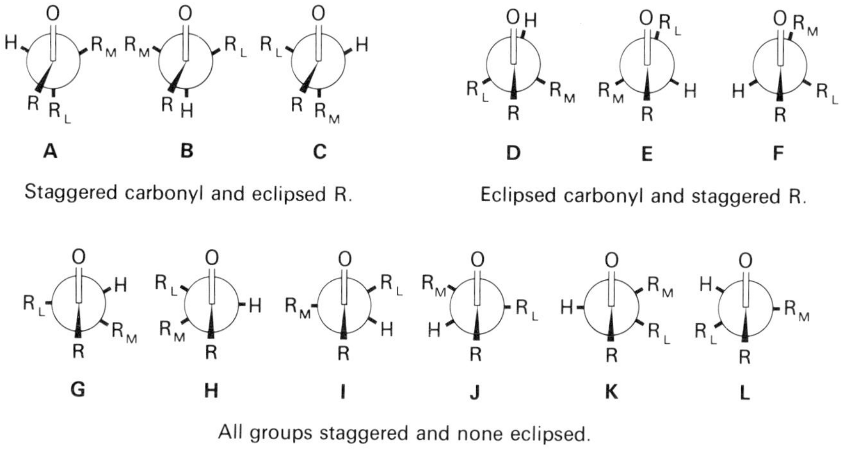

Staggered carbonyl and eclipsed R.

Eclipsed carbonyl and staggered R.

All groups staggered and none eclipsed.

Fig. 3-8. Possible conformations of a chiral carbonyl compound.

* The electron-rich phenyl group in these compounds will certainly be subject to these variables of solvation, complexing, etc., but to a much lesser extent than hetero substituents.

to the carbonyl group from either the right or the left. A summation of the differences in energies of the twelve transition states in which the reagent approaches from the right versus the twelve from the left should give a good approximation of the difference in energies of activation between the formation of one diastereomer and its epimer. Obviously this calculation is not possible at the present time. It might in theory be possible to adapt the Ugi–Ruch type of treatment to such cases (see Sec. **1-4.4b**) but, as it now stands, this treatment is not applicable to systems in which the established and developing chiral centers are adjacent because of specific group interactions which have been ignored in the mathematical development.

Cram,[4] in a logical but highly intuitive choice, made **A** (where $H = R_S$) the basis for his empirical model for predicting the configuration of the preponderant diastereomeric product in such reactions as represented by the examples in Table **3-1**. Thus the assumption was made that the lowest energy conformation would be that one with the two smaller groups staggered with respect to the carbonyl group (considered to be more bulky than R because it was complexed with the reagent). It was then assumed that approach of the reagent from the less hindered side, namely from the side of H (R_S versus R_M), would be controlling and would thus determine the configuration of the predominant product. The success of this model is shown by the fact that there is only one serious exception among the forty-one examples in Table **3-1**. But it is important also to note that the same success would have resulted if **B** had been chosen and the reagent assumed to approach from the side of R_M rather than the side of R_L.

Karabatsos[7] has developed a semiempirical treatment which is a step beyond an empirical model and which considers at least two of the specific group interactions involved in the transition state. His treatment is based upon two initial assumptions. (a) "Little bond breaking and making has occurred at the transition state. Consequently the arrangement of groups around the asymmetric carbon atom with respect to the carbonyl group is similar to that about the sp^3-sp^2 carbon–carbon bond." (b) "The diastereomeric transition states that control product stereospecificity have the smallest group R_S closest to the incoming reagent group."

Karabatsos has considered only those conformations belonging to the second set, namely those with the eclipsed carbonyl (**D**, **E**, **F**), based upon his first assumption and recent nmr and microwave evidence which indicate that the eclipsed carbonyl ground state is generally more stable than the alternate staggered carbonyl conformation.[34] Furthermore, based upon experimentally determined ground state conformational energy differences,

[34] (a) G. J. Karabatsos and N. Hsi, *J. Am. Chem. Soc.*, **87**, 2864 (1965). (b) G. J. Karabatsos and N. Hsi, *Tetrahedron*, **23**, 1079 (1967). (c) G. J. Karabatsos and K. L. Krumel, *ibid.*, **23**, 1097 (1967). (d) G. J. Karabatsos, D. J. Fenoglio, and S. S. Lande, *ibid.*, **91**, 3572 (1969). (e) G. J. Karabatsos and D. J. Fenoglio, *ibid.*, **91**, 1124, 3577 (1969).

quantitative estimates can be made of transition state energy differences based upon thcsc assumptions. Ground state conformational measurements indicate that when R_M is methyl, conformation **F** is favored over **D** by about 800 cal/mol. This is of considerable importance in the Karabatsos treatment since thirty-seven of forty-one examples in Table **3-1** have H for the "small" R_S group.

Karabatsos has reasoned that **E** will be the most stable transition state for approach of the reagent from the right to give one diastereomer (X) and **F** the most stable for approach of the reagent from the left to give the epimer (Y).* Since there is a comparable $H \leftrightarrow R'' \leftrightarrow R$ interaction represented by the approach of the reagent (R″Z) from the right in **E** and from the left in **F**, the evaluation of $\Delta\Delta G^{\ddagger}$ reduces to determining the sum for the energy differences for the interactions in **E** of

$$(R_M \leftrightarrow R) + (R'' \leftrightarrow R_L) + (O \leftrightarrow R_L)$$

versus those in **F** of

$$(R_L \leftrightarrow R) + (R'' \leftrightarrow R_M) + (O \leftrightarrow R_M).$$

The first term in these two sets favors **E**, the second term favors **F**: therefore, to a first approximation, the two are considered to cancel each other. The energy difference represented by the approach of the reagent from the right in **E** and the left in **F** then reduces to the energy difference in the interaction between $O \leftrightarrow R_L$ and $O \leftrightarrow R_M$; i.e.

$$E(O \leftrightarrow R_L) - E(O \leftrightarrow R_M) \cong G^{\ddagger}_X - G^{\ddagger}_Y = -RT \ln \frac{X}{Y}$$

Since the difference $E_{(O \leftrightarrow R_L)} - E_{(O \leftrightarrow R_M)}$ can be estimated from the conformational energy differences of the corresponding ground states (based on the first assumption), a semiquantitative calculation of the stereoselectivity can be made. A reasonably good agreement between experimental and calculated valucs was obtained. This treatment does not take into account the considerable effect on stereoselectivity often observed by a change in reagent. For instance, in the first six cases in Table **3-1**, corresponding to reactions of very similar substrates with aluminum isopropoxide, lithium aluminum hydride, and phenyl and methyl Grignard reagents, the percent stereoselectivities vary between 22 and 60 ($\Delta\Delta G^{\ddagger}$ from 240 to 830 cal/mol).

As indicated in Fig. **3-6**, variations in the achiral R group attached to the carbonyl carbon have a profound effect upon the stereoselectivity which changes from 48 to 96% in substituting *t*-butyl for methyl. Karabatsos' analysis of necessity gives the same stereoselectivity for all these reactions, namely 48% ($\Delta\Delta G^{\ddagger}$, 600 cal/mol). Although this is an important contribution, it is still far from a satisfactory or complete treatment for the generalized case.

* The original reference[7] should be consulted for the rationale of this choice.

Felkin and co-workers[8] have proposed an interpretation for the stereochemical results, which is also consistent with the results encountered in the cyclic series (Sec. **3-4**), that is based on the following four premises. (1) "The transition states for these reactions are in all cases essentially 'reactant-like' rather than 'product-like'." (2) "Torsional strain (Pitzer strain) involving partial bonds in transition states represents a substantial fraction of the strain between fully formed bonds, even when the degree of bonding is quite low." (3) "The important transition state interactions involve R″ (the entering reagent group) and R′ (the achiral group attached to the carbonyl carbon) rather than the interactions of the group on the chiral center with the carbonyl oxygen as envisaged by both Cram and Karabatsos." (4) "Polar effects stabilize those transition states in which the separation between R″ and the electronegative groups (R_L, R_M, R_S) is greatest and destabilize others." These premises lead to the choice of J (Fig. **3-8**) in which the reagent approaches from the left as the lowest transition state as depicted in J‡.

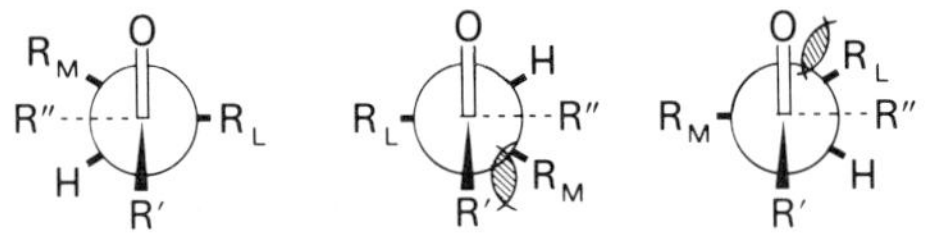

J‡ G‡ I‡

Fig. 3-9. Transition states according to Felkin.[8]

If R_M is a relatively small group there would be little choice between J‡ and G‡, which leads to the epimer, but as the steric bulk of either R_M or R′ increases, the stability of G‡ with respect to J‡ will decrease as a consequence of the steric interaction indicated by the shaded area. A similar situation exists when the steric bulk of R_L is increased. The decreased stabilities of transition states such as I‡, where R_L is adjacent to both R″ and to the carbonyl oxygen, will operate in favor of J‡ even more, with the result that stereoselectivity should rise with increasing steric bulk in R_M, R_L, or R′. This latter effect is substantiated by the results summarized in Fig. **3-6**.

3-4. Steric Control of Asymmetric Induction in Cyclic Ketones

As discussed in Sec. **1-3** in connection with the reduction of cholestanone, each addition reaction of a chiral ketone, such as the naturally occurring keto steroids and terpenes, constitutes an asymmetric synthesis. The number of such reactions is almost limitless and we shall consider only a relatively small number of examples specially chosen to illustrate aspects of the stereochemistry involved. No single generalization, applicable to all systems, has emerged for describing the stereochemical course of the different types of addition reactions to the various classes of cyclic ketones. As a result, discussions of this broad subject generally have been conducted within restricted groups of ketones, namely (a) cyclopentanones, (b) conformationally mobile cyclohexanones, (c) conformationally rigid cyclohexanones, (d) bicyclic ketones, and (e) steroids and terpenes. Within these groups, the observed stereochemistry is considered in terms of several generalizations: Barton's rule,[35] the "steric approach control" versus "product development control" of Dauben, et al.[36] (equivalent to the "steric strain control" and "product stability control" of Brown and Deck[37]), the Wheeler rules,[38] and the transition state analysis of Felkin and co-workers,[8,39] and Eliel.[40] Many of the studies aimed at elucidating the mechanism have been conducted with racemic compounds and thus technically do not constitute asymmetric syntheses. In still other cases, such as the addition reactions of 4-*t*-butylcyclohexanone, the substrate as well as the products are achiral. But the stereochemistry of the cyclic systems cannot be discussed without considering these experiments. Accordingly the choice of material for the following treatment will be quite arbitrary.

3-4.1. Cyclopentanones. In cyclic systems, depending upon the ring size, there may be a certain amount of conformational mobility but there is no possibility of complete rotation about the bond joining the carbonyl carbon and the chiral center as exists in the open-chain systems. If we consider 2-methylcyclopentanone frozen in a planar conformation (**37**), this represents the same framework chosen for the Cram open-chain model. One might reasonably expect therefore that the addition reactions of 2-methylcyclopentanone would conform to this model and give predominantly alcohol (**38A**) by attack from the side of the carbonyl group away from the methyl. Representative results with various reducing agents ($R'' =$ H) are collected in Table **3-9**. Only in the reactions of the bulky aluminum isopropoxide, di(3-methyl-2-butyl)borane and "di-3-pinanylborane" is the less stable **38A** (where $R'' =$ H), which would be predicted on the basis of the Cram rule, produced in excess. With lithium aluminum hydride reduction (No. 1), the ratio of the more stable *trans* 2-methylcyclopentanol to the *cis* compound is approximately 77:23. This same ratio within a few percent

[35] D. H. R. Barton, *J. Chem. Soc.*, 1027 (1953).

[36] W. G. Dauben, G. J. Fonken, and D. S. Noyce, *J. Am. Chem. Soc.*, **78**, 2579 (1956).

[37] H. C. Brown and H. R. Deck, *J. Am. Chem. Soc.*, **87**, 5620 (1965).

[38] (a) D. M. S. Wheeler and J. W. Huffman, *Experientia*, **16**, 516 (1960). (b) O. R. Vail and D. M. S. Wheeler, *J. Org. Chem.*, **27**, 3803 (1962).

[39] M. Chérest and H. Felkin, *Tetrahedron Lett.*, 2205 (1968).

[40] E. Eliel and Y. Senda, *Tetrahedron Lett.*, 6127 (1968) and private communication.

has been obtained in several experiments from different laboratories. The diastereomeric ratio of the thermodynamic equilibrium is approximately 90% *trans* versus 10% *cis* as measured by the sodium in alcohol reductions (Nos. 13 and 14). The Cram rule per se obviously does not apply to this

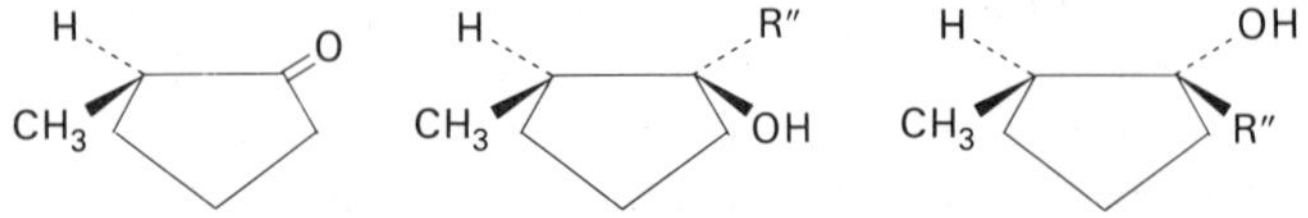

cyclic system, nor did Cram attempt to apply it to cyclic systems. The intriguing question is why should it fail in its application to this system which on the surface appears to be the best possible example for testing the model? The dual concepts of "steric approach control" and "product development control" were proposed[36] to rationalize such results. But especially the product development control concept has come under increasing criticism recently[8,34,38b,39–43] and other more satisfactory explanations have been sought. It is scarcely possible that reductions of open-chain, unhindered cyclic, and hindered cyclic ketones should undergo addition reactions by different mechanisms; the reason for the differences must be sought in the conformational differences in the respective transition states. It is possible that the conformation for 2-methylcyclopentanone as shown in **37** is not correct, that the basis of Cram's rule has no sound mechanistic foundation, or both.

The cyclopentanone ring is, of course, neither planar nor rigid. On the basis of studies on other cyclopentanones[44] one concludes that the 2-methylcyclopentanone ring is most stable in a half-chair conformation represented by **39A** in which the α-carbon is twisted to give an equatorial-like methyl group, while the two β-carbons are twisted with respect to each other so that their hydrogens will be staggered. The position of the methyl group then approaches that which has been shown to be most stable in the open-chain compounds such as propionaldehyde.

Assume that the transition state for the ring resembles this ground state and that the reagent* R″Z will approach the carbonyl bond perpendicularly

[41] A. V. Kamernitzky and A. A. Akhrem, *Tetrahedron*, **18**, 705 (1962).

[42] J. A. Marshall and R. D. Carroll, *J. Org. Chem.*, **30**, 2748 (1965).

[43] (a) J. Richer, *J. Org. Chem.*, **30**, 324 (1965). (b) J. Richer and G. Perrault, *Can. J. Chem.*, **43**, 18 (1965).

[44] E. L. Eliel, N. L. Allinger, S. J. Angyal, and G. A. Morrison, *Conformational Analysis* (New York: Interscience Pub., 1965), pp. 195–205.

* In the metal hydride reductions, the reagent will not be free hydride ion but will be hydride ion associated with the reagent which in turn may be complexed with the oxygen of the carbonyl group. The degree of ion pair formation and of reagent complexing are obviously important variables with different reagents and different solvents.

TABLE 3-9

Stereoselectivities in the Reduction of 2-Methylcyclopentanone

(2-methylcyclopentanone) $\xrightarrow{[H]}$ *trans* and *cis* 2-methylcyclopentanol

No.	Reagent	*trans* Alcohol %	Ref.
1	$LiAlH_4$[a]	79	37
	$LiAlH_4$[b]	76	37
	$LiAlH_4$	75	1
	$LiAlH_4$[c]	77	45
	$LiAlH_4$[c]	82	45
2	$NaBH_4$[d]	74	1
	$NaBH_4$[e]	73	1
3	$LiAlH(O\text{-}t\text{Bu})_3$	72	37
4	$LiAlH(O\text{-Et})_3$	77	37
5	$LiAlH(O\text{-Me})_3$	56	37
6	B_2H_6	69	46
7	$Al(O\text{-}i\text{Pr})_3$[f]	42[f]	45
8	$Al(O\text{-Pr})_3$	24	45
9	$BH(CHMeCHMe_2)_2$	22	46
10	P^*_2BH[g]	6	46
11	H_2 (PtO_2)[h]	72	1
12	H_2 (Pt, HCl)[h]	35	45
13	Na (EtOH)[i]	87	1
14	Na (Moist Et_2O)[i]	90	1

[a] Molar ratio of ketone to $LiAlH_4$ is 1:4.
[b] Molar ratio of ketone to $LiAlH_4$ is 1:1.
[c] Two different samples of $LiAlH_4$; reductions at $-20°$.
[d] In water solvent.
[e] In methanol solvent.
[f] This ratio, obtained after long reaction time, represents the equilibrium ratio of 42:58 of *trans* to *cis* isomer of the aluminum alkoxy salt.
[g] P* represents the 3-pinanyl (isopinocampheyl) group and P^*_2BH represents "di-3-pinanylborane" which almost certainly reacts as the tetra-3-pinanyldiborane (see Sec. **6-2**).
[h] Heterogeneous reaction; added for comparison.
[i] Should be thermodynamic equilibrium mixture.

[45] W. Hückel, M. Maier, E. Jordan, and W. Seeger, *Ann. Chem.*, **616**, 46 (1958).
[46] H. C. Brown and D. B. Bigley, *J. Am. Chem. Soc.*, **83**, 3166 (1961).

from either above or below the plane of the ring as in **39A** or **39B**; then it can be rationalized that a relatively small reagent such as "hydride ion" will require less energy for approach from above the carbonyl group than

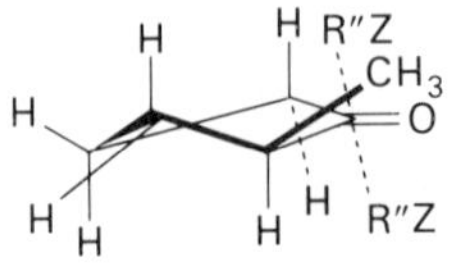

39A 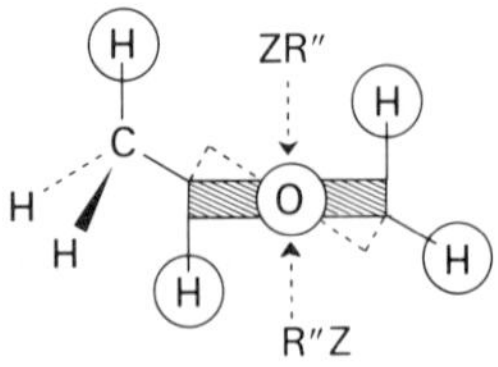 **39B**

from below. The difference in energies of activation for the two modes of approach results from the fact that a pseudoequatorial α-methyl group actually offers less interference to approach of a small reagent (from above in **39**) than do the axial α-hydrogens (from below in **39**). This reasoning in a slightly different form was considered by Umland and Williams[1] but was rejected because it was felt that the carbonyl, when complexed with the reagent, would be too bulky to permit significant contribution to the transition-state by conformation **39**. This objection is hard to assess but would not seem to be borne out by inspection of models. Richer[43a] has used essentially the same explanation, with emphasis on the steric role of the α-axial hydrogens, in rationalizing stereochemical results in the cyclohexanone series.

As the size of the reagent increases, approach from above in **39** will encounter increasing interference with the α-methyl group but approach from below would be confronted with a reasonably constant energy barrier, namely that required to cause the two "axial" hydrogens to rotate (torsional strain) so that the reagent could reach the carbonyl carbon.* On this basis a graded change of stereoselectivities would be expected favoring *cis* attack to give the *trans* alcohol when the reagent is small and favoring *trans* attack to give *cis* alcohol when the reagent is large. Although it is difficult to evaluate the relative size of the reagents (Sec. **3-5.2**), this predicted trend is evident in the first ten examples in Table **3-9**.

In this same way, a graded increase in the size of the α-substituent should raise the energy of activation for approach from above in **39B** but would have no major steric effect upon approach of the reagent from below, although it would contribute, but to a lesser extent, to an increase in torsional strain. The *shape* of the α-substituent (and the reagent) will be important. Thus an ethyl or isopropyl group would cause increased interference from above,

* This is an adaptation to the cyclopentanone series of the explanation advanced by Felkin,[8,39] Richer,[43] Marshall,[42] and Eliel[40] to account for comparable stereochemical results in the cyclohexanone case (Sec. **3-5**).

while a vinyl group might easily cause decreased steric hindrance as compared to methyl. Thc mcager data available on this point (Table **3-10**) support the generalization that the use of either a more hindered reagent or a more hindered α-substituted ketone will decrease the proportion of isomer resulting from approach of the reagent from above in **39**.

3-4.2. Cyclohexanones. Soon after his pioneering paper on conformational analysis in 1950, Barton[35] generalized the observed results of many reductions of cyclohexanone derivatives as follows: "Reduction with sodium borohydride and lithium aluminum hydride in general affords the equatorial epimer if the ketone group is not hindered, the polar (axial) epimer if it is hindered or very hindered. Meerwein–Ponndorf reduction, which is only applicable to relatively unhindered ketones, gives a higher proportion

TABLE 3-10

Effect of α-Substituent on Stereoselectivity of Reductions of the 2-Alkylcyclopentanones[a]

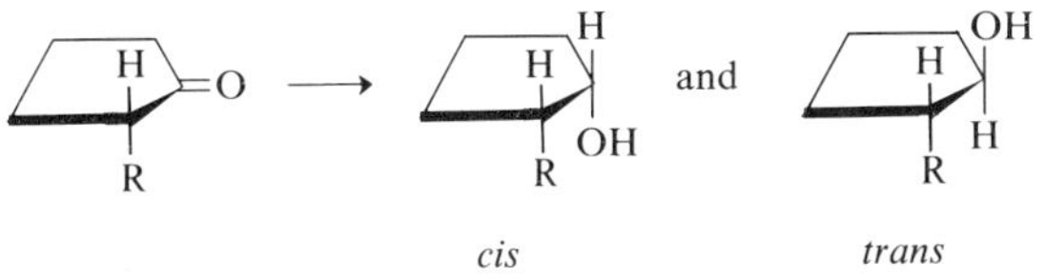

cis *trans*

Reagent	Percent alcohol which is *cis*			
	Substituent			
	α-Me	α-Et	α-*i*-Pr	α-CycloPr
$LiAlH_4$ (Et_2O)	18	"mostly *cis*"[47]	59	53
Al(O-*i*-Pr)$_3$	58		68	72

[a] Data from Reference 44 unless otherwise noted.

of the polar (axial) hydroxyl than do other methods." Kamernitzky and Akhren[41] have collected data from over three hundred examples of various reactions, and Hajós[48] has collected data on all such metal hydride and metal

[47] V. M. Mičovič, *Bull. Soc. chim. Belgrade*, **14**, 181 (1949); *Chem. Abst.*, **46**, 11121 (1952).

[48] A. Hajós, *Komplexe Hydride und Ihre Anwendung in der Organischen Chemie* (Berlin: VEB Deutscher Verlag der Wissenschaften, 1966).

alkoxide reductions. Within the loose boundaries of what is meant by an unhindered and a hindered ketone, this empirical correlation is generally valid. Dauben, Fonken, and Noyce[36] in 1956 were led to propose their "steric approach control" versus "product development control" ideas in order to explain why this generalization was successful. More recently Wheeler and co-workers[38] have enlarged on the Barton correlation and have made the suggestion that the covalently bound complex metal hydrides give greater amounts of the more stable equatorial alcohol but reducing agents with greater ionic character give relatively more of the less stable axial isomer. Recent work has strongly indicated that the stereochemical results of addition reactions in cyclic systems can be accommodated by a more accurate assessment of the steric interactions and torsional strains in the competing transition states. The stereochemistry of the addition reactions of 4-*t*-butylcyclohexanone is informative and is summarized in Table **3-11**.

It is apparent that a substituent in a remote position, i.e., *t*-butyl in the 4-position of the cyclohexane ring, has a profound effect on the steric course of the reaction. This is a conformational effect and not a result of direct steric hindrance of the reagent by the *t*-butyl group. Transition states having the *t*-butyl group axial are of such high relative energy that they are too sparsely populated to contribute significantly to the product mixture. When the substitutuent in the 4-position is methyl instead of *t*-butyl, the energy differences between transition states in which the methyl group is equatorial versus axial are not so great and thus the observed stereoselectivities are not so high although they show the same trends; for example, the lithium aluminum hydride reduction of 4-methylcyclohexanone gives a 81:19 ratio of *trans* to *cis* 4-methylcyclohexanol[49–51] as compared with the 91:9 ratio found with the 4-*t*-butyl homolog. A 4-chloro or 4-methoxy substituent produces an effect in the opposite direction, which has been explained on the basis of dipole interactions in the transition state.[51a,51b] (See Addenda[74]).

The nature of the reagent as well as the substituent has an important role in determining the stereoselectivity. In reductions of 4-*t*-butylcyclohexanone, the percentage of *trans* alcohol in the product can be changed from 96, by use of lithium aluminum hydride in tetrahydrofuran at −40°, to 59 by use of lithium trimethoxyaluminum hydride. In general the reagents which may be considered "large," such as the Grignard reagents, aluminum isopropoxide, and lithium trimethoxyaluminum hydride, give higher yields of the *cis* carbinol, while the reagents which may be considered "small," such as lithium aluminum hydride, sodium borohydride, and lithium acetylide,

[49] (a) E. L. Eliel, *Rec. Chem. Progr.*, **22**, 129 (1961). (b) E. L. Eliel and R. S. Ro, *J. Am. Chem. Soc.*, **79**, 5992 (1957).

[50] D. S. Noyce and D. B. Denney, *J. Am. Chem. Soc.*, **72**, 5743 (1950).

[51] (a) M. G. Combe and H. B. Henbest, *Tetrahedron Lett.*, 404 (1961). (b) H. Kwart and T. Takeshita, *J. Am. Chem. Soc.*, **84**, 2833 (1962).

TABLE 3-11

Stereochemistry of the Addition Reactions of 4-*t*-Butylcyclohexanone

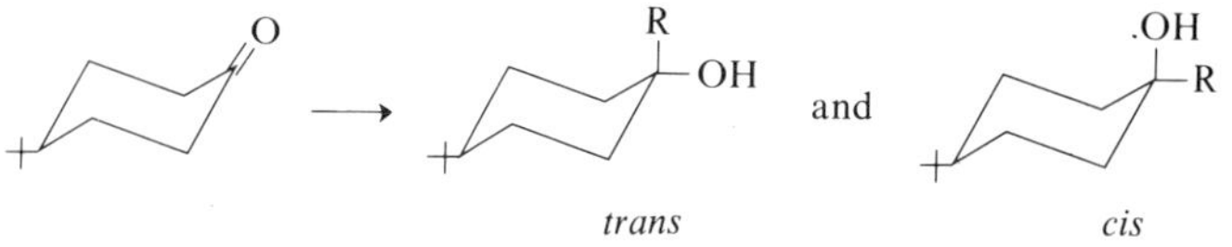

trans and *cis*

No.	Reagent	R	%*trans*	%*cis*	Ref.
1	$LiAlH_4$ (Et_2O)	H	92	8	49
2	$LiAlH_4$ (Et_2O)	H	91	9	43a
3	$LiAlH_4$ (Et_2O)	H	90	10	40
4	$LiAlH_4$ (THF, 27°)	H	92	8	53
5	$LiAlH_4$ (THF, −40°)	H	96	4	53
6	$LiAlH_4$, $AlCl_3$[a]	H	81	19	49
7	$LiAlH(O\text{-}t\text{-}Bu)_3$	H	90	10	43a
8	$LiAlH(OMe)_3$	H	59	41	49
9	$LiC{\equiv}CH$	$C{\equiv}CH$	89	11	43a
10	$NaBH_4$ (MeOH)	H	89	11	40
11	$NaBH_4$ (MeOH, 27°)	H	80	20	53
12	$NaBH_4$ (MeOH, 0°)	H	84	16	53
13	$NaBH_4$ (MeOH, −40°)	H	88	12	53
14	MeMgBr	Me	40	60	52
15	Me_2Mg	Me	35	65	52
16	*n*-PrMgBr	*n*-Pr	27	73	39
17	AllylMgBr	Allyl	51	49	39
18	$Al(O\text{-}iPr)_3$ equilibration	H	79	21	49
19	$Al(O\text{-}iPr)_3$	H	49.1	50.9	51b

[a] With excess $LiAlH_4$-$AlCl_3$ reagent, this appears to give the equilibrium mixture. When excess ketone was used, the product was essentially all *trans*.[49a,54]

give higher proportions of *trans* alcohol. The designation of the effective size of the reagent for stereochemical purposes is at present largely arbitrary and is based upon the kind of data presented in Tables **3-9**, **3-11**, and **3-12.** The assumption is made that the "smaller" reagents give the higher proportion of the equatorial product and the "larger" reagents give the higher proportion of axial products. The nature of the solvent is important; reactions in the more strongly solvating media tend to give an apparent increase in size of reagent, other things being equal. The approximate order, with smallest first, is $LiAlH_4$, $LiAlH(O\text{-}t\text{-}Bu)_3$, $LiC{\equiv}CH < NaBH_4$

[52] H. O. House and W. L. Respess, *J. Org. Chem.*, **30**, 301 (1965).
[53] P. T. Lansbury and R. E. MacLeay, *J. Org. Chem.*, **28**, 1940 (1963).
[54] V. J. Shiner, Jr. and D. Whittaker, *J. Am. Chem. Soc.*, **85**, 2337 (1963).

$< KBH_4 < LiAlH(OCH_3)_3$, and $MeMgX < RMg <$ "di-3-pinanyl-borane." Aluminum isopropoxide generally acts as though it is a little smaller than the Grignard reagents but since it will also cause equilibration of the isomers, the results are often variable. It has been shown that the active reducing agent with aluminum isopropoxide is the trimer.[54] It might therefore be assumed that it would be one of the largest reagents. However, the important factor from the stereochemical standpoint is not the total bulk but the effective size at the site of hydrogen transfer. The fact that sodium borohydride appears larger than lithium aluminum hydride is not expected and has been discussed,[38a] but the more surprising observation is that lithium tri-*t*-butoxyaluminum hydride seems to show the same stereo-selectivity as lithium aluminum hydride. This observation has not been satisfactorily rationalized.[55,56] It has been proposed that a facile equilibrium might give lithium aluminum hydride as the active reducing agent but this explanation is not altogether satisfactory.[37]

It is possible to obtain a *higher* yield of the thermodynamically more stable isomer than is present at equilibrium.[40] Thus in the lithium aluminum hydride reduction of 4-*t*-butylcyclohexanone a 92:8 *trans*:*cis* ratio is obtained, while the equilibrium ratio is 79:21. This situation has been observed as well in the reduction of 3-*t*-butylcyclopentanone[43b] and 2,2-dimethyl-4-*t*-butylcyclohexanone[43a] and constitutes a strong argument against the functioning of a product development control factor.

In the lithium aluminum hydride reduction of α-substituted cyclohexanones, as the bulk of the α-substituent increases, the proportion of *cis* (less stable) isomer increases. The comparable ratios of *trans* to *cis* isomer for α-methyl-, α-isopropyl-, and α-*t*-butyl-cyclohexanones[42] are very nearly 3:1, 2:1, and 1:1, respectively. From the data collected in Table **3-12** it is also seen that the proportion of the *trans* isomer decreases (*cis* increases) upon changing the reducing agent from lithium aluminum hydride (69–82%) to sodium borohydride (69%) to aluminum isopropoxide (50–63%) to potassium borohydride (41%) to lithium trimethoxyaluminum hydride (31%). These results are in accord with the postulated increasing effective bulk of these reagents.

We shall attempt to rationalize these results, without reference to the thermodynamic stability of products, using an analysis of the probable conformation of the transition state and the various steric factors involved as developed by Eliel,[*,40] Felkin,[8,39] Marshall,[42] and Richer.[43] In the

[55] (a) H. Haubenstock and E. L. Eliel, *J. Am. Chem. Soc.*, **84**, 2363 (1962). (b) *ibid.*, 2368 (1962).

[56] D. C. Ayres and R. Sawdaye, *Chem. Commun.*, 527 (1966).

* The authors are deeply indebted to Professor E. Eliel for enlightening discussions concerning salient features of these reactions and interpretations and for information on results prior to publication.

TABLE 3-12

Effect of Reagent on Stereoselectivity of Addition Reactions of 2-Methylcyclohexanone[a]

"*trans*" and "*cis*"

No.	Reagent R″Z	R″	% "*trans*"
1	Na, HOEt	H	72, 88
2	$LiAlH_4$ (Et_2O)	H	69, 70, 82
3	$LiAlH_4$ (THF)	H	76
4	$LiAlH_4$, $AlBr_3$	H	59–71
5	$LiAlH(O\text{-}t\text{Bu})_3$	H	63–70
6	$NaBH_4$ (MeOH, H_2O)	H	69
7	$Al(O\text{-}i\text{Pr})_3$	H	50, 63
8	B_2H_6	H	65[b]
9	KBH_4	H	41
10	$LiAlH(OMe)_3$	H	31
11	$BH(CHMeCHMe_2)_2$	H	23[b]
12	P_2^*BH	H	8[b]
13	HC≡CK (KOH or K-O-*t*Bu)	C≡CH	60
14	KCN	CN	69
15	MeMgI	Me	25

[a] Selected values adapted from Table I in Reference 41.
[b] Data from Reference 45.

2-alkylcyclohexanones, the alkyl group occupies an equatorial conformation which is almost in the plane of the carbonyl group as shown in **40A**. The α-methyl group, being equatorial and symmetrical, will not appreciably hinder axial approach* of the reagent, *if the reagent is small enough*, and in fact the approach from above in **40A** (axial attack) to give the equatorial carbinol appears to be more open than the approach from below (equatorial attack) to give the axial carbinol (cf. also Figs. **39A** and **39B** for analogous cyclopentanone example). The reagent interactions along the reaction coordinate, whether from above or below, are with two axial hydrogens, the

* It is noteworthy that the approach, if it is to be at 90° to the carbonyl double bond, is neither truly axial nor truly equatorial since the final axial or equatorial bond in the product is approximately 55° from the plane of the original carbonyl group. Nevertheless it will be convenient to refer to "axial" attack to give equatorial carbinol and "equatorial" attack to give axial carbinol.

3,5-axial hydrogens from above and the 2,6-axial hydrogens from below; the 2-methyl group is scarcely involved. A major difference in the two

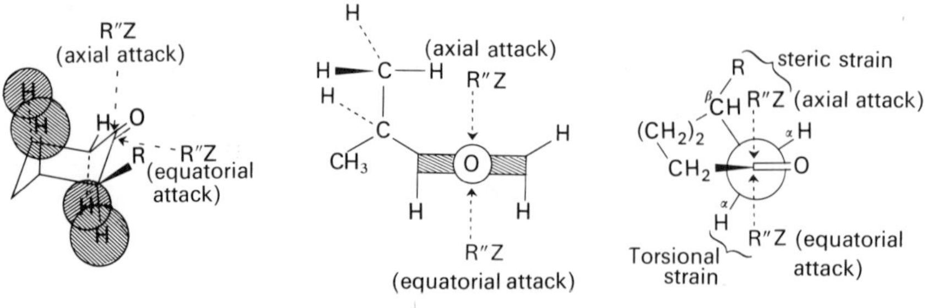

40A **40B** **40C**

approaches is the distance along the reaction coordinate from the carbonyl carbon at which the transition state is reached.[42] The α-axial hydrogens are hindering closer to the carbonyl carbon from below and the β-axial hydrogens, farther out from above. As the reagent approaches along the reaction coordinate, the smaller the reagent is, the closer to the carbonyl carbon it will be at the transition state. Other requirements resulting from the developing $sp^2 \rightarrow sp^3$ change should be much the same in either case. These considerations with a small hydride reagent favor the formation of *trans*-2-methylcyclohexanol (axial attack to give equatorial alcohol), as observed. The *trans* compound happens to be the more stable isomer, but this is not a factor in the steric control of the reaction. When the α-methyl group is replaced with an α-ethyl, α-isopropyl, or α-*t*-butyl group, in which one of the terminal groups of the α-substituent can now extend back over the upper face of the ring to a certain extent but not over the lower face (**40B**), then classical steric hindrance to axial approach of the reagent from above (cf. **40A** and **40B**) will increase while the energy of the transition state for equatorial approach from below will be essentially unchanged. The hindrance to equatorial attack is still primarily due to interference with the pair of α-axial hydrogens, as it is in the 2-methylcyclohexanone case. Felkin[8,39] has proposed that an important portion of the energy requirement for equatorial approach is due to torsional strain resulting from the development of two eclipsed R″ ↔ H interactions with the pair of α-axial hydrogens early in the transition state, while the axial approach involves a much higher proportion of classical steric hindrance. The introduction of either a larger α-equatorial substituent or a β-axial substituent (**40C**) will raise the transition state energy for attack from above by increasing steric hindrance but will have little effect on the energy for equatorial attack where the torsional strain will remain essentially unchanged. Thus the proportion of equatorial attack to produce the axial isomer will increase and can easily become dominant even though the absolute rate for equatorial attack is essentially unchanged from that in the lower homolog. Also, as the size of the *reagent* increases, the α- and β-substituents remaining constant, steric hindrance with the β-axial substituents, which are farther away along the reaction coordinate, will increase while the energy required for equatorial approach past the α-axial hydrogens will remain relatively constant since the latter is primarily a matter of torsional strain. Therefore the formation of the axial carbinol relative to the equatorial isomer will be favored by such an increase in reagent size.

When a methyl group is introduced into the α-equatorial or β-axial position of the cyclohexanone ring as shown in **41** and **42**, the ratio of

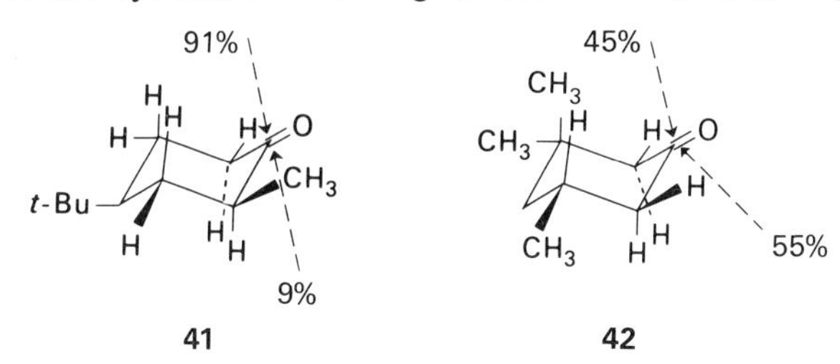

Fig. 3-10. Percent axial versus equatorial attack by LAH resulting from the introduction of α-equatorial or β-axial methyl groups in the cyclohexanone ring.

products from lithium aluminum hydride reduction via axial and equatorial attack are as indicated by the arrows (axial attack from above giving the equatorial carbinol and equatorial approach from below giving the axial carbinol).[40,55] Extensive competitive rate experiments by Eliel and Senda[40] in which 4-*t*-butylcyclohexanone was used as a standard and reductions were conducted on di-, tri-, and tetra-methylcyclohexanones with lithium aluminum hydride, lithium tri-*t*-butoxyaluminum hydride, lithium trimethoxyaluminum hydride, and sodium borohydride have shown that the increase in axial over equatorial carbinols as compared with those from 4-*t*-butylcyclohexanone is primarily the result of a rate retardation of the axial approach pathway while the change in rate of the equatorial approach is relatively small. For example in **42** the rate of the axial approach as compared with 4-*t*-butylcyclohexanone is retarded very greatly while the change in rate of equatorial approach is only minor. These observations are in accord with expectation based upon the analysis presented.

3-4.3. Polycyclic Systems. The conclusions reached by consideration of the basic cyclohexanone derivatives are readily extended to the steroidal ketones and the results are rationalized by similar analyses. The many reductions in the steroid series have been extensively tabulated and reviewed.[41,48,57,58] A wide variety of noncatalytic reductions[38b] of 3-choles-

[57] C. Djerassi, *Steroid Reactions* (San Francisco: Holden-Day, Inc., 1963).

[58] E. Toromanoff in *Topics in Stereochemistry*, ed. E. L. Eliel and N. L. Allinger (New York: Interscience Pub., 1967), Vol. II, pp. 158–198.

tanone (**42A**) have been reported to give product mixtures of 3-β-cholestanol (equatorial OH) versus 3-α-cholestanol (axial OH) of 83–93% to 17–7%. Lithium aluminum hydride reduction of the comparable conformationally biased 4-*t*-butylcyclohexanone gives a 91% *trans*:9% *cis* ratio of 4-*t*-butylcyclohexanols. The larger lithium trimethoxyaluminum hydride reagent causes a decrease in the amount of equatorial alcohol to 82% in the 3-cholestanone case,[38b] and to 59% in the 4-*t*-butylcyclohexanone case (Table **3-8**). The fact that the shift is not as great in the steroid case may be due to the added buttressing effect of the C-10 methyl group.

3-Coprostanone (**42D**) can also be compared directly with 4-*t*-butyl-

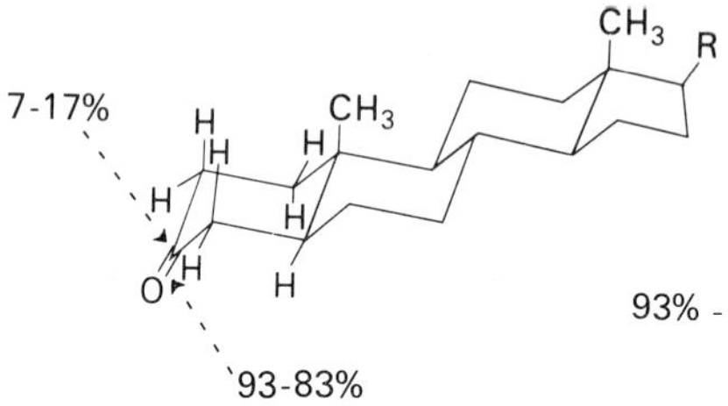

42A, R = C_8H_{17}
42B, R = $COCH_3$
42C, R = =O

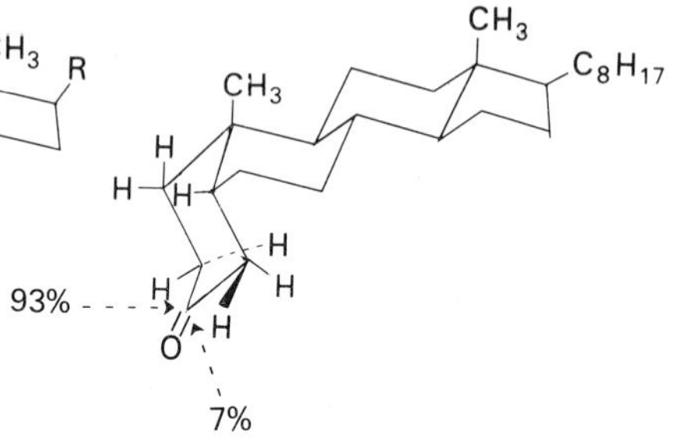

42D Coprostanone

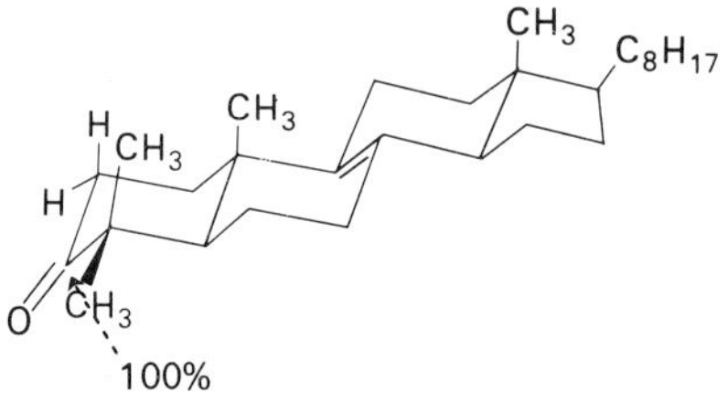

42E Δ^8-Lanostenone

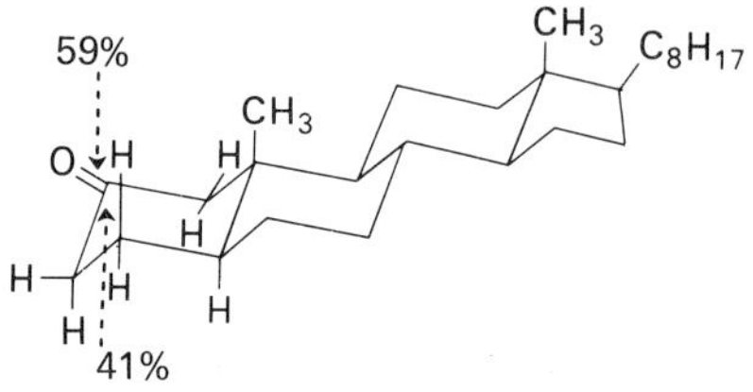

42F Cholestan-2-one

cyclohexanone; the preference for attack by lithium aluminum hydride from the α-side past the pair of axial hydrogens is approximately the same (93:7 in 3-coprostanone, 91:9 in 4-*t*-butycyclohexanone). Δ^8-Lanostenone (**42E**) can be compared to 2,2-dimethyl-4-*t*-butylcyclohexanone. The α-axial methyl group is controlling in each case but in Δ^8-lanostenone (**42E**), either because of the reduced interference for approach from below (due to the double bond at C-8, 9) or the added interference from above (because of the methyl at C-10), the only product isolated is the equatorial Δ^8-β-3-lanostenol. The 2-ketone **42F** is closely analogous to 3,3,5-trimethylcyclohexanone and the stereoselectivities of lithium aluminum hydride reduction are very nearly the same, i.e., 59:41 versus 55:45 in the two cases.* Substituents in a distant ring usually have only a minor effect upon the stereoselectivity as shown by the lithium aluminum hydride reduction of **42B** and **42C** which give, respectively, 88:12 and 93:7 ratios of β- and α-forms.[57] Under some circumstances a polar substituent at a distance may have a significant effect upon the stereoselectivity.[59] Thus sodium borohydride reduction of the 12-oxotigogenin derivative **42G** gives a 22:78 ratio of the 12-α to 12-β carbinol but with a 3-α-chlorine present, **42H**, the ratio was 33:67.

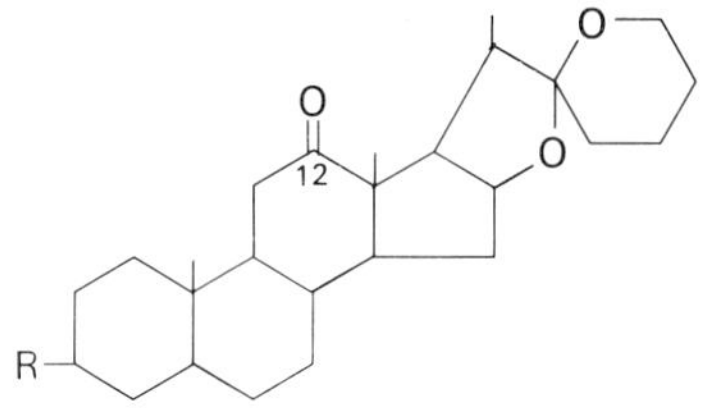

42G, R = H
42H, R = Cl

3-4.4. Bicyclic Ketones. The results from addition reactions to many bicyclic ketones are illustrated by a consideration of norcamphor and camphor which have been studied extensively.[37,50,60] Lithium aluminum hydride reduction of norcamphor (**43A**) gives an 8:92 ratio of *exo* to *endo* isomer; but reduction of camphor (**43B**) gives a 90:10 ratio of *exo* (iso-

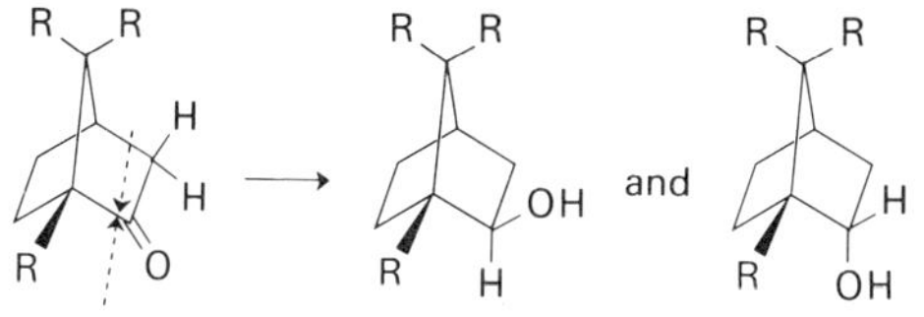

	44	**45**
43A, Norcamphor, R = H →	8%	92%
43B, (+)-Camphor, R = CH_3 →	90%	10%

borneol) to *endo* (borneol) isomers. In norcamphor, a reagent approaching from the *exo* side, i.e., from above, in **43A** passes through a transition state in which the reagent encounters one hydrogen from the methylene bridge and a pseudoequatorial α-hydrogen. Approach from the *endo* side, i.e., from

* Extensive pertinent data on other steroidal ketones have been collected in Tables 4, 5, and 6 of Reference 41.

[59] H. B. Henbest, *Proc. Chem. Soc.*, 159 (1963).
[60] C. H. DePuy and P. R. Story, *J. Am. Chem. Soc.*, **82**, 627 (1960).

below, in **43A** involves reagent interactions with three pseudoaxial hydrogens from the six-membered ring. It is very clear from all considerations that the former approach is less sterically hindered in accordance with the observation that **45** predominates. But when the hydrogens on the methylene bridge are replaced with methyl groups as in camphor, the transition state for *exo* approach from above in **43B** is now greatly hindered in comparison with that for the *endo* approach from below with the result that isoborneol **44B** is the major product even though under thermodynamic equilibrium borneol **45B** is favored to almost 100%.

Lithium tri-*t*-butoxyaluminum hydride gives a 93–95% yield[37,1] of *endo* isomer, thus showing only a slightly greater stereoselectivity than lithium aluminum hydride itself while lithium trimethoxyaluminum hydride gives 99% *endo* isomer. From these data one may conclude that this latter reagent is not only "larger" than lithium aluminum hydride but also "larger" than lithium tri-*t*-butoxyaluminum hydride (Sec. **3-4.2**). The Meerwein–Ponndorf reduction of camphor[50] gives a 70:30 ratio of *endo* to *exo* isomer which is out of line with its generally considered large steric requirement. However, in this case, the reaction is slow and it appears that there is concomitant thermodynamic equilibration of the aluminum salts of borneol and isoborneol.

We have considered almost entirely the reductions by metal hydrides in this section because of the extensive data available. The much less systematic data available on the addition of Grignard reagent indicates that these addition reactions parallel the metal hydride reductions with the provision that the Grignard reagents are large as compared to metal hydride reagents.

1969 Addenda

Sec. 3-2.3. It has been shown[61] that the most stable ground state rotamer of chloroacetaldehyde and bromoacetaldehyde has the halogen–carbon bond eclipsing the carbonyl group, contrary to the conformation postulated for the transition state in the dipolar model.

Sec. 3-2.4. Addition reactions to carbonyl compounds containing an α-oxygen substituent have been extended to include the reactions of phenyllithium and phenylmagnesium bromide with 2,3-O-dibenzyl- and 2,3-O-isopropylidene-D-glyceraldehyde.[62] The stereoselectivities were only moderate (maximum *threo*:*erythro* ratio, 1:1.9). The stereoselectivities were higher and were reversed when diphenylmagnesium reacted with the com-

[61] G. J. Karabatsos and D. J. Fenoglio, *J. Am. Chem. Soc.*, **91**, 1124 (1969).

[62] Y. Ohgo, J. Yoshimura, M. Kono, and T. Sato, *Bull. Chem. Soc. Japan*, **42**, 2957 (1969); cf. also M. H. Delton and G. U. Yuen, *J. Org. Chem.*, **33**, 2473 (1968).

parable imine, N-benzyl-2,3-O-isopropylidene-D-glyceraldimine (maximum *erythro*:*threo* ratio, 1:5.5 in the presence of two molar equivalents of $MgBr_2$).[63]

Sec. 3-2.5. High stereoselectivities have been recorded in the asymmetric reductions of some β-amino ketones with chiral centers in both the α and β positions.[64] The results have been rationalized in terms of a six-membered ring, intra-complex, hydrogen transfer which is pertinent to the discussions on pages 106 to 108.

Sec. 3-2.6. Further data on 1,3-asymmetric inductions (cf. Table **3-6**) are given by the results of a study of the LAH reduction of twenty-two β-amino ketones.[65] With tertiary amino ketones the asymmetric syntheses ranged up to 54% with the *threo* diastereomer predominating; with secondary amines the asymmetric syntheses were either zero or low with the *erythro* isomer predominating.

Sec. 3-4.1. The stereochemistry of the addition of sixteen different organometallic reagents to 2-methylcyclohexanone (cf. Table **3-9**) has been studied.[66] The ratio of attack from the face of the carbonyl group *cis* to the methyl substituent versus attack of the face *trans* to the methyl group varied from 99:1 (phenylmagnesium bromide in THF) to 0.1:1 (sodium acetylide in benzene-toluene). The results can be rationalized by the considerations given in *Secs.* **3-4.1** and **3-4.7**.

Sec. 3-4.1. As part of a more extensive study, the addition of methyllithium to (*R*)-(+)-2-methyl-2-phenylcyclopentanone was observed to give 75% of the (1*R*,2*R*)-*cis* and 25% of the (1*S*,2*R*)-*trans*-1,2-dimethyl-2-phenylcyclopentanol.[67]

Sec. 3-4.2. Additional studies on the stereoselectivity of the reduction of 4-*t*-butylcyclohexanone and 3,3,5-trimethylcyclohexanone with lithium, sodium, and potassium isopropoxides have been reported.[68]

Sec. 3-4.2. Five publications[69–73] have appeared during 1969 which deal with the stereochemistry of addition reactions of organometallic reagents

[63] Y. Ohgo, J. Yoshimura, and T. Sato, *Bull. Chem. Soc. Japan*, **42**, 728 (1969).
[64] M. J. Lyapova and B. J. Kurtev, *Chem. Ber.*, **102**, 3739 (1969).
[65] M. J. Brienne, C. Fouquey, and J. Jacques, *Bull. Soc. Chim. Fr.*, 2395 (1969).
[66] J.-P. Battioni, M.-L. Capmau, and W. Chodkiewicz, *Bull. Soc. Chim. Fr.*, 976 (1969).
[67] T. D. Hoffman and D. J. Cram, *J. Am. Chem. Soc.*, **91**, 1000 (1969).
[68] D. N. Kirk and A. Mudd, *J. Chem. Soc.* (*C*), 804, 968 (1969).
[69] G. Chauvière, Z. Welvart, D. Eugène, and J.-C. Richer, *Can. J. Chem.*, **47**, 3285 (1969).
[70] J.-C. Richer and D. Eugène, *Can. J. Chem.*, **47**, 2387 (1969).
[71] M. Perry and Y. Maroni-Barnaud, *Bull. Soc. Chim. Fr.*, 2372 (1969).
[72] P. R. Jones, E. J. Goller, and W. J. Kauffman, *J. Org. Chem.*, **34**, 3566 (1969).
[73] M. R. Harnden, *J. Chem. Soc.* (*C*), 960 (1969).

to the carbonyl group of substituted cyclohexanone derivatives. These studies include examples which should be considered along with those in Tables **3-11** and **3-12**. The data in these papers have been interpreted in accord with the presentation on pages 124 to 127. Kirk[74] has reinterpreted the ratio of axial versus equatorial attack on 4-chlorocyclohexanone in terms of conformational equilibria rather than charge-dipole interactions.

Sec. 3-4.3. A study has been made of the stereoselectivity of the reductions of a series of steroidal ketones by lithium, sodium, and potassium isopropoxides, by various metal hydrides,[72] by alkali metal-ammonia reducing agents[68] and by the chloroiridic acid-triphenylphosphite reagent.[75]

[74] D. N. Kirk, *Tetrahedron Lett.*, 1727 (1969).
[75] P. A. Browne and D. N. Kirk, *ibid.*, 1653 (1969).

4

Various Asymmetric Addition Reactions of Carbonyl Compounds

4-1. The Cyanohydrin Reaction

The cyanohydrin reaction occupies a key position in the field of asymmetric synthesis. The reaction of benzaldehyde with hydrogen cyanide to give mandelonitrile, which was hydrolyzed to mandelic acid, was discovered in 1832.[1] This reaction was first applied to the sugar series by Kiliani in 1885[2]; E. Fischer soon thereafter utilized it for progressing from one sugar to the next higher homolog. In applying this reaction to L-arabinose (**1**), without isolating the nitriles (**2** and **3**), he first isolated only L-mannonic acid (**4**) as the lactone but later also found about one-third as much L-gluconic acid (**5**) in the mother liquors.[3] He subsequently commented[4] that "To my knowledge these observations furnish the first definitive evidence that further synthesis with asymmetric systems proceeds in an asymmetric manner." The cyanohydrin reaction when applied to a variety of sugars

[1] E. W. Winckler, *Ann. Chem.*, **4**, 246 (1832).
[2] H. Kiliani, *Ber.*, **18**, 3066 (1885); **19**, 221, 767, 3029 (1886); **20**, 282, 339 (1887); **21**, 915 (1888); **22**, 521 (1889).
[3] E. Fischer, *Ber.*, **23**, 2611 (1890).
[4] E. Fischer, *Ber.*, **27**, 3210 (1894).

```
CHO                 CN         CN                 COOH        COOH
HCOH                HCOH       HOCH               HCOH        HOCH
HOCH   --HCN-->   [ HCOH and   HCOH ]  ------>    HCOH  and   HCOH
HOCH                HOCH       HOCH               HOCH        HOCH
CH2OH               HOCH       HOCH               HOCH        HOCH
                    CH2OH      CH2OH              CH2OH       CH2OH
 1                   2          3                  4           5
```

generally gives as the predominant product that one in which the group on the newly formed chiral center at C-2 and the hydroxyl group on the established chiral center at C-4 are opposite to each other when written in the Fischer convention[5] as illustrated in the reaction of **1** to give predominately **4**. However, the ratio of the diastereomeric products can vary, depending upon conditions. Thus Isbell and co-workers[6] found that starting with D-arabinose and using hydrogen cyanide in sodium bicarbonate buffer solution, D-mannonitrile and D-gluconitrile were formed in 3:1 ratio but that the ratio was reversed when hydrogen cyanide in sodium carbonate solution was used.

Kuhn and co-workers[7] have used the analogous reaction of hydrogen cyanide and aniline to develop a synthesis for the homologous α-amino sugars. The reaction with the four D-aldopentoses gives as the predominant product that one in which the anilino group at C-2 and the adjacent hydroxyl group at C-3 are on opposite sides when written according to the Fischer convention as shown in going from **6** to **7** or from **9** to **10**. Thus the simple

```
   ¹CH=O                               CN                      CN
HO—²C—H        HCN              H—¹C—NHPh           PhNH—¹C—H
 (CHOH)2     -------->         HO—²C—H       and      HO—²C—H
 CH2OH        PhNH2             (CHOH)2                (CHOH)2
              EtOH              CH2OH                  CH2OH
    6                              7                       8
D-Arabinose   ------>      Gluco-, 65%       and     Manno-, 10%
D-Lyxose      ------>      Galacto-, 51%     and     Talo-, 12%
```

[5] C. S. Hudson, "The Fischer Cyanohydrin Synthesis and the Configuration of Higher Carbon Sugars and Alcohols" in *Adv. in Carbohydrate Chem.*, **1**, 1 (1945).

[6] (a) H. S. Isbell, J. V. Karabinos, H. L. Frush, N. B. Holt, A. Schwebel, and T. T. Galkowski, *J. Res. Nat. Bur. Stand.*, **48**, 163 (1952); *Chem. Abst.*, **47**, 3244 (1953). (b) H. S. Isbell, H. L. Frush, and N. B. Holt, *J. Res. Nat. Bur. Stand.*, **53**, 325 (1954); *Chem. Abst.*, **49**, 9512 (1955).

[7] R. Kuhn, W. Bister, and H. Fischer, *Ann.*, **617**, 109 (1958).

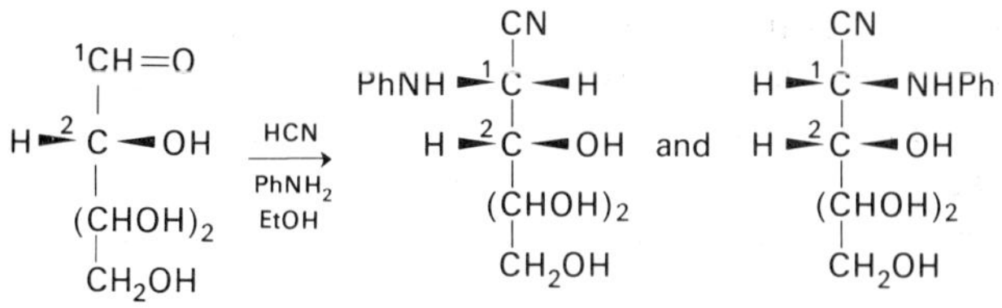

9		10		11
D-Ribose	⟶	*Altro*-, 67%	and	*Allo*-, 13%
D-Xylose	⟶	*Ido*-, 80%	and	*Gulo*-, 9%

cyanohydrin reaction and the modification using aniline do not follow parallel stereochemical courses and no doubt there are mechanistic differences. The free aldehyde is almost certainly involved in the first case and the Schiff base or its iminium ion in the second.

If in the cyanohydrin reaction on a chiral aldehyde the newly formed asymmetric center could be separated from the original chiral moiety, then an asymmetric synthesis in the Marckwald sense would be accomplished, as illustrated for the reaction of the phenyl Grignard reagent with a carbohydrate derivative (Sec. **1-3**). Something similar to this was achieved[8] by carrying out a cyanohydrin reaction on the glucoside of salicylaldehyde (helicin, **12**) as shown in Fig. **4-1**. The formation of (+)-*o*-hydroxymandelic acid, (**14**), $[\alpha]_D^{20}$ +1.9°, represents an asymmetric synthesis of about 3%.*

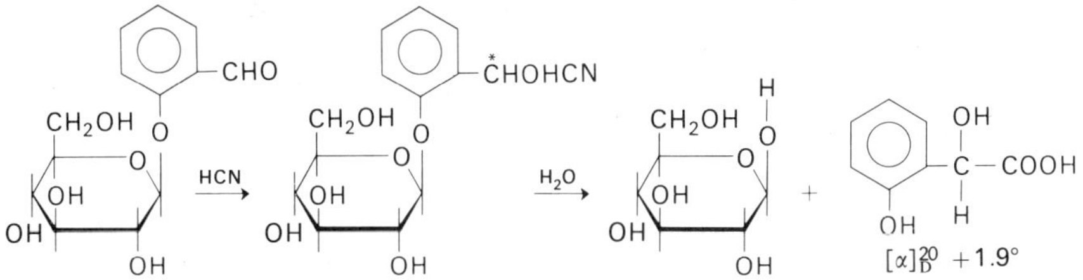

$[\alpha]_D^{20}$ +1.9°

Fig. 4-1. Cyanohydrin reaction of helicin.

In order to draw meaningful conclusions concerning the factors controlling the steric course of these cyanohydrin reactions in the carbohydrate series, the mechanism for the reaction must be considered. Lapworth[9]

[8] E. Fischer and M. Slimmer, *Ber.*, **36**, 2575 (1903).

[9] A. Lapworth, *J. Chem. Soc.*, **83**, 995 (1903); **85**, 1206 (1904).

* Based upon the now known maximum rotation of *o*-hydroxymandelic acid [A. McKenzie and P. A. Stewart, *J. Chem. Soc.*, 104 (1935)]. It seems probable that this constitutes a valid asymmetric synthesis; however, Fischer was very cautious in concluding this, since the rotation of the product was low and the product was an oil.

showed that the cyanohydrin reaction of benzaldehyde and hydrogen cyanide in aqueous alcohol was reversible and that the rate was first order with respect to both aldehyde and cyanide ion. These facts are compatible with a slow, rate-determining, direct reaction (Eq. 1) of cyanide ion with aldehyde followed by a rapid protonation to give the cyanohydrin. Since

$$R{-}CH{=}O + CN^- \underset{}{\overset{\text{Slow}}{\rightleftarrows}} R{-}\overset{\overset{\displaystyle O^-}{|}}{C}HC{\equiv}N \qquad (1)$$

$$R{-}\overset{\overset{\displaystyle O^-}{|}}{C}HC{\equiv}N + HB \overset{\text{Fast}}{\rightleftarrows} R{-}\overset{\overset{\displaystyle OH}{|}}{C}HC{\equiv}N + B^- \qquad (2)$$

hydrogen cyanide is a weak acid, the rate is very dependent upon the pH of the solution. Subsequent kinetic studies have also shown that the reaction may be subject to general acid catalysis[10] which would permit interpretation in terms of the following equations.

$$R{-}CH{=}O + HB \overset{\text{Fast}}{\rightleftarrows} R{-}CH{=}O\cdots HB \qquad (3)$$

$$R{-}CH{=}O\cdots HB + CN^- \overset{\text{Slow}}{\rightleftarrows} R{-}\overset{\overset{\displaystyle OH}{|}}{C}HC{\equiv}N + B^- \qquad (4)$$

In either case one would expect the attack of cyanide ion to be governed by the same steric factors which were discussed in connection with the Cram models (Sec. **3-2**). The cyanide ion would presumably act as a relatively small reagent (see Table **3-9**, No. 11).

It is difficult to know how the Cram model should be applied in the carbohydrate series (cf. Sec. **3-2.2**). A naïve application of the Cram open-

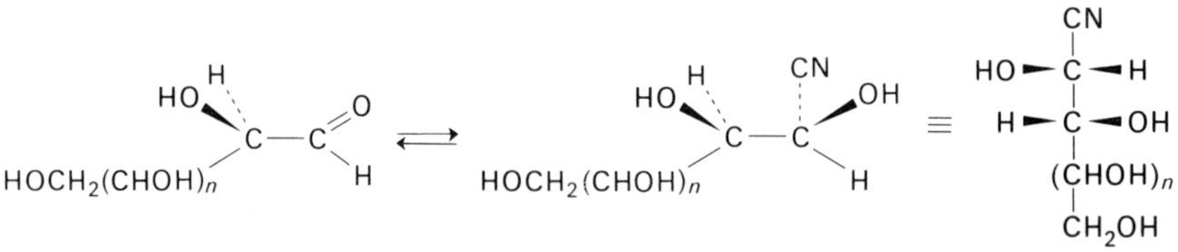

HO H HOCH₂(CHOH)ₙ C—C O H ⇄ HO H HOCH₂(CHOH)ₙ C—C CN OH H ≡ CN / HO—C—H / H—C—OH / (CHOH)ₙ / CH₂OH

Fig. 4-2. Unwarranted application of Cram's open-chain model to Kiliani synthesis.

chain model to this system (Fig. **4-2**), considering $R_S = H$, $R_M = OH$, and $R_L = (CHOH)_nCH_2OH$, leads to the prediction that the predominant isomer will have the hydroxyl group on the original α-chiral center and the hydroxyl group on the newly established chiral center opposite to each other, when

[10] W. J. Svirbely and J. F. Roth, *J. Am. Chem. Soc.*, **75**, 3106 (1953).

written according to the Fischer convention. This in fact is not generally observed but, instead, the major product in the Kiliani synthesis is usually the isomer which has the hydroxyls at C-2 and C-4 opposite to each other when written in the Fischer convention. On the other hand, application of Cram's open-chain model on the same basis correctly predicts the predominant isomer in the modified cyanohydrin reactions (that have been studied) using aniline[7] or ammonia in the Strecker amino acid synthesis (Sec. **7-4.1**). In these cases, the step which determines the stereochemistry probably is the addition of cyanide ion to the Schiff base[10,11] of the acyclic form of the reducing sugar. The stereochemistry is further complicated by the fact that the intermediate Schiff base may exist in *syn* and *anti* forms. This reaction might be expected to have quite different stereochemical requirements from the addition of cyanide to the carbonyl group of the corresponding substrate. In view of the difference in mechanism, the many possibilities for hydration and internal hydrogen bonding, and the multiple asymmetric centers in these carbohydrate substrates, the lack of correspondence in the stereochemical course of these reactions is not surprising. Since the cyanohydrin reaction in the sugar series has been reviewed,[5,6,11] the stereochemical course of the many known examples will not be summarized in detail here.

In 1908 Rosenthaler[12] showed that a crude enzyme preparation of emulsin from bitter almonds catalyzed the reaction of benzaldehyde and hydrogen cyanide to give, after hydrolysis, (*S*)-(+)-mandelic acid (9% e.e.). Subsequent work, which has been summarized,[13] extended this enzymatic asymmetric synthesis to many aldehydes and the results were interpreted in terms of simultaneous cyanohydrin reactions, one of which was enzyme-catalyzed and stereospecific and another which was nonenzymatic and nonstereospecific.

$$\underset{\mathbf{15}}{PhCHO} + HCN \xrightarrow{\text{Emulsin}} Ph\overset{*}{C}H(OH)C{\equiv}N \xrightarrow{H_2O} \underset{R(+)\text{-}\mathbf{16}}{H\text{—}C(COOH)(Ph)\text{—}OH} + \underset{S(-)\text{-}\mathbf{16}}{HO\text{—}C(COOH)(Ph)\text{—}H}$$

This enzyme-catalyzed cyanohydrin synthesis recently has been made a practical synthesis of optically active α-hydroxy nitriles.[14,15] The flavoprotein D-oxynitrilase was absorbed on "ECTEOLA"* cellulose ion exchange

[11] H. S. Isbell and H. L. Frush, *J. Org. Chem.*, **23**, 1309 (1958).
[12] L. Rosenthaler, *Biochem. Z.*, **14**, 238 (1908).
[13] L. Rosenthaler, *Fermentforsch.*, **5**, 334 (1922).
[14] W. Becker and E. Pfeil, *J. Am. Chem. Soc.*, **88**, 4299 (1966).
[15] W. Becker, H. Freund, and E. Pfeil, *Angew. Chem.*, **77**, 1139 (1965).

* This is an epichlorohydrin-triethanolamine modified cellulose; see E. A. Peterson and H. A. Sober, *Biochem. Prep.*, **8**, 43 (1961).

resin, and an aqueous methanol mixture of highly purified benzaldehyde and hydrogen cyanide was passed slowly through the column; upon evaporation of the effluent, there was obtained a 95% yield of mandelonitrile consisting of 97% of the *R*-(+) isomer and 3% of the *S*-(−) enantiomer. Kilogram quantities of the (+)-mandelonitrile can be made by the continuous operation of the column since the enzyme is very stable. Yields of 84–100% of optically active cyanohydrins from a variety of methyl, hydroxy, and methoxy substituted benzaldehydes as well as furfural, 2-thiophene aldehyde, pentanal, butanal, 2-methylpropanal, and crotonaldehyde were reported;[15] only the yields from ethanal and propanal were low: 28 and 45%, respectively.

After Rosenthaler demonstrated the enzymatic cyanohydrin reaction, Bredig and Fiske[16] found that optically active mandelonitrile was formed in low enantiomeric purity when the reaction was catalyzed with either quinine or quinidine and carried out in nonaqueous solution (chloroform or toluene). Because of the equilibrium nature of this reaction, the observed rotation of the reaction mixture first rose to a maximum and then fell. These experiments were successfully extended to other aldehydes.[17] It was also shown that a catalyst made by incorporating the diethylamino group into cotton or wool promoted an asymmetric synthesis,[18] but again the enantiomeric purity of the products was low. Extensive experiments have been conducted by Tsuboyama[19] on the benzaldehyde cyanohydrin reaction using a basic chiral catalyst made by polymerizing (*S*)-isobutylethyleneimine (**17**). A maximum asymmetric synthesis of the (*S*)-(−)-mandelonitrile (*ca.* 20% e.e.) was obtained when the catalyst was the benzene-soluble secondary polyamine, **18**, R = H. Treating this same polymer with toluene-2,4-diisocyanate

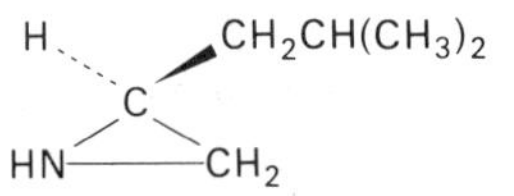

H, $CH_2CH(CH_3)_2$, C, HN, CH_2

17

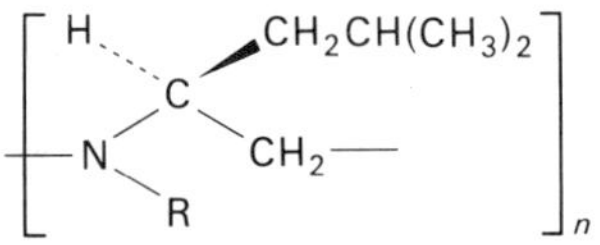

H, $CH_2CH(CH_3)_2$, C, N, CH_2—, R, *n*

18

gave an insoluble urethane cross-linked polymer which catalyzed the formation of the enantiomeric (*R*)-(+)-mandelonitrile (1–3% e.e.). Since the configuration at the chiral carbon center is unaffected by the cross-linking, the steric course of the reaction must be controlled by the overall conformation of the polymer and not by the configuration of the chiral carbon center

[16] G. Bredig and P. S. Fiske, *Biochem. Z.*, **46**, 7 (1912).

[17] G. Bredig and M. Minaeff, *Biochem. Z.*, **249**, 241 (1932).

[18] G. Bredig, F. Gerstner, and H. Lang, *Biochem. Z.*, **282**, 88 (1935).

[19] (a) S. Tsuboyama, *Bull. Chem. Soc. Jap.*, **35**, 1004 (1962); *Chem. Abst.*, **57**, 11096 (1962). (b) *ibid.*, **38**, 354 (1965); *Chem. Abst.*, **62**, 16100 (1965). (c) *ibid.*, **39**, 698 (1966); *Chem. Abst.*, **65**, 3710 (1966).

alone. Upon conversion to a tertiary amine, the same polymer (**18**, R = CH_3, C_2H_5, and $CH_2C_6H_5$) was a less stereoselective catalyst for the cyanohydrin reaction (0–2% asymmetric synthesis).

These results have been compared to those obtained with a diethylaminoethylcellulose catalyst (6% asymmetric synthesis), a diethylaminoethyl wool catalyst (2% asymmetric synthesis), and a cross-linked polystyrene to which triacetylglucosamine units had been attached (1–2% asymmetric synthesis). These reactions serve as interesting models for the enzyme-catalyzed cyanohydrin synthesis but fall far short with respect to extent of asymmetric synthesis and give very limited information concerning the structural features which control the stereoselectivity.

The most thorough stereochemical study of the asymmetric cyanohydrin reaction has been conducted by Prelog and Wilhelm[20] whose experiments were carried out in nonaqueous solvent (chloroform containing 1% ethanol). It had been shown previously[21] that the rate of the benzaldehyde-hydrogen cyanide reaction in nonaqueous solvent which was catalyzed by hydrocinchonidine was first order with respect to benzaldehyde, second order with respect to the alkaloid or other added base and practically independent of the hydrogen cyanide concentration. This was interpreted by Prelog and Wilhelm in terms of the following three steps, where B may be alkaloid or

$$\mathrm{HCN + B \xrightleftharpoons{\text{Fast}} (HB^+CN^-)} \tag{5}$$

$$\mathrm{R{-}CH{=}O + (HB^+CN^-) \xrightleftharpoons{\text{Fast}} (R{-}CH{=}O \cdots HB^+CN^-)} \tag{6}$$

$$\mathrm{(R{-}CH{=}O \cdots HB^+CN^-) + (HB^+CN^-) \xrightleftharpoons{\text{Slow}} R{-}\overset{\displaystyle OH}{\overset{|}{C}}HCN + (HB^+CN^-) + B} \tag{7}$$

other added base. The important difference in the reaction in chloroform versus aqueous solvent is that the base and hydrogen cyanide react to form a closely bound ion pair in the former (Eq. 5) but give independent hydrated ions in the latter. According to this scheme, the stereochemical course of the reaction is determined by the process represented by Eq. 7.

Because of the reversibility of the reaction, the true stereoselectivity must be determined by extrapolating to zero time. This was done for three examples and it was found that the percent stereoselectivities for the reaction of cinnamaldehyde and hydrogen cyanide in chloroform solution catalyzed by quinine, cinchonidine, or quinine bisiodomethylate in the presence of diethylamine were approximately 10, 2, and 25% respectively.

The last mentioned reaction is of particular interest. In the presence of quinine bisiodomethylate, cinnamaldehyde does not react with hydrogen

[20] V. Prelog and M. Wilhelm, *Helv. Chim. Acta*, **37**, 1634 (1954).

[21] H. Albers and E. Albers, *Z. Naturforsch.*, **9b**, 122, 133 (1954).

cyanide until a trace of diethylamine is added; reaction then proceeds with the highest stereoselectivity observed for any of the cinchona alkaloids tested. The greater the relative concentration of the nonchiral amine catalyst, the lower the stereoselectivity. With quinine alone, however, the stereoselectivity is essentially independent of the base concentration. These results clearly demonstrate the dual nature of the catalyst; one molecule reacts as an ammonium ion and one as a proton-donating base in the transition state.

Prelog and Wilhelm[20] determined the percents enantiomeric excess of the cyanohydrins produced by the reaction of hydrogen cyanide and cinnamaldehyde (represented in Fig. **4-3**, R = CH=CHPh) catalyzed by twenty-five different cinchona alkaloids of known configuration. In addition, the results from literature reports and original experiments using benzaldehyde (Fig. **4-3**, R = Ph) and twenty-nine cinchona bases were tabulated and compared.

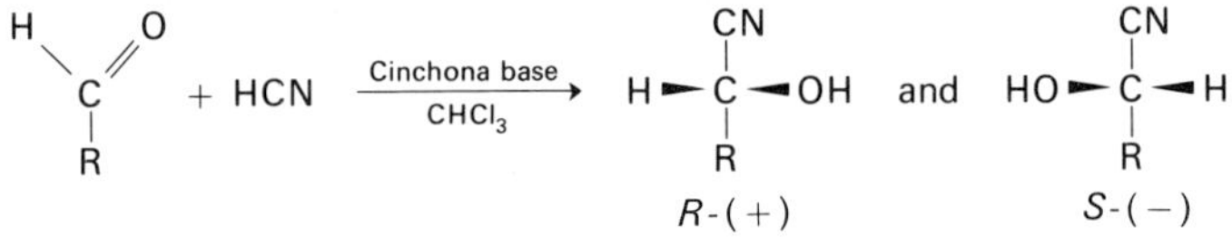

H(R)C=O + HCN —(Cinchona base, $CHCl_3$)→ H–C(CN)(R)–OH *R*-(+) and HO–C(CN)(R)–H *S*-(−)

Fig. 4-3. Asymmetric cyanohydrin reaction with benzaldehyde and cinnamaldehyde catalyzed by cinchona alkaloids; R = Ph and CH=CHPh.

The cinchona alkaloids represent four configurational types; formulas **19-22** illustrate these without any intention of showing the preferred conformations. The effect on stereoselectivity caused by variations in con-

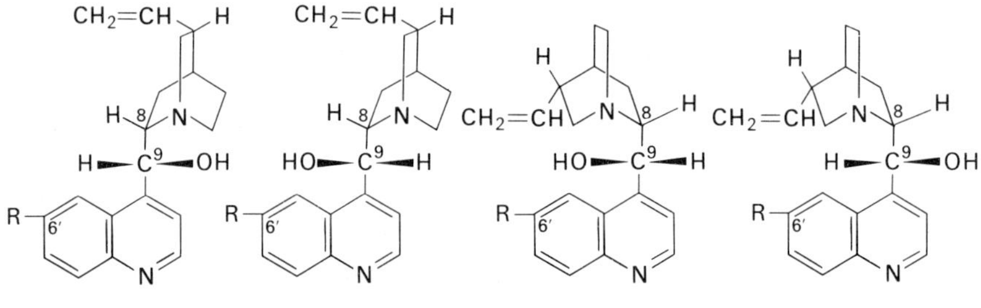

	19	**20**	**21**	**22**
R = OMe,	Quinine	9-Epiquinine	Quinidine	9-Epiquinidine
R = H,	Cinchonidine	9-Epicinchonidine	Cinchonine	9-Epicinchonine

figurations at chiral centers C-8 and C-9 have been studied. The complexity of the stereochemical control of this asymmetric synthesis is illustrated by the following. Quinine and cinchonidine, which differ only in the substitution of a methoxy group for hydrogen at position 6′ of the quinoline nucleus, catalyze the formation of enantiomeric (*R*)-(+)- and (*S*)-(−)-mandelonitrile (2.6 and 8.7% e.e., respectively), but these same two catalysts *both* catalyze the formation of (*R*)-(+)-cinnamaldehyde cyanohydrin (5.7 and 1.1% e.e.). This difference is rather surprising considering the distance between the 6′ substituent and the basic nitrogen atoms. The hydro derivatives of quinine, quinidine, cinchonine, or cinchonidine, in which the vinyl group is reduced, show about the same stereoselectivities as the parent alkaloids. The catalysts in which the asymmetric center at C-9 is destroyed by substituting hydrogen for hydroxyl have reduced stereoselectivities (0–2% e.e. in four different cases). The catalysts with chlorine instead of hydroxyl at C-9 generally showed lower stereoselectivities and often, but not invariably, produced the cyanohydrin of opposite configuration. Cinchona alkaloids* which have the *R*-configuration at C-9 (**19, 22**) gave (*R*)-cinnamaldehyde cyanohydrin in excess and alkaloids with the *S* configuration at C-9 (**20, 21**) gave (*S*)-cinnamaldehyde cyanohydrin. However, this regularity was *not* observed in benzaldehyde cyanohydrin syntheses.

Based on these observations and some assumptions concerning the conformation at C-9 in the cinchona alkaloids, Prelog and Wilhelm advanced a working hypothesis, namely that the stereochemical course of the reaction in nonaqueous solvent was determined principally by two aldehyde-ammonium ion "intermediates" as represented in **23** and **24**. They proposed that the cyanide ion–ammonium ion pair can attack each of these (Eq. 7) from one side preferentially because of the shielding of the quinoline nucleus, thus leading to two diastereomeric transition states which must then give enantiomers in unequal amounts. In order to explain the influence of C-6′ substituents, it was postulated that the methoxy group in the quinine series (**23** and **24**, R = OCH_3) was sterically important when dealing with

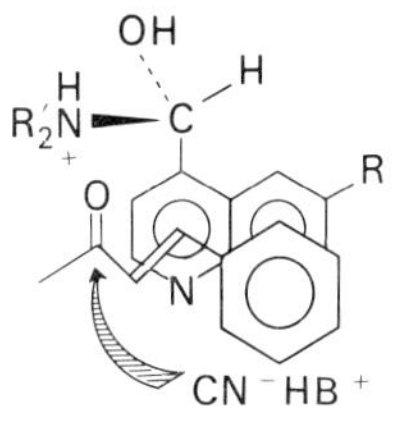

OH, H, $R'_2\overset{+}{N}H$—C, O, R, N, $CN^- HB^+$

23

OH, H, $R'_2\overset{+}{N}H$—C, O, R, N, $CN^- HB^+$

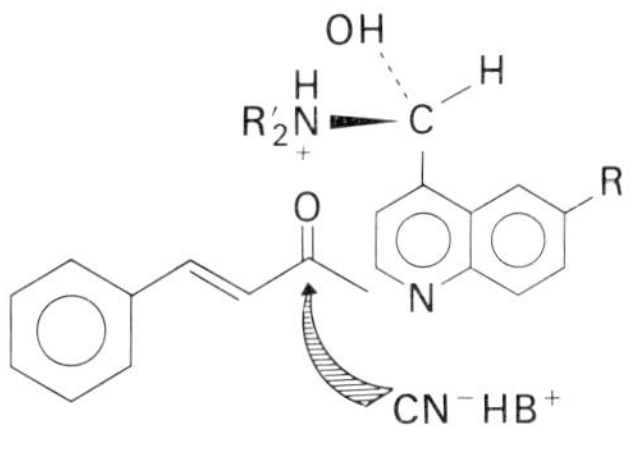

24

cinnamaldehyde as a substrate, but it was relatively unimportant when the substrate was benzaldehyde which would not "reach" to the 6′ position. Many of the experimental observations could be rationalized with these highly intuitive models.

* Two exceptions were tetrahydro derivatives in which the pyridine moiety of the quinoline was reduced and which therefore had an additional chiral center introduced at C-4′.

4-2. Aldol-Type Addition Reactions

There are numerous possibilities for asymmetric syntheses which could make use of aldol-type reactions. For example, according to the broad definition we have adopted, the reaction of an aldehyde with the C-17 acetyl group of progesterone to give an α-hydroxy ketone with a new chiral center constitutes an asymmetric synthesis. However, there are few examples in which the inducing chiral center can be separated conveniently from the product and thus the aldol reaction has not been studied extensively for the purpose of demonstrating asymmetric synthesis per se.

General mechanistic and stereochemical aspects of base-catalyzed aldol-type reactions have been reviewed[22] and we shall consider only those cases which have been designed specifically for elucidating the nature of the asymmetric synthesis process. Several formally related reactions are represented by the reaction sequence: **25** + **26** → **27**, in which R* is a chiral group such as (−)-menthyl. All of these asymmetric syntheses have been studied and

$$\underset{\mathbf{25}}{\mathrm{Ph{-}\overset{\overset{O}{\|}}{C}{-}H}} + \underset{\mathbf{26}}{\mathrm{H_2\overset{\overset{X}{|}}{C}{-}COOR^*}} \xrightarrow[2)\ H_2O]{1)\ Z} \underset{\mathbf{27}}{\mathrm{Ph{-}\overset{\overset{OH}{|}}{\underset{\underset{H}{|}}{C}}{-}CH_2COOH}} + \mathrm{R^*OH}$$

1, Malonic ester-type reaction: X = COOH, Z = pyridine
2, Aldol-type addition: X = H, Z = Et_2NMgBr
3, Reformatsky reaction: X = Br, Z = Zn
4, Darzens condensation: X = Cl, Z = $^{-}$O-*t*-Bu
5, For a Wittig reaction which is related, see Sec. **9-2.2**

the information relating to one system should be of value in interpreting the data from the others.

An isolated example of this type of reaction is the pyridine-catalyzed reaction of benzaldehyde (**25**) with (−)-menthyl hydrogen malonate (**28**) followed by decarboxylation and hydrolysis to give (−)-3-hydroxy-3-phenylpropanoic acid (**30**, 21% e.e., 55% crude yield).[23] In this sequence

$$\underset{\mathbf{25}}{\mathrm{PhCH{=}O}} + \underset{\mathbf{28}}{\mathrm{\overset{\overset{COOR^*}{|}}{\underset{\underset{COOH}{|}}{CH_2}}}} \xrightarrow{C_5H_5N} \underset{\mathbf{29}}{\mathrm{Ph\overset{\overset{OH}{|}}{C}H{-}\overset{\overset{COOR^*}{|}}{\underset{\underset{COOH}{|}}{C}}H}} \xrightarrow[+H_2O]{-CO_2} \underset{\mathbf{30}}{\mathrm{Ph\overset{\overset{OH}{|}}{C}H{-}\overset{\overset{COOH}{|}}{C}H_2}} + \mathrm{R^*OH}$$

R* = (−)-menthyl

[22] D. J. Cram, *Fundamentals of Carbanion Chemistry* (New York and London: Academic Press, 1965).

[23] E. B. Abbot, E. W. Christie, and A. McKenzie, *Ber.*, **71B**, 9 (1938).

there is the possibility for concentration of one isomer over the other via kinetic resolution by either incomplete decarboxylation or incomplete hydrolysis. Nevertheless this appears to be an authentic case of asymmetric synthesis. Recrystallization of the crude product (**30**), which removed some cinnamic acid and also may have concentrated the (−) isomer of **30**, gave (−)-3-hydroxy-3-phenylpropanoic acid (58% e.e.). Since other examples are lacking, it is not possible to draw any general conclusions; however, this would appear to be an interesting system for further study.

The highest stereoselectivity encountered in the aldol-type condensation is reported for the reaction of (−)-menthyl acetate and acetophenone (Table **4-1**, No. 1, 93% e.e.)[24] using the powerful condensing agent diethylaminomagnesium bromide, prepared by adding an equivalent amount of diethylamine to ethylmagnesium bromide. The stereoselectivities for all the examples using propiophenone were lower (Nos. 2–7, 20–70% e.e.).[25] In comparable cases, the extent of asymmetric synthesis is generally higher when the inducing chiral moiety is menthyl rather than bornyl.[26] In two exceptions to this generalization (Nos. 10–11 and 12–13) the reported extents of asymmetric syntheses are so low that the deviations are of doubtful meaning, but Nos. 18 and 19 constitute clear exceptions. In the benzoylformate-atrolactate asymmetric synthesis system, the (−)-menthyl and (−)-bornyl esters, which have a common chiral order of the R_S, R_M, and R_L groups at the carbinol centers (see **31** and **32**), induce the formation of the *same* enantiomer in excess. In the present system, (−)-menthyl and (+)-bornyl esters

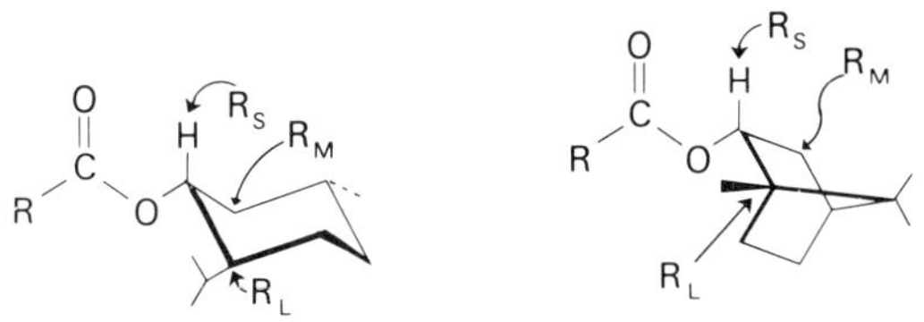

31 **32**

Fig. 4-4. Comparison of R_S, and R_M, and R_L in (−)-menthyl and (−)-bornyl esters. These structures are not intended to imply preferred ester conformations. Note that the chiral ester carbon in both **31** and **32** has the *R* configuration while the R_L carbon in both has the *S* configuration.

generally induce corresponding chirality in the product although this is not always so (Nos. 14–15 and 16–17). Thus the Prelog model, which so

[24] S. Mitsui, K. Konno, I. Onuma, and K. Shimizu, *J. Chem. Soc. Jap.*, **85**, 437 (1964); *Chem. Abst.*, **61**, 13167 (1964).

[25] S. Mitsui and Y. Kudo, *Tetrahedron*, **23**, 4271 (1967).

[26] K. Sisido, K. Kumazawa, and H. Nozaki, *J. Am. Chem. Soc.*, **82**, 125 (1960).

successfully correlates the configuration of the products in the atrolactic asymmetric synthesis, cannot be applied directly to these aldol-type condensations. It was at first suggested[27] that in these aldol-type asymmetric syntheses it was the chiral center in the position *beta* to the carbinol oxygen (R_L in **31** and **32**) which was stereochemically controlling. This has been refuted by the results of experiments (Nos. 4–7) in which only a single α-chiral center is present in the alcohol moiety of the ester and yet the extents of asymmetric synthesis in these cases are still substantial. A more likely explanation[25] is that the conformations of the transition states of these

TABLE 4-1

Asymmetric Aldol-Type Addition Reactions[a,b]

$$\text{Ph(R)C=O} + CH_3\overset{O}{\overset{\|}{C}}-OR^* \xrightarrow{Et_2NMgBr} \xrightarrow{H_2O} \text{Ph(R)C(OH)}-CH_2\overset{O}{\overset{\|}{C}}-OH + R^*OH$$

No.	R	R*	Product		
			Config.	Asym. syn. % e.e. or $[\alpha]_D$	Yield %
1	Me	(−)-Menthyl	*S*	93	53
2	Et	(−)-Menthyl	*S*	70[c]	37
3	Et	(+)-Bornyl	*S*	36[c]	35
4	Et	(+)-1-Phenylethyl	*S*	39[c]	50
5	Et	(−)-1-Phenylethyl	*R*	46[c]	36
6	Et	(+)-1-(1′-Naphthylethyl)	*S*	20[c]	37
7	Et	(−)-1-(1′-Naphthylethyl)	*R*	20[c]	18
8	*p*-MeO-Ph	(−)-Menthyl	(+)	16	72
9	*p*-MeO-Ph	(+)-Bornyl	(+)	7	72
10	*p*-Tolyl	(−)-Menthyl	(+)	$[\alpha]_D$ +0.3°	69
11	*p*-Tolyl	(+)-Bornyl	(+)	$[\alpha]_D$ +1.0°	70
12	*p*-ClC_6H_4	(−)-Menthyl	(+)	$[\alpha]_D$ +0.0°	72
13	*p*-ClC_6H_4	(+)-Bornyl	(+)	$[\alpha]_D$ +0.2°	75
14	*o*-Tolyl	(−)-Menthyl	(−)	48	72
15	*o*-Tolyl	(+)-Bornyl	(+)	4	73
16	2,4-$Me_2C_6H_3$	(−)-Menthyl	(−)	$[\alpha]_D$ −74°	74
17	2,4-$Me_2C_6H_3$	(+)-Bornyl	(+)	$[\alpha]_D$ +7.7°	72
18	α-Naphthyl	(−)-Menthyl	(+)	21	79
19	α-Naphthyl	(+)-Bornyl	(+)	30	82

[a] In all experiments, a toluene solution of the ketone and chiral acetate was added to an ether solution of the catalyst.

[b] Experiments No. 1 at −15 to −20°, Reference 24; Nos. 2–7 at −5 to −10°, Reference 25; Nos. 8–19 at −5°, Reference 26.

[c] Isolated as the methyl ester by distillation.

[27] K. Sisido, O. Nakanisi, and H. Nozaki, *J. Org. Chem.*, **26**, 4878 (1961).

reactions are such that the stereochemical control is exercised by both the *alpha* and *beta* chiral centers in the menthyl and bornyl groups. In the menthyl group these two chiral centers may work to reinforce each other, while in the bornyl group the two centers may work in opposition, with the overall effect that the extent of asymmetric synthesis is reduced or in some cases the sense of the asymmetric synthesis may even be reversed.

It is mechanistically significant that an asymmetric synthesis of 16% was obtained when (−)-menthyl acetate was condensed with *p*-methoxybenzophenone (Table **4-1**, No. 8) in contrast to the undetectable asymmetric synthesis observed in the reduction of this same ketone by a chiral Grignard reducing agent[28a] (Sec. **5-2**) or chiral LAH-quinine complex[28b] (Table **5-12**). The difference between the *p*-methoxyphenyl and phenyl groups resides in the *para* positions remote from the developing chiral center. If a purely steric interpretation for the control of asymmetric synthesis in these cases is assumed, then there must be some interactions between the menthyl group of the ester and the *para* positions of the *p*-methoxyphenyl and phenyl groups. Inspection of molecular models tends to discount such interactions, although they cannot be unequivocally eliminated from consideration. Alternatively, the cause of the stereoselectivity may reside in the different electronic interactions of the phenyl versus *p*-methoxyphenyl groups. It should be quite possible to collect the relevant information for a Hammett *sigma-rho* correlation which should allow a decision between these alternatives.

The asymmetric synthesis achieved when *o*-tolyl phenyl ketone is the substrate (Table **4-1**, No. 14), is 48%, indicating a much higher degree of stereoselectivity when the substituent is closer to the site of the developing chiral center. This result is in accord with a primarily steric explanation for the control of these asymmetric syntheses. It would be of interest to obtain results using other substrates such as mesityl phenyl ketone and 2,4,6-trimethylbenzaldehyde. Further data must be collected concerning the stereochemical effect of many other substrates and chiral inducing agents before the factors controlling this type of asymmetric synthesis can be properly elucidated. Some comparisons between the results obtained in this system (Table **4-1**) and those obtained in the asymmetric Reformatsky reaction (Table **4-2**) are made in Sec. **4-3**.

4-3. The Reformatsky Reaction

The older chemical literature concerned with the Reformatsky reaction has been reviewed[29] from a preparative standpoint but without regard to

[28] (a) H. S. Mosher and E. D. Parker, *J. Am. Chem. Soc.*, **78**, 4081 (1956). (b) O. Červinka and O. Bělovský, *Coll. Czech. Chem. Commun.*, **32**, 3897 (1967).

[29] R. L. Shriner, *Organic Reactions*, **1**, 1 (1942).

stereochemistry. In a formal sense, the Reformatsky reaction may be compared with the Grignard reaction in which the organozinc reagent prepared from an α-haloester replaces the Grignard reagent. In both reactions, a

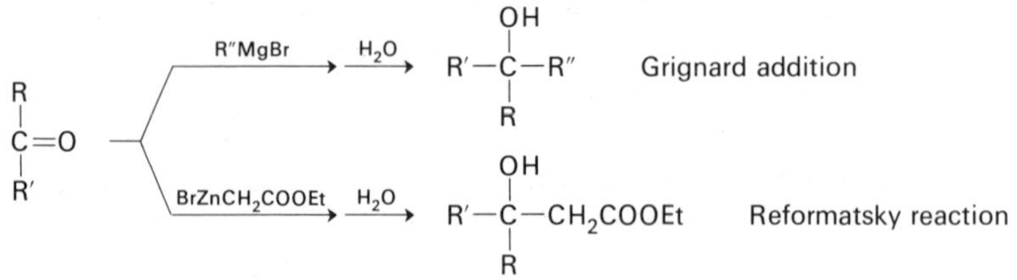

$$R(R')C{=}O \xrightarrow{R''MgBr} \xrightarrow{H_2O} R'{-}C(OH)(R){-}R'' \quad \text{Grignard addition}$$

$$R(R')C{=}O \xrightarrow{BrZnCH_2COOEt} \xrightarrow{H_2O} R'{-}C(OH)(R){-}CH_2COOEt \quad \text{Reformatsky reaction}$$

Fig. 4-5. Formal comparison of Grignard addition and Reformatsky reaction.

new chiral center is created if the R groups are different, as represented in Fig. **4-5**. Two distinct situations can lead to asymmetric synthesis. If the carbonyl compound is chiral, there will be an asymmetric synthesis of the type covered by Cram's rule. Although there may be a few scattered examples of such cases, there have been no systematic studies of this type (the Reformatsky reagent does not appear in Tables **3-1** to **3-7**). If the Reformatsky reagent is chiral, there will be an asymmetric synthesis at the newly formed chiral center; most of the examples are of the type in which the alcohol moiety of the ester is chiral.

Although the Grignard and Reformatsky reactions appear to be similar in a formal sense, the Grignard reagent has a true organometallic bond, while the Reformatsky reagent is best represented as an enolate ion, **33**, or as a complex in which the halogen (X) and zinc atoms are held in a cyclic structure, **34**. The Reformatsky reaction is therefore more closely related to the diethyl-

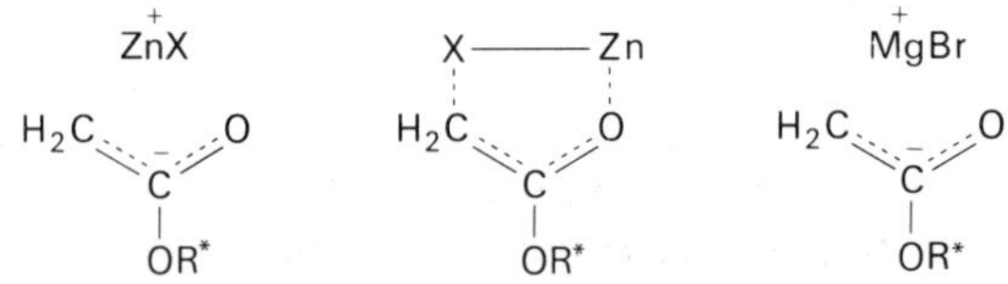

$\overset{+}{ZnX}$ — $H_2C{\cdots}\bar{C}(OR^*){\cdots}O$ **33**

$X{-}Zn$ — $H_2C{\cdots}C(OR^*){\cdots}O$ **34**

$\overset{+}{MgBr}$ — $H_2C{\cdots}\bar{C}(OR^*){\cdots}O$ **35**

aminomagnesium bromide-catalyzed aldol-type condensation (Sec. **4-2**), in which the reagent can be represented as **35**, than it is to the Grignard reaction. This point cannot be considered settled, however, in view of the experiments[30] using optically active methyl α-bromopropionate (**36**) in the reaction with either benzaldehyde (**37A**) or acetophenone (**37B**). Four

[30] J. Canceill, J. Gabard, and J. Jacques, *Bull. Soc. Chim. Fr.*, 231 (1968).

isomers comprising two diastereomeric sets of enantiomers, **38** and **39**, are formed in each reaction. In the one employing benzaldehyde, the ratio of **38A** to **39A** was found to be 37:63; both products exhibited optical activity. In the predominant diastereomeric product (**39A**), the *S*,*S* enantiomer was found to be present in 2.2% excess over the *R*,*R* enantiomer. Thus while the major course of the reaction was racemization, a small excess of inversion had occurred. However, in the reaction of the same optically active reagent

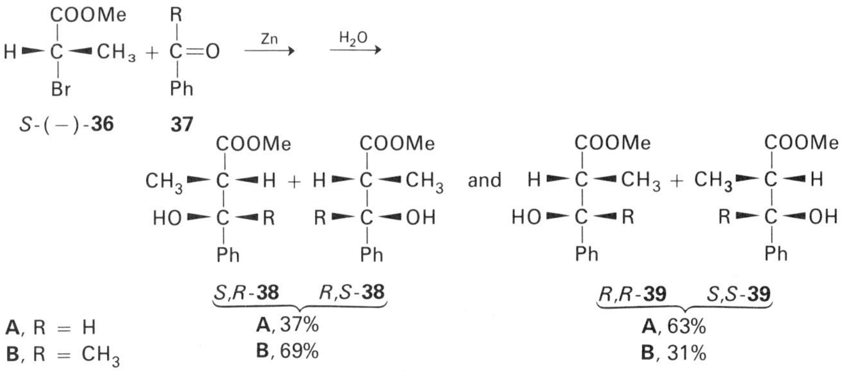

COOMe R
H—C—CH_3 + C=O $\xrightarrow{Zn}$ $\xrightarrow{H_2O}$
Br Ph

S-(−)-**36** **37**

COOMe COOMe COOMe COOMe
CH_3—C—H + H—C—CH_3 and H—C—CH_3 + CH_3—C—H
HO—C—R R—C—OH HO—C—R R—C—OH
Ph Ph Ph Ph

S,*R*-**38** *R*,*S*-**38** *R*,*R*-**39** *S*,*S*-**39**

A, R = H **A**, 37% **A**, 63%
B, R = CH_3 **B**, 69% **B**, 31%

Fig. 4-6. Comparison of the Reformatsky reactions of benzaldehyde and acetophenone with optically active methyl α-bromopropionate.

with acetophenone (**37B**), the diastereomeric pairs **38B** and **39B** were produced in 69:31 ratio and these products showed no optical activity. The presumption is that the organozinc reagent is first formed with retention of configuration but that this goes over to the symmetrical enolate anion which cannot lead to optically active products.*

Using racemic ethyl α-bromopropionate and four different ketone substrates, approximately the same "*threo*:*erythro*" isomer ratios were obtained,[31,32] namely, acetophenone, 70:30; 2-acetylnaphthalene, 70:30; 2-propionylnaphthalene, 75:25; 2-acetyl-6-methoxynaphthalene, 75:25.

In addition to the stereoselective formation of the *R*,*R*-*S*,*S* enantiomeric pair over the *R*,*S*-*S*,*R* pair or vice versa (Fig. **4-6**), a distinct type of stereoselectivity which fits the Marckwald classical asymmetric synthesis scheme is illustrated by the reaction of (−)-menthyl α-bromoacetate with benzaldehyde

* R. Rétey (private communication; see also footnote in Reference 30) has found that the reagent made from optically active methyl α-bromopropionate gave inactive 2-methyl-3-hydroxypropanoic acid upon treatment with formaldehyde.

[31] M. Mousseron, M. Mousseron, J. Neyrolles, and Y. Beziat, *Bull. Soc. Chim. Fr.*, 1483 (1963).

[32] Y. Beziat and M. Mousseron-Canet, *Bull. Soc. Chim. Fr.*, 1187 (1968).

or acetophenone (Fig. **4-7**). This asymmetric synthesis was discovered by Reid and Turner[33,34] and extended by Palmer and Reid.[35,36] When

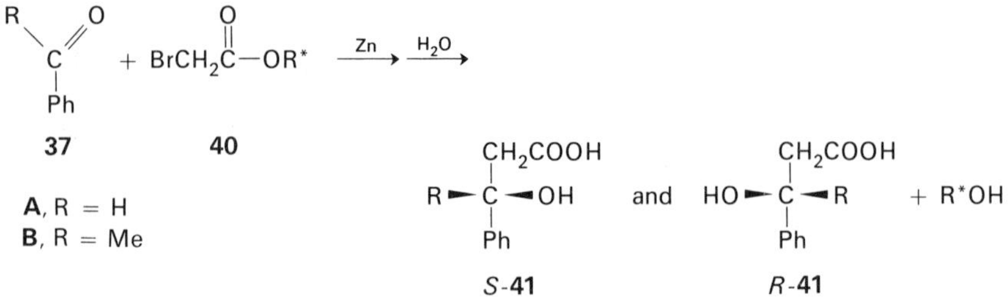

$$\underset{\substack{\mathbf{37}\\ \mathbf{A},\ R = H\\ \mathbf{B},\ R = Me}}{R(Ph)C{=}O} + \underset{\mathbf{40}}{BrCH_2C(=O){-}OR^*} \xrightarrow{Zn} \xrightarrow{H_2O} \underset{S\text{-}\mathbf{41}}{R{-}C(CH_2COOH)(OH)(Ph)} \text{ and } \underset{R\text{-}\mathbf{41}}{HO{-}C(CH_2COOH)(R)(Ph)} + R^*OH$$

Fig. 4-7. Asymmetric Reformatsky synthesis using (−)-menthyl α-bromoacetate; i.e. R* = (−)-menthyl.

benzaldehyde (**37A**) is used as the carbonyl compound, (*S*)-(−)-3-hydroxy-3-phenylpropanoic acid (*S*-(−)-**41A**) predominates over the *R*-(+) isomer by approximately 15%. When acetophenone (**37B**) is treated in the same fashion, the (*S*)-(+)-3-hydroxy-3-phenylbutanoic acid (*S*-(+)-**41B**) predominates over the *R*-(−)-isomer by approximately 30%. The data for reactions with other chiral α-bromoacetates are collected in Table **4-2**.

Despite the distance of three atoms separating the established chiral center in the alcohol moiety and the developing chiral center on the β-carbon atom, stereoselectivities are substantial. They are generally higher in the reactions of acetophenone (Nos. 10–20) than those of benzaldehyde (Nos. 1–9). The maximum percent asymmetric synthesis was 61 when a cadmium reagent, generated with di-*n*-propyl cadmium, was employed instead of the zinc reagent (No. 14).

It is of historical interest that Reid and Turner[33] pointed out in 1949 that an "asymmetric induction" in the Marckwald sense [caused by an asymmetric force "inducing" dissymmetry in the carbonyl group (Sec. **1-1**)] cannot operate here since the carbonyl group and inducing center are in different molecules.

It has been proposed[37] that the asymmetric synthesis in the Reformatsky reaction is controlled on the surface of the zinc metal as the reagent is generated. Palmer's and Reid's results[35,36] have effectively contradicted this proposal by showing that use of a homogeneous Reformatsky reagent prepared by using iodine-activated zinc followed by treatment with benzaldehyde

[33] J. A. Reid and E. E. Turner, *J. Chem. Soc.*, 3365 (1949).
[34] J. A. Reid and E. E. Turner, *J. Chem. Soc.*, 3694 (1950).
[35] M. H. Palmer and J. A. Reid, *J. Chem. Soc.*, 931 (1960).
[36] M. H. Palmer and J. A. Reid, *J. Chem. Soc.*, 1762 (1962).
[37] C. L. Arcus and D. G. Smyth, *J. Chem. Soc.*, 34 (1955).

(No. 3) gives approximately the same extent of asymmetric synthesis as observed in the classical "one stage" procedure using metallic zinc. In the case of acetophenone (No. 13) the homogeneous reaction gives a higher asymmetric synthesis than the heterogeneous reaction.

The chiral reagents in the Reformatsky reaction and the aldol-type condensation with esters (Sec. **4-2**) are probably closely related (cf. **33–35**); therefore, the stereoselectivities observed for analogous reactions should be similar. A comparison of the data in Tables **4-1** and **4-2** shows that, as far as the examples overlap, the same isomer predominates in both cases (compare Table **4-1**, No. 1 with Table **4-2**, Nos. 11–16). Thus, in the Reformatsky reaction, (−)-menthyl and (−)-bornyl acetates, which have a common chiral order of R_S, R_M, and R_L at the carbinol carbon (Fig. **4-4**), give enantiomeric products; in the aldol reaction, (−)-menthyl and (+)-bornyl acetates give the *same* enantiomer in excess.

TABLE 4-2

Asymmetric Reformatsky Reaction with Optically Active Bromoacetate Esters[a]

$$\underset{R'}{\overset{R}{\Big\rangle}}C{=}O + BrCH_2COOR^* \longrightarrow R{-}\underset{R'}{\overset{OH}{\overset{|}{\underset{|}{C}}}}{-}CH_2COOH + R^*OH$$

No.	R	R′	R*	β-Hydroxy Acid: Stereo-selectivity[b] % e.e.	Config.	Yield %	Conditions	Ref.
1	H	Ph	(−)-Menthyl	15	*S*	74	—	36
2	H	Ph	(−)-Menthyl	17	*S*	64	$ZnBr_2$	36
3	H	Ph	(−)-Menthyl	18	*S*	40	Activated Zn	36
4	H	Ph	(−)-Menthyl	25	*S*	63	Homogeneous	36
5	H	Ph	(−)-Menthyl	30	*S*	54	Zn(*n*-Pr)$_2$	36
6	H	Ph	(−)-Menthyl	18	*S*	79	Iodoacetate	36
7	H	Ph	(−)-Bornyl	15	*R*	76	—	36
8	H	Ph	(−)-α-Fenchyl	16	*R*	88	—	36
9	H	Ph	(+)-*s*-Octyl[c]	2	*S*	84	—	36
10	CH_3	Ph	(−)-Menthyl	29[d]	*S*	64	—	33, 35
11	CH_3	Ph	(−)-Menthyl	34	*S*	64	Benzene, ether	35
12	CH_3	Ph	(−)-Menthyl	9	*S*	58	Toluene	35
13	CH_3	Ph	(−)-Menthyl	34	*S*	37	Homogeneous	35
14	CH_3	Ph	(−)-Menthyl	61	*S*	68	Cd(*n*-Pr)$_2$	35
15	CH_3	Ph	(−)-Menthyl	44	*S*	66	Zn(*n*-Pr)$_2$	35
16	CH_3	Ph	(−)-Menthyl	31	*S*	63	$MgI_2 \cdot Mg$	35
17	CH_3	Ph	(−)-Bornyl	12	*R*	79	—	35

TABLE 4-2—continued

Asymmetric Reformatsky Reaction with Optically Active Bromoacetate Esters[a]

				β-Hydroxy Acid				
No.	R	R′	R*	Stereo-selectivity[b] %e.e.	Config.	Yield %	Conditions	Ref.
18	CH_3	Ph	(−)-Fenchyl	7	*R*	56	—	35
19	CH_3	Ph	(+)-*s*-Butyl[e]	5	*S*	82	—	35
20	CH_3	Ph	(+)-*s*-Octyl[c,f]	0	—	80	—	35
21	H	c-C_6H_{11}[g]	(−)-Menthyl	—[h]	—	69	—	36
22	H	n-C_3H_7	(−)-Menthyl	—[i]	—	58	—	36
23	H	n-C_4H_9	(−)-Menthyl	—[i]	—	71	—	36
24	H	n-C_6H_{13}	(−)-Menthyl	—[i]	—	84	—	36
25	H	n-C_8H_{17}	(−)-Menthyl	—[i]	—	75	—	36

[a] The reactions were carried out in benzene solvent under classical conditions.[29]

[b] Based on $[\alpha]_D^{25}$ −19.0° (EtOH) for (*S*)-(−)-3-phenyl-3-hydroxypropanoic acid and $[\alpha]_D^{25}$ +8.9° (EtOH) for (*S*)-(+)-3-phenyl-3-hydroxybutanoic acid.[24,34]

[c] *s*-Octyl represents the 1-methylheptyl group.

[d] Reaction mixture refluxed for 1 hr. When it was refluxed for 20 hr, the yield dropped to 55 % and the stereoselectivity to 26 %. The latter was unchanged by the addition of $CoCl_2$ or $CdBr_2$ but was lowered by the addition of $MgBr_2$ (23 %) or $ZnBr_2$ (25 %).[33]

[e] Di-*n*-propylzinc in homogeneous ether solution gave a 77 % yield of *S*-(+) isomer which was 5 % enantiomerically pure.

[f] Also no detectable asymmetric synthesis when di-*n*-propylcadmium was used as the condensing agent.

[g] c-C_6H_{11} represents the cyclohexyl group.

[h] Specific rotation $[\alpha]_D$ − 1.20° (c 5.7, EtOH). Stereoselectivity unknown since this compound has not been resolved.

[i] The observed rotations were 0.05° or less but since these products have not been resolved, the percent asymmetric synthesis cannot be calculated.

It is quite clear that the Prelog generalization developed for the atrolactic acid asymmetric synthesis (Sec. **2-2.1**) cannot be applied directly to these Reformatsky asymmetric syntheses. Although both (−)-menthyl and (−)-bornyl esters induce asymmetric synthesis of the same order of magnitude in the atrolactic and Reformatsky reactions, in the former both esters lead to the formation of (*R*)-(−)-atrolactic acid in excess while in the latter they give (*R*)-(−)- and (*S*)-(+)-3-hydroxy-3-phenylpropanoic acids, respectively. In this regard the asymmetric Reformatsky reaction resembles the asymmetric aldol condensation which shows a similar stereochemical correlation pattern (Sec. **4-2**).

Zimmerman and Traxler[38] proposed a cyclic mechanism to account for the 24:76 "*erythro*:*threo*" ratio of diastereomers from the Ivanov reaction

[38] H. E. Zimmerman and M. D. Traxler, *J. Am. Chem. Soc.*, **79**, 1920 (1957).

of the magnesium enolate of phenylacetic acid with benzaldehyde. This mechanism was applied[31,32] to the reaction of the Reformatsky reagent with aldehydes, ketones, imines, and nitriles as shown in Fig. **4-8** for the reaction of α-bromophenylacetic ester and acetophenone. In the transition state represented in **43** there is a confrontation of the methyl and phenyl groups

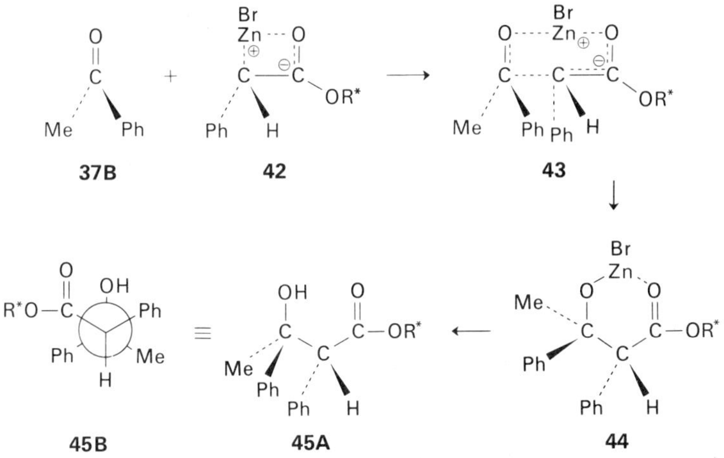

Fig. 4-8. Cyclic mechanism for Reformatsky reaction on acetophenone. Only one diastereomeric transition state and product are shown.

of the acetophenone with the hydrogen and phenyl groups of the Reformatsky reagent. On the basis of this model one would expect an excess of the *R,S* + *S,R* diastereomeric pair, rather than the *R,R* + *S,S* pair. This is what is observed with acetophenone, but Jacques and coworkers[30] have shown that the opposite is observed when benzaldehyde is the substrate (Fig. **4-6**). Furthermore, according to this explanation, the stereoselectivity should change in the reactions of α-alkyl substituted α-bromopropionates as the bulk of the α-alkyl group is changed in going from α-methyl to α-*t*-butyl, while in fact there is only a small increase in the "*threo*:*erythro*" ratio, from 63:37 to 69:31. The steric confrontation of the various substituents will not be as direct as the planar representation **43** would make it appear. The conformation of the transition state will adjust itself to minimize the various interactions. In a chair conformation the reagent could attack the carbonyl carbon at somewhat less than a 90° angle and with rotation about the developing C—C bond, could minimize steric interactions as shown in **46**.[38] However, the factors which would favor **46** over **47**, which are isomeric at the developing α- and β-chiral centers, are difficult to assess. In the face of so many uncertainties it is unwise to speculate further at the present time concerning the steric and electronic interactions in the transition state which control the steric course of these asymmetric Reformatsky reactions.

[Developing chiral centers]

[Established chiral moiety]

46 **47**

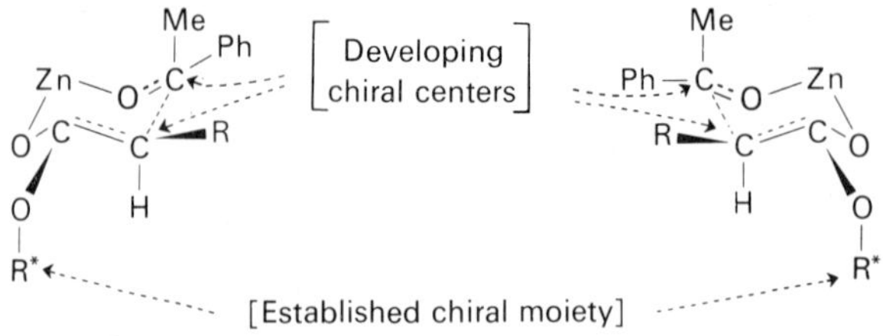

4-4. The Darzens Reaction

As indicated (Sec. **4-2**), the Darzens reaction (glycidic ester condensation) bears a certain resemblance to the aldol-type addition and Reformatsky reactions and it is appropriately considered in close connection with these. The synthetic aspects of the Darzens condensation have been reviewed[39] but without reference to the stereochemistry involved. Although the published work on the mechanism of the reaction is not extensive, the classical stereochemical course of the reaction is fairly well understood,[40,41] based primarily upon the one study which has been designed specifically to investigate the asymmetric synthesis aspects of this reaction.[27]

Acetophenone (**37B**) and (−)-menthyl chloroacetate (**48**) react on treatment with potassium *t*-butoxide to give the glycidic ester (**50**) which is not isolated but is reduced with lithium aluminum hydride to give (*S*)-(−)-3-hydroxy-3-phenyl-1-butanol (**51**) (35% yield, 14–15% e.e.) which is configurationally related to (*S*)-(+)-3-hydroxy-3-phenylbutanoic acid (**41B**).

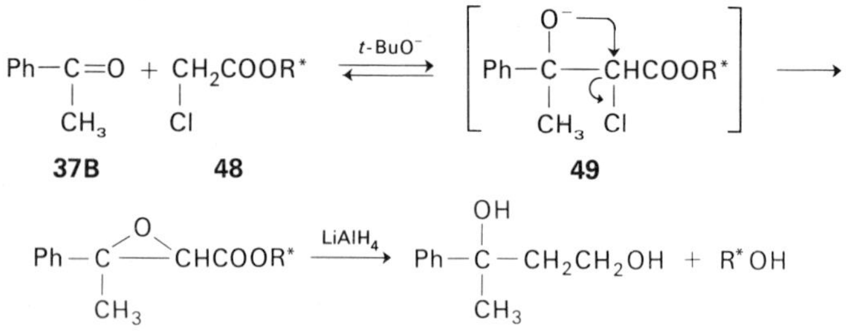

$$\underset{\mathbf{37B}}{Ph{-}C(CH_3){=}O} + \underset{\mathbf{48}}{ClCH_2COOR^*} \underset{}{\overset{t\text{-}BuO^-}{\rightleftharpoons}} \left[\underset{\mathbf{49}}{Ph{-}C(CH_3)(O^-){-}CHCl{-}COOR^*}\right] \longrightarrow$$

$$\underset{\mathbf{50}}{Ph{-}C(CH_3)\overset{O}{\frown}CHCOOR^*} \xrightarrow{LiAlH_4} \underset{S\text{-}(-)\text{-}\mathbf{51}}{Ph{-}C(OH)(CH_3){-}CH_2CH_2OH} + R^*OH$$

[39] M. S. Newman, *Org. Reactions*, **5**, 413 (1949).

[40] H. Dahn and L. Loewe, *Chimia*, **11**, 98 (1957).

[41] H. E. Zimmerman and L. Ahramjian, *J. Am. Chem. Soc.*, **82**, 5459 (1960).

In the same reaction sequence, (+)-bornyl chloroacetate also gives (*S*)-(−)-**51** (33% yield, 4–5% e.e.). Since (−)-menthol and (+)-borneol have opposite chiral orders of R_S, R_M, and R_L around the carbinol carbon, the stereochemistry of this reaction is not compatible with a Prelog-type model but follows the same pattern as the Reformatsky- (Sec. **4-3**) and aldol-type reactions (Sec. **4-2**), at least as far as these two experiments are concerned.

The diastereomeric intermediates represented by **49** are formed rapidly and reversibly;[40,41] the rate-determining step, and therefore the one which controls the asymmetric synthesis, is the conversion of **49** to **50**. The isomer composition of epoxide **50** was not investigated, but from other studies it seems likely that it is mainly a mixture of the *R,S* and *S,R* diastereomers. Which diastereomeric pair predominates has been discussed in terms of "overlap control"[41] of the transition state for the conversion of **49** into **50**, but whether the *R,S* enantiomer predominates over the *S,R* isomer is determined by the chiral R* group. Transition states **49A** and **49B** may be taken as representations of two low energy pathways, **49A** leading to the *cis*-epoxide and **49B** to the *trans*. Whether the reaction leading to *cis* is faster or slower than that leading to *trans* will be determined by the difference in the sum of the interaction energies of (Ph ↔ COOR* + CH_3 ↔ H) in **49A** versus (CH_3 ↔ COOR* + Ph ↔ H) in **49B**. It is intuitively reasonable to expect that the latter should be lower and that the *trans* phenyl-carboxyl product should predominate. Another diastereomeric transition state, **49C**, will also lead to the *trans*-epoxide but with the opposite configurations at the α- and β-carbon atoms. Which of these two, **49B** or **49C**, will have the lower energy depends upon the interactions of the chiral R* group with the rest of the molecule. The (−)-menthyl group occupies mirror-image environments in **49B** and **49C** and the energies of these competing transition state

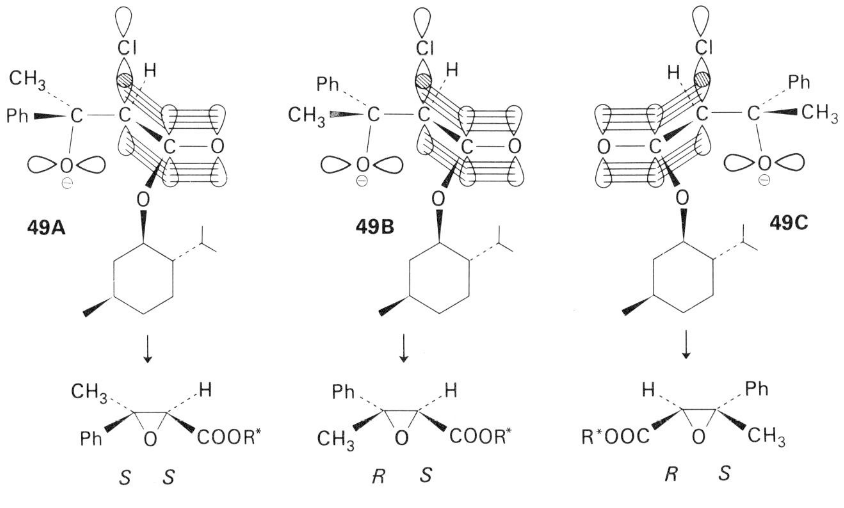

49A → CH3 / H / Ph / O / COOR* — *S S*

49B → Ph / H / CH3 / O / COOR* — *R S*

49C → H / Ph / R*OOC / O / CH3 — *R S*

models will not be the same. From the observed stereochemistry of the product one can deduce that if this is a correct approximation for the competing transition states, **49B** is preferred.

Based on such a picture, one can see that the chirality at positions other than that of the carbinol carbon of the ester should exert a significant influence on the asymmetric synthesis. Therefore, it is perhaps not surprising that (−)-menthol and (+)-borneol induce the same chirality even though they have opposite chiral orders of R_S, R_M, and R_L at the carbinol carbon atom.

4-5. Intramolecular Oxidation-Reduction

The transformation of α-keto aldehydes (**52**) to α-hydroxy acids (**53**) can be carried out both enzymatically and nonenzymatically. This has often been referred to as an internal Cannizzaro reaction, but it also can be considered a variant of the benzilic acid rearrangement[42] in which a hydrogen

$$R{-}\overset{O}{\overset{\|}{C}}{-}\overset{O}{\overset{\|}{C}}{-}H + H_2O \longrightarrow R{-}\underset{H}{\overset{OH}{C}}{-}COOH \text{ and } R{-}\underset{OH}{\overset{H}{C}}{-}COOH$$

52 *R*-**53** *S*-**53**

A, R = CH_3
B, R = Ph
C, R = $HOOCCH_2CH_2$
D, R = 2-thienyl (thiophene ring, S)
E, R = CH_3–C_6H_4– (p-tolyl)

migrates instead of an aryl group. The enzymatic transformation was discovered in 1913 and was explored by Neuberg and co-workers[43] and others over the following 20 years. The results of experiments involving various substrates as well as numerous microbiological systems and plant and animal enzymatic materials have been reviewed[43,44] and it would not be profitable to summarize in any detail the extensive older literature here. A wide variety of microbiological fermentations using bacteria, yeast, cell-free extracts; animal tissue such as muscle, liver, or nerve; vegetable materials such as soybean meal, and extracts of whole green plants were found to catalyze

[42] W. E. Knox in *The Enzymes*, eds. P. D. Boyer, H. Lardy, and K. Myrbäck (New York and London: Academic Press, 1960), Vol. II, pp. 271–281.

[43] (a) C. Neuberg and E. Simon, *Abderhaldens Handbuch, der biologische Arbeitsmethoden*, Abteilung IV, Teil 2, 2225–2248 (1935). (b) C. Neuberg, *Adv. Carbohydr. Chem.*, **4**, 75 (1949).

[44] P. D. Ritchie, *Asymmetric Synthesis and Asymmetric Induction* (London: Oxford Univ. Press, 1933), pp. 38–39.

the conversion of methylglyoxal (**52A**) and phenylglyoxal (**52B**) to lactic and mandelic acids, respectively. Depending upon the system employed, either enantiomer may predominate and the excess of one enantiomer may vary from essentially 0 to 100%, although the results were often in the 70 to 90% range. In most cases the lactic acid was isolated by precipitation and crystallization of the zinc salt, while the mandelic acid was also purified by crystallization. Since fractionation of isomers may have taken place during the crystallizations, the percent asymmetric syntheses based upon rotations of crystallized materials are of very limited value.

In addition to the two most common substrates, methylglyoxal and phenylglyoxal (**52A** and **52B**), β-carboxyethylglyoxal (**52C**),[45,46] α-thienylglyoxal,[47] and *p*-tolylglyoxal[48] have been used. Since the latter two are unnatural substrates, it is evident that the enzyme system involved is not very specific and it is presumed that almost any substituted glyoxal would be converted to the corresponding optically active α-hydroxy acid (for instance, by actively fermenting yeast).

It is now recognized that two enzymes, glyoxalase I and glyoxalase II are responsible for these transformations. Furthermore glutathione acts as a coenzyme in this system. It has also been shown by means of isotope experiments that both the enzymatic[49,50] and nonenzymatic[51–53] internal Cannizzaro reactions take place without incorporating hydrogen from the solvent into the *alpha* position of the product. It is therefore established that this rearrangement in both enzymatic and nonenzymatic processes involves an internal 1,2-hydrogen shift and that enolic intermediates (which would lead to deuterium or tritium incorporation) are not involved. Based on this evidence, the enzymatic process can be formulated to involve an initial reaction of the substituted glyoxal with reduced glutathione (GSH) to give the thiohemiacetal **54**. This intermediate (**54**) undergoes a 1,2-hydrogen shift under the influence of glyoxalase I to give the α-hydroxythioester **55**, which is hydrolyzed in the presence of the thioesterase glyoxalase II to give the α-hydroxy acid **53** and reduced glutathione. By controlling the conformation of the substrate, the enzyme system determines to which diastereotopic face of the carbonyl group the hydrogen will be transferred in going from **54** to **55**. There must be different forms of this enzyme to account for the

[45] S. Fujise, *Biochem. Z.*, **236**, 237 (1931).
[46] C. Neuberg and H. Collatz, *ibid.*, **225**, 242 (1930).
[47] S. Fujise, *ibid.*, **236**, 241 (1931).
[48] C. Neuberg and Cl. Ostendorf, *Biochem. Z.*, **279**, 459 (1935).
[49] V. Franzen, *Chem. Ber.*, **89**, 1020 (1956).
[50] I. A. Rose, *Biochim. Biophys. Acta*, **25**, 214 (1957).
[51] H. Frendenhagen and K. F. Bonnhoeffer, *Z. Phys. Chem.*, **A181**, 379 (1938).
[52] W. E. Doering, T. I. Taylor, and E. F. Schoenewaldt, *J. Am. Chem. Soc.*, **70**, 455 (1948).
[53] V. Franzen, *Chem. Ber.*, **88**, 1361 (1955).

52 + GSH ⇌ 54 —Glyoxalase I→ 55 —Glyoxalase II→ 53 + GSH

GSH = [$HOOCCH(NH_2)CH_2CH_2CONHCH(CH_2SH)CONHCH_2COOH$]

reduced glutathione

different stereochemical results obtained with enzyme preparations from various sources.

Franzen[53] established that β-aminothiols ($R_2NCH_2CH_2SH$) would catalyze the nonenzymatic conversion of phenylglyoxal (**52B**) to mandelic acid (**53B**). The SH group was necessary since the corresponding methyl thioether or dimethyl sulfonium iodide showed no catalytic effect. It was proposed that these catalysts operated by first forming the thiohemiacetal, **56**, which has the β-amino group favorably situated to assist in the removal of a proton from the hydroxyl group, to give the alkoxide, **57**. The formation of the alkoxide will favor the internal shift of a hydrogen as a hydride ion to give the product **58**. This thioester (**58**) is readily hydrolyzed in the basic

56 → 57 → 58

reaction mixture to regenerate the catalyst. Furthermore, two *chiral amino thiols* (Table **4-3**, Nos. 1 and 9) were found to catalyze an asymmetric conversion of phenylglyoxal to methyl mandelate.[54] These experiments

[54] V. Franzen, *Chem. Ber.*, **90**, 2036 (1957).

TABLE 4-3

Asymmetric Internal Cannizzaro Reaction

$$R-\overset{O}{\overset{\|}{C}}-\overset{O}{\overset{\|}{C}}-H + CH_3OH \xrightarrow{R^*SH} R-\underset{H}{\underset{|}{\overset{OH}{\overset{|}{C}}}}-\overset{O}{\overset{\|}{C}}-OCH_3$$

No.	R	R*SH	Product Config.	Product Asym. syn. % e.e.	Ref.
1[a]	Ph	(+)-$PhCH_2CH(CH_3)N(CH_3)CH_2CH_2SH$	*S* (+)	11	54
2[a]	Ph	(+)-$PhCH_2CH(CH_3)N(CH_3)CH_2CH_2SH$	—	0	56
3	Ph	(+)-$PhCH(SH)CH_2NMe_2$	*R* (−)	14	55
4	Ph	(−)-$PhCH(SH)CH_2NMe_2$	*S* (+)	12	55
5	Ph	(−)-$PhCH_2CH(NMe_2)CH_2SH$	*S* (+)	1	57
6	Ph	*threo*-(−)-$PhCH(NMe_2)CH(CH_3)SH$	—	—	58
7	Ph	*threo*-(+)-$PhCH(NMe_2)CH(CH_3)SH$	—	—	58
8	Ph	*erythro*-(−)-$PhCH(NMe_2)CH(CH_3)SH$	—	—	58
9	Ph	(+)-2-(Mercaptomethyl)piperidine	*S* (+)	2	54
10	α-Naphthyl	(+)-$PhCH_2CH(CH_3)N(CH_3)CH_2CH_2SH$	—	0	56
11	α-Naphthyl	(−)-$PhCH_2CH(CH_3)N(CH_3)CH_2CH_2SH$	—	0	56
12	α-Naphthyl	(−)-$PhCH(SH)CH_2NMe_2$	*S* (+)	9	55
13	α-Naphthyl	(+)-$PhCH(SH)CH_2NMe_2$	*R* (−)	9	55

[a] Note that the same experiment in the hands of two different workers gave disparate results.

have been extended[55–58] to include four other chiral amino thiols. From the data in Table **4-3** it is seen that the extents of asymmetric syntheses are generally significant but not large. From the limited evidence available, very few conclusions can be drawn concerning the factors which control the stereochemical course of the reaction. The position of the asymmetric center in the catalyst does not appear to be crucial since significant asymmetric syntheses result whether the thiol function, amino function, or both are situated on a chiral center. There are no examples which would indicate whether both functions are absolutely necessary, for instance, whether or not

[55] S. Ose and Y. Yoshimura, *Yakugaku Zasshi*, **77**, 730 (1957); *Chem. Abst.*, **51**, 17856 (1957). In this abstract, phenylglyoxal is mistakenly represented as benzil and 2-dimethylamino-1-phenylethanethiol as 2-dimethylamino-1-phenylethanol.

[56] S. Ose and Y. Yoshimura, *Yakugaku Zasshi*, **77**, 734 (1957); *Chem. Abst.*, **51**, 17856 (1957).

[57] S. Ose and Y. Yoshimura, *Yakugaku Zasshi*, **78**, 687 (1958); *Chem. Abst.*, **52**, 18289 (1958).

[58] Y. Yoshimura, *Yakugaku Zasshi*, **84**, 305 (1964); *Chem. Abst.*, **61**, 560 (1964).

chiral thiols or dithiols in the presence of basic catalysts or chiral alkoxides would also bring about such asymmetric syntheses.

It is disturbing that Franzen found substantial asymmetric synthesis using the primary thiol, $PhCH_2CH(CH_3)N(CH_3)CH_2CH_2SH$, as a chiral catalyst (Table **4-3**, No. 1), while Ose and Yoshimura, using the same catalyst, failed to observe any asymmetric synthesis on the same substrate (No. 2) or on the α-naphthylglyoxal (Nos. 10 and 11).

The stereochemical situation is illustrated in Fig. **4-9**. Formation of the thiohemiacetals, **59**, is a reversible process which establishes a new chiral

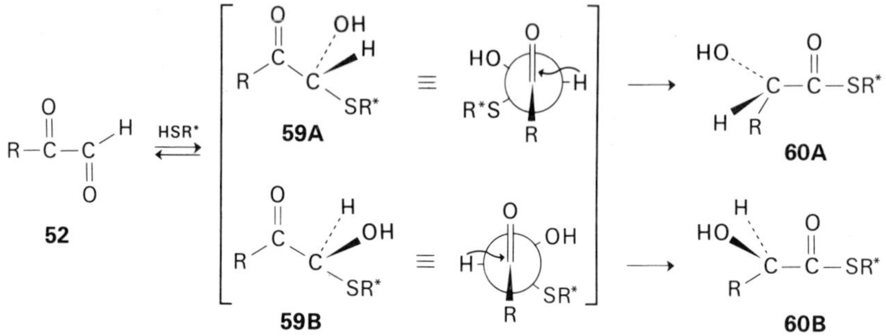

Fig. 4-9. Stereochemical representation of the asymmetric internal Cannizzaro reaction catalyzed by chiral thiols.

center. Thus the diastereomers **59A** and **59B** will not be formed in exactly a 50:50 ratio. But the stereochemically determining step is undoubtedly the internal irreversible hydrogen transfer represented by **59** to **60** (where only one of two or more conformations is shown). Whether the rate of reaction of **59A** is faster or slower than **59B** will depend upon the manner in which the chiral group R* interacts with the chiral thiohemiacetal center and the developing chiral α-hydroxy center in these diastereomeric intermediates.

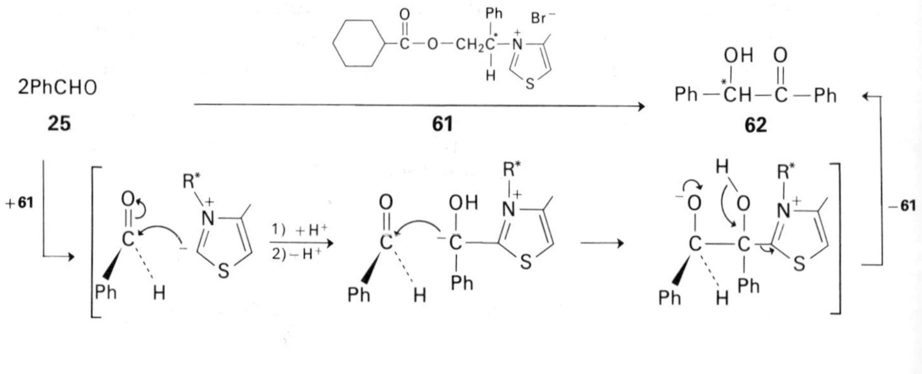

Asymmetric synthesis should occur when the R group of the glyoxal is chiral but there appears to be no such recorded examples.

A similar, but intermolecular instead of intramolecular, reaction is the acyloin condensation. Benzaldehyde (**25**) in the presence of the thiamin analog catalyst **61** gives optically active benzoin[59] **62**. Two crops of crystals were obtained, the first in 50% yield with 0.77% e.e. and the second in 9.4% yield in 22% e.e. There seem to be no chemical examples of mixed condensations such as that between acetaldehyde and benzaldehyde but the biochemical counterpart carried out in the presence of fermenting yeast is well-known.[43b]

1969 Addenda

Chapter 4. Coupling of (−)-menthyl pyruvate to itself in the presence of trimethyl phosphite gave, after hydrolysis, a 35% yield of (−)-(*S*,*S*)-2,3-dimethyltartaric acid, 7% e.e.[60] This can be considered as a special case of addition of a chiral substrate to a chiral carbonyl compound of which no other example has been reported in Chapter 4.

Sec. 4-1. A stereochemical study has been made of the addition of cyanide to *cis*-3,5-dimethyl-, *cis*,*cis*-3,4,5-trimethyl-, and 3,3,5-trimethylcyclohexanones.[61]

[59] J. C. Sheehan and D. H. Hunneman, *J. Am. Chem. Soc.*, **88**, 3666 (1966).
[60] M. Muroi, Y. Inouye, and M. Ohno, *Bull. Soc. Chem. Japan*, **42**, 2948 (1969).
[61] M. R. Harnden, *J. Chem. Soc.* (C), 960 (1969).

5

Hydrogen Transfer from Chiral Reducing Agents to Achiral Substrates

5-1. Meerwein–Ponndorf–Verley Reductions

5-1.1. Introduction. The Meerwein–Ponndorf–Verley (MPV) reduction has been reviewed from a synthetic standpoint[1,2] including the stereochemistry involved in the formation of *erythro* and *threo* isomers. However, it was subsequent to these reviews that the *asymmetric* MPV reaction was discovered independently by Doering and Young[3] and Jackman, Mills, and Shannon.[4] The historical importance of this discovery to the development of the field of asymmetric synthesis has been pointed out (Sec. **1-1**).

[1] A. L. Wilds, *Org. Reactions*, **2**, 178 (1944).

[2] T. Bersin in *Neuere Methoden der Präparativen Organischen Chemie*, ed. W. Foerst (Berlin: Verlag Chemie, GmbH, 1944), pp. 137–154; English translation and revision, E. R. Webster and J. V. Crawford, *Newer Methods of Preparative Organic Chemistry* (New York: Interscience Pub., 1948), pp. 125–153.

[3] W. von E. Doering and R. W. Young, *J. Am. Chem. Soc.*, **72**, 631 (1950).

[4] L. M. Jackman, J. A. Mills, and J. S. Shannon, *J. Am. Chem. Soc.*, **72**, 4814 (1950).

The MPV reaction is known to proceed with direct hydrogen transfer[3–9] and can be summarized by the overall Eq. 1 in which isopropyl alcohol is the solvent and aluminum isopropoxide the catalyst. This overall reaction is

$$R_2C{=}O + MeCHOHMe \underset{}{\overset{Al(OCHMe_2)_3}{\rightleftharpoons}} R_2CHOH + Me{-}\overset{\overset{O}{\|}}{C}{-}Me \quad (1)$$

believed to involve the following individual steps. The first, Eq. 2, is the coordination of the ketone with alkoxide reagent; the second, Eq. 3, the hydrogen transfer; the third, Eq. 4, the decomplexing of the acetone produced, and finally alcoholysis, Eq. 5, of the newly formed alkoxide with isopropyl alcohol to give the original alkoxide and liberate the reduced substrate. Since all these steps are reversible, the series of reactions is complicated

$$R_2C{=}O + Al(OCHMe_2)_3 \rightleftharpoons R_2C{=}O{\rightarrow}Al(OCHMe_2)_3 \quad (2)$$

$$R_2C{=}O{\rightarrow}Al(OCHMe_2)_3 \rightleftharpoons R_2CHOAl(OCHMe_2)_2{\leftarrow}O{=}CMe_2 \quad (3)$$

$$R_2CHOAl(OCHMe_2)_2{\leftarrow}O{=}CMe_2 \rightleftharpoons R_2CHOAl(OCHMe_2)_2 + Me_2C{=}O \quad (4)$$

$$R_2CHOAl(OCHMe_2)_2 + HOCHMe_2 \rightleftharpoons Al(OCHMe_2)_3 + R_2CHOH \quad (5)$$

by the fact that the newly formed alkoxide $R_2CHOAl(OCHMe_2)_2$ can itself become involved in subsequent hydrogen transfer leading to $(R_2CHO)_2AlOCHMe_2$ and ultimately to $(R_2CHO)_3Al$. In these equations aluminum isopropoxide has been represented as a monomer but it has been shown[10] that the reactive species is in fact a trimer (**1**). The usual form, a tetramer[10b] (**2**), reacts much more slowly as a reducing agent. On the other hand the usual form of aluminum tri-*t*-butoxide is a dimer (**3**). Furthermore the rate-determining step *when aluminum isopropoxide is used as a catalyst* is the alcoholysis, Eq. 5, and not the hydrogen transfer, Eq. 3, as is usually assumed. Consequently the reaction proceeds more rapidly when the aluminum alkoxide is used in stoichiometric rather than catalytic amounts. The degree of association of the reactive species in the chiral alkoxides which we shall now discuss has not been studied; thus we shall represent these reagents as monomers but it must be recognized that this is a necessary simplification.

[5] L. M. Jackman and J. A. Mills, *Nature*, **164**, 759 (1949).
[6] A. Streitwieser, Jr., *J. Am. Chem. Soc.*, **75**, 5014 (1953).
[7] V. J. Shiner and D. Whittaker, *J. Am. Chem. Soc.*, **85**, 2337 (1963).
[8] W. von E. Doering and T. C. Aschner, *J. Am. Chem. Soc.*, **75**, 393 (1953).
[9] E. D. Williams, K. A. Kreiger and A. R. Day, *J. Am. Chem. Soc.*, **75**, 2404 (1953).
[10] (a) V. J. Shiner, Jr., D. Whittaker, and V. P. Fernandez, *J. Am. Chem. Soc.*, **85**, 2318 (1963). (b) W. Fieggen, H. Gerdind and N. M. M. Nibbering, *Rec. Trav. Chem. Pays-Bas*, **87**, 377 (1968). Even in the vapor "$Al(O\text{-}i\text{Pr})_3$" is a tetramer.

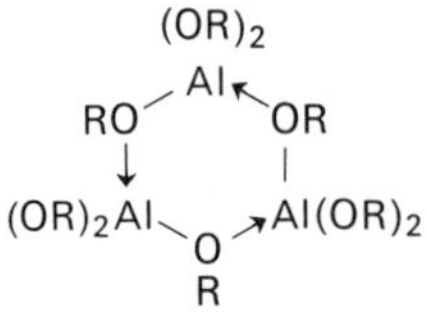

1
$Al(O\text{-}i\text{Pr})_3$
Trimer

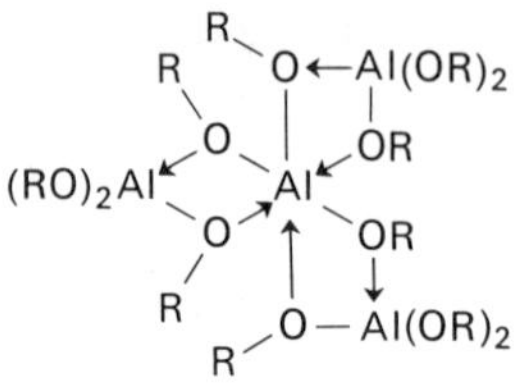

2
$Al(O\text{-}i\text{Pr})_3$
Tetramer

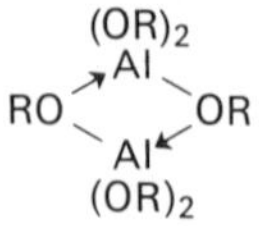

3
$Al(O\text{-}t\text{-Bu})_3$
Dimer

Based on earlier kinetic evidence,[11] it has been postulated that the MPV reaction may take placc by either an intramolecular cyclic hydrogen shift as symbolized by **4** or by a bimolecular process as symbolized by **5**. It is also quite possible that in the polymeric forms hydrogen transfer may take place via an eight-membered ring as shown in **6**. However the kinetic evidence upon which the postulate of a bimolecular process was based did not take into account the polymeric forms of the alkoxide[10] nor the slow alcohol-alkoxide exchange[7] which would constitute serious sources of error in these previous interpretations.

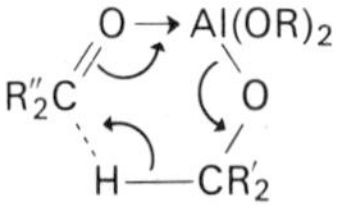

4 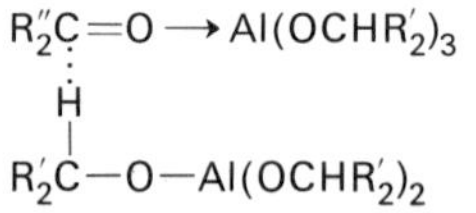 **5** 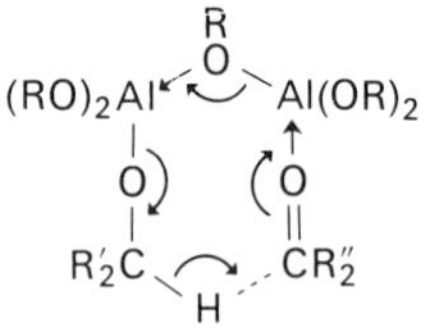 **6**

A Hammett *rho* constant of 1.296 has been found for the MPV reaction between substituted benzophenones and diethylcarbinol.[12] Thus an increased positive charge on the carbonyl carbon atom of the benzophenone facilitates the reaction. This supports the conclusion that the hydrogen is transferred with its pair of electrons as a hydride ion.

All the steps in Eqs. 1-4 are reversible and thus, although the stereochemistry of the *initial* product may be kinetically controlled, equilibration will take place as the reaction proceeds until ultimately a 50:50 ratio of enantiomers will be established. Thus quantitative comparisons of the stereoselectivities of various MPV reductions, in theory, should be made by extrapolating back to the stereochemical composition of the initially formed product. This usually has not been done and so the data available are of reduced value for quantitative comparisons.

[11] (a) W. N. Moulton, R. E. van Atta, and R. R. Ruch, *J. Org. Chem.*, **26**, 290 (1961). (b) L. M. Jackman and A. K. MacBeth, *J. Chem. Soc.*, 3252 (1952).

[12] D. E. Pickhart and C. K. Hancock, *J. Am. Chem. Soc.*, **77**, 4642 (1955).

5-1.2. Stereochemical Model for the Transition State. The stereochemical course of the asymmetric MPV reaction can be understood in terms of the following formulas. In this case the chiral reducing agent (**7**) is prepared from (*S*)-(+)-methylisopropylcarbinol; methyl cyclohexyl ketone (**8**) is the substrate. In the transition state models (**9A** and **9B**) the aluminum atom shown is only one of several (probably three) in the alkoxide polymer; the others, which are not shown, are complexed with the two OR groups.

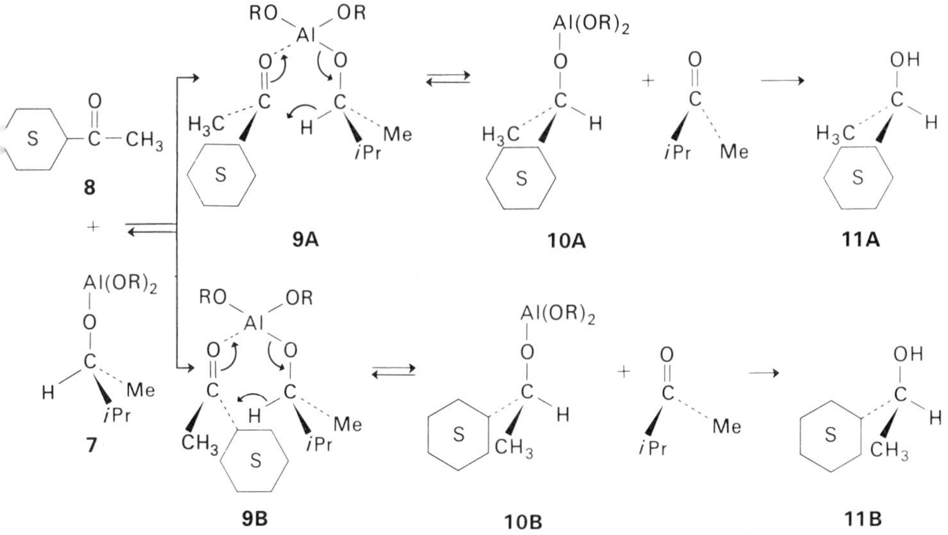

Fig. 5-1. Comparison of transition state models in the asymmetric Meerwein–Ponndorf–Verley reduction.

If the OR groups of the transition states **9A** and **9B** which are not involved specifically in the six-membered ring hydrogen transfer are chiral, they also will have an effect upon the stereochemical outcome of the reaction but this should be a second-order effect compared to the influence of the chiral alkoxide moiety actually involved in the hydrogen transfer as delineated in **9A** and **9B**. It is logical to assume that the transition state symbolized by **9B**, in which the smaller methyl group from the reagent is adjacent to the larger cyclohexyl group of the ketone and the larger isopropyl group of the reagent is adjacent to the smaller methyl group of the ketone, will be favored over the diastereomeric transition state **9A**, in which the two larger groups and the two smaller groups are in juxtaposition. Thus, starting with the aluminum alkoxide from (*S*)-(+)-methylisopropylcarbinol (**7**), one would predict that the predominant enantiomer from the reduction of methyl cyclohexyl ketone (**8**) would be (*S*)-(+)-methylcyclohexylcarbinol (**11B**), a prediction which is in accord with observation[3] (22% e.e.).

The concept that a better fit and therefore a lower energy of activation is obtained by the "dovetailing" of R'_S and R'_L of the reagent with R_L and R_S

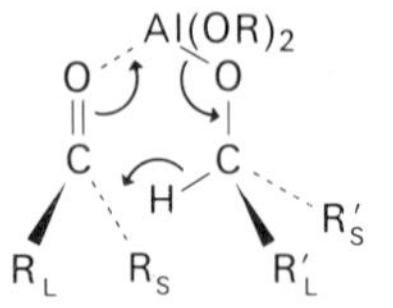

12A
Model of nonpreferred transition state

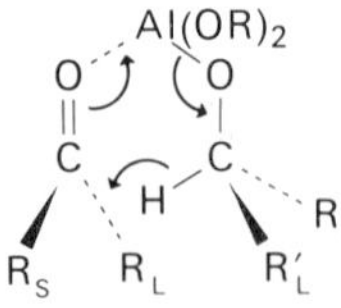

12B
Model of preferred transition state

of the substrate, as in the model represented by **12B** in contrast to **12A** where R_L-R'_L and R_S-R'_S are in juxtaposition, has been verified in all the examples in Tables **5-1** and **5-2**. This is also the guiding principle for predicting the configuration of the predominant product in the Grignard reduction reaction (Sec. **5-2**) and presumably in other systems involving similar 1,3-interactions in a cyclic transition state.

5-1.3. Reductions with Chiral Aluminum Alkoxides. The reductions reported in Table **5-1** were carried out with excess chiral alcohol and catalytic amounts of aluminum alkoxide under equilibrium conditions which are not strictly comparable. Meaningful conclusions concerning the quantitative results therefore are not feasible. It is quite possible that the zero stereoselectivity found in Nos. 3 and 4 is a result of racemization due to equilibration.

Numbers 6 and 7 represent the experiments in which Baker and Linn[13] attempted to achieve the first asymmetric MPV reduction as mentioned in the introduction (Sec. **1-1**). The reducing agent was the aluminum alkoxide of a primary alcohol, (−)-2-methyl-1-butanol, in which the chiral center is in the position *beta* to the oxygen atom. As a consequence the chiral center is not incorporated into the six-membered ring of the hydrogen transfer process and there are diastereotopic hydrogens *beta* to the aluminum, either one of which can be transferred during the reduction step. (Refer to formulas **12A** and **12B** where R'_S would be hydrogen and R'_L would include the chiral group.) This situation is one which is discussed more fully in connection with the analogous Grignard asymmetric reduction reaction (Sec. **5-2**). Asymmetric MPV reductions have been reported with this reagent[14] but need to be confirmed (see Table **5-1**, Nos. 8–10 and Footnote c).

[13] R. H. Baker and L. E. Linn, *J. Am. Chem. Soc.*, **71**, 1399 (1949).
[14] S. Yamashita, *J. Organometal. Chem.*, **11**, 377 (1968).

TABLE 5-1

Reductions with Chiral Aluminum Alkoxides

$$R'_L R'_S C(H)OH + R_L C(=O) R_S \xrightarrow{Al(OR)_3} R'_S C(=O) R'_L + H C(OH) R_L R_S$$

No.	R'_S	R'_L	R_S	R_L	Carbinol product Config.	Asym. syn. % e.e.	Ref.
1	Me	*i*-Pr	Me	*c*-C_6H_{11}[a]	*S* (+)	22	3
2	Me	Et	Me	*i*-C_6H_{13}	*S* (+)	6	3
3	Me	Et	Me	CH_2CH_2Ph	—	0	4
4	Me	Et	Me	*t*-Bu	—	0	4
5	Me	*t*-Bu	Me	*n*-C_6H_{13}	*S* (−)	6	4
6	H	*CHMeEt[b]	Me	Ph	—	0	14
7	H	*CHMeEt[b]	Ph	*p*-ClPh	—	0	14
8	H	*CHMeEt[b]	Me	Et	—[c]	—[c]	14
9	H	*CHMeEt[b]	Me	Ph	—[c]	—[c]	14
10	H	*CHMeEt[b]	3-methylcyclohexanone ring (C=O[d], ~Me)		—[c]	—[c]	14

[a] *c*-C_6H_{11} represents cyclohexyl.

[b] *CHMeEt represents part of the primary active amyl radical, $CH_2CHMeEt$.

[c] Asymmetric synthesis was reported to be high for No. 8 but low for Nos. 9 and 10. However the data are not included here because of discrepancies in the original publication[14] to which the reader is referred. The rotation of the (−)-2-methyl-1-butanol used was reported to be $[\alpha]^{25}_{589}$ −10.13°, while the maximum rotation for this alcohol is $[\alpha]^{25}_{589}$ −5.7° (neat). Furthermore, the rotation of the 2-butanol obtained in No. 8 was reported to be $[\alpha]^{20}_{589}$ +13.08°, $[\alpha]^{20}_{436}$ +89.63°, while the literature values for pure (+)-2-butanol are $[\alpha]^{20}_{589}$ + 13.82° (neat) and $[\alpha]^{20}_{435}$ + 27.23° (neat). Obviously these experiments should be checked.

[d] Represents 3-methylcyclohexanone substrate and 3-methylcyclohexanol product.

5-1.4. Reductions with Chiral Magnesium Alkoxides

5-1.4a. *The Preparation of Chiral α-Deuterio Primary Alcohols.* The most important synthetic application of these alkoxide asymmetric reductions has been in the production of optically active α-deuterio primary alcohols.[6,15–17] Magnesium alkoxides have proved to be more convenient

[15] (a) A. Streitwieser, Jr., and W. D. Schaeffer, *J. Am. Chem. Soc.*, **78**, 5597 (1956). (b) A. Streitwieser, Jr., and J. R. Wolfe, *J. Am. Chem. Soc.*, **79**, 903 (1957). (c) A. Streitwieser, Jr., J. R. Wolfe, and W. D. Schaeffer, *Tetrahedron*, **6**, 338 (1959).

[16] A. Streitwieser, Jr., and M. R. Granger, *J. Org. Chem.*, **32**, 1528 (1967).

[17] B. Belleau and J. Burba, *J. Am. Chem. Soc.*, **82**, 5751 (1960).

reagents for this purpose than the aluminum alkoxides. The most widely studied such reagent has been isobornyloxymagnesium bromide[15–18] (**15**) made by treating (−)-isoborneol (**14**) in ether with a Grignard reagent. Lithium aluminum hydride reduction of (+)-camphor (**13**) gives a good yield of (−)-isoborneol contaminated with about 10% of the epimeric borneol. Since the magnesium bromide salt of borneol is not a reducing agent under these conditions,[18b] it is possible to use the 90:10 isoborneol-

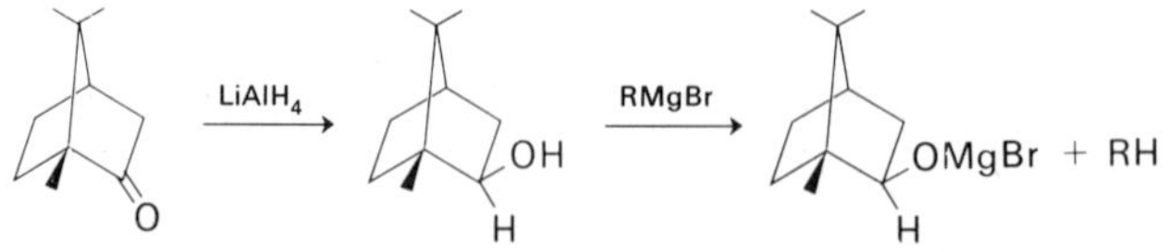

borneol mixture directly for preparation of the magnesium halide salt; alternatively chromatography on alumina removes most of the borneol.[19] In this adaptation of the Meerwein–Ponndorf–Verley reduction, the alkoxide is not used as a catalyst but is present in excess. The reaction with aldehydes proceeds at room temperature and the equilibrium strongly favors the formation of camphor. There is a minimum of racemization due to the reverse reaction when properly carried out. The reduction of deuteriobenzaldehyde (**16**) to give benzyl-α-*d* alcohol in which *R*-(−)-**19** predominates over *S*-(+)-**19** (45% e.e.) is shown in Fig. **5-2**.

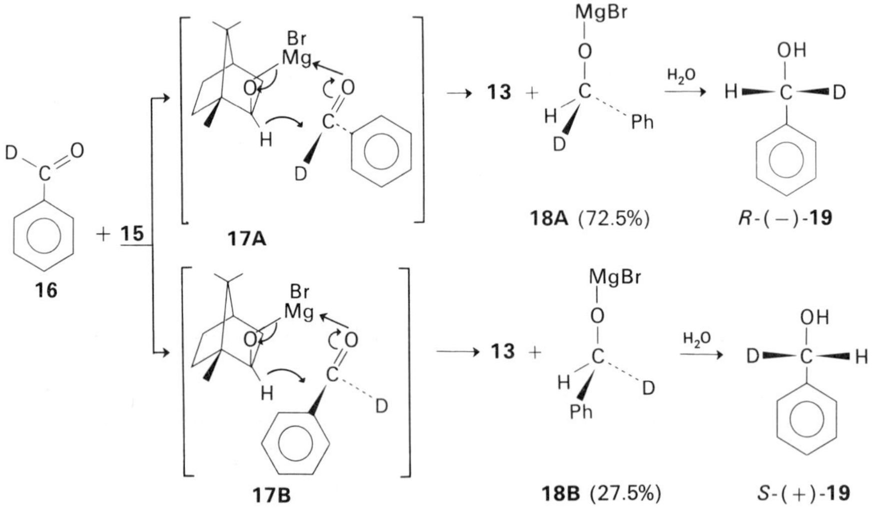

Fig. 5-2. (−)-Isobornyloxymagnesium bromide reduction of deuteriobenzaldehyde.

[18] (a) G. Vavon and A. Antonini, *Compt. rend.*, **230**, 1870 (1950). (b) G. Vavon and A. Antonini, *ibid.*, **232**, 1120 (1951).

[19] H. Gerlach, *Helv. Chim. Acta*, **49**, 1291 (1966).

The carbinol carbon (C-2) of the isobornyloxy reagent is flanked on one side by the quaternary bridge-head carbon (C-1) which acts as R_L and on the other side by the secondary methylene carbon (C-3) which acts as R_S. The lower energy transition state then should be represented by **17A**, in which the deuterium atom of the substrate opposes C-1 and the phenyl opposes C-3, rather than **17B** where the opposite obtains. The *R*-(−) enantiomer which would be formed on hydrolysis of the salt with structure **18A** is the one actually produced in excess (41–45% e.e.) (Table **5-2**, Nos. 5 and 6).

If (+)-camphor is reduced with lithium aluminum deuteride, the deuterated reagent (−)-isoborneol-2-*d* (**15**-2*d*) which is formed can be used to transfer deuterium instead of hydrogen. Thus the hydrogen and the deuterium in the reagent and substrate represented in Fig. **5-2** can be interchanged. Such deuterated reagents have been used in several of the examples of Table **5-2**. Whether one uses the deuterated substrate and isotopically normal reagent or vice versa is a matter of convenience and economics. Interchanging hydrogen and deuterium in substrate and reagent will of necessity

TABLE 5-2

Use of Chiral Alkoxy Magnesium Reagent to Produce Optically Active α-Deuterio Primary Alcohols

$$R-\overset{\overset{\displaystyle O}{\|}}{C}-H(D) + [R'R''\overset{\overset{\displaystyle D(H)}{|}}{C}-O]MgX \longrightarrow R\overset{*}{C}HDOH$$

No.	Alcohol from which the reagent was made	Substrate	Alcohol Produced: Config.	Asym. syn. % e.e.	Ref.
1	*S*-(+)-2-Octanol-2-*d*	*i*-PrCHO	*R*-(−)	8–15	6
2	*R*-(−)-2-Octanol-2-*d*	*i*-PrCHO	*S*-(+)	6	6
3	*R*-(−)-2-Octanol	*i*-PrCDO	*R*-(−)	10	15a
4	(−)-Isoborneol-2-*d*	*i*-PrCHO	*S*-(+)	19	15c
5	(−)-Isoborneol	PhCDO	*R*-(−)	45	15c
6	(−)-Isoborneol	PhCDO	*R*-(−)	41	15b
7	(−)-Isoborneol-2-*d*	CH_3CHO	*S*-(−)	44[a]	16
8	(−)-Isoborneol-2-*d*	*p*-$MeOPhCH_2CH_2CHO$	*S*-(−)	—[b]	17
9	(−)-Isoborneol-2-*d*	*cis*-4-Me-$C_6H_{10}CHO$	*S*-(+)	—[c]	19
10	(−)-Isoborneol-2-*d*	*trans*-4-Me-$C_6H_{10}CHO$	*S*-(+)	—[c]	19

[a] The precision of this value is ±9% based on the rotation of enzymically produced ethanol-1-*d*, according to H. R. Levy, F. A. Loewus, and B. Vennesland, *J. Am. Chem. Soc.*, **79**, 2949 (1957).

[b] In Reference 17 it was assumed that this material was the pure enantiomer but this conclusion was based on a false assumption,[15b] which has now been corrected.[16]

[c] Deuterium incorporation was less than expected because of a Tischenko condensation which gave undeuterated benzyl benzoate. C_6H_{10} is the disubstituted cyclohexane nucleus.

produce enantiomeric products in excess (Table **5-2**, Nos. 2 and 3). The absolute stereoselectivities of the matched reactions in which hydrogen and deuterium have been interchanged will not necessarily be the same. Nevertheless the quantitative difference observed between Nos. 2 and 3 cannot be taken as a measure of the inherent differences in stereoselectivities since these are equilibrium reactions which were not necessarily conducted under comparable conditions. There is a greater variation between the several runs represented by Nos. 1 and 2 (which on symmetry principles must lead, under identical conditions, to the same absolute stereoselectivities) than between Nos. 2 and 3.

An interesting use of asymmetric reduction with isobornyloxy-2-*d*-magnesium bromide (**15**-2*d*) has been in establishing the stereochemical course of the enzymatic amino acid decarboxylation reaction.[17] Reduction of *p*-methoxyphenylacetaldehyde (**20**) with chiral reagent (**15**-2*d*) via the transition state represented by **21** gave (−)-*p*-methoxyphenylethanol-1*d* (**22**) which according to the accepted interpretation of this reaction will have the absolute *S* configuration. By the indicated series of reactions, which includes one inversion on treatment of the tosylate with sodium azide, the *S*-(−) carbinol (**22**) was converted to the *R*-tyramine-α-*d* (**24**). By comparison of the rates of amine oxidase-catalyzed oxidation of this isomer with that of the amine obtained from enzymatic decarboxylation of tyrosine (**25**) in

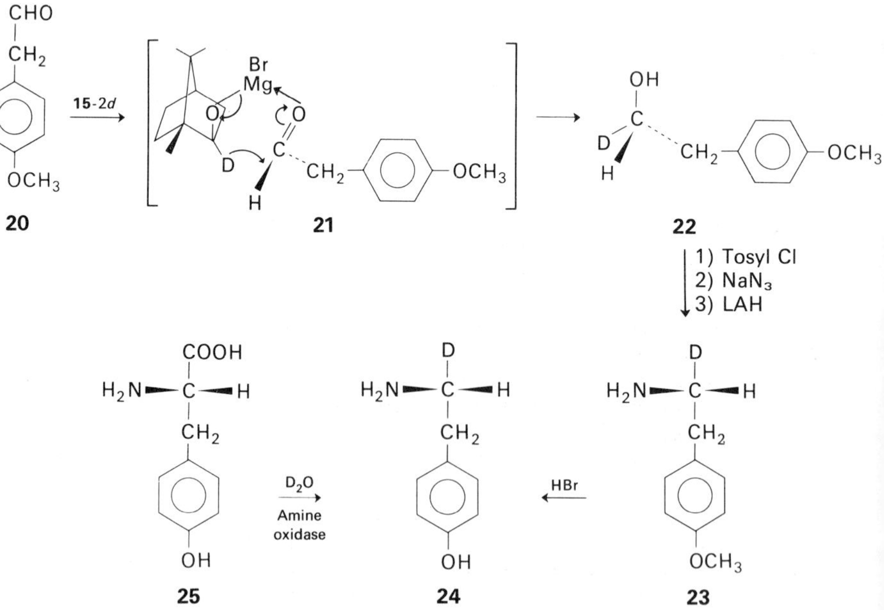

CHO
CH_2
OCH_3
20
15-2*d*
Br
Mg
O
O
C
D
H
CH_2
OCH_3
21
OH
C
D
H
CH_2
OCH_3
22
1) Tosyl Cl
2) NaN_3
3) LAH
COOH
H_2N C H
CH_2
OH
25
D_2O
Amine oxidase
D
H_2N C H
CH_2
OH
24
HBr
D
H_2N C H
CH_2
OCH_3
23

Fig. 5-3. Determination of stereochemistry of amine oxidase reaction.

D_2O, it was concluded that the two were identical and that the decarboxylation proceeds with retention of configuration.

A beautiful application of this asymmetric reduction is the proof of configuration of chiral *cis*- and *trans*-4-methylcyclohexylidene-acetic acids[19] (cf. Sec. **9-2.3**, Fig. **9-8**).

5-1.4b. *The Reduction of Glyoxylic Acids.* Table **5-3** summarizes results of Vavon and Antonini[18] in which phenylglyoxylic acid and some related derivatives were reduced with an excess of several chiral, terpenoid, magnesium alkoxy reagents. The stereochemistry of Nos. 1, 2, and 4, in which the product is mandelic acid, is correctly predicted based on the analysis represented in Fig. **5-4** in which COOH (actually COOMgBr) is assumed to act as a smaller group than phenyl.

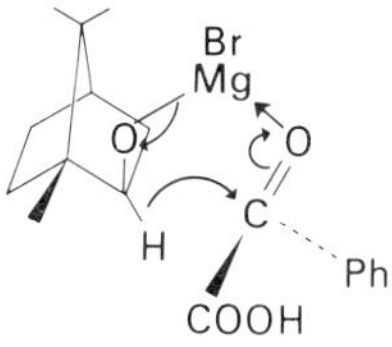

(−)-Isobornyloxy
26A

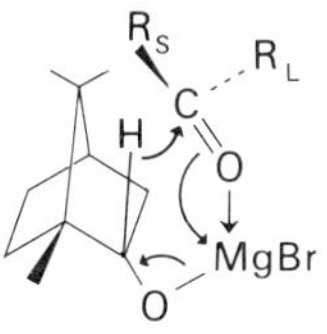

(−)-Bornyloxy
26B

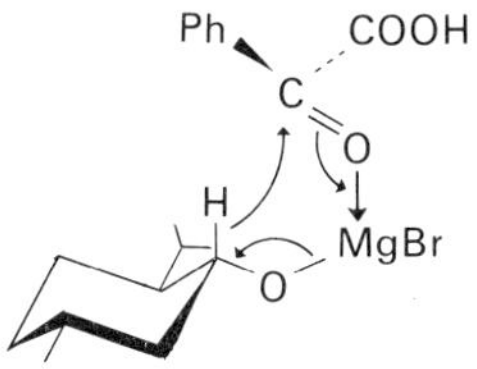

(−)-Menthoxy
26C

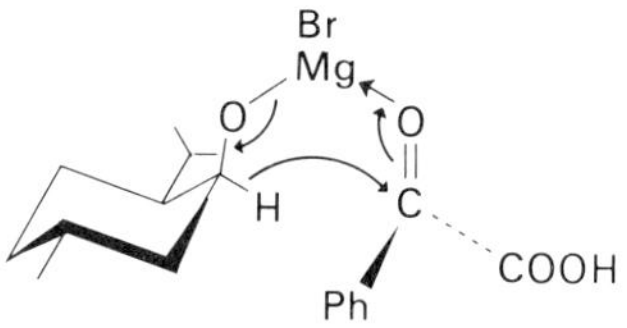

(+)-Neomenthoxy
26D

Fig. 5-4. Preferred transition states for hydride transfer in MPV reductions by bromomagnesium alkoxides of (−)-isoborneol, (−)-borneol, (−)-menthol, and (+)-neomenthol.

There is essentially no reduction by the bornyl alkoxide (**26B**) presumably because the hydrogen is too highly shielded by the C-9 methyl group. Comparable results have been observed in the asymmetric Grignard reduction using an analogous system (Sec. **5-2**). The absolute configuration of α-hydroxy-α-naphthylacetic acid has not been firmly established;* thus

* In his review Klabunovskii [E. I. Klabunovskii, *Asymmetric Synthesis* (Moscow: Zoachnyi Inst. Sovet. Torgovli, 1960), p. 61; German translation by G. Rudakoff, (Berlin: VEB Deutscher Verlag der Wissenschaften, 1963)] states that α-hydroxy-α-naphthylacetic acid and α-hydroxyphenylacetic acid of the same sign of rotation have opposite configurations. The evidence for this was not given and the statement may be based upon rationalization of the data in Table **5-3** rather than independent evidence.

conclusions concerning the stereochemical course of examples Nos. 5, 6, and 7 can only be tentative. The very low order of asymmetric synthesis in No. 5 precludes any meaningful conclusion based on the sign of rotation. That this stereoselectivity is only 1–2% is surprising since one would anticipate an increase in stereoselectivity when the phenyl group was replaced with a naphthyl group. It is seen from reductions Nos. 2 and 6 as well as Nos. 1 and 7 that replacement of phenyl with α-naphthyl results in α-hydroxy-α-naphthylacetic acid with sign of rotation opposite to that of the α-hydroxyphenylacetic acid. If one assumes that these two acids with opposite signs of rotations have related absolute configurations, then the results of Nos. 5 and 6 can be rationalized by the same models used in Fig. **5-4** in which phenyl is replaced with α-naphthyl.

Numbers 8 and 9 of Table **5-3** are especially interesting since both reagent and substrate are chiral. One might expect that the asymmetric synthesis which is controlled by the (−)-menthyl moiety of the ester would proceed

TABLE 5-3

Reduction of Aryl Glyoxylic Acid and Derivatives with Chiral Magnesium Alkoxides[a]

$$\mathrm{R{-}\overset{O}{\overset{\|}{C}}{-}\overset{O}{\overset{\|}{C}}{-}O{-}R'} \xrightarrow[\text{2) } H_2O]{\text{1) } R^*OMgBr} \mathrm{R{-}\underset{OH}{\underset{|}{C}}HCOOH}$$

No.	R	R′	R*	α-Hydroxy acid Config.	$[\alpha]_D$ °	Asym. syn. % e.e.
1	Ph	H	(−)-Menthyl	*S*-(+)	+54	33
2	Ph	H	(+)-Neomenthyl	*R*-(−)	−27.3	17
3	Ph	H	(+)-Bornyl	—[b]	—[b]	—[b]
4	Ph	H	(−)-Isobornyl	*S*-(+)	+24	15
5	α-Naphthyl	H	(−)-Menthyl	(−)[c]	−2.8 −3.5	1–2
6	α-Naphthyl	H	(+)-Neomenthyl	(+)[c]	+30.2	15
7	α-Naphthyl	Et	(−)-Menthyl	(−)[c]	−21	6–11
8	α-Naphthyl	(−)-Menthyl	(−)-Menthyl[d]	(−)[c]	−12.4	6
9	Ph	(−)-Menthyl	(−)-Menthyl[d]	*R*-(−)	−66	41

[a] Data from References 18a and 18b.

[b] The synthetic yield was essentially zero and thus no asymmetric synthesis could be determined.

[c] If one assumes that (+)-α-hydroxyphenylacetic acid and (−)-α-hydroxy-α-naphthylacetic acid with opposite signs are both of the *S* configuration, then the data in Nos. 5, 6, and 7 are consistent with those from Nos. 1 and 2. (See footnote on page 169.)

[d] See Sec. **5-3.1** for a discussion of these "double asymmetric induction" reactions.

according to the Prelog generalization to favor the formation of (*S*)-(+)-mandelic acid (see examples in Table **2-4**). Also, since the action of (−)-menthoxymagnesium bromide on the achiral benzoylformic acid gives the *S*-(+) isomer (Table **5-3**, No. 1) as predicted by the model represented by **26C** in Fig. **5-4**, one could reasonably expect that these two controlling factors would reinforce each other to give an especially high degree of stereoselectivity. As seen by the data reported in Table **5-3**, No. 9, the asymmetric synthesis was the highest reported in this series (41 % e.e.), but it was the unexpected *R*-(−)-isomer which was formed in excess! It can be argued that the carbomenthoxy group is no longer smaller than the phenyl group and that this has caused the reversal of the sense of asymmetric synthesis, but this would mean that the two controlling factors were operating in opposition to each other; under these circumstances one would anticipate a reduced extent of stereoselectivity. Quite apparently these simple models are inadequate to handle the more complicated situation in which both substrate and reagent are chiral (cf. also Table **5-12**, Nos. 28–29 and Sec. **5-3.1** for other "double asymmetric inductions").

5-1.4c. *The Reduction of dl-2-(o-Tolyl)cyclohexanone.* The (−)-isobornyloxymagnesium bromide reduction of 2-(*o*-tolyl)cyclohexanone is of special interest because of the chirality of both the substrate and reducing agent.[20] This serves as an example for the reduction of chiral cyclic ketones by chiral reagents, although in fact a racemic ketone was used as substrate. The product, 2-(*o*-tolyl)-cyclohexanol was a mixture of 92 % *cis* and 8 % *trans*

O, Me — 15 → — OH, H, Me — and — H, H, O, Me

RS — *R,R* and *S,S*, 92% (2% e.e. *R,R*) — *R,S* and *S,R*, 8% (% e.e. unknown)

isomers. The *cis* diastereomer had an estimated 2 % excess of the *R,R* over the *S,S* isomer.* Assuming 100 % reaction, if one started with pure (*R*)-2-(*o*-tolyl)-cyclohexanone, one should obtain a 94 % yield of the *R,R* isomer and a 6 % yield of the *R,S* isomer. Thus the overpowering stereochemically controlling factor for formation of the *cis* isomer is the axial versus equatorial attack (Sec. **3-4**) and not the chirality of the reducing agent. It would be interesting

[20] (a) A. C. Huitric and T. R. Newell, *J. Pharm. Sci.*, **52**, 608 (1963). (b) A. C. Huitric and J. B. Carr, *J. Org. Chem.*, **26**, 2648 (1961). (c) D. R. Galpin and A. C. Huitric, *J. Pharm. Sci.*, 447 (1968). Additional details concerning the resolution of *o*-tolylcyclohexanol are reported in *J. Org. Chem.*, **33**, 921 (1968).

* The percent asymmetric synthesis can only be estimated since the published rotation data for the resolution[20c] was given for methanol, while that from the asymmetric synthesis was for ethanol solvent.

to know the % e.e. for the *trans product* which may be quite different from that for the *cis*.

5-1.5. MPV-Type Reduction Using a Chiral Alkoxyaluminodichloride. A reagent prepared from 4 moles of a chiral alcohol, 1 mole of lithium aluminum hydride, and 3 moles of aluminum chloride is another type of asymmetric reducing agent.[21] Presumably this reagent is an alkoxyaluminodichloride ($AlCl_2OR^*$) which is formed according to the following equations. This is a more facile reducing agent than the standard Meerwein–Ponndorf–Verley reagent. It is distinct from the reagent prepared from 1 mole of lithium aluminum hydride and less than 4 moles of chiral alcohol

$$LiAlH_4 + 4R\overset{*}{O}H \longrightarrow LiAl(OR^*)_4 + 4H_2$$

$$LiAl(OR^*)_4 + 3AlCl_3 \longrightarrow LiCl + 4AlCl_2OR^*$$

which is discussed in Sec. **5-3**. The stereoselectivity of this alkoxyaluminodichloride reagent was not significantly altered by preparing it by first mixing aluminum chloride with ethereal LAH and then introducing the chiral alcohol (Method A) or by adding LAH to the chiral alcohol to form the lithium tetra-alkoxide followed by the addition of an ethereal solution of aluminum chloride (Method B). When the ketone to be reduced was added to preformed lithium aluminum tetra-alkoxide, there was no reduction under the usual reaction conditions; when aluminum chloride was added to such a mixture (Method C), reduction started but the stereoselectivity was quite different from that observed by Method A or B. The results in Table **5-4** are for Method A. The predominate products reported in this table have the configuration which would be expected by considering the same steric factors as discussed for the Meerwein–Ponndorf–Verley reduction (cf. formulas **12**, **17**, **26A**). The extents of asymmetric synthesis range up to a maximum of 70% for reduction of phenyl isopropyl ketone by the isobornyl reagent. The extents of asymmetric synthesis of the phenyl alkyl ketones somewhat parallel those for the related Grignard asymmetric reductions (Table **5-7**). It must be significant that in both of these asymmetric reductions the stereoselectivity increases in going from phenyl ethyl ketone to phenyl *i*-propyl ketone but decreases in going from the latter to phenyl *t*-butyl ketone. Attempts to rationalize this behavior[21d] are discussed in Sec. **5-2.3a**.

5-1.6. MPV-Type Asymmetric Reductions Using Chiral Alkali Metal Alkoxides. In an experiment which preceded the discovery of the asymmetric MPV aluminum alkoxide reduction, Doering and Aschner*

[21] (a) D. Mea-Jacheet and A. Horeau, *Bull. Soc. Chim. Fr.*, 3040 (1966). (b) D. Nasipuri and G. Sarkar, *J. Indian Chem. Soc.*, **44**, 165 (1967). (c) D. Nasipuri and C. K. Ghosh, *ibid.*, **44**, 556 (1967). (d) D. Nasipuri, G. Sarkar, and C. K. Ghosh, *Tetrahedron Lett.*, 5189 (1967).

* Paper presented at the September 17, 1947, 112th Meeting of the American Chemical Society, Abstracts, p. 216; referred to in Reference 3 but not otherwise published.

TABLE 5-4

Carbonyl Reductions with Chiral Alkoxyaluminodichlorides[a]

$$R-\overset{\overset{\displaystyle O}{\|}}{C}-R' \xrightarrow{AlCl_2OR^*} RCHOHR'$$

No.	R* of Reagent	R—COR′	Carbinol % e.e.	Carbinol Config.	Ref.
1	(−)-Isobornyl[b]	MeCOEt	—	*R*	21a
2		MeCOEt	1–3	*R*	21b
3		MeCO-*i*-Bu	5–6	*R*	21b
4		MeCO-*i*-Pr	12–16	*R*	21b
5		MeCO-*t*-Bu	22	*R*	21a
6		MeCO-*t*-Bu	12–18	*R*	21b
7		PhCOMe	19–25	*R*	21b
8		PhCOEt	35–38	*R*	21b
9		PhCO-*i*-Bu	45–67	*R*	21b
10		PhCO-*i*-Pr	66–70	*R*	21b
11		PhCO-*t*-Bu	23	*R*	21d
12		PhCO-*c*-Hex[c]	40	*R*	21d
13		PhCOCOOH	7	*S*	21c
14	(−)-Bornyl	PhCOCOOH	52	*S*	21c
15	(−)-Menthyl	PhCOCOOH	1	*S*	21c
16	(−)-Isobornyl	PhCOCOOEt	14–17	*S*	21c
17	(−)-Bornyl	PhCOCOOEt	9	*S*	21c
18	(−)-Menthyl	PhCOCOOEt	4	*S*	21c
19	(−)-Isobornyl	PhCOCN	2–4	*S*	21c
20	(−)-Bornyl	PhCOCN	9	*S*	21c

[a] Reduction carried out according to Method A, Reference 21b.
[b] This reagent is a mixture of approximately 90% isobornyl and 10% bornyl; the alcohol was obtained by the reduction of camphor with LAH.
[c] *c*-Hex represents the cyclohexyl group.

reported that in benzene the sodium alkoxide made from (−)-2-methyl-1-butanol caused the reduction of phenyl isopropyl ketone to (−)-phenylisopropylcarbinol (2% e.e.). The presumption is that in benzene the sodium alkoxide is a tight ion pair and acts like aluminum or magnesium in forming a cation bridge in a six-membered transition state analogous to **4** and **17**; however, the aluminum alkoxide of (−)-2-methyl-l-butanol failed to give asymmetric reduction.[13] The reagent from the primary alcohol (−)-2-methyl-1-butanol is a special case because there are diastereotopic hydrogens on the α-carbon atom, of which either can be transferred. Thus the chiral center of the reducing agent is not directly incorporated into the ring of the six-membered transition state. A comparable situation is discussed in detail in Secs. **5-1.3** and **5-2.3d**.

Červinka and co-workers[22a,22b] found that asymmetric reductions were consistently obtained using potassium alkoxides of (*S*)-(+)-methyl-*t*-butylcarbinol, (*S*)-(−)-methylphenylcarbinol, (−)-quinine, and (+)-quinidine in benzene solvent. In Nos. 1, 2, 3, and 6, Table **5-5**, where the configurations

TABLE 5-5

Potassium Alkoxide Asymmetric MPV Reductions[a,b]

$$R_SCOR_L \xrightarrow{R^*OK} R_S\overset{*}{C}HOHR_L$$

No.	Ketone		Optical activity of product from R*OK reductions[a,b]			
	R_S	R_L	*S*(+) Me–CHOK[a]–CMe_3 $[\alpha]_D$ °	*S*(−) Me–CHOK[a]–Ph $[\alpha]_D$ °	(−)-Quinine[b] and KO-*t*-Bu $[\alpha]_D$ °	(+)-Quinidine[b] and KO-*t*-Bu $[\alpha]_D$ °
1	Me	Cyclohex	*S*(+) 0.52[c]	*S*(+) 0.42[c]	—	—
2	Me	α-Naphth	*S*(−) 1.70[d]	*S*(−) 1.05[d]	—	—
3	*t*-Bu	Ph	*S*(−) 4.70[e]	*S*(−) 4.03[e]	—	—
4	Ph	*o*-Tolyl	*S*(+) 1.50[f]	*R*(−) 1.38[f]	*S*(+) 1.79[f]	*R*(−) 1.78[f]
5	Ph	α-Naphth	*S*(−) 4.95[f]	*R*(+) 3.18[f]	*S*(−) 4.72[f]	*R*(+) 1.53[f]
6	Cyclohex	Ph	*S*(−) 5.77[g]	*S*(−) 4.55[g]		

[a] Data from Reference 22c, rotations in ethanol.
[b] Data from Reference 22b, rotations in ethanol.
[c] Maximum rotation $[\alpha]_D^{20}$ 5.6° (neat).
[d] Maximum rotation $[\alpha]_D^{20}$ 76° (ethanol).
[e] Note that Ph is represented as R_L and *t*-Bu as R_S contrary to assumption in Reference 22c where it was incorrectly concluded that the configuration was *R*-(−). Maximum rotation is $[\alpha]_D$ 36.2° (ether); 30.6° (acetone); 27.3° (benzene); unreported in ethanol.
[f] Maximum rotation unknown; absolute configuration inferred from these and other empirical correlations (Table **5-12**).
[g] Maximum rotation is $[\alpha]_D^{20}$ 37.6° (ether); $[\alpha]_D^{20}$ 22° (benzene); unreported in ethanol.

of the product carbinols are known, the enantiomer produced in excess is the one which is predicted by the considerations of models **12A** and **12B** [with K replacing $Al(OR)_2$]. However in Nos. 4 and 5 the purely aliphatic methyl-*t*-butyl reagent gave *S* carbinols while the methyl-phenyl reagent gave the *R* carbinols. Apparently there is an added electronic interaction

[22] (a) O. Červinka, V. Suchan, and B. Masař, *Coll. Czech. Chem. Comm.*, **30**, 1693 (1965). (b) O. Červinka, O. Bělovský, A. Fabryová, V. Dudek, and K. Grahman, *Coll. Czech. Chem. Commun.*, **32**, 2618 (1967).

introduced when the aromatic rings are incorporated into both substrate and reagent (cf. Sec. **5-2.3a**). Potassium alkoxides of (−)-quinine and (+)-quinidine, with opposite configurational relationships at C-8 and C-9, gave enantiomeric products as expected. The extents of asymmetric synthesis appear to vary from 1 to 25% but cannot be determined accurately since the observed rotations are recorded in different solvents from those for which the rotations of the pure enantiomers are known.

5-1.7. Use of the Asymmetric MPV Reduction to Determine the Configuration of Biphenyl Derivatives. A discussion of a closely related MPV reduction, which is a partial kinetic resolution of a racemate and not an asymmetric synthesis, is very pertinent to the present subject. Mislow and co-workers[23] have used the chiral alkoxide reduction of racemic bridged biphenyl ketones such as (+)-4′,1″-dinitro-1,2,3,4-dibenz-1,3-cycloheptadiene-6-one (**27**) to establish the absolute configuration of the bridged biphenyl and its derivatives. The *R* and *S* forms of both the ketone **27** and the corresponding carbinol **28** are shown in Fig. **5-5**.

It is important to note that the chirality of the carbinol **28** is determined by the axial dissymmetry of the biphenyl system and not by the carbinol center.

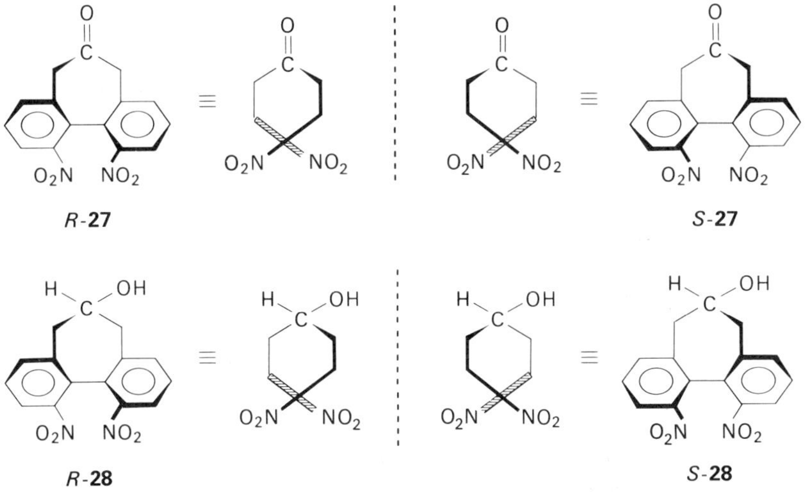

Fig. 5-5. Configurational representations of enantiomeric (*R*)- and (*S*)-4′,1″-dinitro-1,2,3,4-dibenz-1,3-cycloheptadiene-6-one and -6-ol.

[23] (a) P. Newman, P. Rutkin, and K. Mislow, *J. Am. Chem. Soc.*, **80**, 465 (1958). (b) K. Mislow and F. A. McGinn, *ibid.*, **80**, 6036 (1958). (c) K. Mislow, R. E. O'Brien, and H. Schaefer, *ibid.*, **84**, 1940 (1962). (d) K. Mislow, R. Graeve, A. J. Gordon, and G. H. Wahl, *ibid.*, **86**, 1733 (1964).

This is readily seen by looking at models; interchanging the positions of H and OH in the "end-on" view of *R*-**28** is equivalent to viewing the structure from behind the plane of the page instead of from in front as shown in Fig. **5-5** and *does not* invert the configuration of the molecule. Thus reduction of *R*-**27** produces only one alcohol, namely *R*-**28**, regardless of which face of the carbonyl group is attacked. It is established that reduction of the pure (+) isomer of **27** gives (+)-**28**. Thus the ketone and carbinol with the same absolute axial dissymmetry have the same signs of rotation.

The partial reduction of racemic ketone *R*,*S*-**27** by (*S*)-(+)-methyl-*t*-butylcarbinol in the presence of aluminum *t*-butoxide produces a carbinol **28** which is levorotatory and leaves behind unreduced ketone which is dextrorotatory. The established nature of the MPV reduction permits a clear-cut decision concerning which isomer will be reduced more rapidly by aluminum (*S*)-(+)-methyl-*t*-butylcarbinylalkoxide **29**. In **30A**, Fig. **5-6**, the

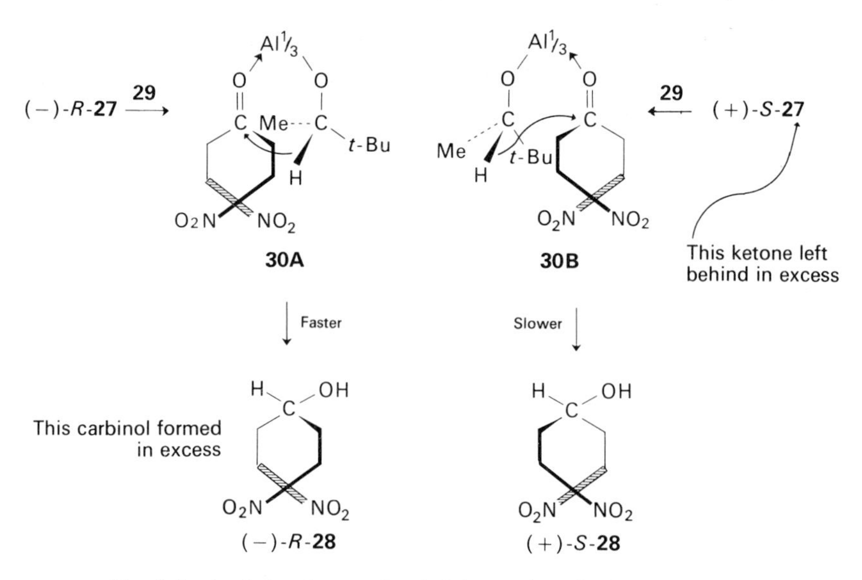

Fig. 5-6. Analysis of stereochemical interactions in the MPV reduction of (+)-4′,1″-dinitro-1,2,3,4-dibenz-1,3-cycloheptadiene-6-one (**27**) to the -6-ol (**28**) by aluminum *S*-(+)-methyl-*t*-butylcarbinylalkoxide (**29**).

bulky *t*-butyl group opposes a hydrogen of the methylene group while the smaller methyl group interferes with the edge of the phenyl ring. In **30B** the reverse is true. There is considerably more steric interaction with the protruding edge of the phenyl ring than with the methylene hydrogen (best seen by inspecting models) and thus the transition state represented by **30A** is clearly favored. Accordingly ketone *R*-**27** will be consumed faster than *S*-**27** at the beginning of the reaction and carbinol *R*-**28** will be formed preferentially. Since incomplete reduction in this reaction gave a product enriched in the (−) carbinol and left behind unreacted ketone enriched in the (+) isomer, it follows that the configurations are *R*-(−)-**28** for the carbinol and *S*-(+)-**27** for the ketone. These assignments have now proved to be self-consistent with other evidence.[23]

5-2. The Asymmetric Grignard Reduction Reaction

5-2.1. Introduction and Mechanism. The major reaction of a Grignard reagent with a carbonyl compound is often *reduction** rather than *addition*, as illustrated in the following equation:

$$R_2C{=}O + R'_2CHCH_2MgX \longrightarrow \begin{cases} R_2C(OMgX){-}CH_2CHR'_2 & \text{(addition)} \\ R_2CH(OMgX) + R'_2C{=}CH_2 & \text{(reduction)} \end{cases}$$

The reduction reaction was recognized by Grignard in his original investigations,[24] and in 1929 Conant and Blatt[25] observed that reduction was favored in reactions with sterically hindered ketones (**31**). Whitmore and George[26] noted a correlation between the extent of reduction of diisopropyl ketone and the availability of hydrogens on the *beta* carbon atom of the Grignard reagent (**32**). Particularly revealing was the fact that neopentyl Grignard reagent with no β-hydrogens gave no reduction, while *t*-butyl Grignard reagent, with nine β-hydrogens, gave 95% reduction. Based on these observations, Whitmore proposed† that the Grignard reduction reaction proceeds by the reversible formation of a complex (**33**) which then undergoes a direct transfer of hydrogen from reagent to substrate. When the carbonyl group in **31** is sterically hindered, the addition reaction is repressed and transfer of the small hydrogen on the *beta* carbon atom of the Grignard reagent is favored as represented in **34B** to give the magnesium alcoholate **35** and the olefin **36**. The overall result, after hydrolysis, is the reduction of the ketone

[24] V. Grignard, *Ann. chim.*, (7) **24**, 433 (1901).

[25] J. Conant and A. H. Blatt, *J. Am. Chem. Soc.*, **51**, 1227 (1929).

[26] F. C. Whitmore and R. S. George, *J. Am. Chem. Soc.*, **64**, 1239 (1942).

* The reduction reaction of Grignard reagents with carbonyl compounds has been reviewed by M. S. Kharasch and O. Reinmuth, *Grignard Reactions of Nonmetallic Substances* (Englewood, N.J.; Prentice-Hall, Inc., 1954) pp. 147–160.

† Presented at the 143rd Meeting of the American Chemical Society, Atlantic City, April, 1943. Details are to be found in the Ph.D. thesis of R. S. George, The Pennsylvania State College, July, 1943.

to the carbinol by the Grignard reagent, which is oxidized to the olefin. The production of equivalent amounts of olefin and carbinol has been demonstrated repeatedly. Furthermore, exclusive β-hydrogen transfer has been established by deuterium tracer experiments.[27] The similarities of the

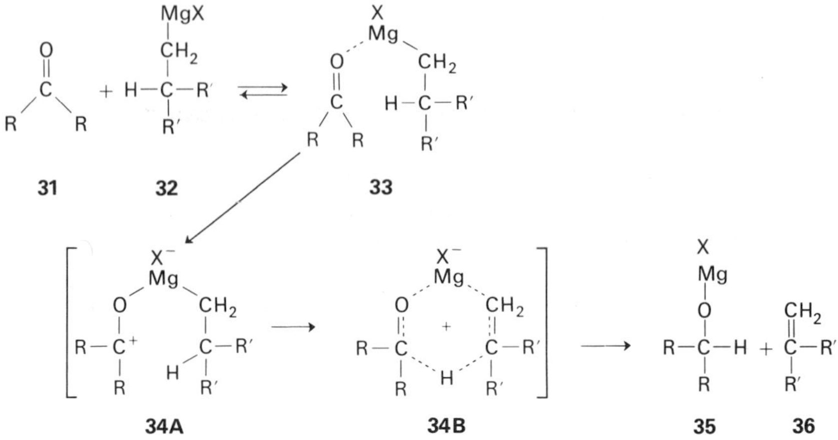

Fig. 5-7. Whitmore mechanism for the Grignard reduction reaction.

mechanisms for the Meerwein–Ponndorf–Verley reduction and the Grignard reduction, which was proposed prior to the developments discussed in Sec. **5-1**, are apparent. An important difference, however, is that the Grignard reaction is essentially irreversible in contrast to the equilibrium of the alkoxide hydrogen transfer process.

Vavon and co-workers were the first to observe an asymmetric Grignard reduction in the reaction of acetophenone (**38**) with the reagent (**37**) prepared from "pinene hydrochloride"[28] to give *R*-(+)-methylphenylcarbinol (**40**, 30% e.e.). This reagent consists of a mixture of the bornyl- and isobornylmagnesium chlorides; presumably the isobornyl species (**37**) is the active

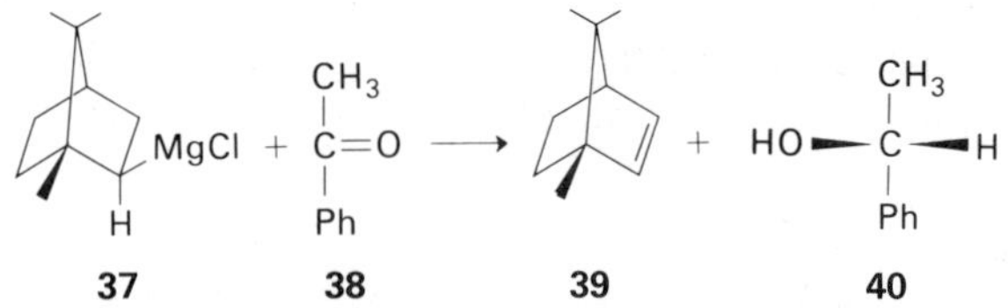

reducing agent (cf. Sec **5-2.3c**). This reaction was extended to a series of alkyl phenyl ketones,[29] but no attempt was made to interpret these reactions

[27] G. E. Dunn and J. Warkentin, *Can. J. Chem.*, **34**, 75 (1956).

[28] G. Vavon, C. Rivière, and B. Angelo, *Compt. rend.*, **222**, 959 (1946).

[29] G. Vavon and B. Angelo, *Compt. rend.*, **224**, 1435 (1947).

in terms of a cyclic hydrogen transfer mechanism nor to analyze the factors leading to the asymmetric synthesis other than to point out that there was a relationship between the steric hindrance of the ketone and the extent of asymmetric reduction.

The first of a series of investigations specifically designed to study the nature of this reaction and to employ it as a tool for studying the factors involved in the transition state for the asymmetric reduction reaction was published in 1950.[30] In these initial studies the Grignard reagent from *S*-(+)-1-chloro-2-methylbutane, which has a single asymmetric center at the position *beta* to the magnesium atom, was employed. The reduction of methyl *t*-butyl ketone (**41**) by this reagent (**42**) is represented in Fig. **5-8**.

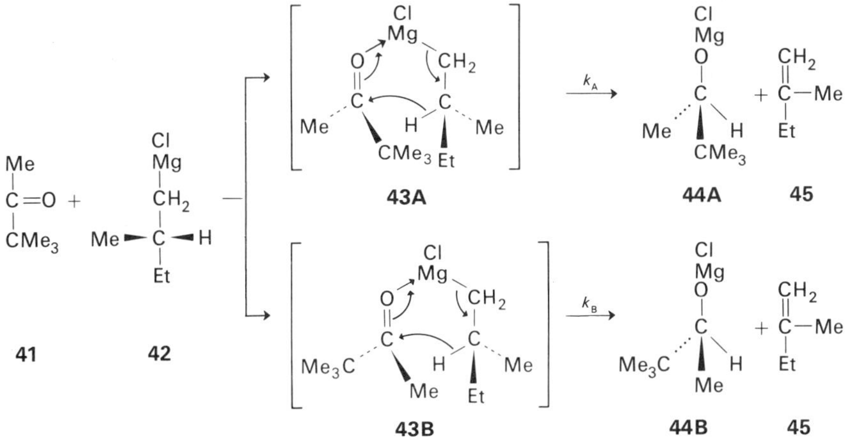

Fig. 5-8. Competing reactions in the asymmetric reduction of methyl *t*-butyl ketone by the Grignard reagent from *S*-(+)-1-chloro-2-methylbutane.

The reagent can attack either enantiotopic face of the carbonyl group via transition states represented by **43A** and **43B**. Since the transition states are diastereomeric, the rates of the two reactions to give either **44A** or **44B** must be different. Common starting materials are used in both processes and the products differ only in the chirality of one of the products. Thus the ground state energies of starting materials for, as well as products from, the competing processes are identical. Therefore from an energy standpoint, the processes differ solely in their energies of activation. In **43A**, the two larger groups are in juxtaposition in front of the plane of the six-membered cyclic transition state as written, while the two smaller groups are in juxtaposition behind the plane of the transition state. On the other hand, in **43B**, one large and one small group from the substrate and reagent oppose each other both in front

[30] H. S. Mosher and E. La Combe, *J. Am. Chem. Soc.*, **72**, 3994 (1950).

and in back. Since this latter situation offers a better "fit" of the groups during the hydrogen transfer process, one predicts that **43B** represents the transition state of lower energy and therefore, *S*-(+)-methyl-*t*-butylcarbinol from **44B**, would be the preferred product. This prediction is realized (12% e.e. *S*-(+) isomer).

This picture for the hydrogen transfer process is incomplete in several details. The ether molecule which is certainly coordinated with the magnesium atom is not shown. If an ether molecule and chloride ion are both coordinated to the magnesium atom, then the magnesium constitutes another chiral center whose effect on the asymmetric synthesis has not been taken into consideration in **43A** and **43B**. Also, the transition state has been represented as essentially planar, while it must in fact be able to minimize the various interactions by assuming more of a chair-like conformation. Depending upon how far the bond making and breaking has progressed before the transition state is reached, it will have more or less the equivalent of two "half double bonds" (cf. **34B**). To the extent this is so, the transition state cannot be considered a cyclohexane chair conformation.

There is still some uncertainty as to the reactive species in the Grignard reduction. Is it monomeric as represented or more highly associated as is now rather well established in the case of the aluminum alkoxide reductions* (Sec. **5-1.2**)? The interpretation of the stereochemistry of the reaction is further complicated by the fact that the Grignard reagent may react via an R_2Mg species, and as the reaction progresses, via an RMgOR′ species. It must also be kept in mind that the magnesium alkoxide product [**44** or RMgOR′ or $Mg(OR')_2$] has the potential for reacting in a MPV-type reduction.[31] These points will now be considered under a discussion of the experimental conditions and limitations of the reaction.

5-2.2. Experimental Considerations. Studies have been made, primarily on the reduction of methyl *t*-butyl ketone with the Grignard reagents from (+)-1-halo-2-methylbutanes (Fig. **5-8**), to test the effect on the stereoselectivity caused by certain experimental variables. Although the synthetic yield of reduction product was quite dependent on the halide

[31] R. Hamelin, *Compt. rend.*, **250**, 1081 (1960).

* The associations of pure methyl, ethyl, and *t*-butyl Grignard reagents in diethyl ether have been measured [E. C. Ashby and F. Walker, *J. Organometal. Chem.* **7**, 17 (1967)]. The reagent from *t*-butyl chloride has an *i* value of approximately 2 in the concentration range of 0.2–2.0 molar. The association of ethyl Grignard shows a higher *i* value than this, and that for the reagents which have been used in the asymmetric synthesis have not been studied but presumably are comparable. More important however is the established fact that the association of the species in a Grignard solution goes up rapidly in the presence of oxygen or alkoxy species which of course are formed as the reaction progresses [M. S. Singer, R. M. Salinger, and H. S. Mosher, *J. Org. Chem.*, **32**, 3821 (1967) and further references in Footnote 12 therein].

(chloride, 29%; bromide, 14%; iodide 0.5%), the effect on asymmetric synthesis was minor (chloride, 13%; bromide, 12%; iodide, 8%).[32] Based on these results, the chlorides generally have been preferred for the Grignard asymmetric reduction studies.

The stereoselectivity of this reaction is not particularly temperature dependent. The reduction of methyl *t*-butyl ketone by the reagent from (+)-1-chloro-2-methylbutane in ether solvent[30] shows an increase in stereoselectivity from 13 to 16% when the temperature is lowered from 35 to −75°. The reduction of phenyl isopropyl ketone by the Grignard reagent from (+)-1-chloro-2-phenylbutane in tetrahydrofuran (THF) solvent also shows an increase in stereoselectivity from 66 to 78% when the temperature is lowered from 66 to 0°.[33] The asymmetric reduction at 25° in ether solvent was 82%. The extent of asymmetric synthesis is favored in ether versus THF solvent not only in this case but also in the reduction of phenyl *t*-butyl ketone by the reagent from (+)-1-chloro-2-phenylpropane, where the stereoselectivity dropped from 22 to 4% with this solvent change.[34] There has been no systematic study of the effect of solvent on the stereoselectivity of these asymmetric reductions.

The problem of the change in reducing species as the reaction progresses is illustrated by the following equation where the Grignard reagent reacts with the initially produced alkoxide (such as **35**) to give either **46** or **47**.

$$\underset{\textbf{35}}{R^*MgX + R'OMgX} \longrightarrow \underset{\textbf{46}}{R^*MgX \leftarrow \overset{R'}{\overset{|}{O}}-MgX} \quad \text{or} \quad \underset{\textbf{47}}{R^*MgOR' + MgX_2}$$

These new chiral reagents (**46** and **47**) might have appreciably different steric requirements than the initial Grignard reagent and might therefore lead to a progressive change in the stereoselectivity of the reaction as it progresses. This was first tested by reversing the order in which the reagents were mixed. When the ketone was added to the Grignard reagent (normal addition, Grignard always in excess), the stereoselectivity was the same, within experimental error, as when the Grignard was added to the ketone (reverse addition, ketone in excess).[30] Furthermore the stereoselectivity was essentially unchanged by adding one equivalent of ketone to two equivalents of reagent or by first adding one equivalent of acetaldehyde to two equivalents of reagent followed by one equivalent of the ketone substrate to be reduced.[35]

[32] W. M. Foley, F. J. Welch, E. M. La Combe, and H. S. Mosher, *J. Am. Chem. Soc.*, **81**, 2779 (1959). The analysis of the iodide is reported by J. A. Dale, D. L. Dull, and H. S. Mosher, *J. Org. Chem.*, **34**, 2543 (1969).

[33] J. S. Birtwistle, K. Lee, J. D. Morrison, W. A. Sanderson, and H. S. Mosher, *J. Org. Chem.*, **29**, 37 (1964).

[34] C. Aaron, Unpublished results, Stanford University, 1964.

[35] D. O. Cowan and H. S. Mosher, *J. Org. Chem.*, **28**, 204 (1963).

Thus if the nature of the reducing agent does change during the course of the reduction, the stereoselectivity of the various reducing species toward methyl *t*-butyl ketone are very nearly the same.

The extent of asymmetric synthesis obtained using the *dialkylmagnesium* reagent prepared from (+)-1-chloro-2-methylbutane was almost the same (11–12%) as that obtained with the usual Grignard reagent (13%).[36]

In studies of the asymmetric Grignard reaction, the synthetic yield of reduction product often becomes a practical consideration since these reactions always involve the competition of the addition as well as the enolization reaction versus reduction. Thus addition to give a secondary alcohol is the major reaction in the treatment of benzaldehyde-1-*d* with the Grignard reagent from (+)-1-chloro-2-methylbutane and only a 21% yield of reduction product, benzyl-α-*d* alcohol,[37] is obtained. This inherently low yield becomes a practical limitation for extension to other deuterio-aldehydes with the exception of highly hindered ones such as trimethylacetaldehyde.[38] In still other cases the rate of reduction may be abnormally slow because of excessive steric hindrance toward transfer of the *beta* hydrogen as in the homobornyl system **48**.[39] Also this is undoubtedly the reason why bornyloxymagnesium bromide **49** usually does not act as a reducing agent in the MPV-type reduction in contrast to isobornyloxymagnesium bromide **15**. Electronic factors

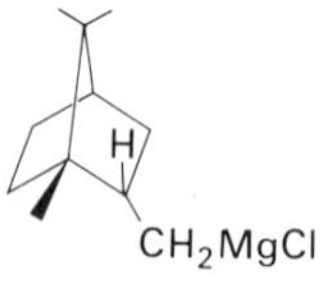

H CH_2MgCl **48** H OMgBr **49** CF_3 Ph—C—CH_2MgCl H **50**

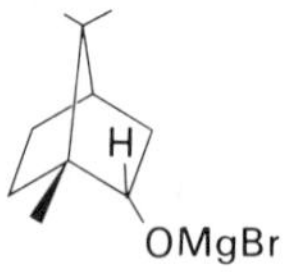

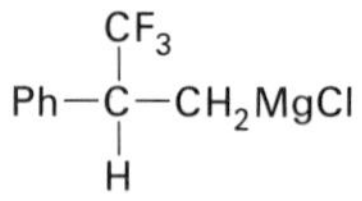

may also retard the α-hydrogen transfer. For this reason, the asymmetric reduction studies making use of the Grignard reagent from (*S*)-(+)-1-chloro-2-phenyl-3,3,3-trifluoropropane (**50**) could not be extended to include a complete series of reductions.[40]

5-2.3. The Effect of Various Substrates and Chiral Reducing Agents in Asymmetric Grignard Reductions.

5-2.3a. *Asymmetric Reductions with the Grignard Reagent from* (+)-1-*Chloro-2-methylbutane.* The largest number of asymmetric Grignard reductions (summarized in Table **5-6**) have been performed with the aliphatic

[36] H. S. Mosher and P. K. Loeffler, *J. Am. Chem. Soc.*, **78**, 4959 (1956).

[37] V. E. Althouse, D. M. Feigl, W. A. Sanderson, and H. S. Mosher, *J. Am. Chem. Soc.*, **88**, 3595 (1966).

[38] V. E. Althouse, E. Kaufmann, P. Loeffler, K. Ueda, and H. S. Mosher, *J. Am. Chem. Soc.*, **83**, 3138 (1961).

[39] R. Koehler, M.S. thesis, Stanford University, 1957.

[40] D. Dull, Ph.D. thesis, Stanford University, 1967; *Chem. Abst.*, **70**, 3115 (1969).

TABLE 5-6

Asymmetric Reductions by the Grignard Reagent from (+)-1-Chloro-2-methylbutane

$$R_S(R'_L)C{=}O + Me\text{—}C(H)(Et)\text{—}CH_2\text{—}MgCl \longrightarrow [\text{52}] \longrightarrow R'_L(R_S)C(H)OH + CH_2{=}C(Me)\text{—}Et$$

51 **42** **52** **53** **45**

52: Predominant transition model

53: Predominant isomer

No.	R_S	A: R'_L = *t*-butyl[a] %e.e.	Config.	B: R'_L = cyclohexyl[b] %e.e.	Config.	C: R'_L = phenyl[c] %e.e.	Config.
1	D	12[d]	$S(+)$		—	19[e]	$S(+)$
2	Me	13	$S(+)$	4	$S(+)$	4	$S(-)$
3	Et	11	$S(-)$	9	$S(-)$	6	$S(-)$
4	*n*-Pr	11	$S(-)$	9	$S(-)$	6	$S(-)$
5	*n*-Bu	11	$S(-)$	9[f]	$S(-)$	6[f]	$S(-)$
6	*i*-Bu	6	$S(-)$	16	$S(-)$	10	$S(-)$
7	*i*-Pr	0	$S(-)$	2	$S(+)$	24	$S(-)$
8	C_6H_{11}	2[b]	$S(-)$		—[g]	25	$S(-)$
9	*t*-Bu		—[g]	[2][h]	$S(-)$	16	$S(-)$

(Percent asymmetric reductions with R′ equal to: A, B, C.)

[a] Data in Reference 32, unless otherwise noted.
[b] Data in Reference 41.
[c] Data in Reference 42, unless otherwise noted.
[d] See Reference 38.
[e] See Reference 43.
[f] These values are slightly lower than reported in the original publications[41,42] because the value for the maximum rotation of *n*-butylphenylcarbinol has been revised upward.[44]
[g] These are degenerate cases in which the two R groups are identical and thus lead to an achiral product.
[h] Brackets indicate configuration opposite to that labeled "predominant isomer" at the top of the table. Note that cyclohexyl is now smaller (R_S) than *t*-Bu (R'_L).

[41] E. P. Burrows, F. J. Welch, and H. S. Mosher, *J. Am. Chem. Soc.*, **82**, 880 (1960).
[42] R. MacLeod, F. J. Welch, and H. S. Mosher, *J. Am. Chem. Soc.*, **82**, 876 (1960).
[43] J. Keiner, Unpublished results, Stanford University, 1966.
[44] A. Horeau, J. P. Guetté, and R. Weidmann, *Bull. Soc. Chim. Fr.*, 3513 (1966).

reagent from *S*-(+)-1-chloro-2-methylbutane (**42**). The simplest series is represented by the reductions of alkyl *t*-butyl ketones (**51A**, R_S = alkyl, R'_L = *t*-butyl) to give optically active alkyl-*t*-butylcarbinols (**53A**, R_S = alkyl, R'_L = *t*-butyl). In the examples in Column A, the enantiomer with the absolute *S* configuration is produced in excess via the preferred transition state model represented by **52** at the head of the table. Since the R groups in Nos. 1–8 clearly are sterically smaller (R_S) than *t*-butyl (R'_L), these results can be generalized in the form of models **54A** and **54B** for the transition states in which **54A** will be preferred because it represents a better fit of the two pairs of interfering groups, i.e., (R_S versus R''_L) + (R'_L versus R'''_S) in **54A** as opposed to (R_L versus R''_L) + (R'_S versus R'''_S) in **54B**.

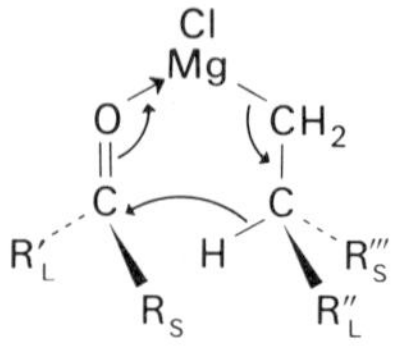

54A

Preferred transition state model

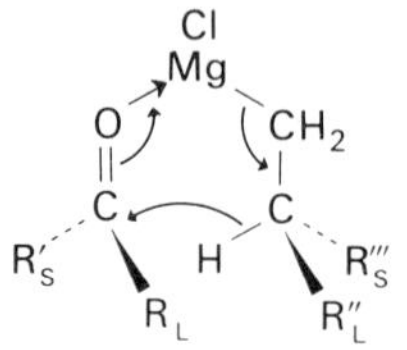

54B

Nonpreferred transition state model

Fig. 5-9. Generalized models for preferred and nonpreferred transition states in the asymmetric Grignard reduction reaction.

It seems logical to deduce from a quantitative standpoint that the greater the difference in steric bulk between R'_L and R_S of the carbonyl compound, the greater will be the difference in energy of activation, $\Delta\Delta G^{\ddagger}$, between transition state models **54A** and **54B**, and therefore the higher the percent asymmetric synthesis. For the reduction of the alkyl *t*-butyl ketones (Table **5-6**, Column A, Nos. 2–8), this deduction is verified. As the steric bulk of the alkyl group R_S increases, so that the bulk of the alkyl group approaches that of the *t*-butyl group, the asymmetric synthesis decreases from 13 to 2%. However the very first example in Table **5-6**, the reduction of trimethylacetaldehyde-1-*d*, on this basis should be higher than the others by a significant amount, since there is the maximum possible difference in group bulk (D versus *t*-butyl). In fact, the asymmetric synthesis observed (12% e.e.) is essentially the same as that for reduction of methyl *t*-butyl ketone (13% e.e.). It can be argued that a certain critical bulk is needed before there is significant steric interaction and thus going below methyl in size is no advantage. There is no significant difference between the asymmetric reduction observed when R_S is ethyl, *n*-propyl, or *n*-butyl. This is true for the data reported in Columns A, B, and C and indicates that the critical steric bulk is that which is manifest by branching on the α-carbon atom. For this reason in other investigations, the *n*-propyl and *n*-butyl homologs usually have not been studied. Thus when the groups involved are purely aliphatic, the transition state model **54** not only correlates the absolute configuration but also satisfactorily predicts the relative order of extents of asymmetric synthesis. It should be noted that the conformation of the alkyl group itself cannot be ignored. Thus, if the ethyl and isopropyl groups were spinning rapidly with respect to the rate of hydrogen transfer, they would sweep through the volume of a cone which would be the same volume as occupied by the *t*-butyl group; therefore, their effective steric bulks and consequently their asymmetric syntheses should be comparable, contrary to observation.

In Column B of Table **5-6** the asymmetric reductions of various alkyl cyclohexyl ketones by the same reagent are reported. In each case the model (**54A** versus **54B**) correctly predicts the absolute *configuration* of the carbinol produced in excess but there is no such clear quantitative *trend in the extents* of asymmetric reduction as was apparent in the alkyl *t*-butyl series. In fact, in going from methyl to ethyl to *i*-butyl, the extents of asymmetric synthesis increase in contrast to the results in Column A. But with isopropyl cyclohexyl ketone, which has the same extent of branching on each side of the carbonyl group, the asymmetric synthesis does decrease (to 2%) as expected. In the case of cyclohexyl *t*-butyl ketone, the *t*-butyl substituent is now larger than cyclohexyl with respect to branching on the α-carbon atom and one observes as predicted that the predominant enantiomeric carbinol has the configuration opposite to that of **52** given at the top of the table, although the extent of asymmetric synthesis is smaller than anticipated (2% e.e.; data in brackets to indicate opposite stereochemistry to **52**). Since *t*-butyl takes precedence over cyclohexyl in the Cahn–Ingold–Prelog nomenclature scheme, the predominant product here has the same *S* configurational designation as the other carbinols produced in this series of asymmetric reductions.

The configuration of the predominant product from the reduction of the phenyl alkyl ketones shown in Column C of Table **5-6** is correctly predicted in every case by the proposed model with the assumption that phenyl is larger than all alkyl groups, even *t*-butyl. With the exception of the first and last examples in Column C, increasing the bulk of the alkyl group of the alkyl phenyl ketone increases the extent of asymmetric synthesis in a manner *quite the opposite of that shown in the data for the alkyl t-butyl ketones* of Column A. The fact that α-deuteriobenzaldehyde gives a much higher asymmetric synthesis than acetophenone is what would be expected intuitively since the difference in steric bulk between D and phenyl is obviously much larger than that between methyl and phenyl. But this is not what is observed in the comparable *t*-butyl series.

The case of phenyl *t*-butyl ketone presents a special situation and embodies conformational possibilities not inherent in the purely aliphatic ketones. The methyl and *t*-butyl groups possess C_{3v} symmetry and, like a spinning top, can be considered to offer about the same steric interaction regardless of the specific orientation of the group. But the phenyl group with C_{2v} symmetry must exhibit quite different steric and electronic interactions depending upon its orientation. Interaction with the adjacent group will be quite different when the flat, π electron face of the phenyl ring is involved as in **55A** as compared with the situation in which the hydrogen-bearing edge is involved as in **55B**. In **55A** the carbonyl group and phenyl

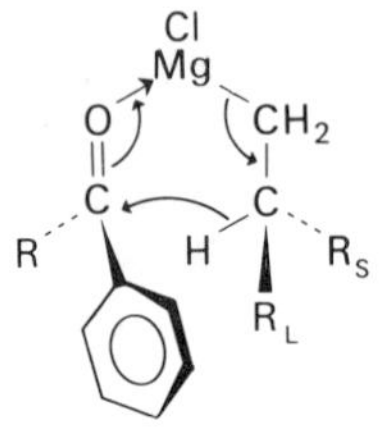

55A 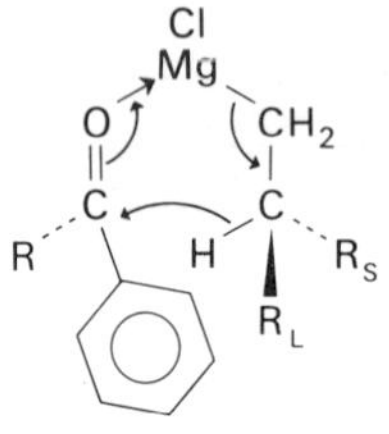 **55B** 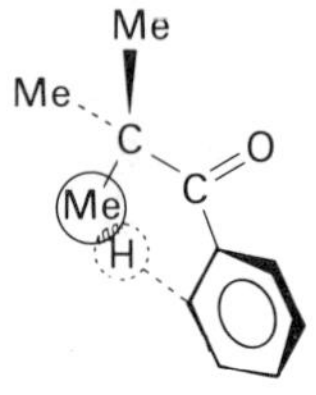 **56**

ring are coplanar and are stabilized thereby. The noncoplanar situation in **55B** would not be expected to be important unless some factor prevented coplanarity. Such a factor is present in phenyl *t*-butyl ketone and may explain why the upward trend in extent of asymmetric synthesis (shown in Table **5-6**, Column C) does not continue through the *t*-butyl homolog. The ultraviolet spectra of the alkyl phenyl ketones clearly show that the *t*-butyl homolog is anomalous, indicating that the *t*-butyl group prevents the carbonyl and phenyl groups becoming coplanar[45] as indicated in **56**. This noncoplanarity is clearly reflected in the relative rates of reduction of these alkyl phenyl ketones using sodium borohydride.[46] It is difficult to assess the effect of this on the extent of asymmetric reduction. In the transition state the carbonyl carbon is progressing from the sp^2 to sp^3 state. The steric hindrance to coplanarity should assist this change but the crucial question is, will it assist the pathway leading to one enantiomer more than it will assist that leading to the other?

It is necessary to emphasize that the energy differences being considered here are small; $\Delta\Delta G^{\ddagger}$ for a reaction giving 17% asymmetric reduction at 35° corresponds to about 200 calories (0.2 kcal) per mole. Therefore the interplay of the steric interactions which control these small differences being considered may be very subtle, such as precisely how well the two pairs of groups

[45] G. D. Hedden and W. G. Brown, *J. Am. Chem. Soc.*, **75**, 3744 (1953).
[46] H. C. Brown and K. Ichikawa, *J. Am. Chem. Soc.*, **84**, 373 (1962).

slide by each other during the change of the carbonyl hybridization from sp^2 to sp^3, when one or the other enantiotopic face of the carbonyl is attacked.

The results from the reductions of 2,2-dimethylcyclohexanone and 3,3-dimethylcyclohexanone by the Grignard reagent from *S*-(+)-1-chloro-2-methylbutane are pertinent;[47,48] the 2,2-dimethyl isomer gives the *S*-(+) isomer (18% e.e.) as predicted with slightly better stereoselectivity than in the case of methyl *t*-butyl ketone (13% e.e.) which is the open-chain analog. On the other hand, 3,3-dimethylcyclohexanone gives an asymmetric reduction of less than 2%. It is thus apparent that the important interactions during the hydrogen transfer are those close to the developing chiral center and that in this case the interaction with the β-axial methyl group is not very prominent.

5-2.3b. *Chiral Reagents with Phenyl Groups.* When a phenyl group is incorporated in both substrate and reagent, both conventional steric forces as well as dipole interactions are operative in the transition state. The data dealing with these examples are summarized in Table **5-7**. With one exception (the reduction of phenyl *t*-butyl ketone shown in brackets in the first column) the configurations of the products are those predicted on the basis of the transition state model **54** with the assumption that phenyl acts as R_L with respect to the alkyl groups, methyl through *t*-butyl. The observed configurations of the products also coincide with those predicted on the assumption that there is a strong electronic repulsion between phenyl groups so that the transition state with phenyl groups in juxtaposition would be especially unfavorable. Thus in the examples of Table **5-7**, it would be predicted that the interaction of steric and electronic effects should reinforce each other. This prediction is supported by the generally higher stereoselectivities observed in these reactions as compared with those reported in Table **5-6**.

The reduction of phenyl *t*-butyl ketone by the Grignard reagent from (+)-1-chloro-2-phenylpropane gives the predominant enantiomer opposite to that predicted on the basis of model **54** *assuming t-butyl is smaller than phenyl.* This result is at variance with the previous finding with the Grignard reagent from (+)-1-chloro-2-phenylbutane and serves as a warning that there are some limitations to the use of model **54** for this asymmetric synthesis.

From a quantitative standpoint there is a general trend of increasing stereoselectivity with increasing size of the alkyl substituent of the ketone with the exception of the reduction of phenyl *t*-butyl ketone in the series represented in Columns A and B. The stereoselectivity dropped from 59 to 22% in going from phenyl isopropyl ketone to phenyl *t*-butyl ketone in Column A and from 82 to 16% in Column B. But in the third column the

[47] E. La Combe, Ph.D. thesis, Stanford University, 1950.

[48] M. G. Edelstein, M.S. thesis, Stanford University, 1963.

TABLE 5-7

Asymmetric Reduction of Alkyl Phenyl Ketones by Grignard Reagents from (+)-β-Alkyl-β-phenylethyl Chlorides[a]

$$R(Ph)C{=}O + R'(Ph)(H)C{-}CH_2MgCl \longrightarrow [\text{transition state}] \longrightarrow H{-}C(R)(Ph){-}OH + CH_2{=}C(Ph){-}R'$$

		Percent asymmetric synthesis with *R'* equal to:		
		A[b]	B[c]	C[d]
No.	R	R' = methyl	R' = ethyl	R' = isopropyl
1	Me	38	47	—
2	Et	38	52	66
3	*i*-Bu	—	53	—
4	*i*-Pr	59	82[e]	80
5	*t*-Bu	[22][f]	16	91

[a] In some experiments the enantiomeric reagent was employed and the enantiomeric carbinol was formed from that shown in the equations. The results are presented as though reagents of related configurations were used throughout. All products on this basis had the absolute *S*-(−) configuration with the one exception noted in brackets.[f]

[b] See Reference 34.

[c] See Reference 33.

[d] See Reference 49.

[e] Three runs in THF solvent at 0, 35, and 66° gave percent asymmetric syntheses of 78, 75, and 69%.

[f] This product has the *R*-(+) configuration opposite to that shown in the heading of the table. When the reduction was carried out in THF solvent, *R*-(+) isomer was formed with 4% asymmetric synthesis.

stereoselectivity rose from 80 to 91% with the same change. Since the trend is not uniform, it is difficult to speculate as to its cause. Before the results from the reductions using the reagent from (+)-1-chloro-2-phenyl-3-methylbutane were known (Table **5-7**, Column C), it was proposed that this drop in stereoselectivity with phenyl *t*-butyl ketone was associated with the noncoplanarity of the phenyl ring and carbonyl group in this particular substrate. In effect, for the nonpreferred transition state **57A** there is an increase in the steric interactions of the two phenyl rings but a decrease in the electronic repulsions over that to be expected when the two phenyl rings

[49] J. L. Schmiegel, Ph.D. thesis, Stanford University, 1967; *Dissert. Abstr.*, **28**, 4507 (1967).

could be face to face as in **57C**. The electronic repulsion would not be involved in the preferred transition state **57B** where the two phenyl rings are far apart. Therefore the overall effect might be a balance of those forces which would greatly reduce the difference in activation energies between **57A** and **57B** from that expected if the phenyl ring could be coplanar as in **57C**. But the fact that the stereoselectivity rose to a maximum with the *t*-butyl example in Column C cannot be readily accounted for on the same basis and implies some very specific and subtle differences in the group interactions during the hydrogen transfer step.

Nasipuri and co-workers[50] have attempted a semiquantitative rationalization of the data on the asymmetric reduction of alkyl phenyl ketones. The first assumption is that as the steric bulk of the alkyl group increases, other things being equal, the stereoselectivity should decrease. The second assumption is that the slower the rate of reduction, other things being equal, the higher the stereoselectivity. Since the greater the bulk of the alkyl group, the slower the reduction these two factors are acting in opposition to each other. Finally, it is specifically assumed that the conformation of the alkyl group in the transition state is as delineated by Karabatsos for reactions of ketones with an α-chiral center (Sec. **3-3**). There is no sound theoretical basis for the second assumption in spite of the empirical evidence cited in its favor. Furthermore this proposal is subject to the same difficulties as encountered in the discussion of **57A**, **57B**, and **57C** (and the data in Tables **5-7** and **5-8** which were not available in their entirety to these authors).

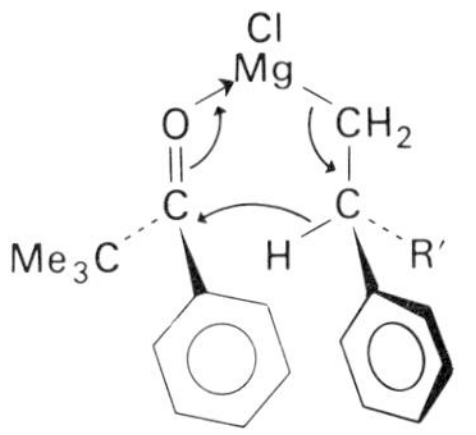

Cl Mg O CH$_2$ C C Me$_3$C H R'

57A

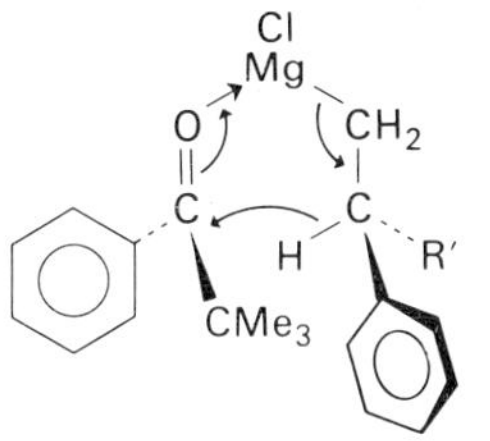

Cl Mg O CH$_2$ C C H R' CMe$_3$

57B

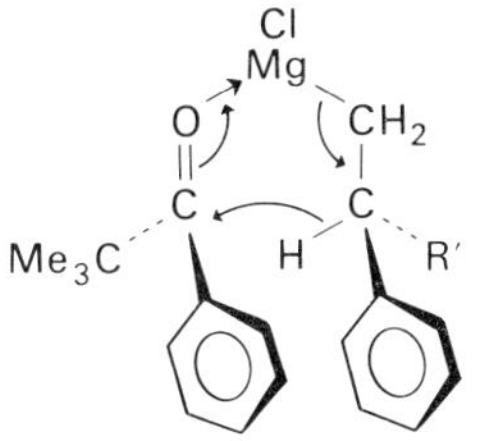

Cl Mg O CH$_2$ C C Me$_3$C H R'

57C

The experiments reported in Table **5-7** suggest that the degree of stereoselectivity of these asymmetric Grignard reductions may be influenced as much by electronic as by steric factors namely, phenyl-phenyl interactions. But these specific examples are such that the two factors reinforce each other and so a clear evaluation of the relative importance of steric and electronic interactions can hardly be made. The reduction of *p*-chloro-, *p*-methoxy-, and *o*-chloro-benzophenones by the Grignard reagent from (+)-1-chloro-2-methylbutane gave, within experimental error, optically inactive products

[50] D. Nasipuri, G. Sarkar, and C. K. Ghosh, *Tetrahedron Lett.*, 5189 (1967).

in the first two cases and an asymmetric synthesis of 10% in the third.[51] This is in accord with a steric explanation for the control of these asymmetric syntheses since the groups in the *para* position are too distant from the reaction site to exert any steric control, although they could exert an inductive effect which should be manifest if one of the groups of the chiral reducing agent were phenyl or another electronegative group.

In an attempt to obtain further data bearing on the question of electronic versus steric effects, several trifluoromethyl ketones have been reduced to secondary trifluoromethylcarbinols by chiral Grignard reagents.[52–54] Electronic interactions should be a maximum with the trifluoromethyl substituent. Furthermore it has the advantage of possessing C_{3v} symmetry so that, in common with the methyl and *t*-butyl groups, specific rotational conformations of the group itself need not be considered. The relative size of the trifluoromethyl group is also known; van der Waal diameters of methyl, trifluoromethyl, and *t*-butyl groups, considered as spinning tops, are, respectively, 4.0, 5.1, and 6.2 Å; the energy barriers as measured by rotation about the single bond in ethane or substituted ethanes are 2.9, 3.3, and 4.3 kcal/mole and the conformational energy barriers in the cyclohexane ring chair to boat conversion are about 1.7, 2.4, and 5.3 kcal/mole, respectively.

The asymmetric reductions of methyl trifluoromethyl ketone, *t*-butyl trifluoromethyl ketone, and phenyl trifluoromethyl ketone by several different chiral Grignard reagents are reported in Table **5-8**. Since there are no special polar differences between the methyl and ethyl groups, one would anticipate that the factors controlling the asymmetric reduction of methyl trifluoromethyl ketone by the purely aliphatic Grignard reagent from (+)-1-chloro-2-methylbutane would be primarily steric in nature. Thus, of the two transition state models **58A** and **58B**, in which R = CH_3, the one of lower energy should be **58A**, where the methyl and ethyl groups eclipse each other in front and the trifluoromethyl and methyl eclipse each other behind

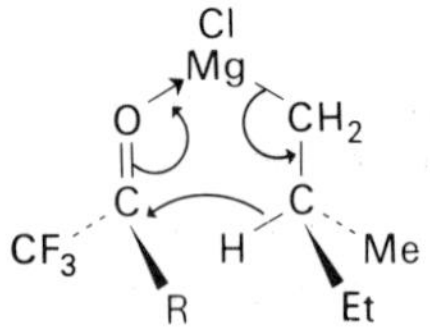

58A

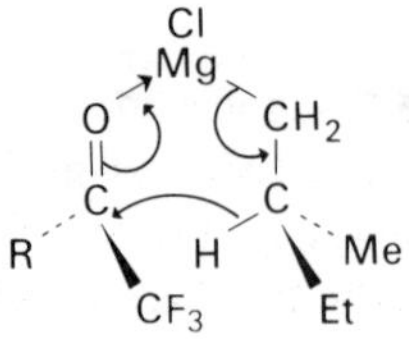

58B

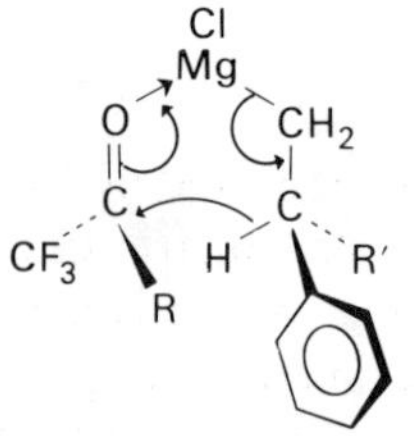

59A

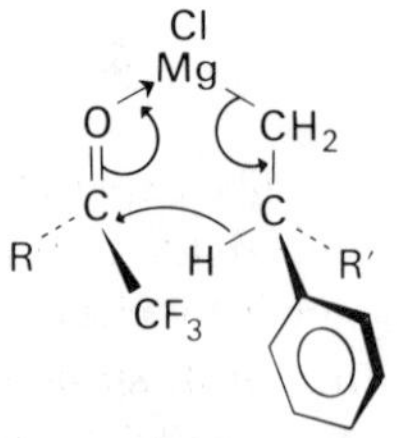

59B

[51] H. S. Mosher and E. D. Parker, *J. Am. Chem. Soc.*, **78**, 4081 (1956).

[52] H. S. Mosher, J. E. Stevenot, and D. O. Kimble, *J. Am. Chem. Soc.*, **78**, 4374 (1956).

[53] D. M. Feigl, Ph.D. thesis, Stanford University, 1965.

[54] H. Peters, D. M. Feigl, and H. S. Mosher, *J. Org. Chem.*, **33**, 4245 (1968).

TABLE 5-8

Asymmetric Reduction of Trifluoromethyl Ketones[a]

$$\underset{R}{\overset{CF_3}{C}}=O + H\blacktriangleright\underset{R''}{\overset{R'}{C}}\blacktriangleleft CH_2MgCl \longrightarrow\longrightarrow HO\blacktriangleright\underset{R}{\overset{CF_3}{C}}\blacktriangleleft H + R'R''C=CH_2$$

		Percent asymmetric reduction			
		A	B	C	D
No.	R	R′ = methyl R″ = ethyl % e.e.	R′ = methyl R″ = phenyl % e.e.	R′ = ethyl R″ = phenyl % e.e.	R′ = isopropyl R″ = phenyl % e.e.
1	Me	3[b]	67	64	—
2	*t*-Bu	0	74	63	—
3	Ph	22	47	38	22[b]

[a] Data taken from Reference 53 unless otherwise noted.
[b] Data taken from Reference 49.

the plane of the six-membered transition state, rather than as shown in **58B** where the two larger groups oppose each other in front and the two smaller groups behind. From the known configuration of methyltrifluoromethylcarbinol[54] this prediction is confirmed although the observed value (3% e.e.) is perhaps lower than expected. When the R group in **58A** and **58B** is *t*-butyl, the situation should be reversed since the *t*-butyl group is larger than the trifluoromethyl group. A low order of asymmetric synthesis would be predicted on steric grounds with the product from the transition state model **58B** being preferred. In fact a negligible asymmetric synthesis was observed.

The asymmetric reduction of phenyl trifluoromethyl ketone by the aliphatic Grignard (as represented in **58A** and **58B**, R = phenyl) is of special interest. One must conclude, based on several lines of evidence, that the trifluoromethyl group has a smaller steric bulk than phenyl. In asymmetric synthesis, phenyl repeatedly acts as R_L with respect to *t*-butyl and trifluoromethyl is smaller than *t*-butyl. This last conclusion derives not only from van der Waals radii, rotational barriers in substituted ethanes, and from conformational energy values, but also from the fact that uv spectral measurements show that the phenyl ring in phenyl *t*-butyl ketone is twisted out of coplanarity with the carbonyl group (**56**) while this is not so with phenyl trifluoromethyl ketone. Therefore, when the R group in **58A** and **58B** is phenyl, one predicts *on steric grounds alone* that the latter represents the preferred transition state. Contrary to this prediction, a 22% asymmetric synthesis of the enantiomer (from **58A**) whose configuration is shown at the top of Table **5-8** (third entry in column A) was obtained. This is a clear indication that steric factors alone are not controlling the stereochemistry of this reaction. There would seem to be no simple way of rationalizing this observation by invoking electronic factors. Postulating that trifluoromethyl acts as though it were larger than phenyl could resolve this specific case, but it would be incompatible with other evidence.

The experimental results from the reduction of the three trifluoromethyl ketones by the Grignard reagents possessing a phenyl group in the *beta* position (Table **5-8**, Columns B, C, and D) indicate that the lower energy transition state model is represented by **59A** rather than **59B** regardless of whether R is methyl, *t*-butyl, or phenyl. In **59A** the phenyl and trifluoromethyl groups are oriented away from each other. When R is methyl, a consideration of both steric and dipolar interactions leads to the conclusion that transition state **59A** should be preferred as observed. But when R is either phenyl or *t*-butyl, steric considerations alone would predict that **59B** should be preferred. The experimental results clearly show that this is not the case, and we must conclude that the electronic repulsion between trifluoromethyl and phenyl is the overriding consideration which determines that the lower energy transition state is represented by **59A**.

From the data in Table **5-7**, it was seen that the preferred transition state model had the two phenyl groups distant from each other and that stereoselectivities up to 91% were achieved. In spite of this, the unfavorable interactions of two phenyl groups is overcome in the reduction of phenyl trifluoromethyl ketone (Table **5-8**, last example in Columns B, C, and D) so that transition state model **59A** in which the two phenyl groups are in juxtaposition is preferred. This must be attributed to the greater repulsion of trifluoromethyl and phenyl versus that of two phenyl groups. For all combinations of steric and electronic factors reported in Table **5-8**, therefore, the electronic factors appear to be controlling. It is important to reemphasize that an analysis of these data *based only on steric interactions* leads to a prediction of the wrong configuration in about half of the cases in Table **5-8**.

A study in which the trifluoromethyl group was incorporated into the Grignard reagent could not be completed because of the poor yield of reduction products.[40] Only phenyl trifluoromethyl ketone was successfully reduced by the Grignard reagent from *S*-(+)-1-chloro-2-phenyl-3,3,3-trifluoropropane to give *R*-(−)-phenyltrifluoromethylcarbinol (71% e.e., 16% synthetic yield). The absolute configuration of the product is known[54] and there is very strong circumstantial evidence for the configuration of the trifluoromethyl Grignard reagent.[40,55] Of the two transition state models shown in **60A** and **60B**, the former is the one which correctly predicts the

[55] C. Aaron, D. Dull, J. L. Schmiegel, D. Jaeger, Y. O. Ohashi, and H. S. Mosher, *J. Org. Chem.*, **32**, 2797 (1967).

observed result (71% e.e.). In this transition state, the trifluoromethyl groups oppose each other as do the two phenyl groups. In previous examples involving the interactions of two phenyl groups, the transition state where

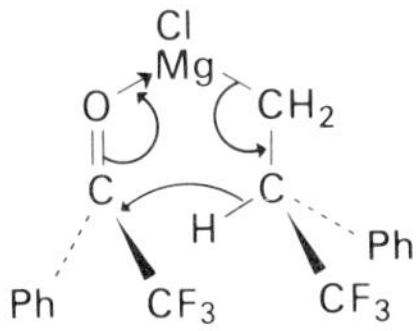

60A
Preferred

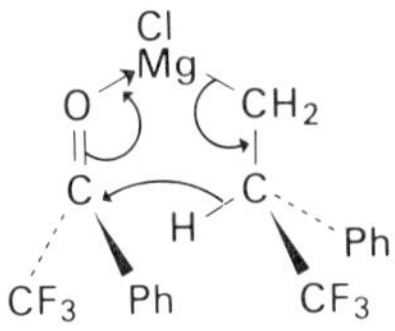

60B
Nonpreferred

these were far apart and not eclipsed was invariably preferred. In previous examples involving the interaction of a phenyl and trifluoromethyl group, the transition state (**59A**) in which these were distant from each other was preferred. This unexpected experimental result shows that the interaction of (CF_3 versus CF_3) + (Ph versus Ph) is more favorable than two (CF_3 versus Ph) interactions. Alternatively, some different mechanism is operative. More data must be obtained before these results can be systematized.

One factor ignored in the previous discussion is the possibility of attractive forces in contrast to repulsive forces between groups in very close proximity. Intermolecular attractive forces in perfluorocarbons are minimal, suggesting that intratransition state attractive forces are not involved in CF_3 versus CF_3 interactions. Although such weak attractive forces could exist for Ph versus Ph, such does not seem to have been evident in the other asymmetric syntheses where two phenyl groups were involved; therefore intergroup attractive forces among the nonfunctional alkyl and aryl groups we have been considering must be relatively unimportant. In other systems, especially enzymatic reactions, where the groups involved can be involved in hydrogen bonding, salt formation and hydrophobic bonding, attractive forces may be the overriding factor.

Formulas **61A** and **61B** represent a situation in which the same group interactions are involved in the transition state but in a different sequence. The first (**61A**) represents the preferred transition state for the reaction of the Grignard reagent from 1-chloro-2-phenylbutane with methyl phenyl ketone,

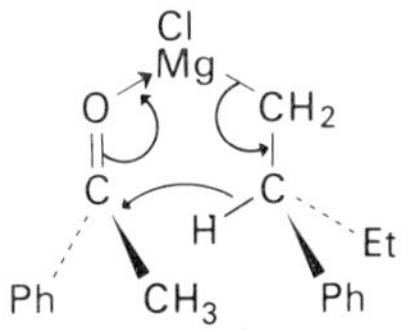

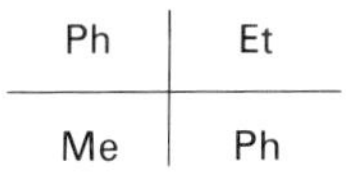

61A

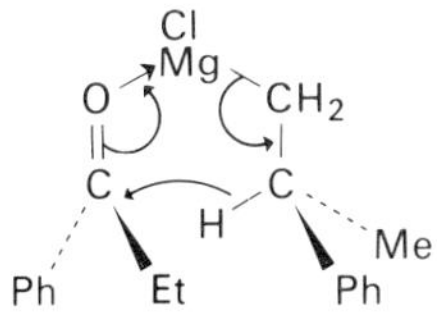

61B

Ph	Et
Me	Ph

61A′

Ph	Me
Et	Ph

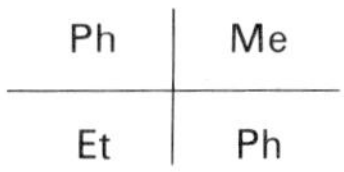

61B′

while the second (**61B**) represents that for the reaction of the Grignard reagent from 1-chloro-2-phenylpropane with ethyl phenyl ketone. The interactions are represented diagrammatically in **61A**′ and **61B**′ where the groups to the left are from the carbonyl compound and to the right are from the Grignard reagent. In these reactions primary nonbonded steric interactions with the same groups are involved. Although the transition states are not identical, albeit very similar, one would expect very nearly the same extents of asymmetric synthesis in the two cases. Experimentally the reaction represented by **61A**′ leads to an asymmetric synthesis of 47%, while that from **61B**′ gave 38%. Diagrams **62A** to **63B** represent the other known cases where such asymmetric syntheses can be matched in this manner. The correspondence in percent asymmetric synthesis is only fair. This must mean that the simplified concept in which the differences in energies of activations is based solely on the group interactions involved is a first-order approximation which has ignored other important variables.

Ph	*i*-Pr
Et	Ph

62A
66% e.e.

Ph	Et
i-Pr	Ph

62B
82% e.e.

Ph	*t*-Bu
CF_3	Ph

63A
4% e.e.

Ph	CF_3
t-Bu	Ph

63B
8% e.e.

Fig. 5-10. Comparison of Grignard asymmetric reductions in which the same group interactions are involved in the transition state. Each diagram represents only the model of the preferred transition state and the % e.e. is the enantiomeric excess resulting from this transition state.

5-2.3c. *Reagents with Multiple Chiral Centers.* A series of alkyl phenyl ketones was reduced with the Grignard reagent from "pinene hydrochloride" as outlined in Table **5-9** (Nos. 1–11). This reagent is a mixture of bornyl- (*endo*) and isobornyl- (*exo*) magnesium chloride **37** in which the magnesium is attached to a chiral center. From partial carbonation studies, Riviére[56] concluded that it was the isobornyl (*exo*) reagent which was the active reducing agent. Whether one starts with bornyl or isobornyl chloride, presumably the same reagent mixture will result. From nmr studies on the norbornyl system[57] it was also concluded that it is the *exo* compound which is the active reducing agent. These studies also established that the interconversion of *exo* and *endo* norbornyl Grignard reagents is relatively slow (half-life of a few hours at room temperature). Direct extrapolation of stereochemical

[56] C. Riviére, *Ann. chim.*, **1**, 157 (1946).

[57] F. R. Jensen and K. L. Nakamaye, *J. Am. Chem. Soc.*, **88**, 3437 (1966); see also A. G. Davies and B. P. Roberts, *J. Chem. Soc.* (*B*), 317 (1969).

TABLE 5-9

Asymmetric Reduction of Ketones by Grignard Reagents with Multiple Chiral Centers

No.	Reagent	Ketone	Carbinol configuration	Asymmetric synthesis % e.e.	Ref.
1	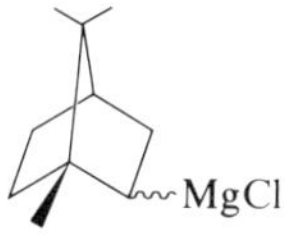 ~MgCl [Grignard reagent prepared from "pinene hydrochloride" made from (+)-α-pinene]	PhCOMe	*R* (+)	36	28
2		PhCOEt	*R* (+)	19	28
3		PhCO-*n*Pr	*R* (+)	46	28
4		PhCO-*n*Bu	*R* (+)	52	28
5		PhCO-*i*Pr	*R* (+)	55	28
6		PhCO-*t*Bu[a]	*R* (+)	72[a]	28
7		PhCO-*t*Bu[a]	*R* (+)	40–72[a]	58
8		PhCO-*t*Bu[b]	*R* (+)	68	60a
9		PhCO-*t*Bu[c]	*R* (+)	52	60a
10		PhCO-*t*Bu	*R* (+)	14	61
11		*p*-Cl-PhCOMe	(−)[d]	35	59
12		*o*-Cl-PhCOMe	(−)[d]	45	59
13		2,4,6-Cl_3PhCOMe	(−)[d]	79	59
14	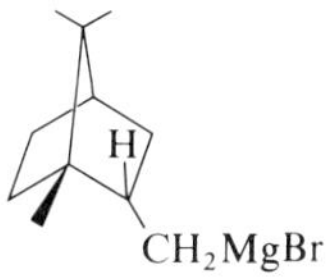 H, CH_2MgBr	$PhCOCF_3$	*S* (+)	44	61
15		PhCO-*t*Bu	—[e]	—[e]	61
16		Me_3CCOMe	—[e]	—[e]	39
17	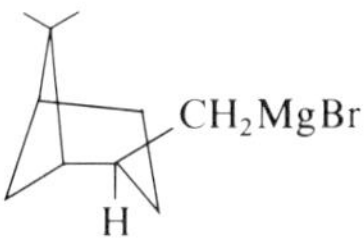 CH_2MgBr, H	PhCOMe	*R* (+)	6	61
18		PhCOEt	*R* (+)	28	61
19		PhCO-*i*Pr	*R* (+)	35	61
20		PhCO-*t*Bu	*R* (+)	41	61
21		$PhCOCF_3$	*R* (−)	8	61
22	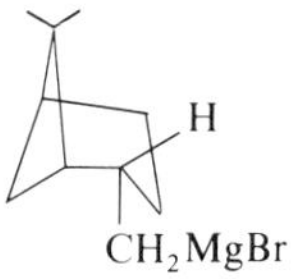 H, CH_2MgBr	PhCOMe	*S* (−)	40[f]	61
23		$PhCOCF_3$	*S* (+)	10[f]	61
24	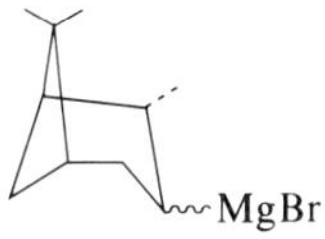 ~MgBr	PhCO-*t*Bu	—	0	60b
25	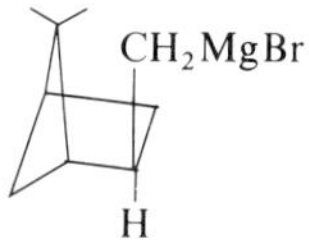 CH_2MgBr, H	PhCO-*i*Pr	*R* (+)	57	61

TABLE 5-9—continued

No.	Reagent	Ketone	Carbinol configuration	Asymmetric synthesis % e.e.	Ref.
26	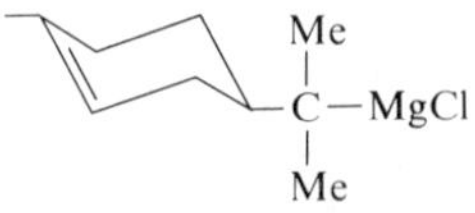 Me–C(Me)(MgCl)– (cyclohexenyl)	PhCO-*n*Bu	*R* (+)	5	60c
27		*p*-Br-PhCOMe	(+)[g]	—[g]	60c
28	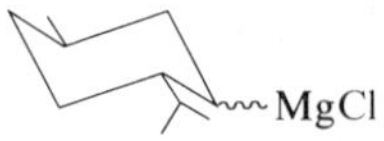 (menthyl)MgCl	PhCO-*t*Bu	*S* (−)	27	60c
29	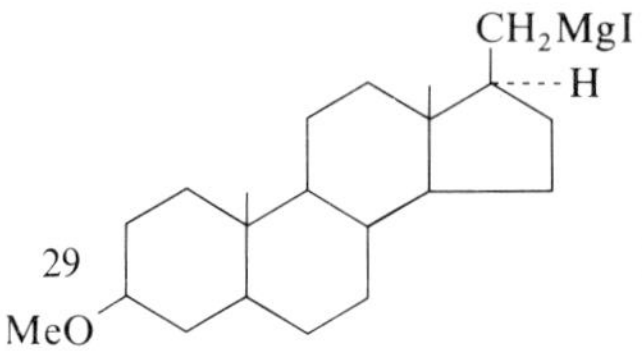 MeO-steroid–CH_2MgI (H)	PhCO-*i*Pr	*S* (−)	28[f]	61

[a] In References 28 and 58 this is described as pseudobutyl phenyl ketone. *Chemical Abstracts* translated this "secondary butyl". Presumably the ketone is *t*-butyl as here represented, although the original reports are not sufficiently explicit to be positive.
[b] Grignard reagent prepared from purified bornyl chloride.
[c] Grignard reagent prepared from purified isobornyl chloride.
[d] Although these carbinols have been resolved, their absolute configurations are unknown.
[e] There was no detectable yield of reduction product, presumably because the β-hydrogen is so severely hindered by the *gem*-dimethyl group.
[f] The synthetic yields in these reductions were very low.
[g] Observed $[\alpha]_D + 3.2°$; the maximum rotation and absolute configuration are unknown.

results from the *exo*-norbornyl to the *exo*-bornyl system may not be justified since in the former the substituent on the *exo*-position is the less hindered while in the latter the *gem*-dimethyl group renders the *exo* substituent the more hindered. In the assumed transition state correlation models represented by **64A** and **64B**, the *exo* or *endo* nature of the reagent is not designated. Neither is it specified whether it is the *exo* or *endo* β-hydrogen, or a mixture of both, which is transferred. One predicts on the basis of these empirical models that **64A**, in which the small R group of the ketone is on the side of the

[58] R. Bousett, *Bull. Soc. Chim. Fr.*, 210 (1955).
[59] G. Vavon, M. Lewi, and S. Lewi, *Pharmazie* (Sofia), 26 (1956); *Chem. Zentralblatt*, 909 (1957).
[60] (a) M. F. Tatibouet, *Bull. Soc. Chim. Fr.*, 868 (1951). (b) *ibid.*, 871 (1951). (c) *ibid.*, 867 (1951).
[61] B. J. G. McFarland, Ph.D. thesis, Stanford University, 1966; *Dissert. Abstr.*, **27**, 111 (1966).

sterically hindering *gem*-dimethyl group of the reagent, would be favored over **64B**.

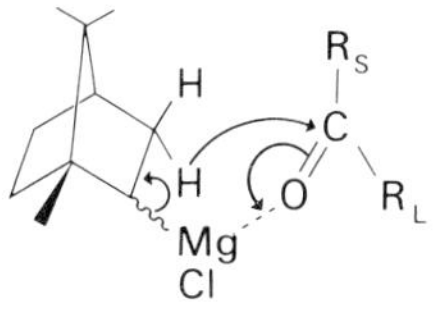

64A
Preferred

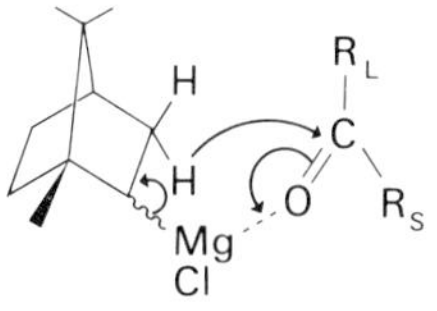

64B
Nonpreferred

The alkylphenylcarbinols produced from the asymmetric reduction of the corresponding ketones all belong to the *R*-(+) series in consonance with this prediction based on the assumption that all the alkyl groups including *t*-butyl act as though they are smaller than phenyl. But the reduction of phenyl trifluoromethyl ketone does not fit this pattern. The configuration of *R*-(−)-phenyltrifluoromethylcarbinol is "opposite"* to that of the *R*-(+)-alkylphenylcarbinols. If we assume the validity of models **64A** and **64B** and if we interpret the results only on a steric basis, we are forced to conclude that CF_3 is larger than phenyl, a deduction which is hard to justify with other evidence as pointed out previously (Sec. **5-2.3b**).

Although the extents of asymmetric synthesis are generally substantial in these reductions using reagents with multiple chiral centers (Table **5-9**), they are not higher than those with reagents possessing a single chiral center (compare results in Table **5-7**). The highest asymmetric reductions were observed with the reagent from "pinene hydrochloride" (Table **5-9**, Nos. 1–11) and the most highly hindered ketones, phenyl *t*-butyl ketone and 2,4,6-trichlorophenyl methyl ketone. The extents of asymmetric reduction of acetophenone and *p*-chloroacetophenone by this reagent were essentially the same while that of *o*-chloroacetophenone was significantly higher as one would predict solely on steric considerations.

A comparison of the reductions with the reagent from (−)-*exo*-10-bromopinane (Table **5-9**, Nos. 17–21) and from the (−)-*endo* reagent (Nos. 22 and 23) shows that these diastereomeric reagents give enantiomeric products in excess as would be predicted based upon the proposed models **65A** and **65B** for the preferred transition state. Again the absolute configurations for the alkylphenylcarbinols are correctly predicted based on the assumption that phenyl is larger than all the alkyl groups. But to rationalize the results on

* (+)-Phenyltrifluoromethylcarbinol is configurationally related to (+)-phenylmethylcarbinol[53,54] but the former is designated *S* and the latter *R* since CF_3 takes nomenclature preference over C_6H_5 in the Cahn–Ingold–Prelog system.

the reduction of phenyl trifluoromethyl ketone, one must postulate that the trifluoromethyl group is larger than phenyl.

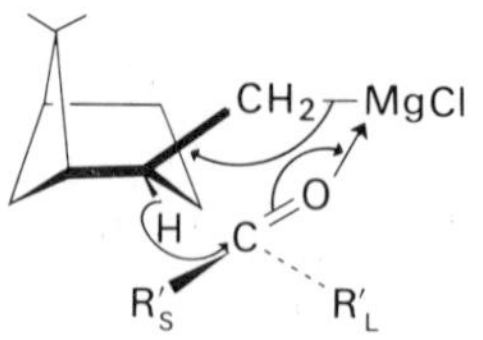

65A 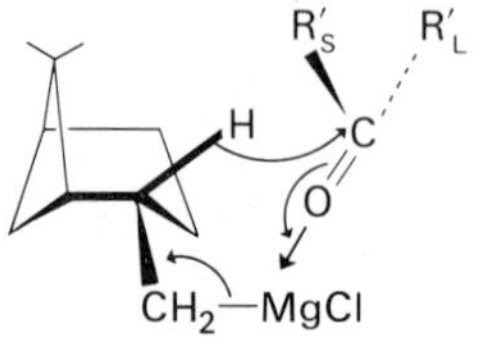 **65B**

The reduction of phenyl isopropyl ketone by the Grignard reagent from 10-bromopinane (Table **5-9**, No. 19) gave a 35% excess of the *R*-(+) carbinol, whereas reduction by comparable reagent from the ring contracted norpinane (No. 25) gave a 57% excess of the same carbinol. This observation leads to the conclusion that the greater rigidity in the latter case favors higher stereoselectivity, other things being equivalent.

Finally it is worth commenting that in these reductions the important steric effects are those centered around the developing chiral center and that a reagent with multiple chiral centers, such as those prepared from 3β-methoxy-17β-bromomethyl-Δ^5-androstene (No. 29) is no better a priori than one with a single chiral center properly situated.

5-2.3d. *Reagents with the Chiral Center in the gamma Position.* In all the examples of the Grignard reduction considered up to this point, the chiral center has been situated on the *beta* position of the reagent so that the hydrogen transferred is from the chiral center which is part of the six-membered ring transition state.* When methyl *t*-butyl ketone is reduced with the Grignard reagent from (+)-1-chloro-3-methylpentane, in which the chiral center is on the γ-position of the reagent, the methyl-*t*-butylcarbinol which is formed, in very low yield, is optically inactive within experimental limits.[62] It has now been shown,[63] however, that the reduction of phenyl *i*-propyl ketone by the Grignard reagents from (*R*)-(−)-1-chloro-3- phenylbutane and (*R*)-(−)-1-chloro-3-phenylpentane (homologs of **61A** and **61B**) gives (*R*)-(+)-phenylisopropylcarbinol (25 and 29% e.e., respectively). The situation may be analyzed by considering the following four representations of the many competing transition states for this reaction. Either of the diastereotopic hydrogens labeled H_A (*pro-S*) or H_B (*pro-R*) can be transferred

[62] H. S. Mosher and E. La Combe, *J. Am. Chem. Soc.*, **72**, 4991 (1950).

[63] J. D. Morrison, D. L. Black, and R. W. Ridgway, *Tetrahedron Lett.*, 985 (1968).

* An exception is the reagent prepared from "pinene hydrochloride" (Table **5-9**, Nos. 1–13) in which the α-chiral center is in the six-membered ring transition state but the hydrogen is transferred from an achiral β-center.

to one or the other enantiotopic faces of the carbonyl group. Since H_A and H_B are diastereotopic, they will be transferred at different rates; however realization of an asymmetric synthesis does not require that H_A and H_B be transferred at different rates. Consider only the transfer of H_A as shown in

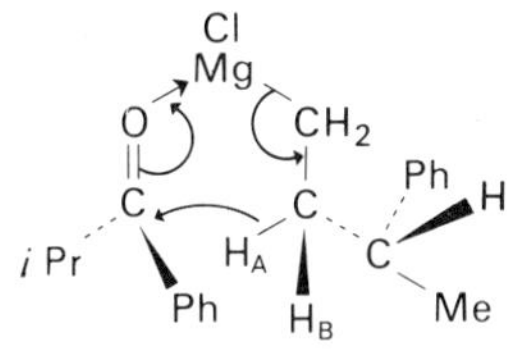

66A 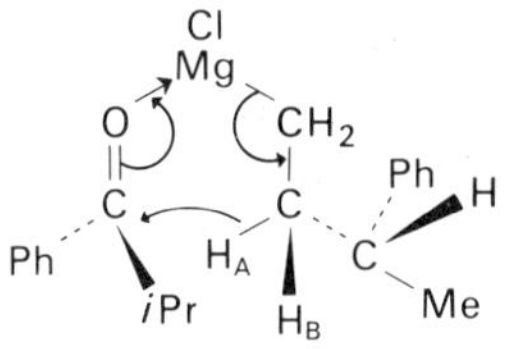 **66B**

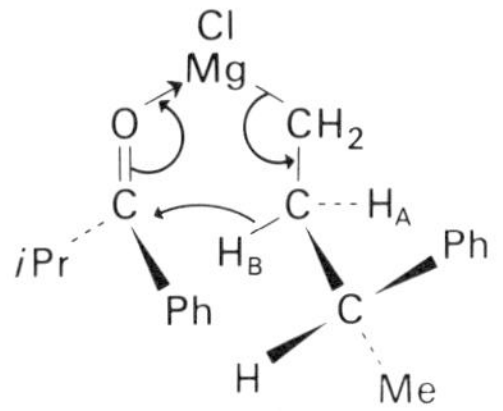

66C 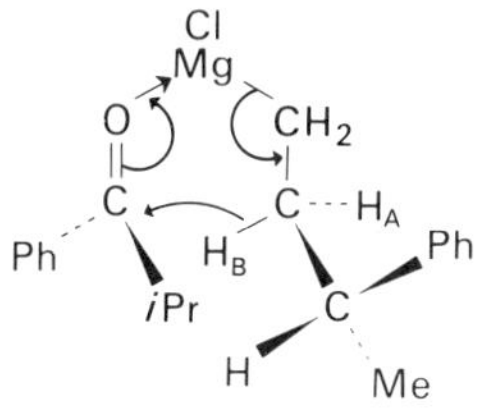 **66D**

66A and **66B**, i.e., the hypothetical situation which would pertain if the rate of transfer of H_A were many times faster than that of H_B. *In the conformation written*, **66A** presumably represents a lower energy transition state than **66B** since the larger phenyl group of the ketone is in juxtaposition to the hydrogen (H_B) rather than to the larger chiral α-methylbenzyl group as in **66B** where H_A is transferred to the opposite face of the keto group. Thus one predicts for this conformation a preponderance of the *R* carbinol resulting from transfer of H_A as shown in **66A**. A similar analysis for the transfer of H_B as represented by **66C** and **66D** leads to the conclusion that the transfer of H_B favors the *S* carbinol as represented by **66D**. If **66A** and **66D** were mirror-image transition states and if **66B** and **66C** were also enantiomeric, then one would predict a racemic product from this reaction since the stereoselectivity for transferring H_B would be equal and opposite to that for transferring H_A. But because of the chiral center in the α-phenyl group, A and D constitute diastereomeric, not enantiomeric, transition states and therefore the summation of all the steric interactions in the transition states will not cancel; therefore symmetry principles require the formation of a chiral product. Intuitively one would not expect a large difference since the established chiral center is not adjacent to the developing chiral center; thus the observed 25% asymmetric synthesis is surprisingly high, especially in view of the analogous aliphatic example.[62]

5-2.3e. *Reduction of 1-Deuterioaldehydes.* In studying steric effects in asymmetric hydrogen transfer to carbonyl compounds, the smallest possible R_S group is deuterium. Thus by studying the asymmetric reduction of 1-deuterioaldehydes to give chiral 1-deuterio primary alcohols, the difference between R_S and R_L can be maximized.[37,42,64] The results from such experiments using the asymmetric Grignard reduction are collected in Table **5-10** along with the values for reduction of the corresponding methyl ketone for comparison. In all cases the predominate isomer has the expected configuration predicted based on the model represented by **54**.

TABLE 5-10

Comparison of Asymmetric Reduction of *1*-Deuterioaldehydes with Corresponding Methyl Ketones

$$R'C(D){=}O + ClMgCH_2C(H)(R'')(R''') \longrightarrow H{-}C(D)(R'){-}OH + CH_2{=}C(R'')R'''$$

Carbonyl reagent \ Grignard reagent	$ClMgCH_2C(H)(Me)(Et)$ [Carbinol % e.e.]	$ClMgCH_2C(H)(Et)(Ph)$ [Carbinol % e.e.]
t-BuCDO	12[a]	29[f]
(*t*-BuCOMe)[d]	(13)[c]	(7)[d,e]
PhCDO	19[b]	67[f]
(PhCOMe)[d]	(4)[c]	(47)[d]

[a] See reference 64.

[b] Reported in Reference 37 as 19 ± 2% e.e. An error was made in the conversion of the observed to specific rotation in one of the two experiments (p. 3598, Column 2, line 4). The value is more properly reported as 17 ± 4% e.e.

[c] Corresponding methyl ketones included for comparison (cf. Table **5-6**).

[d] See Table **5-7**.

[e] This value is for the asymmetric reduction by 2-phenyl-1-propylmagnesium chloride instead of 2-phenyl-1-butylmagnesium chloride.

[f] See Reference 43.

[64] V. E. Althouse, K. Ueda, and H. S. Mosher, *J. Am. Chem. Soc.*, **82**, 5938 (1960).

It is surprising that the percent asymmetric reduction of 1-deuterio-trimethylacetaldehyde by the Grignard reagent from 1-chloro-2-methyl-butane (12%) is essentially no different than that of methyl *t*-butyl ketone (13%) while that of α-deuteriobenzaldehyde (29%) is substantially greater than that of acetophenone (4%). The fact that these stereoselectivities are not higher indicates that factors other than the *difference* in steric requirement between R_S and R_L are important. One of these factors must be the overall steric crowding in the transition state. The transition states must accommodate the steric interactions in a manner that minimizes the energy of activation. If it is reasonably flexible and open and if the crowding is not too severe, the apparent differences between two groups such as deuterium and *t*-butyl may not be much greater than the differences between methyl and *t*-butyl. But with more highly crowded situations the effect of the differences in group sizes may be more pronounced. Obviously further data bearing on this question are desirable.

Asymmetric reductions to give 1-deuterio-primary alcohols can be achieved by either transferring hydrogen to a deuterioaldehyde or transferring deuterium to an isotopically normal aldehyde. Models for the comparable preferred transition states are shown in **67A** and **67B** for the formation

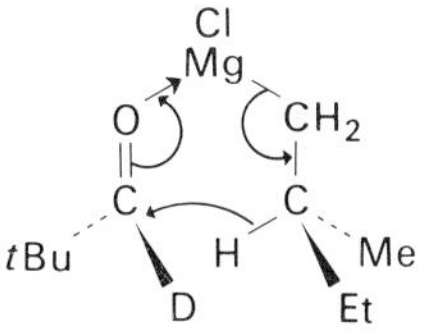

Cl Mg O CH2 C C tBu H Me D Et

67A

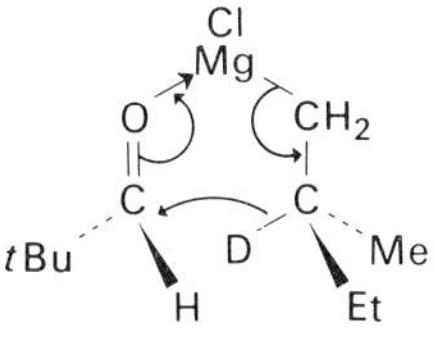

Cl Mg O CH2 C C tBu D Me H Et

67B

of neopentyl-1-*d* alcohol. The transition states represented by these models are very closely related but not identical. In **67A**, a C—H bond is being broken and reformed while in **67B** it is a C—D bond which is being broken and reformed; in the former the interaction is between deuterium and ethyl and in the latter between hydrogen and ethyl. Although the deuterium transfer will be appreciably slower (Dunn and Warkentin[27] observed a deuterium isotope effect of 2.0–2.2 in the reduction of benzophenone with 2-methyl-propyl-2-*d*-magnesium chloride), one would think a priori that the stereoselectivities of the two reactions should be approximately the same. In fact a 12% asymmetric reduction was observed when hydrogen was transferred and 36% when deuterium was transferred.* It has also been observed† that the $LiAlH_4 \cdot$ quinine reduction (Sec. **5-3.1**) of benzaldehyde-

* The observed rotations in these experiments were not large[38] and the work was done before gas chromatographic purification techniques. The results should be repeated and extended to other related systems.

† Research by A. Horeau, H. Lapin, D. Meá, and A. Nouaille, as quoted by J. Mathieu and J. Weill-Raynal, *Bull. Soc. Chim. Fr.*, 1211 (1968).

1-*d* gave (−)-benzyl-α-*d* alcohol, 30% e.e., while $LiAlD_4$ · quinine reduction of isotopically normal benzaldehyde gave 39% e.e. of the enantiomer.

Assuming the correctness of these results, the only reasonable explanation would seem to be one based on operation of the tunneling effect.[65] The longer de Broglie wavelength for hydrogen can allow it to be transferred from reagent to substrate without going over the energy barrier. The transfer of hydrogen could therefore take place between the carbon atoms of the reagent and substrate when they are appreciably further apart than in the case of the transfer of deuterium which shows negligible tunneling. The necessity for the carbon atoms to be closer together for the transfer of deuterium versus hydrogen would contribute to the crowding of the transition state and would thereby accentuate the steric interactions involved.

The reduction of phenyl *t*-butyl ketone by (*S*)-2-phenyl-1-butylmagnesium chloride (Table **5-7**) versus the reduction of the same ketone by the reagent bearing a deuterium in the β-position showed the same extent of asymmetric synthesis within experimental error (16% versus 15%).[66] The same is true for the reduction of phenyl ethyl ketone by the Grignard reagents from (+)-1-chloro-2-phenylbutane and (+)-1-chloro-2-phenylbutane-2-*d* (54 and 55% asymmetric synthesis, respectively).[66] These two reactions are not exactly comparable to the reduction of trimethylacetaldehyde represented by **67A** and **67B** but would suggest that the important factor was the difference in steric interactions between H and D and not the difference in H versus D transfer.

5-3. Reduction by Chiral Metal Hydride Complexes

That lithium aluminum hydride (LAH) could be converted into a chiral lithium aluminum alkoxy hydride which could be used in asymmetric reductions was conceived by Bothner-By[67] who reported that such a reagent, prepared from LAH and (+)-camphor (1:2 molar ratio) reduced methyl ethyl ketone and methyl *t*-butyl ketone to the corresponding secondary carbinols which were optically active. These results were shown to be erroneous;[68,69] the optical activity was traced to contamination of the products by camphor. It was subsequently shown[69] that reduction of methyl *t*-butyl ketone by a reagent prepared from LAH and (−)-menthol also failed to give detectable asymmetric reduction. These initial failures

[65] Harold S. Johnston, in *Advances in Chemical Physics*, ed. I. Prigogine (New York: Interscience Pub. Inc., 1961), vol. **3**, pp. 131–170.

[66] J. D. Morrison and H. S. Mosher, Unpublished results.

[67] A. A. Bothner-By, *J. Am. Chem. Soc.*, **73**, 846 (1951).

[68] P. S. Portoghese, *J. Org. Chem.*, **27**, 3359 (1962).

[69] S. R. Landor, B. J. Miller, and A. R. Tatchell, *Proc. Chem. Soc.*, 227 (1964); *J. Chem. Soc.*, 1822 (1966).

may have been due to the facile disproportionation of the intermediate lithium aluminum alkoxy hydride compounds (cf. Sec. **3-4.2**) according to the following equations in which OR* represents the chiral isobornyloxy or menthyloxy group. These equilibrium reactions can give LAH which is the more active reducing agent, which would bring about achiral reduction.

$$LiAlH_4 + HOR^* \longrightarrow LiAlH_3OR^* + H_2$$

$$2LiAlH_3OR^* \rightleftarrows LiAlH_2(OR^*)_2 + LiAlH_4$$

$$2LiAlH_2(OR^*)_2 \rightleftarrows LiAl(OR^*)_4 + LiAlH_4$$

Such a scheme may be operative in some cases but certainly not in all. It has now been shown that (−)-menthol-LAH and (+)-camphor-LAH complexes give asymmetric reduction of methyl and ethyl benzoylformates[70] (Table **5-12**, Nos. 26 and 27), that (+)-borneol-LAH and (−)-menthol-LAH complexes give asymmetric reduction of a group of imines[71] (cf. Tables **5-13** and **5-14**), and that conjugated enynols[72] are reduced to chiral allenic carbinols by a (−)-menthol-LAH complex (cf. Table **9-1**). Finally it has now been reported that methyl ethyl ketone is in fact reduced asymmetrically by a (+)-camphor-LAH complex.[73]

Červinka[74] discovered that complexes formed from LAH and carbinol amines (quinine, quinidine, cinchonine, cinchonidine, ephedrine) gave asymmetric reductions. These results have been amply verified and extended[75–81] (cf. Tables **5-11** through **5-15**). It was reasoned that complexes of aminoalcohols would show less tendency to dissociate and would be more stereoselective than those of monohydric alcohols such as (−)-menthol, (+)-borneol, or (−)-isoborneol.

Landor and co-workers[72,81] showed that carbonyl compounds were reduced asymmetrically with a reagent prepared from LAH and dihydroxy-

[70] A. Horeau, H. B. Kagan, and J. P. Vigneron, *Bull. Soc. Chim. Fr.*, 3795 (1968).

[71] O. Červinka, V. Suchan, O. Kotýnek, and V. Dudek, *Coll. Czech. Chem. Commun.*, **30**, 2484 (1965).

[72] (a) R. J. D. Evans, S. R. Landor, and J. P. Regan, *Chem. Commun.*, 397 (1965). (b) S. R. Landor, B. J. Miller, J. P. Regan, and A. R. Tatchell, *Chem. Commun.*, 585 (1966).

[73] Y. Minoura and H. Yamaguchi, *J. Polym. Sci., Part A-1*, **6**, 2013 (1968).

[74] O. Červinka, *Chimia (Aarau)*, **13**, 332 (1959).

[75] O. Červinka and O. Bělovský, *Coll. Czech. Chem. Commun.*, **32**, 3897 (1967).

[76] O. Červinka, *Coll. Czech. Chem. Commun.*, **30**, 1684 (1965).

[77] A. Pohland, Private communication.

[78] O. Červinka and O. Bělovský, *Coll. Czech. Chem. Commun.*, **30**, 2487 (1965).

[79] H. Christol, D. Duval, and G. Solladié, *Bull. Soc. Chem. Fr.*, 4151 (1968).

[80] O. Červinka, *Coll. Czech. Chem. Comm.*, **26**, 673 (1961); **30**, 2403 (1965).

[81] (a) S. R. Landor and A. R. Tatchell, *J. Chem. Soc. (C)*, 2280 (1966); (b) S. R. Landor, B. J. Miller, and A. R. Tatchell, *ibid.*, 197 (1967).

monosaccharide derivatives. Červinka and Fabryová[82] demonstrated that monohydroxy sugar derivatives also formed LAH complexes which gave asymmetric reductions (cf. Table **5-17**).

The convenience of the experimental procedure and the availability of chiral alcohol amines as well as monosaccharide derivatives make this type of asymmetric reduction attractive for the preparation of optically active secondary carbinols. By using a basic chiral agent followed by extraction from acidic solutions, there is positive assurance that the neutral product will not be inadvertently contaminated with optically active basic impurities. The enantiomeric purities of products from these reductions usually have been low, but a few[72,76,81] have approached 50% e.e., and one[81b] was 71%. Consequently there is considerable incentive to search for systems which might develop even higher stereoselectivities.

5.3-1. Asymmetric Reduction of Ketones. The data in Table **5-11** were collected in order to explore the effects of various solvents and chiral reagents upon the extent of asymmetric reduction of acetophenone. Ether uniformly promotes higher stereoselectivities than THF and the other solvents tried, although benzene in two examples gave reasonably good results (22 and 24% e.e.). It is especially interesting that changing the solvent from ether to THF reversed the stereochemical course of the reduction in two cases (Nos. 1 and 10) but not uniformly.

Of the alkaloids tested, quinine led to the highest stereoselectivities observed (48% e.e.), although the drug Darvon was almost as effective (45% e.e.). Structures showing possible stereochemical relationships of the reagents are depicted in **68** and **69**. The possibility for the electron pair on nitrogen to assist in displacement of the hydride ion in the transition state via a five- or six-membered ring may allow a more rigid conformation for the

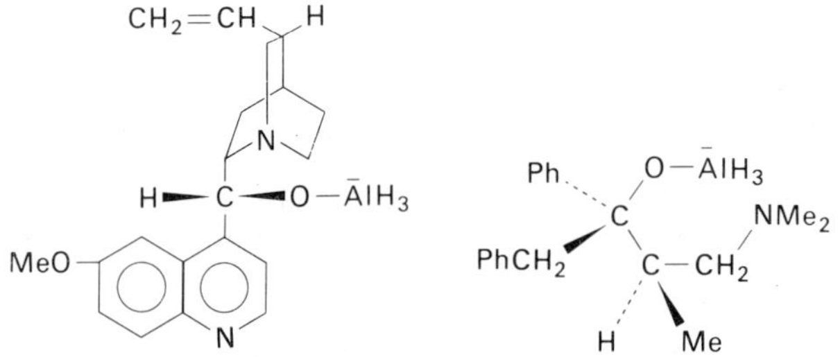

68
Quinine-LAH reagent

69
Darvon-LAH reagent

[82] O. Červinka and A. Fabryová, *Tetrahedron Lett.*, 1179 (1967).

TABLE 5-11

Asymmetric Reduction of Acetophenone by Lithium Aluminum Hydride Modified by Chiral Amino Carbinols in Various Solvents[a]

$$LiAlH_3(OR^*)^b + PhC(=O)Me \longrightarrow \longrightarrow HO-C(Me)(Ph)-H \text{ and } H-C(Me)(Ph)-OH + R^*OH$$

R-(+) and *S*-(−)

No.	Chiral agent R*OH	Ether % e.e.	Ether config.	THF % e.e.	THF config.	Other solvents	% e.e. config.	Ref.
1	(−)-Quinine	48	*R*	9[c]	*S*	*i*Pr ether	3 *R*	75
2	(−)-Quinine	—	—	—	—	*i*Bu ether	3 *R*	75
3	(−)-Quinine	—	—	—	—	Anisole	12 *R*	75
4	(−)-Quinine	—	—	—	—	Dioxane	11 *R*	75
5	(−)-Quinine	—	—	—	—	Benzene	24 *R*	75
6	(−)-Ephedrine	13	*R*	7	*R*	—	—	75
7	(−)-N-Ethylephedrine	25	*R*	10	*R*	Dioxane	16 *R*	75
8	(−)-N-Ethylephedrine	—	—	—	—	Benzene	22 *R*	75
9	(+)-4-Ephedrine	3	*R*	2	*R*	—	—	75
10	(+)-Cinchonine	18	*S*	1	*R*	—	—	75
11	(+)-Quinidine	23	*S*	—	—	—	—	76
12	(−)-Cinchonidine	12	*R*	—	—	—	—	76
13	(+)-Cinchonine	18	*S*	—	—	—	—	76
14	Darvon[d]	45	—	—	—	—	—	77
15	(−)-$PhCHOHCH_2NEt_2$	0.5	*R*	—	—	—	—	75
16	(−)-$MeCHOHCMe_3$	3.9	*S*	—	—	—	—	75

[a] The reagent prepared from LAH and chiral alcohol-amine is first prepared in the indicated solvent and the ketone is then added. The molar ratio of LAH, chiral reagent, and ketone is 1.1 : 1.1 : 1.0.

[b] The reagent is symbolized as $LiAlH_3OR^*$; this is meant only to reflect the molar ratio of LAH and chiral reagent R*OH used in its preparation.

[c] Asymmetric synthesis dropped to 0.3 % when $NaBH_4$ was substituted for $LiAlH_4$.

[d] Darvon is 1,2-diphenyl-3-methyl-4-dimethylamino-2-butanol (cf. **69**).

hydride transfer step thereby increasing the stereoselectivity of the process.

Conceptually there are several distinct ways in which the asymmetric reduction of benzoylformate esters to give optically active mandelic acid can be achieved: (a) by the reduction of a chiral ester with an achiral reducing agent; (b) by the reduction of an achiral ester with a chiral reducing agent; (c) by a combination of a chiral ester and chiral reducing agent. Horeau and co-workers[70] have investigated these possibilities with (−)-menthyl benzoylformate (**70A**) and ethyl benzoylformate (**70B**) (Table **5-12**, Nos. 26–29). Either a simple asymmetric reduction of **70A** with an achiral reducing agent LAH-cyclohexanol, process (a), or of **70B** with a chiral reducing agent LAH-(+)-camphor, process (b), gives (*R*)-(−)-mandelic acid

TABLE 5-12

Asymmetric Reduction of Ketones by LAH-Alkaloid Complexes[a]

$$LiAlH_3OR^* + R{-}\overset{\overset{\displaystyle O}{\|}}{C}{-}R' \longrightarrow\longrightarrow RCHOHR' + R^*OH$$

No.	Ketone R—COR′	Chiral inducing agent: (−)-Quinine % e.e./config.		(+)-Quinidine % e.e./config.		(−)-Cinchonidine % e.e./config.		(+)-Cinchonine % e.e./config.		Other % e.e./config.		Ref.
1	MeCOEt	0	—	—	—	—	—	—	—	2[b]	—	76, 73
2	MeCO-*i*-Pr	6	*S*	—	—	—	—	2	*R*	14	*S*[c]	76
3	MeCO-*t*Bu	11	*S*	—	—	8	*S*	7	*R*	10	*S*[c]	76
4	$MeCOC_6H_{13}$	6	*S*	—	—	6	*S*	—	—	—	—	76
5	$MeCOC_6H_{13}$	4	*S*	3	*S*	—	—	3	—	—	—	78
6	$MeCOCH_2Ph$	2	*S*	3	*R*	—	—	—	—	—	—	76
7	MeCOPh[d]	48	*R*	23	*S*	12	*R*	18	*S*[d]	13	*R*[c]	76
8	MeCO-*p*-MeOPh[e]	+19°	—	−8°	—	—	—	−10°	—	+10°[c]	—	76
9	MeCOMesityl	41	*R*	3	*S*	—	—	1	*S*	3	*R*[c]	76
10	MeCO-α-Naph	1	*R*	—	—	—	—	—	—	—	—	76
11	MeCOTricyhexPh[f]	16	*R*	—	—	—	—	—	—	—	—	76
12	PhCO-*p*-MeOPh	0	—	—	—	—	—	—	—	—	—	75
13	PhCO-p-Tolyl	0	—	—	—	—	—	—	—	0	—	22b
14	PhCO-*o*-Tolyl[e]	+3°	—	−3°	—	—	—	−3°	—	—	—	22b
15	PhCO-*o*-ClPh[e]	−1°	—	+2°	—	—	—	—	—	—	—	22c
16	PhCO-*p*-NMe_2Ph[e]	+0.1°	—	—	—	—	—	—	—	—	—	75
17	PhCOMesityl	39	*S*[g]	13	*R*	—	—	24	*R*	8	*S*[c,g]	22b,76
18	PhCO-α-Naph[e]	−10°	—	+3°	—	—	—	+1°	—	—	—	22b,76
19	PhCO-9-Anthryl[e]	−14°	—	+13°	—	—	—	+1°	—	—	—	22b,76

20	PhCO-*t*Bu	2	*S*[h]	—	—	—	—	—	—	—	—	22c
21	$PhCOC_6H_{13}$	26	—	10	—	—	—	4	—	—	—	76,78
22	*o*-TolylCOC_6H_{13}[e]	+3°	—	−1°	—	—	—	−0.3°	—	—	—	76,78
23	α-NaphCOC_6H_{13}[e]	0°	—	+8°	—	—	—	0°	—	—	—	76,78
24	Spiro[4.4]nonanone-1	5	*R*[i]	—	—	2	*R*[i]	1	*S*[i]	12	*S*[i,j]	79
25	Poly-3-butenone[e,k]	—	—	—	—	—	—	—	—	−2-12°	—[k]	73
26	PhCOCOOMe	17	*S*	—	—	5	*R*	—	—	4[l]	*R*[l]	70
27	PhCOCOOEt	—	—	—	—	—	—	—	—	5[m]	*R*[m]	70
28	PhCOCOO(−)-menthyl[n]	7	*R*	—	—	26	*R*	—	—	40[l]	*R*[l]	70
29	PhCOCOO(−)-menthyl[n]	—	—	—	—	—	—	—	—	49[m]	*R*[m]	70

[a] The LAH:chiral agent:substrate ratio was 1.1:1.1:1.0 unless otherwise noted. This complex is symbolized as $LiAlH_3OR^*$ without prejudice as to its real structure.

[b] This reagent was prepared by allowing (+)-camphor and LAH to react in ether. An 18% e.e. of methyl-*t*-butylcarbinol was reported[68] using this reagent; this was reportedly[68,69] due to contamination with camphor, while Reference 73 reports in three runs (with molar ratios of 1:1, 1:2 and 1:3 of LAH and camphor) gas chromatographically pure (−)-isomer (2% e.e.) was formed.

[c] The chiral agent is (−)-ephedrine.

[d] In THF solvent *R* isomer (9% e.e.) formed in quinine-LAH reduction and *R* isomer (1% e.e.) formed in (+)-cinchonine reduction (cf. Table **5-11**).

[e] Maximum rotation unknown; therefore, specific rotations recorded.

[f] TricyhexPh refers to the 2,4,6-tricyclohexylphenyl group.

[g] Using (−)-quinine in THF instead of ether, *S* isomer formed (2% e.e.); using (−)-N-ethylephedrine, *S* isomer (17% e.e.); using (+)-4-ephedrine, *R* isomer (27% e.e.); using 1-phenyl-2-diethylaminoethanol, *S* isomer (2% e.e.); using *S*-(−)-methylphenylcarbinol, *R*(+)-isomer formed (0.2% e.e.).

[h] Reported[22c] as *R*(−) based on the assumption that *t*-butyl is larger than phenyl but the configuration has now been proved to be *S*(−) as here given.

[i] The LAH:chiral agent:ketone molar ratio was 1:2:1.

[j] The chiral agent was (−)-α-methylbenzylamine; with (+)-α-methylbenzylamine, *R* isomer was formed (14% e.e.). Note that this agent does not have an OH group in contrast to all the others in this table.

[k] Poly-3-butenone was reduced to poly-methylvinylcarbinol with $[\alpha]_D^{15}$ (ethanol) of −1.6°, −9.2°, and −12.3°, depending upon the LAH-camphor molar ratio of 1:1, 1:2, or 1:3.

[l] A 1:3 LAH-(−)-menthol reagent, cf. Table **5-3** for reductions of benzoylformate esters by chiral Meerwein-Ponndorf-Verley reagents.

[m] A 1:3 LAH-(+)-camphor reagent.

[n] Cf. Sec. **2-3.4** (Table **2-4**) for reductions of menthyl benzoylformate by achiral reducing agents.

after hydrolysis (10 and 4% e.e., respectively). But the "double asymmetric reduction" using both chiral ester (**70C**) and chiral reducing agent, process (c), results in 49% asymmetric synthesis. This "double asymmetric induction" is higher than would be anticipated on the basis of any simple additive effect.

On the other hand, reduction of methyl benzoylformate by an LAH-quinine complex gives (*S*)-mandelic acid (17% e.e.). Since the reduction of (−)-menthyl benzoylformate by an achiral reducing agent as in process (a)

PhC(=O)—C(=O)—OR (**70**) → (a) LAH/3 Cyclohexanol; (b) LAH/3 (+)-Camphor; (c) LAH/3 (+)-Camphor → H—C(COOH)(Ph)—OH

	% e.e.	Config.
(a)	10	*R*
(b)	4	*R*
(c)	49	*R*

(A) R = (−)-menthyl
(B) R = ethyl
(C) R = (−)-menthyl

or by LAH alone gives the enantiomeric (*R*)-mandelic acid (6–10% e.e., Sec. **2-3.4**, Table **2-4**), the stereochemical control of these two processes operate in opposition to each other. Reduction of the chiral (−)-menthyl ester **70C** with the chiral reducing agent, LAH-quinine, gives (*R*)-mandelic acid with low stereoselectivity (7% e.e., Table **5-12**, No. 28) in accord with expectations for such an opposing "double asymmetric induction" (cf. also Table **5-3**, No. 9).

Reductions of enamines (Table **5-13**) have shown that a 1:1 molar ratio of chiral agent and LAH gives a reagent which results in higher stereoselectivities as well as higher synthetic yields. Based on this evidence, the experiments reported in Table **5-11** and most of the other reductions (Tables **5-12** through **5-16**) were conducted with a LAH:reagent:substrate ratio of 1.1:1.1:1.0. The reduction of acetophenone by the LAH-quinine reagent was studied extensively (Table **5-11**, Nos. 1–5). In all experiments except one carried out in the THF solvent, the (*R*)-(+)-phenylmethylcarbinol isomer was produced in excess.* Furthermore, (−)-cinchonidine and (−)-ephedrine, which have comparable *S,R*-"erythro" configurations on the amino alcohol centers, (**71**, cf. also **68**) also gave *R*-(+)-methylphenylcarbinol while (+)-quinidine and (+)-cinchonine with opposite *R,S*-"erythro" arrangements at the amino alcohol centers (**72**) gave the enantiomeric *S*-(−)-methylphenylcarbinol in excess. The reagent from (−)-quinine gave higher stereoselectivities than that from (−)-cinchonidine, which differs only in

* This is the enantiomer of phenylmethylcarbinol obtained in a MPV-type reduction using (−)-quinine and potassium *t*-butoxide. However, it was convincingly demonstrated using deuterium labeling[22c] that the quinine-LAH reductions were *not* the MPV-type; i.e., hydrogen was not transferred from the carbinol carbon atom.

having H instead of OCH_3 at C_6, cf. **68**, and that from (+)-quinidine or (+)-cinchonine (**72**), which have enantiomeric relationships at C_8 and C_9.

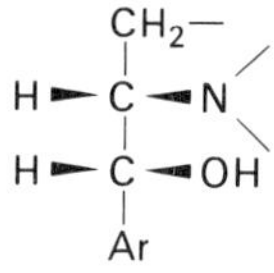

71

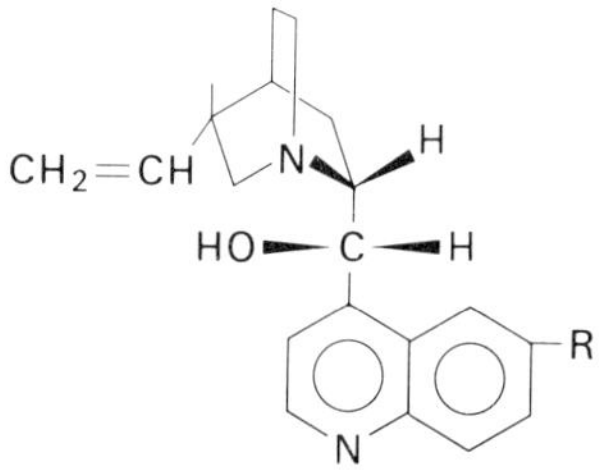

72
(+)-Quinidine, R = OMe
(+)-Cinchonine, R = H

These experiments have been used as the basis for the correlation of configuration of carbinols. Considerable caution must be exercised in this regard, however, since the stereochemical balance of the reaction appears to be quite delicate. Thus in several reductions (Table **5-11**, No. 1 and Table **5-12**, Nos. 3 and 6) there is a reversal of the stereochemical course of the reaction in changing from ether to THF solvent. There was also such a reversal with temperature in dioxane solvent. Also in the reduction of enamines (Table **5-13**) there is a reversal in stereochemistry with a change in the LAH-quinine ratio. Furthermore methyl alkyl ketones give *S*-carbinols while methyl aryl ketones give *R*-carbinols under the same conditions (Table **5-12**, Nos. 1–10). Little confidence can be held for configurational assignments based on this empirical method unless the compounds being studied belong to a closely related series or until considerable additional evidence has been amassed. It is difficult to rationalize any quantitative relationships depending upon the size of the substituents R_S and R_L attached to the carbonyl group. For instance, the percent asymmetric reduction decreases as the difference in steric bulk increases in the series methyl phenyl ketone (48%), methyl mesityl ketone (41%), and methyl 2,4,6-tricyclohexylphenyl ketone (16%). At the same time the percent asymmetric reduction decreases as the bulk of the carbonyl substituents approach each other in size in the series methyl phenyl ketone (48%), cyclohexyl phenyl ketone (26%), and *t*-butyl phenyl ketone (2%). There are certainly factors in addition to difference in steric bulk which are important in determining the extent of stereoselectivity in these reactions.

The steric argument is supported however by the fact that *p*-methyl- and *p*-methoxy-benzophenone, where the substituent which determines the chirality of the carbinol center is in fact remote from it, gave optically inactive reduction products; while *o*-methyl-benzophenone, where the methyl group is close to the developing chiral center, gave an optically active product (3% e.e.). On the other hand the 6-methoxy group in (−)-quinine or (+)-quinidine, which is remote from the chiral center of the reagents, has a large effect on the stereoselectivity in comparison to the unsubstituted (−)-cinchonidine or (+)-cinchonine.

With a change in temperature from 0 to 60° the asymmetric reduction of acetophenone by the LAH-quinine reagent in THF increased from 2 to 9%, from which $\Delta\Delta G^{\ddagger}$ for the competing transition states was calculated to be −2.05 cal/mol/degree. But in dioxane solvent the reverse trend was noted in going from 60 to 105° with a change in formation of the predominate isomer from *R* (11% e.e.) to *S* (2% e.e.).

5-3.2. Asymmetric Reduction of Enamines and Imines. Červinka[22a,74] has shown that enamines[80] and imines[71] can be reduced to optically active amines by LAH which first has been treated with a chiral alcohol such as (−)-menthol. Contamination of the basic asymmetric reduction product by a neutral, chiral inducing agent can be excluded by its extraction from acid solution. This asymmetric reaction was first explored in the reduction of 2-substituted N-methyl-Δ^1-tetrahydropyridinium perchlorates (**73**, Table **5-13**) and then was extended to the dihydropyrrole analogs (**75**, Table **5-14**). These enamine salts are not ether soluble and thus these reactions may be largely heterogeneous. From the data in Table **5-14**, it is evident that the ratio of LAH to (−)-menthol in the chiral reducing agent is critical for determining both the sense and quantitative amount of asymmetric reduction. Since the highest stereoselectivities as well as the highest synthetic yields were generally found with a 1:1 molar ratio of LAH to menthol, subsequent reactions were standardized using a reagent of this composition which can be symbolized as $LiAlH_3OR^*$. (−)-Menthol gave higher stereoselectivity in the reduction of the 2-phenyl compound (Table **5-13**, No. 5) than did (+)-borneol. The reactions in ether solvent gave higher stereoselectivities than in THF. A sodium borohydride-(−)-menthol reagent gave essentially optically inactive product in methanol solvent but in ether the product, $[\alpha]_D$ +3.11°, had the opposite sign to that obtained by LAH-(−)-menthol reduction.

Of the examples in Table **5-13**, only the maximum rotation of N-methyl-2-propylpiperidine (N-methylconiine) is known; the rotation of $[\alpha]_D$ +9.49° represents 12% asymmetric synthesis of the *S* isomer. It is known that the dextrorotatory 2-methyl, 2-*n*-propyl, and 2-β-phenylethyl N-methylpiperidines reported in Table **5-13** have the configuration indicated in the equation at the top of the table and the same is assumed for the other 2-alkyl derivatives. No such deduction can be made with any confidence concerning the (−)-2-phenyl and (+)-2-α-naphthyl derivatives formed under the same conditions because of the uncertainties of electronic versus steric effects and

TABLE 5-13

Asymmetric Reduction of 2-Substituted N-Methyl-Δ^1-tetrahydropyridinium Perchlorates by LAH-(−)-Menthol Reagent[a]

$$\mathbf{73}\ (\text{N-Me } ClO_4^-) \xrightarrow[2)\ H_2O]{1)\ LiAlH_3(OR^*)} \mathbf{74}$$

No.	R	2-Substituted N-methylpiperidine, **74**			
		Et_2O solvent		THF solvent	
		$[\alpha]_D°$	config.	$[\alpha]_D°$	config.
1	Me	+0.69	*S*	—	—
2	Et	+2.20	—	—	—
3	*n*-Pr	+9.49[b]	*S*	+1.24[c]	*S*
4	*n*-Amyl	—	—	+1.83	—
5	Ph[d,e]	−1.55[d]	—	−0.89	—
6	CH_2Ph	+1.02	—	—	—
7	CH_2CH_2Ph	—	—	+0.48	*S*
8	α-Naphthyl	+0.48	—	—	—

[a] Data from Reference 80. The ratio of LAH to (−)-menthol to substrate was 1:1.1:0.5 except in the *n*-Pr example where it was 1:1.1:1.

[b] 12 % e.e.

[c] 1.6 % e.e.

[d] A rotation of +7.34° was observed when the LAH:(−)-menthol molar ratio was 1:2.

[e] With LAH:(+)-borneol (1:1) reagent, the rotation was −0.71° and with LAH:(+)-camphor (1:1) reagent, the rotation was −1.88°.

the apparent relative sizes of these groups with respect to the piperidine nucleus.

Of these examples reported in Table **5-14**, only the maximum rotation of *N*-methyl-2-benzylpyrrole is known; the value of $[\alpha]_D$ +4.75° represents a 6% asymmetric synthesis. Meaningful stereochemical interpretation of these results is not possible because one knows very little concerning the configuration of the reagent and the enantiomeric compositions of most of the products.

The asymmetric reduction of imines (**77**, Table **5-15**) to amines (**78**) by LAH which has been treated with a chiral alcohol such as (−)-menthol is a homogeneous reaction.[70] The ketimines can be made by the reaction of a Grignard reagent on a nitrile and used *in situ.* The stereoselectivities using this reagent on alkyl phenyl ketimines are below 10%; they are not known for the aryl phenyl ketimines since the maximum rotations of the corresponding primary amines have not been determined. This is a potentially useful

TABLE 5-14

Effect of LAH-Menthol Ratio on the Stereoselectivity of 1-Methyl-2-alkyl-Δ^1-dihydro-pyrrolinium Perchlorate Reductions[a]

$$\mathbf{75}\ (\text{N-Me pyrrolinium, 2-R},\ ClO_4^-) \xrightarrow[Et_2O]{LiAlH_m(OR^*)_n} \mathbf{76}\ (\text{N-Me pyrrolidine, 2-H, 2-R})$$

Ratio m:n	Reagent[c]	R = Et $[\alpha]_D$ °	R = *n*Pr $[\alpha]_D$ °	R = *i*Pr $[\alpha]_D$ °	R = CH_2Ph[b] $[\alpha]_D$ °	config.	% e.e.
3:1	$[LiAlH_3OR^*]$	+6.18	+1.32	−1.65	+4.75	*R*	6
2:2	$[LiAlH_2(OR^*)_2]$	+3.38	+1.10	+1.28	−1.32	*S*	1.5
1:3	$[LiAlH(OR^*)_3]$	−0.40	—	—	−0.38	*S*	0.5

[a] Data taken from Reference 80.
[b] Based on $[\alpha]_D^{24}$ −78.5° (neat) maximum rotation obtained by resolution[80] of the dibenzoyl tartrate.
[c] These formulas are intended only to symbolize the ratio of LAH and (−)-menthol (HOR*) without implication concerning the real structure of these complexes.

TABLE 5-15

Asymmetric Reductions of Imines by LAH-Chiral Alcohol Reagents[a]

$$\underset{\mathbf{77}}{R-C(=NH)-R'} \xrightarrow[2)\ H_2O]{1)\ LiAlH_3OR^*} \underset{\mathbf{78}}{R-C^*H(NH_2)-R'}$$

No.	Ketimine 77 R	R′	(−)-Menthol $[\alpha]_D$ °	% e.e.	(+)-Borneol $[\alpha]_D$ °	% e.e.	(−)-Quinine $[\alpha]_D$ °	% e.e.
1	Ph	Me	+0.82	2.2 *R*	−0.67	1.8 *S*	+0.95	2.5 *R*
2	Ph	Et	+0.94	9.9 *R*	−0.89	9.4 *S*	—	—
3	Ph	*n*Pr	+0.63	—	−0.42	—	—	—
4	Ph	α-Naph	+4.13[b]	—	−1.44[c]	—	—	—
5	Ph	*o*-Tolyl	+4.56	—	—	—	—	—
6	Ph	Mesityl	0.0	—	—	—	—	—
7	*o*-Tolyl	α-Naph	−1.81[d]	—	—	—	—	—

(Columns 4–9: Amine 78 produced by)

[a] Data taken from Reference 71.
[b] The hydrochloride on reduction gave $[\alpha]_D$ +3.19°.
[c] The hydrochloride on reduction gave $[\alpha]_D$ −2.28°.
[d] Reduction of imine hydrochloride.

reaction for the empirical correlation of configuration of the resulting amines. It seems justified to conclude that the dextrorotatory amines where R′ is an alkyl group (Nos. 1–3) are configurationally related. It is known that the first two are *R*-(+) and thus the third one should be *R* as well. Similar extension to the phenyl aryl ketimine examples (Nos. 4–7) is premature. Note that phenyl *o*-tolyl ketimine gave an optically active reduction product while phenyl mesityl ketimine inexplicably did not.

5-3.3. LAH-Monosaccharide Complexes in Asymmetric Reductions. Reductions of ketones by LAH which has been modified by treatment with monosaccharide derivatives (**79**–**83**) give optically active secondary carbinols (Table **5-16**).[69,72,81,82] Although the extents of asymmetric reduction generally have been below 10% with the reagents from **79**, **80**, **82**, and **83**, the reagent prepared by the reaction of 1 : 1 molar amounts of

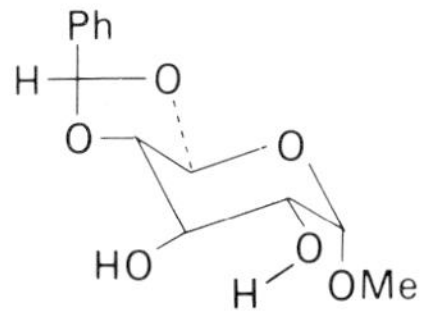

79 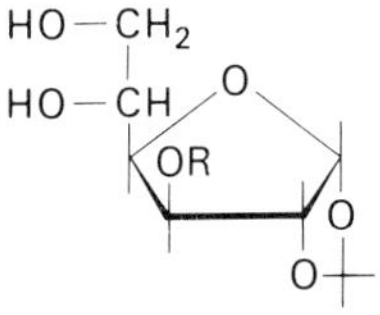 **80** 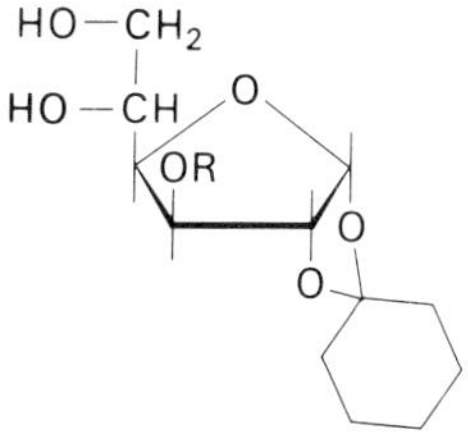 **81**

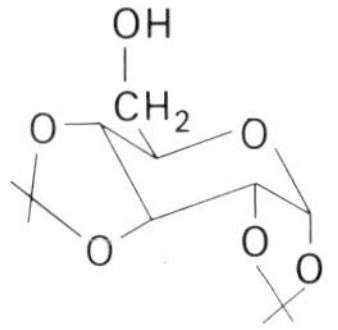

82 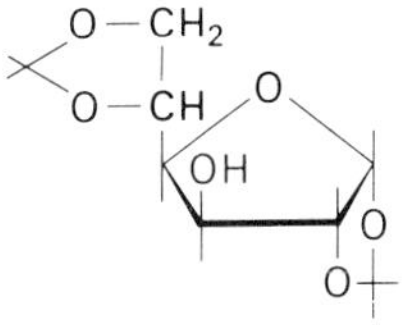 **83**

LAH and 3-O-benzyl-1,2-cyclohexylidene-α-D-glucofuranose (**81**, R = CH_2Ph) gave asymmetric reductions of several ketones in the 30% range (Table **5-16**, column three from the right). This reagent has two active hydrogens and is symbolized by $LiAlH_2(OR^*)_2$).[81a] Landor and coworkers[81b] were able to increase the stereoselectivity of this reagent by added ethanol to give a 1 : 1 : 1 molar reaction mixture of LAH : **81**(R = CH_2Ph) : EtOH (Table **5-16**, next to last column). This reagent now has one active hydrogen and can be symbolized by $LiAlH(OR^*)_2OEt$. Several ketones were reduced with 30–50% stereoselectivity and acetophenone yielded methylphenylcarbinol of 71% e.e.

TABLE 5-16

Asymmetric Ketone Reductions Using LAH-Monosaccharide Complexes[a]

$$R{-}\overset{O}{\overset{\|}{C}}{-}R' \xrightarrow{LiAlH_x(OR^*)_y} R{-}CHOH{-}R'$$

No.	$R{-}\overset{O}{\overset{\|}{C}}{-}R'$	**79**[b]	**80**[b] R = H	**80**[b] R = Me	**80**[b] R = CH_2Ph	**81**[b] R = H	**81**[b] R = Me	**81**[b] R = Et	**81**[b] R = CH_2Ph	**81**[c] R = CH_2Ph	**82(83)**
		% e.e.	% e.e.	% e.e.	% e.e.	% e.e.	% e.e.	% e.e.	% e.e.	% e.e.	% e.e.
1	Me_2CHCH_2COMe	7.4 *R*	2.1 *R*	0	1.3 *S*	3.0 *S*	0.9 *S*	3.3 *S*	30 *S*	17 *R*	6(2) *R*
2	Me_3CCOMe	4.9 *R*	1.3 *R*	3.4*S*	—	5.1 *S*	7.3 *S*	1.7 *R*	4 *S*	18 *R*	2(4) *R*
3	PhCOMe	7.7 *S*	8.9 *S*	0.7 *R*	6.1 *S*	14.3 *S*	1.1 *R*	3.7 *S*	33 *S*	71 *R*	9(4) *R*
4	2-NaphthCOMe	7.3 *S*	—	—	—	6.2 *S*	—	—	29 *S*	40 *R*	—
5	$Me_2C{=}CHCOMe$	4.7 *R*	0	—	1.6 *S*	11.4 *S*	4.1 *R*	1.6 *S*	30 *S*	31 *R*	—
6	$HC{\equiv}CCOMe$[d]	9.9 *S*	—	—	—	14.4 *S*	—	—	—	—	—
7	$HC{\equiv}CCO$*n*Pr[d,e]	−1.00°	−1.12°	+0.56°	—	−0.90°	−0.24°	−0.20°	+0.4°	—	—
8	$HC{\equiv}CCO$*i*Pr[d,e]	−1.73°	−0.23°	—	+0.16°	−0.25°	+0.15°	−0.40°	+0.9°	—	—
9	$HC{\equiv}CCOPh$[d,e]	+1.84°	—	—	—	+1.18°	—	—	—	—	—
10	MeCOEt	—	—	—	—	—	—	—	—	45 *R*	—
11	MeCO*n*Bu	—	—	—	—	—	—	—	12 *S*	13 *R*	—
12	$MeCOCH_2CMe_3$	—	—	—	—	—	—	—	19 *S*	4 *R*	—
13	MeCO*n*Hex	—	—	—	—	—	—	—	12 *S*	25 *R*	—
14	EtCOPh	—	—	—	—	—	—	—	38 *S*	46 *R*	—

[a] See p. 213 for structures **79**–**83**; reductions using **79**–**80**, Reference 69; using **81**, Reference 81; using **82**–**83**, Reference 82.
[b] The chiral reagent and LAH were used in 1:1.1 molar ratio and can be symbolized by $LiAlH_2(OR^*)_2$ since the chiral reagent has two hydroxyl groups.
[c] The chiral reagent made from LAH, **81** (R = CH_2Ph), and EtOH in 1:1:1 molar ratio is symbolized by $LiAlH(OR^*)_2OEt$.
[d] The acetylenic ketones were reduced to acetylenic carbinols which however contained 5–8 % of the corresponding vinylic carbinol as an impurity.
[e] The maximum rotations and configurations are unknown and thus the results are given in terms of $[\alpha]_D$° (neat).

The 1:1 reagent of LAH and **81** (R = CH_2Ph) gave the *S* carbinol in excess in the nine examples reported to date while the 1:1:1 LAH:**81**(R = CH_2Ph):EtOH reagent gave the *R* carbinol in excess, frequently with enhanced stereoselectivity. Such systematic behavior indicates that this reagent may be used for configurational correlations; the configuration of 2-naphthylmethylcarbinol was assigned on this basis. Demonstration of this reversal in stereoselectivity upon the addition of ethanol to the 1:1 reagent may be taken as evidence that the ketone being reduced is behaving according to the standard pattern for this system. For such predictions to be reliable, care must be taken to use standardized reagent (filtered LAH solutions are preferred[81a]) and carefully controlled molar ratios of reactants.

It is significant that the sugar derivative **81**, R = CH_2Ph, was ether soluble and the reductions using it were homogeneous in contrast to most of the other reagents from **79**–**82** which were ether insoluble and gave reductions which were heterogeneous, at least in part.

A convincing transition state correlation model which accounts for the stereochemistry and its observed reversal upon the addition of ethanol has been proposed. This method has great potential and should be extended.

5-3.4. Asymmetric Reductions of Carbonyl and Imino Compounds by Di-3-pinanylborane. The uses of di-3-pinanylborane[83,84] for asymmetric synthesis are considered in Sec. **6-2** where the nature of this reagent and its mechanism of action are discussed in detail in connection with its addition to olefins. However, this reagent* has been used as well for the reduction of carbonyl compounds[84–87] (Table **5-17**) and some imines[88] (Table **5-18**).

Significant asymmetric reduction of carbonyl compounds occurs with this reagent but the percent asymmetric synthesis is not as pronounced as is observed in the use of the same reagent with *cis* olefins (Table **6-1**). The reduction of isobutyraldehyde-1-*d* by the isotopically normal reagent, P_2^*BH, gives (+)-isobutyl-1-*d* alcohol (29% e.e.). Treatment of the isotopically

[83] H. C. Brown and G. Zweifel, *J. Am. Chem. Soc.*, **83**, 486 (1961).
[84] H. C. Brown and D. B. Bigley, *ibid.*, **83**, 3166 (1961).
[85] K. R. Varma and E. Caspi, *Tetrahedron*, **24**, 6365 (1968).
[86] E. Caspi and K. R. Varma, *J. Org. Chem.*, **33**, 2181 (1968).
[87] S. Wolfe and A. Rauk, *Can. J. Chem.*, **44**, 2591 (1966).
[88] D. R. Boyd, M. F. Grundon, and W. R. Jackson, *Tetrahedron Lett.*, 2102 (1967).

* Di-3-pinanylborane, commonly referred to as diisopinocampheylborane, will be represented as P_2^*BH. It is prepared by the reaction of 4 moles of α-pinene with 1 mole of diborane.

$$4\ \alpha\text{-Pinene} + B_2H_6 \rightleftharpoons 2P_2^*BH \rightleftharpoons P_4^*B_2H_2$$

The reagent is present primarily as the dimer, tetra-3-pinanyldiborane. (+)-*R*,*R*-α-Pinene gives a levorotatory reagent while (−)-*S*,*S*-α-pinene gives a dextrorotatory reagent (Sec. **6-2**).

normal aldehyde with the deutero reagent, P_2^*BD, gives the enantiomeric (−)-isobutyl-1-*d* alcohol (27–28% e.e.) as seen in Table **5-17**, Nos. 1–3. This reduction has a comparable stereoselectivity to that observed for the reduction of the same substrate by the isobornyloxymagnesium reagent

TABLE 5-17

Asymmetric Reduction of Carbonyl Compounds by Di-3-pinanylborane[a]

$$R_S-\overset{\overset{\displaystyle O}{\|}}{C}-R_L' \xrightarrow{P_2^*BH} \xrightarrow[H^+]{H_2O} R_S-\overset{\overset{\displaystyle OH}{|}}{C}H-R_L'$$

Ketone					Product Carbinol		Ref.
No.	R_S	R_L'	Reagent	Solvent	% e.e.	config.	
1	H	*i*-Pr	P_2^*BD	THF	27	*R*-(−)	85
2	H	*i*-Pr	P_2^*BD[b]	THF	28	*R*-(−)[b]	85
3	D	*i*-Pr	P_2^*BH[b]	THF	29	*S*-(+)[b]	85
4	H	Ph	P_2^*BD	THF	6	*R*-(−)	85
5	H	Ph	P_2^*BD	THF	30	*S*-(+)	87
6	Me	Et	P_2^*BH[b]	THF	7	*S*-(+)[b]	85
7	Me	Et	P_2^*BH	Diglyme	11	*R*-(−)	84
8	Me	*i*-Pr	P_2^*BH[b]	THF	20	*S*-(+)[b]	85
9	Me	*i*-Pr	P_2^*BH	Diglyme	16	*S*-(+)	85
10	Me	*i*-Pr	P_2^*BH	Diglyme	17	*R*-(−)	84
11	Me	*t*-Bu	P_2^*BH	Diglyme	30	*S*-(+)[c]	84
12	Me	*t*-Bu	P_2^*BH	THF	12	*S*-(+)[c]	85
13	Me	Ph	P_2^*BH	Diglyme	14	*R*-(+)	84
14	Et	*i*-Pr	P_2^*BH	THF	62	*S*-(−)	85
15	RCH_2CH_2[d]	*i*-Pr	P_2^*BH	THF	—[e]	*S*-(−)	86

[a] The chiral reagent is made by mixing, in either diglyme or THF, $NaBH_4$, α-pinene, and BF_3 in a ratio to give P_2^*BH. Although both (+)-α-pinene and (−)-α-pinene were used, the data in this table have been adjusted as though only (+)-*R*,*R*-α-pinene were used to give the (−)-P_2^*BH reagent.

[b] Data are given for (−)-reagent, although (+)-reagent from (−)-α-pinene actually reported.

[c] The configuration of pinacolyl alcohol is now known to be *S*-(+) [J. Jacobus, Z. Majerski, K. Mislow, and P. V. R. Schleyer, *J. Am. Chem. Soc.*, **91**, 1998 (1969)].

[d] RCH_2CH_2 represents the 2-tetrahydropyranyloxyethyl group; i.e., R = tetrahydropyranyloxy.

[e] Maximum rotation unknown, $[\alpha]_D^{25}$ −0.80° from (+)-P_2^*BH reagent and $[\alpha]_D^{25}$ +0.82° from the (−)-P_2^*BH reagent.

(19% e.e., cf. Table **5-2**, No. 4) but the stereoselectivity was much less for the reduction of benzaldehyde-1-*d* with P_2^*BH (6% e.e.) than with the isobornyloxymagnesium reagent (41–45%). The reaction of P_2^*BH with terminal olefins followed by oxidation (Tables **6-3** and **6-4**) and the reaction of P_2^*BH with aldehydes (Table **5-17**) are formally similar processes, but there is no obvious quantitative correlation of the results from these two types of substrates. The former seems to be more stereoselective in most cases.

Both the extent of asymmetric reduction and the configuration of the predominate isomer differ in the reports from two separate laboratories on the asymmetric reduction of benzaldehyde by P^*_2BD. (Table **5-17**, Nos. 4 and 5). Furthermore it should be noted that the stereoselectivities for experiments Nos. 6 and 7 and Nos. 9 and 10 (Table **5-17**) from different laboratories agree fairly well in extent of asymmetric synthesis but unaccountably give enantiomeric products in excess. On the other hand experiments Nos. 11 and 12 do not agree in extent of asymmetric synthesis but do agree in the configuration of the predominant enantiomer. It is possible that this irreproducibility of results is connected with the "age" of the reagent as noted in another study[89] (cf. Sec. **6-2.25**).

It also seems inconsistent that the stereoselectivity in the reduction of ethyl isopropyl ketone should be 62% while it was only 12–30% for methyl *t*-butyl ketone. It is apparent that the experimental variables in this system should be studied further.

Three chiral reducing agents have been used to reduce 2-alkyl-Δ^1-piperideines[88] (Table **5-18**): "di-3-pinanylborane" [P^*_2BH, made from 4

TABLE 5-18

Asymmetric Reduction of Δ^1-Piperideines by Hydrido Boranes[88]

No.	Δ^1-Piperideine R =	Reagent[a]	Predominant amine config.	Predominant amine % e.e.[b]
1	Me	P^*_2BH	*S*	2.0–3.3
2	Me	$P^*_3B_2H_3$	*S*	2.2–3.2
3	*n*Pr	$P^*_3B_2H_3$	*S*	2.9–10.7
4	Me	$P^*_2nBuBH^-Li^+$	*R*	19.5–24.0
5	*n*Pr	$P^*_2nBuBH^-Li^+$	*R*	4.0–4.3

[a] P* represents the 3-pinanyl radical. See Sec. **6-2.1** for a discussion of the composition of these reagents. The reagent P^*_2BH referred to as di-3-pinanylborane (or di-isopinocampheylborane) is probably the dimer, tetra-3-pinanyldiborane.

[b] These values are not corrected for the isomeric purity of the (+)-pinene, used in preparation of the reagent, the rotation of which was 74% of the maximum literature value.

[89] D. J. Sandman, K. Mislow, W. P. Giddings, J. Dirlam, and G. C. Hanson, *J. Am. Chem. Soc.*, **90**, 4877 (1968).

moles of (+)-α-pinene per mole of diborane], "tri-3-pinanyldiborane" [$P_3^*B_2H_3$, made from 1:3 molar ratio of diborane and (+)-α-pinene], and lithium hydrido-1-butyl-di-3-pinanylborate [$(P_2^*n\text{-BuBH})^-Li^+$, prepared from "di-3-pinanylborane" by adding *n*-butyl lithium. The extents of asymmetric reductions with the first two reagents were generally low (2–10.7% e.e. of the *S* isomer) and it is possible that the same reducing species is acting in both reagents (cf. Sec. **6-2.1**). The asymmetric bias in the reduction of 2-methyl-Δ^1-piperideine with the reagent to which *n*-butyl lithium had been added was substantially higher (19.5–24% e.e., *R* isomer). Treatment of this reagent with 1-butene prior to reduction of the piperideine did not alter the extent of asymmetric reduction. This was interpreted to mean that chiral boranes are not responsible for asymmetric reduction in this system.

1969 Addenda

Sec. 5-1.1. It has been shown[90] that chiral lithium (*R*)-(+)-(α-methylbenzyl)anilide reduces phenyl α-naphthyl ketone to the optically active carbinols in a manner analogous to that of the MPV reduction. The rotation of the carbinol was $[\alpha]_D + 35.5°$ (c 1, acetone) when the reaction was conducted in ether solvent, but $[\alpha]_D + 3.6°$ (c 1, acetone) in THF solvent. The maximum rotation is unknown. No asymmetric reduction of phenyl *p*-biphenyl ketone was observed, indicating that steric hindrance in the *para* position is unimportant in analogy with the Grignard asymmetric reduction (cf. p. 189).

Sec. 5-2.3d. Using stereospecifically labeled deuterium reagents, it was demonstrated that the reduction of phenyl *t*-butyl ketone by the Grignard reagent from (*R*)-1-chloro-3-phenylbutane involves preferential transfer of the *pro-S* hydrogen.[91]

Sec. 5-2.3e. 1-Adamantanecarboxaldehyde-1-*d* has been reduced by actively fermenting yeast to the corresponding 1-adamantylcarbinol-1-*d* which was shown to be at least 97% enantiomerically pure by nmr analysis of its O-methylmandelate[92] (cf. Table **5-9**).

Sec. 5-3.1. The chiral agent obtained by addition of (*S*)-amphetamine or *S*-desoxyephedrine to diborane reduces acetophenone to (*S*)-(+)-phenylmethylcarbinol, 4% e.e. or 5% e.e. (cf. Table **5-17**).[93]

[90] G. Wittig and U. Thiele, *Ann. Chem.*, **726**, 1 (1969).
[91] J. D. Morrison and R. W. Ridgway, *J. Am. Chem. Soc.*, **91**, 4601 (1969).
[92] S. H. Liggero, R. Sustmann, and P. R. Schleyer, *J. Am. Chem. Soc.*, **91**, 4571 (1969).
[93] J. C. Fiaud and H. B. Kagan, *Bull. Soc. Chim. Fr.*, 2743 (1969).

6

Asymmetric Additions to Alkenes

6-1. Introduction

Examples of asymmetric additions to carbon-carbon double bonds fall into two general categories. Either the reagent is chiral and the alkene has enantiotopic faces or the alkene has diastereotopic faces and the reagent is achiral. As an example of the first type, (+)-monoperoxycamphoric acid (**1**) reacts with styrene (**2**) to yield optically active styrene oxide[1] (**3**) (Fig. **6-1**). As an example of the second type, diphenyldiazomethane (**5**) converts optically active menthyl acrylate (**6**) to a cyclopropanecarboxylate (**7**) that

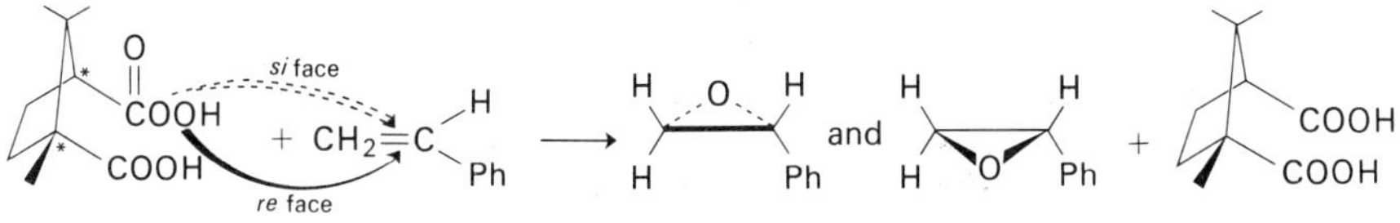

Fig. 6-1. Reaction of a chiral reagent with an achiral alkene having enantiotopic faces.

[1] (a) H. B. Henbest in *Organic Reaction Mechanisms*, Special Publication No. 19, The Chemical Society, London, 1965, pp. 83–92. (b) R. C. Ewins, H. B. Henbest, and M. A. McKervey, *Chem. Commun.*, 1085 (1967).

can be hydrolyzed to optically active 2,2-diphenylcyclopropanecarboxylic acid[2] (**8**) (Fig. **6-2**).

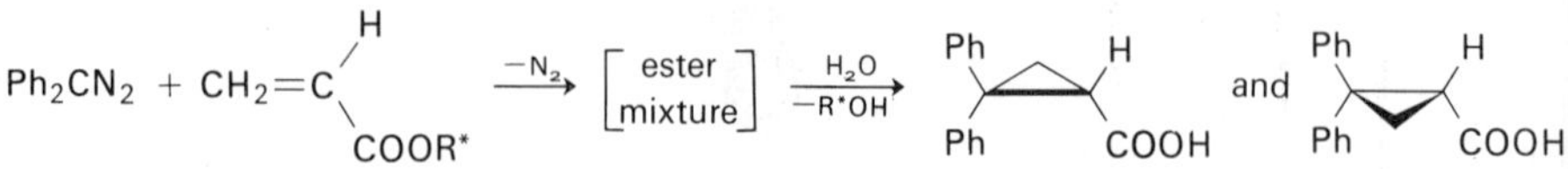

$$Ph_2CN_2 + CH_2{=}C(H)COOR^* \xrightarrow{-N_2} \begin{bmatrix}\text{ester}\\ \text{mixture}\end{bmatrix} \xrightarrow[-R^*OH]{H_2O} \mathbf{8A}\ (51\%) \text{ and } \mathbf{8B}\ (49\%)$$

5 **6** **7** **8A** (51%) **8B** (49%)

Fig. 6-2. Reaction of an achiral reagent with a chiral alkene having diastereotopic faces. R* is (−)-menthyl.

Many asymmetric syntheses of these types are useful for preparative purposes, for correlation of configurations, and for mechanism studies. A number of chiral additions to alkenes will now be discussed including the two general types represented by Figs. **6-1** and **6-2** as well as asymmetric additions under the influence of chiral catalysts.

6-2. Hydroborations with "Di-3-pinanylborane"*

Perhaps the most versatile chiral reagent readily available for laboratory use is "di-3-pinanylborane" (**10**), prepared by reaction of diborane with optically active α-pinene (**9**) at 0°.[3–6] The product of this reaction is usually named as the monomer, although it has been established[5,6] that it exists mainly as a dimer (**12**) in solution. Furthermore the dimer is about 5–10% dissociated into tri-3-pinanyldiborane (**13**) and α-pinene (Fig. **6-3**), the exact degree of dissociation being dependent upon the concentration of dimer, the amount of excess α-pinene present, and the solvent. It is dissociated to a greater extent in tetrahydrofuran (THF) than in diethylene glycol dimethyl ether (diglyme) and partially precipitates in the latter solvent when prepared according to recommended procedures.[7]

[2] H. M. Walborsky and C. G. Pitt, *J. Am. Chem. Soc.*, **84**, 4831 (1962) and references therein.
[3] H. C. Brown and G. Zweifel, *J. Am. Chem. Soc.*, **81**, 247 (1959).
[4] H. C. Brown and G. Zweifel, *J. Am. Chem. Soc.*, **83**, 2544 (1961).
[5] H. C. Brown and G. J. Klender, *Inorg. Chem.*, **1**, 204 (1962).
[6] G. Zweifel and H. C. Brown, *J. Am. Chem. Soc.*, **86**, 393 (1964).
[7] G. Zweifel and H. C. Brown, *Organic Reactions*, (New York: John Wiley & Sons, Inc., 1963), Vol. 13, p. 34.

* This is the *Chemical Abstracts* name for diisopinocampheylborane[3] (**10**). We shall identify a solution of the products from the hydroboration of 4 moles of α-pinene with 1 mole of diborane (2 moles of borane) as "di-3-pinanylborane," but we shall *not* append the quotation marks when we wish to refer exclusively to the monomeric form. The reasons for this are clarified in the subsequent discussion.

Oxidation of (−)-"di-3-pinanylborane" [from (+)-α-pinene] with hydrogen peroxide yields (−)-isopinocampheol (**11**), the absolute configuration of which is known.[6] Since peroxide oxidation of the carbon-boron bond proceeds with retention of configuration,[8] the configuration of the borane precursor is thus established.

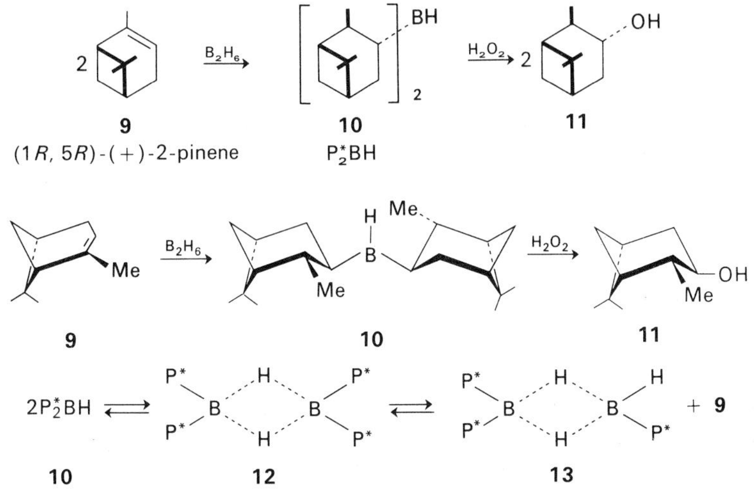

2 **9** (1*R*, 5*R*)-(+)-2-pinene $\xrightarrow{B_2H_6}$ [ ]$_2$BH **10** P^*_2BH $\xrightarrow{H_2O_2}$ 2 OH **11**

9 Me $\xrightarrow{B_2H_6}$ **10** H–B, Me, Me $\xrightarrow{H_2O_2}$ **11** OH, Me

$2P^*_2BH$ ⇄ $P^*_2B(\mu\text{-}H)_2BP^*_2$ ⇄ $P^*_2B(\mu\text{-}H)_2BHP^*$ + **9**

10 **12** **13**

Fig. 6-3. Structure and equilibria for "di-3-pinanylborane" reagent; P* represents the 3-pinanyl (isopinocampheyl) group

Relatively unhindered *cis*-olefins are readily hydrobrated by "di-3-pinanylborane." The resulting diastereomeric trialkylboranes in which boron is attached to a new asymmetric carbon atom are produced in unequal amounts. Stereospecific cleavage of the carbon-boron bonds thus produces a chiral derivative of the original olefin.* For example, from the trialkylborane obtained by hydroborating *cis*-2-butene with "(−)-di-3-pinanylborane," one can synthesize (*R*)-(−)-2-butylamine or (*R*)-(−)-2-butanol by reaction with hydroxylamine-0-sulfonic acid or alkaline hydrogen peroxide, respectively. When prepared from the same solution of epimeric trialkylboranes, the amine **14** and alcohol **15** possess identical enantiomeric purities (*ca.* 75%) within experimental error[9] (Fig. **6-4**, cf. also Table **6-1**, Footnote b).

[8] A. G. Davies and B. P. Roberts, *J. Chem. Soc.* (*C*), 1474 (1968).

[9] L. Verbit and P. J. Heffron, *J. Org. Chem.*, **32**, 3199 (1967).

* Since subsequent cleavage reactions are conducted on a mixture of diastereomers which will react at different rates, there is the possibility of selectively cleaving one isomer unless the reaction is carried to completion. Therefore, one must be certain that the cleavage is essentially complete if the results are to be interpreted from a quantitative standpoint.

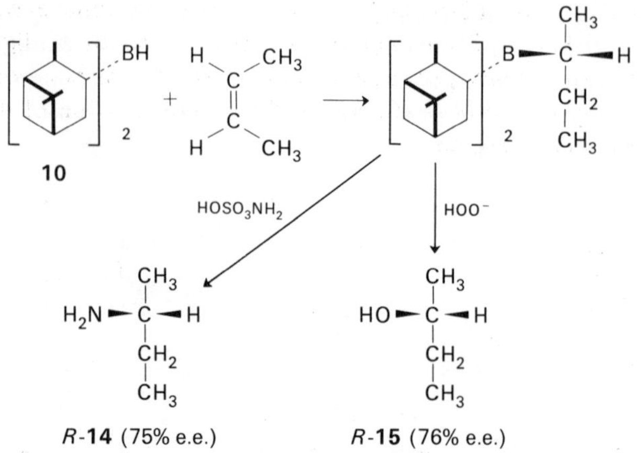

R-**14** (75% e.e.) *R*-**15** (76% e.e.)

Fig. 6-4. Asymmetric synthesis of 2-butanol and 2-aminobutane via hydroboration.

6-2.1. Asymmetric Synthesis of Alcohols via Hydroboration-Oxidation of Olefins.* Relatively unhindered *cis* olefins are hydroborated rapidly by "di-3-pinanylborane." Reaction of the resulting epimeric trialkylboranes with alkaline hydrogen peroxide yields optically active alcohols.[3–11] Relatively hindered *cis* olefins and *trans* olefins are also hydroborated, but much more slowly, and the alcohols resulting from subsequent peroxide oxidation have lower enantiomeric purities than those obtained from less hindered *cis* olefins. The structure of the principal hydroborating species probably depends upon the steric requirements of the olefin. It could be di-3-pinanylborane (**10**), tetra-3-pinanyldiborane (**12**), tri-3-pinanyldiborane (**13**), or other less substituted boranes such as *sym*- or *unsym*-di-3-pinanyldiborane. Hydroboration of *cis* hindered alkenes is usually discussed in terms of **10** or **12** as the effective reagent. Tri-3-pinanyldiborane appears to be the hydroborating species in the case of certain alkenes which are more hindered.[12]

Table **6-1** summarizes the results obtained with a number of relatively unhindered *cis* olefins. The rotation of both the (−)- and (+)-α-pinene used in these experiments was 93–95% of the maximum reported value. Assuming that the α-pinene was chemically pure and that the lower rotation was due to the presence of racemate, the enantiomeric purities of the alcohols

[10] H. C. Brown and D. B. Bigley, *J. Am. Chem. Soc.*, **83**, 486 (1961).

[11] H. C. Brown, N. R. Ayyangar, and G. Zweifel, *J. Am. Chem. Soc.*, **86**, 397 (1964).

[12] H. C. Brown, N. R. Ayyangar, and G. Zweifel, *J. Am. Chem. Soc.*, **86**, 1071 (1964).

* Reduction of unsymmetrical ketones with "di-3-pinanylborane" and other chiral reducing agents also yields optically active alcohols (cf. Sec. **5-3.4**).

TABLE 6-1

Hydroboration-Oxidation of *cis* Olefins with "Di-3-pinanylborane" in Diglyme[a]

No.	α-Pinene	Olefin	Product	Predominant isomer	Asymmetric synthesis % e.e.
1	+	*cis*-2-Butene	2-Butanol	*R*	87[b]
2	−	*cis*-2-Butene	2-Butanol	*S*	86
3	−	*cis*-2-Pentene	2-Pentanol, 76% 3-Pentanol, 24%	*S*	82[c]
4	+	*cis*-3-Hexene	3-Hexanol	*R*	91
5	−	*cis*-4-Methyl-2-pentene	4-Methyl-2-pentanol, 96% 2-Methyl-3-pentanol, 4%	*S*	76[d]
6	+	Norbornene	*exo*-Norborneol	1*S*, 2*S*	67–70
7	−	Norbornene	*exo*-Norborneol	1*R*, 2*R*	65–68
8	+	Bicycloheptadiene[e]	*exo*-Dehydronorborneol	1*R*, 2*S*	48–51

[a] Reference 11, unless otherwise noted. Both the (−) and (+) α-pinene used in preparing "di-3-pinanylborane" were apparently only 93–95% enantiomerically pure, in which case the %e.e. reported in the last column should be increased by a factor of 1.05–1.08 to reflect the %e.e. to be expected with enantiomerically pure reagent.

[b] *cis*-2-Butene gave 2-butanol, 78% e.e., when hydroborated in THF rather than diglyme. Note also that Reference 9 reports 75% e.e. for this reaction and that when the reaction mixture stood for 48 hr before hydrolysis, the stereoselectivity dropped to 70% e.e.

[c] Based on the rotation of the 2- and 3-pentanol mixture.

[d] Based on $[\alpha]_D$ max for 4-methyl-2-pentanol.

[e] K. Mislow and J. G. Berger, *J. Am. Chem. Soc.*, **84**, 1956 (1962).

listed can be multiplied by a factor of 1.05–1.08 to obtain the percent asymmetric synthesis possible with enantiomerically pure-α-pinene.

The observation that in THF the degree of asymmetric synthesis appears to be slightly lower than it is in diglyme (Table **6-1**, Footnote b) may be significant. In diglyme the dimer (**12**) is known to be less dissociated into tri-3-pinanyldiborane (**13**) than it is in THF.[6]

There are some important differences between the reactions of certain isomeric olefins with "di-3-pinanylborane." For example, whereas the reaction of *cis*-2-butene with "di-3-pinanylborane" at 0° is virtually complete in 2 hr, *trans*-2-butene requires 24 hr for essentially complete reaction. Furthermore, hydrogen peroxide oxidation of the trialkylborane from the *cis* isomer yields 2-butanol that is 75–87% enantiomerically pure,[9,12] whereas that obtained by oxidation of borane from *trans* olefin is only 13% enantiomerically pure.[12] The *trans*-2-butene also *displaces 1 mole of α-pinene from the reagent for every 2 moles of olefin hydroborated*, whereas *cis*-butene does not displace α-pinene. Likewise, *trans*-4-methyl-2-pentene reacts much more slowly with "di-3-pinanylborane" than does *cis*-4-methyl-2-pentene. About eight times as much 2-methyl-3-pentanol was obtained from the *trans* olefin as from the *cis* isomer. Again 1 mole of α-pinene was displaced for every 2 moles of olefin hydroborated.[12]

Hydroboration of the more hindered olefin 2-methyl-2-butene with "di-3-pinanylborane" is extremely slow; it proceeds with displacement of *1 mole of α-pinene for each mole of olefin hydroborated.*[13] 3-Methyl-2-butanol (8–14% e.e.) is obtained upon peroxide oxidation.[12,13] Likewise, 1-methylcyclopentene and bicyclo[2.2.2]oct-2-ene also hydroborate very slowly with "di-3-pinanylborane," with the displacement of 1 mole of α-pinene per mole of olefin, to give oxidation products with relatively low enantiomeric purities.

A lower hydroboration rate, a change in directive effect, lower enantiomeric purities of alcohols from oxidation, and particularly displacement of α-pinene are observations which suggest that when the olefin is sufficiently hindered the hydroborating agent is not di-3-pinanylborane or its dimer, tetra-3-pinanyldiborane, but tri-3-pinanyldiborane formed by elimination of α-pinene from the dimer. This conclusion is supported by the observation that the reagent prepared with a 1:3 molar ratio of diborane and α-pinene hydroborates a trisubstituted olefin at a reasonable rate, and alcohols obtained by oxidation have about the same enantiomeric purities as those obtained from slow hydroboration with "di-3-pinanylborane." Even 1-methylcyclohexene, which reacts extremely slowly with dialkylboranes, is hydroborated in 4 hr with the reagent prepared from a 1:3 molar ratio of diborane and α-pinene, which presumably is tri-3-pinanyldiborane.

With certain hindered olefins (e.g., 1-methylcyclopentene) it appears that a second mole of α-pinene is displaced after the tri-3-pinanyldiborane reacts with the first mole of olefin and then a second hydroboration occurs. A reasonable rationalization of hydroboration sequences that proceed with displacement of α-pinene is shown in Figs. **6-5** and **6-6**; some representative experimental results are summarized in Table **6-2**.

Hydroboration of 2-substituted 1-alkenes with di-3-pinanylborane provides a synthetic route to optically active 2-substituted 1-alkanols. The enantiomeric excesses of alcohols obtained in this way are not as high as

$$R_1R_2C{=}CH_2 \xrightarrow[2)\ ^-OOH]{1)\ \text{"}P_2^*BH\text{"}} R_1R_2{}^*CHCH_2OH$$

those of alcohols from the hydroboration of *cis* unhindered olefins. However, considering the fact that the new asymmetric carbon is not that to which boron becomes attached and is therefore more remote from the disymmetric 3-pinanyl moiety in the transition state, the asymmetric syntheses observed

[13] D. K. Shumway and J. D. Barnhurst, *J. Org. Chem.*, **29**, 2320 (1964).

TABLE 6-2

Hydroboration-Oxidation of Hindered Olefins with 3-Pinanylboranes[12] at 0°

Reagent mmoles[a]	Olefin mmoles	Olefin reacted (mmoles)	H_2 evolved on hydrolysis (mmoles)	α-Pinene displaced (mmoles)	Alcohol	Asym. syn. % e.e.
Moderately hindered olefins						
P_2^*BH (50)	*trans*-2-Butene (55)	23[b]	31	11.6	(−)-2-Butanol	13
P_2^*BH (25)	4-Methyl-2-pentene (25)	25	10.8	12.0	4-Methyl-2-pentanol, 68%; 2-Methyl-3-pentanol, 32%	—[c]
P_2^*BH (25)	4-Methyl-2-pentene (50)	42.5	1.6	20.5	4-Methyl-2-pentanol, 78%; 2-Methyl-3-pentanol, 22%	—[c]
P_2^*BH (25)	Cyclohexene (25)	21.7	—[c]	11.4	Cyclohexanol	Achiral product
Hindered olefins						
P_2^*BH (100)	2-Methyl-2-butene (100)	72	—	60	(−)-3-Methyl-2-butanol	14
$P_3^*B_2H_3$ (50)	2-Methyl-2-butene (50)	—[c]	—	—[c]	(−)-3-Methyl-2-butanol	17
P_2^*BH[d] (100)	1-Methylcyclopentene (100)	—	—	63	(+)-*trans*-2-Methylcyclopentanol	22
$P_3^*B_2H_3$ (100)	1-Methylcyclopentene (100)	—	—	—	(−)-*trans*-2-Methylcyclopentanol	17.5
$P_3^*B_2H_3$ (50)	1-Methylcyclohexene (50)	—	—	—	(−)-*trans*-2-Methylcyclohexanol	18
P_2^*BH (50)	Bicyclo[2.2.2] oct-2-ene (50)	38	—	34	(+)-Bicyclo[2.2.2]octan-2-ol	17

[a] P_2^*BH = "di-3-pinanylborane"; $P_3^*B_2H_3$ = tri-3-pinanyldiborane, the presumed product from reaction of α-pinene and diborane in a 3:1 mole ratio. Prepared from (−)-α-pinene unless otherwise noted.

[b] VPC indicated complete reaction of the 2-butene. Some olefin is probably lost during the reaction.

[c] — indicates data not reported.

[d] Prepared from (+)-α-pinene.

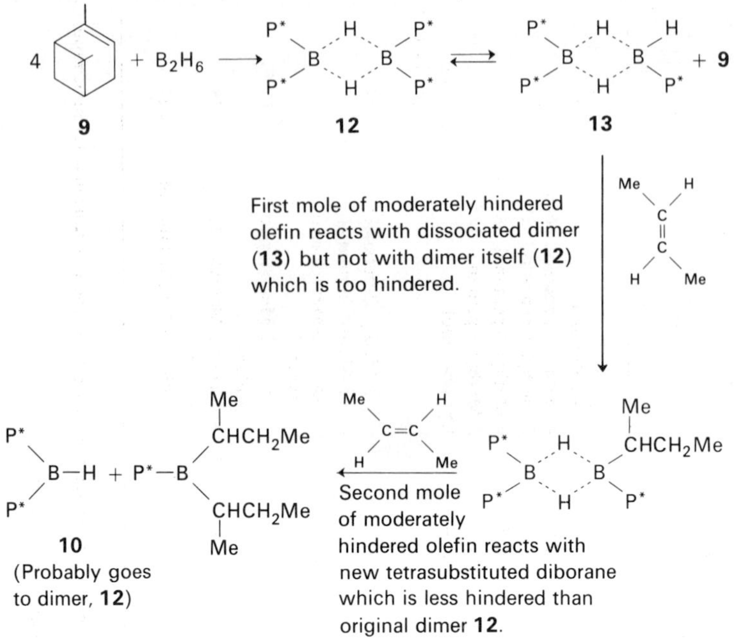

First mole of moderately hindered olefin reacts with dissociated dimer (**13**) but not with dimer itself (**12**) which is too hindered.

Second mole of moderately hindered olefin reacts with new tetrasubstituted diborane which is less hindered than original dimer **12**.

10 (Probably goes to dimer, **12**)

Fig. 6-5. Rationalization of the hydroboration of a moderately hindered olefin with di-3-pinanylborane. The net result is the displacement of 1 mole of α-pinene per 2 moles of olefin hydroborated (cf. Fig. **6-6**).

TABLE 6-3[14]

Hydroboration of 2-Substituted Terminal Alkenes with (−)-"Di-3-pinanylborane"

$$R_1R_2C{=}CH_2 \xrightarrow[\text{2) HOO}^-]{\text{1) "Di-3-pinanylborane"}} R_1R_2CHCH_2OH$$

No.	Alkene R_1	R_2	Alcohol product	Asm. syn. % e.e.
1	Me	Et	*R*-(+)-2-Methyl-1-butanol	21
2	Me	*i*Pr	*R*-(−)-2,3-Dimethyl-1-butanol	30
3	Me	*t*Bu	*R*-(−)-2,3,3-Trimethyl-1-butanol[a,b]	Unknown[b]
4	Me	Ph	*R*-(+)-2-Phenyl-1-propanol	5

[a] About four times the normal amount of (+)-α-pinene was present at termination of the reaction due to dissociation of (−)-"di-3-pinanylborane".

[b] The $[\alpha]_D$ max of (−)-2,3,3-trimethyl-1-butanol has not been determined. The *R* configuration has been assigned on the basis of reasonable stereochemical arguments.

[14] G. Zweifel, N. R. Ayyangar, T. Munekata, and H. C. Brown, *J. Am. Chem. Soc.*, **86**, 1076 (1964).

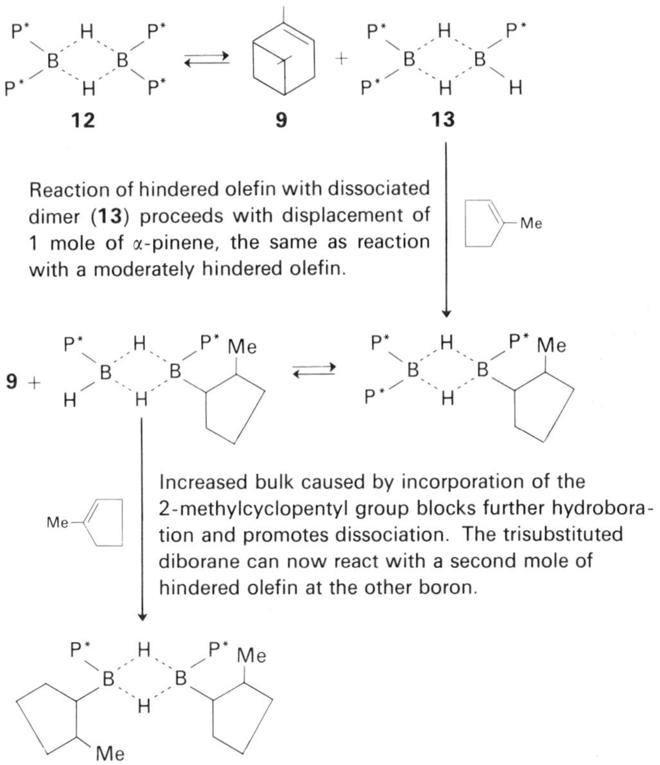

Reaction of hindered olefin with dissociated dimer (**13**) proceeds with displacement of 1 mole of α-pinene, the same as reaction with a moderately hindered olefin.

Increased bulk caused by incorporation of the 2-methylcyclopentyl group blocks further hydroboration and promotes dissociation. The trisubstituted diborane can now react with a second mole of hindered olefin at the other boron.

Fig. 6-6. Rationalization of the hydroboration of a more highly hindered olefin with "di-3-pinanylborane" (cf. Fig. **6-5**). The net result is the displacement of 1 mole of α-pinene per mole of olefin hydroborated.

are substantial (Table **6-3**).[14] The enantiomeric composition of the product alcohol does not depend in an obvious way upon the apparent steric bulk difference between the groups R_1 and R_2. Hydroboration of 2-methyl-1-butene gives an alcohol with four times the enantiomeric purity of that from similar treatment of 2-phenylpropene. Since the displacement of α-pinene is nearly the same with these two olefins, it would appear that the same alkyl borane is the hydroborating species (presumably di-3-pinanylborane or its dimer). The reagent apparently discriminates between methyl and ethyl more effectively than it does between methyl and phenyl* (21 versus 5% e.e.).

* This reaction provides another example of the uncertain relationship of apparent steric bulk differences between sigma bonded ligands at a trigonal carbon to the asymmetric bias of an addition reaction. In certain cases replacement of a simple alkyl group, such as methyl or ethyl, by phenyl increases the asymmetric bias, while in other cases such substitution results in a decrease. Reasonable rationalizations might be developed to explain the direction of the effect for a particular type of substrate and reagent, but there does not seem to be any generalization that accommodates all reagents.

Of the terminal alkenes investigated, only 2,3,3-trimethyl-1-butene displaced significant amounts of α-pinene. In an essentially complete reaction of 100 mmoles of olefin and 100 mmoles of "di-3-pinanylborane," 62.8 mmoles of α-pinene were displaced. This suggests that, with this particular alkene, the hydroboration is mechanistically complex, involving both addition and displacement pathways.

Asymmetric hydroboration has been applied to the synthesis of alcohols and other compounds that are optically active by virtue of deuterium substitution. The results of a number of such syntheses are summarized in Table **6-4**. The experiments represented by Nos. 2 and 3 reveal a dramatic isotope effect.[15] Hydroboration of (*Z*)-1-hexene-1-*d* (*cis*-D and *n*-Bu groups)* gave, after hydrogen peroxide oxidation, (*R*)-(−)-1-hexanol-1-*d*

TABLE 6-4

Summary of Asymmetric Synthesis of Deuterium Labeled Alcohols

$$R_1R_2C{=}CR_3R_4 \xrightarrow[\text{from }(+)\text{-}\alpha\text{-pinene}]{P_2^*B{-}H \text{ or } P_2^*B{-}D} \xrightarrow{H_2O_2} R_1R_2C(OH){-}CH(D)R_3R_4$$

No.	R_1	R_2	R_3	R_4	Reagent	Product	Asym. syn. % e.e.	Ref.
1	D	H	Et	H	P_2^*B-H	(*R*)-(−)-1-Butanol-1-*d*	56	17
2	D	H	*n*-Bu	H	P_2^*B-H	(*R*)-(−)-1-Hexanol-1-*d*	42	15
3	H	D	*n*-Bu	H	P_2^*B-H	(*S*)-(+)-1-Hexanol-1-*d*	86	15
4	H	Me	H	Me	P_2^*B-D	(2*S*,3*R*)-2-Butanol-3-*d*	64	18
5	H	H	H	*n*-Pr	P_2^*B-D	(*R*)-(−)-1-Pentanol-2-*d*	48±7	19

(42% e.e.), whereas the same treatment of (*E*)-1-hexene-1-*d* gave (*S*)-(+)-1-hexanol-1-*d* (86% e.e.). The observed reversal in sense of chirality is expected (cf. Sec. **6-2.5**), but one might have anticipated comparable extents of asymmetric bias, since the two alkenes differ only in the orientation of hydrogen and deuterium relative to the *n*-butyl group. The isotope effect may reflect a "size" difference between hydrogen and deuterium, but the possibility of partial bonding of these atoms giving some primary isotope character to

[15] H. Weber, P. Loew, and D. Arigoni, *Chimia*, **19**, 595 (1965). See also A. Horeau and A. Nouaille, *Tetrahedron Lett.*, 3953 (1966).

* The *E-Z* nomenclature [J. E. Blackwood, C. L. Gladys, K. L. Loening, A. E. Petrarca, and J. E. Rush, *J. Am. Chem. Soc.*, **90**, 509 (1968)] will be used where the *cis-trans* designation is equivocal.

their interactions in the transition state has also been mentioned.[16] Alternatively this may only be a reflection of the uncontrolled "age" of the reagent in the two experiments (cf. Fig. **6-13**). Hydroboration of (*Z*)-1-butene-1-*d*

$$\underset{Z}{\text{H}(n\text{-Bu})\text{C}{=}\text{C}(\text{H})\text{D}} \xrightarrow[2)\ \text{HOO}^-]{1)\ \text{P}_2^*\text{BH}} n\text{-BuCH}_2\text{—C(H)(D)OH} \qquad 42\%\ \text{e.e.}$$

$$\underset{E}{\text{H}(n\text{-Bu})\text{C}{=}\text{C}(\text{D})\text{H}} \xrightarrow[2)\ \text{HOO}^-]{1)\ \text{P}_2^*\text{BH}} n\text{-BuCH}_2\text{—C(D)(H)OH} \qquad 86\%\ \text{e.e.}$$

with P_2^*B—H (Table **6-4**, No. 1) gave (*R*)-(−)-1-butanol-1-*d* (56% e.e.),[17] but there are no comparison figures for hydroboration of (*E*)-2-butene-1-*d*.

Experiment No. 4 (Table **6-4**) suggests that there is also an isotope effect associated with the transfer of hydrogen versus deuterium from "di-3-pinanylborane" (or deuterioborane) to the same unlabeled alkene. When *cis*-2-butene was treated with "di-3-pinanyldeuterioborane"[18,*] and then oxidized with peroxide, *erythro*-2-butanol-3-*d* was obtained (mixture of 2*S*,3*R* and 2*R*,3*S* enantiomers). The enantiomeric composition was determined by degrading the product to (*S*)-(−)-ethanol-1-*d* (64% e.e.). Since the hydroboration is known to proceed by *cis* addition, this constitutes an elegant proof of the configuration of (−)-ethanol-1-*d*. Assuming no racemization at the labeled carbon during the degradation sequence (Fig. **6-7**), this enantiomeric composition reflects the asymmetric bias of the deuterioboration, which is lower than that observed for the comparable hydroboration of *cis*-2-butene with isotopically normal reagent (75–87% e.e.,[9,11] Table **6-1**). It is possible that these experimental differences in stereoselectivity with hydrogen versus deuterium transfer may only reflect differences in preparation or age of the reagent (cf. Sec. **6-2.5** and Fig. **6-13**).

The last experiment (No. 5) summarized in Table **6-4**[19] indicates that the asymmetric bias for the reaction of "P_2^*B-D" with 1-pentene [48 ± 7% e.e. (*R*)-(−)-1-pentanol-2-*d*] is higher than that of the reaction of "P_2^*B-H" with a number of 2-substituted-1-alkenes (Table **6-3**). Unfortunately 1-pentene-2-*d* has not been subjected to asymmetric hydroboration with "P_2^*B-H"

[16] D. R. Brown, S. F. A. Kettle, J. McKenna, and J. M. McKenna, *Chem. Commun.*, 667 (1967).

[17] A. Streitwieser, Jr., L. Verbit, and R. Bittman, *J. Org. Chem.*, **32**, 1530 (1967).

[18] H. Weber, J. Seibl, and D. Arigoni, *Helv. Chim. Acta*, **49**, 741 (1966).

[19] A. Streitwieser, Jr., I. Schwager, L. Verbit, and H. Rabitz, *J. Org. Chem.*, **32**, 1532 (1967).

* Prepared from (+)-α-pinene, LiD, and BF_3 etherate.

for direct comparison. Isotope effects on stereoselectivity have also been noted in the asymmetric Grignard reduction (Sec. **5-2.3e**). All of these results are intrinsically intriguing. More experimentation is needed to confirm them, to provide additional examples, and to stimulate theoretical consideration of their origin.

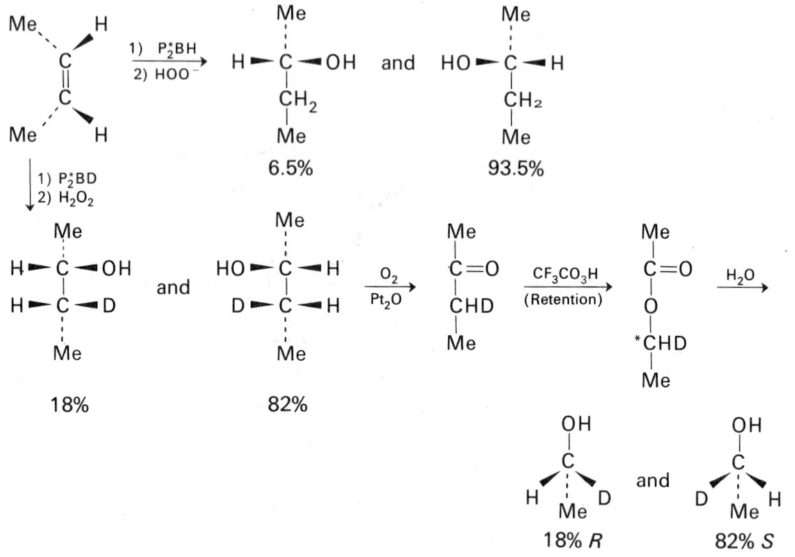

Fig. 6-7. Isotope effect on the asymmetric hydroboration of *cis*-2-butene and configurational proof for (−)-ethanol-1-*d*.

6-2.2. Kinetic Resolution of Racemic Olefins.

Incomplete hydroboration with a chiral hydridoborane is an example of a kinetically controlled asymmetric transformation (Sec. **1-4.4a**). Although such processes do not qualify as asymmetric syntheses, since a new asymmetric center is not produced, it is appropriate at this point to illustrate the utility of "di-3-pinanylborane" for this purpose.

Treatment of 3-methylcyclopentene with half the amount of "di-3-pinanylborane" required for complete reaction results in optically active residual olefin (45% e.e.). Increasing the amount of borane to 80% of that required for complete reaction increases the stereoselectivity (65% e.e.).[11] Partial kinetic resolutions of 3-ethylcyclopentene,[11] 1-methylnorbornene,[11] *trans,trans*-2,8-*trans*-bicyclo[8.4.0]tetradecadiene,[20] trimethylcyclodecatriene,[21] and 4-methylcyclohexene[22] have been reported using this technique.

[20] P. S. Wharton and R. A. Kretchmer, *J. Org. Chem.*, **33**, 4258 (1968).

[21] J. Furukawa, T. Kakuzen, H. Morikawa, R. Yamamoto, and O. Okuno, *Bull. Chem. Soc. Jap.*, **41**, 155 (1968).

[22] S. I. Goldberg and F. L. Lam, *J. Org. Chem.*, **31**, 240 (1966).

The method is somewhat limited by the fact that only relatively unhindered olefins will react with "di-3-pinanylborane."

Racemic olefin 0.15 mole + "P_2^*BH" 0.12 mole → BP_2^* + R-(+) 82.5% and S-(−) 17.5% (0.03 mole); CrO_3 → Optically active + Optically active

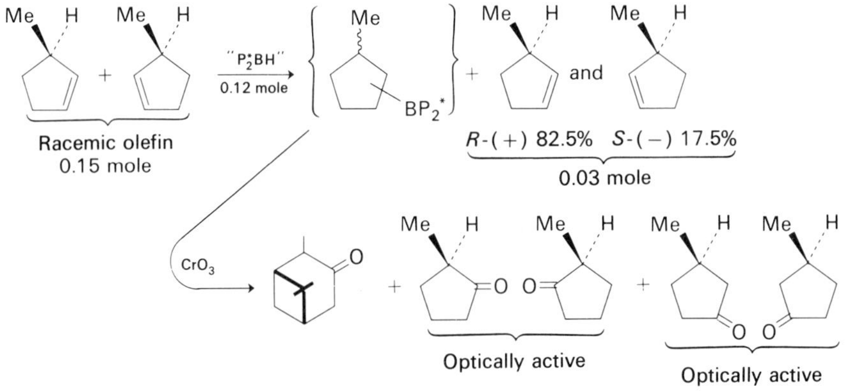

Fig. 6-8. Partial kinetic resolution of a racemic olefin with "di-3-pinanylborane."

6-2.3. Synthesis of Optically Active Ketones. Since oxidation of boranes can yield ketones,[23] this constitutes a method of synthesis of chiral ketones via reaction of certain olefins with a chiral borane. The hydroboration of the symmetrical olefin norbornene (**16**) with "di-3-pinanylborane" serves as an example.[24] After chromic acid oxidation of the chiral intermediate mixture of boranes **17-18**, optically active (1*S*,4*R*)-(+)-norcamphor (**21**) (21% e.e.) is obtained. Alternatively the intermediate can

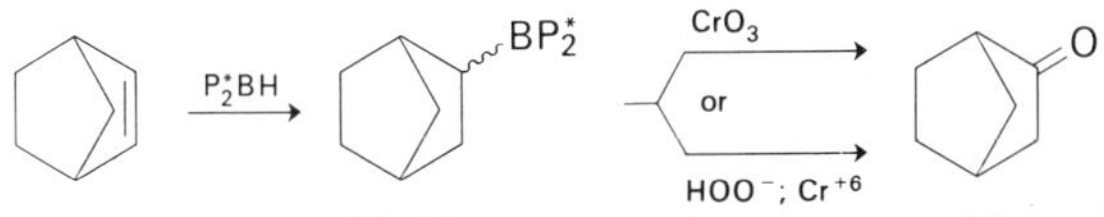

16 —P_2^*BH→ **17–18** (BP_2^*) —CrO_3 or HOO^-; Cr^{+6}→ **21**

be oxidized first with peroxide to give the alcohols followed by further oxidation with Sarett reagent.[25] The stereoselectivity of the reaction is determined by the composition of the mixture of diastereomeric boranes, **17–18**, which is more fully illustrated in Fig. **6-9**. The four diboranes (**17A** and **17B** + **18A** and **18B**) on peroxide oxidation give two sets of enantiomeric

[23] (a) R. Pappo, *J. Am. Chem. Soc.*, **81**, 1010 (1959). (b) H. C. Brown and C. P. Garg, *ibid.*, **83**, 2951 (1961).

[24] R. K. Hill and A. G. Edwards, *Tetrahedron*, **21**, 1501 (1965).

[25] (a) G. I. Poos, G. E. Arth, R. E. Beyler, and L. H. Sarett, *J. Am. Chem. Soc.*, **75**, 422 (1953). (b) L. F. Fieser and M. Fieser, *Reagents for Organic Synthesis* (New York: John Wiley & Sons, Inc., 1967) p. 145.

alcohols **19A** and **19B** and **20A** and **20B** but further oxidation destroys the carbinol chiral center giving the enantiomeric ketones **21A** and **21B**.

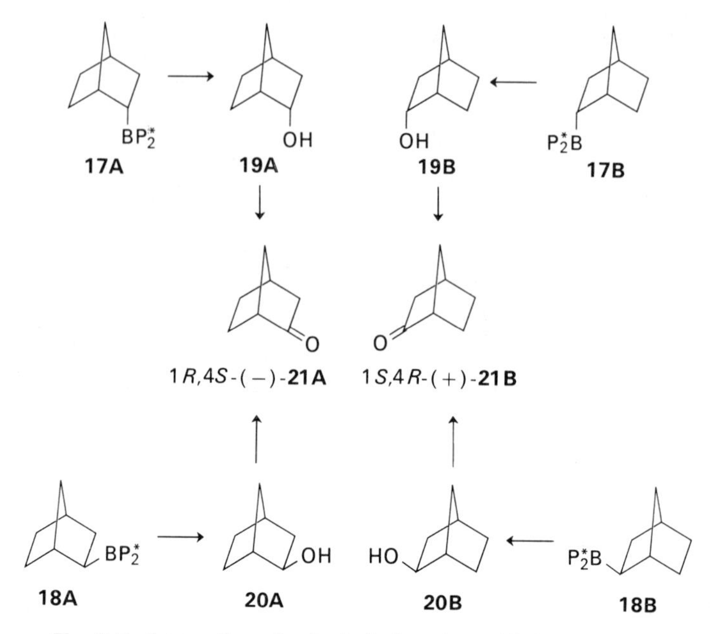

Fig. 6-9. Intermediates in the hydroboration-oxidation of norbornene with "di-3-pinanylborane."

A rather different situation is illustrated by the chromic acid oxidation of the mixture of boranes from the reaction of 3-methylcyclopentene (**22**) with "di-3-pinanylborane." *If the hydroboration is carried to completion* and if there is no kinetic resolution in the oxidation, the ketone formed (Fig. **6-10**, **27**) must be racemic; for although the rate of formation of the *cis* intermediates, **24A** + **24B**, may not equal the rate of formation of the *trans* intermediates, **23A** + **23B**, the amount of **23A** + **24A** must equal the amount of **23B** + **24B**. Since the chiral center at the point of boron substitution is ultimately destroyed, the yield of *S*-(−)-**27** must equal the yield of *R*-(+)-**27** and a racemate will result. But if racemic 3-methylcyclopentene is *incompletely* hydroborated with "di-3-pinanylborane," the enantiomeric composition of the 3-methylcyclopentanone (*S*-(−)-**27A** and *R*-(+)-**27B**) would reflect the relative rates of reaction of the enantiomeric olefins via pairs of diastereomeric transition states leading to unequal amounts of **27A** and **27B**. Furthermore, even though the olefin was completely hydroborated, optically active 3-methylcyclopentanone could result by kinetic resolution if the oxidation was not carried to completion.

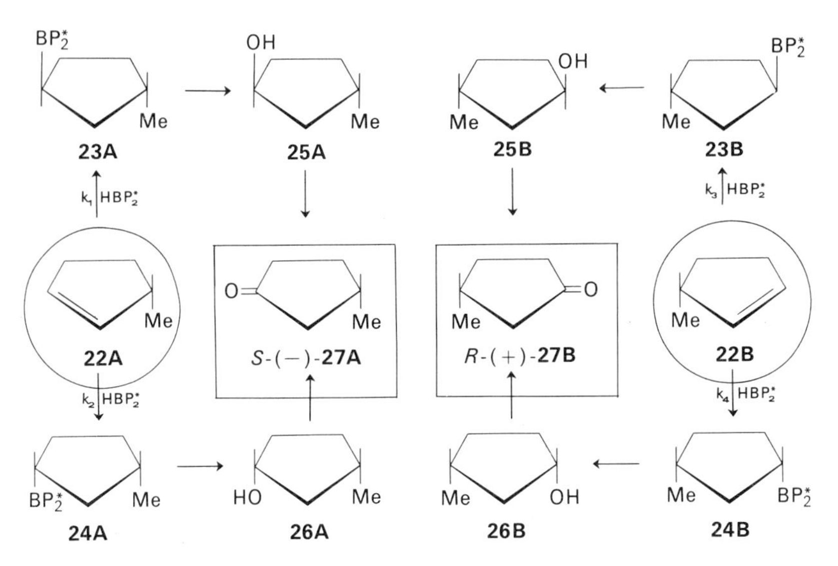

Fig. 6-10. Intermediates in the hydroboration-oxidation of racemic 3-methylcyclopentene with "di-3-pinanylborane." Circled formulas are the enantiomers of the racemic starting materials; boxed formulas are the enantiomers of the product.

6-2.4. Other Asymmetric Syntheses via Hydroboration with "Di-3-pinanylborane."

Up to the present time stereospecific (with retention) oxidation of the borane to an alcohol has been the most frequently used reaction subsequent to asymmetric hydroboration. In a few other cases, protonolysis[19] and aminolysis[9] have been used. A number of recently reported reactions of trialkyboranes have yet to be applied to the products from asymmetric hydrobration with "di-3-pinanylborane."

For example, it would appear that one might be able to convert the mixture of epimeric boranes (**29**) from the asymmetric hydroboration of certain achiral olefins (**28**) to the methylol derivative (**30**) of the olefin by

$$\underset{\mathbf{28}}{R_2C{=}C(H)(R')} \xrightarrow{HBP^*_2} \underset{\mathbf{29}}{R_2CH{-}\overset{H}{\underset{R'}{C^*}}{-}BP^*_2} \xrightarrow[\substack{2)\ BH_4^- \\ 3)\ H_2O}]{1)\ CO} \underset{\mathbf{30}}{R_2CH{-}\overset{H}{\underset{R'}{C^*}}{-}CH_2OH}$$

reaction with carbon monoxide in the presence of sodium or lithium borohydride.[26] Similarly, reaction of intermediate boranes with dimethyloxo-

[26] M. W. Rathke and H. C. Brown, *J. Am. Chem. Soc.*, **89**, 2740 (1967).

sulfonium methylide,[27a] dimethylsulfonium methylide,[27b] substituted ylides,[27c] trimethylammonium methylide,[27d] triphenylphosphonium methylide,[27e] phenyl(bromodichloromethyl)mercury,[28] diazomethane,[29] or isonitriles,[30] followed by appropriate oxidative or hydrolytic cleavage, would appear to offer additional possibilities for study. A large number of new and novel reactions of organoboranes have been reported recently.[31] Obviously, as new reactions of boranes are discovered the utility of "di-3-pinanylborane" as an asymmetric hydroborating agent will be extended.

6-2.5. Transition State Models for Hydroboration with "Di-3-pinanylborane." Several models have been devised to rationalize stereochemical consequences of asymmetric hydroboration. These models represent empirical correlations and for the most part do not represent the actual transition state, although they are believed to incorporate some of the features which may characterize it. One important simplification of most of them is that di-3-pinanylborane is pictured as the hydroborating agent. As pointed out (Sec. **6-2**), there is evidence that either tetra-3-pinanyldiborane or, in the case of more hindered olefins, tri-3-pinanyldiborane are actually involved. H. C. Brown and co-workers[11] have made use of an empirical model which nicely accommodates the results observed with unhindered *cis* olefins (those that do not displace α-pinene). However, when displacement of α-pinene occurs in stoichiometric amounts in the hydroboration of olefins with "di-3-pinanylborane," the product has been found to be the enantiomer of that predicted on the basis of the original Brown model. The model is based upon a three-dimensional representation of a presumed lowest energy conformation for di-3-pinanylborane, **31**.* As shown in **32** and **33**, a *cis* alkene (2-butene) can be positioned so that either enantiotopic face may undergo hydroboration. The premise of the Brown model is that approach as shown in **32A** would be favored since the R

[27] (a) J. J. Tufariello and L. T. C. Lee, *J. Amer. Chem. Soc.*, **88**, 4757 (1966). (b) J. J. Tufariello, P. Wojtkowski, and L. T. C. Lee, *Chem. Commun.*, 505 (1967). (c) J. J. Tufariello, L. T. C. Lee, and P. Wojtkowski, *J. Am. Chem. Soc.*, **89**, 6804 (1967). (d) W. K. Musker and R. R. Stevens, *Tetrahedron Lett.*, 995 (1967). (e) R. Köster and B. Rickborn, *J. Am. Chem. Soc.*, **89**, 2782 (1967).

[28] D. Seyferth and B. Prokai, *J. Am. Chem. Soc.*, **88**, 1834 (1966).

[29] C. E. H. Bawn and A. Ledwith, *Progr. Boron Chem.*, **1**, 345 (1964).

[30] (a) G. Hesse and H. Witte, *Angew. Chem.*, **75**, 791 (1963). (b) J. Casanova, Jr. and R. E. Schuster, *Tetrahedron Lett.*, 405 (1964). (c) J. Casanova, Jr., H. Kiefer, D. Kuwada, and A. Boulton, *ibid.*, 703 (1965).

[31] (a) H. C. Brown and M. W. Rathke, *J. Am. Chem. Soc.*, **89**, 2737, 2738, 4528 (1967). (b) H. C. Brown, G. W. Kabalka, and M. W. Rathke, *ibid.*, **89**, 4530 (1967). (c) H. C. Brown and C. D. Pfaffenberger, *ibid.*, **89**, 5475 (1967). (d) H. C. Brown and E. Negishi, *ibid.*, **89**, 5285, 5477 (1967). (e) H. C. Brown, et al., *ibid.*, **89**, 5708, 5709 (1967).

* In this and all other models in this section, the illustrations are based on di-3-pinanylborane from *R*,*R*-(+)-α-pinene.

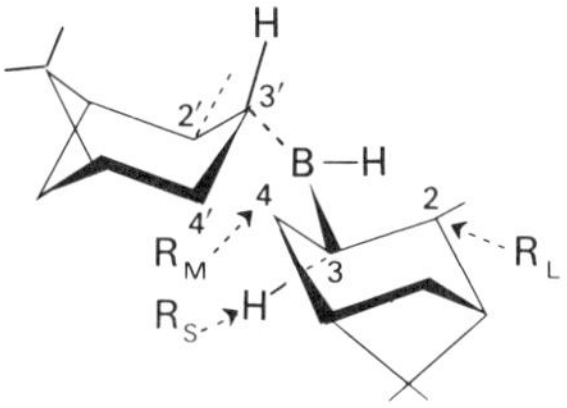

31A 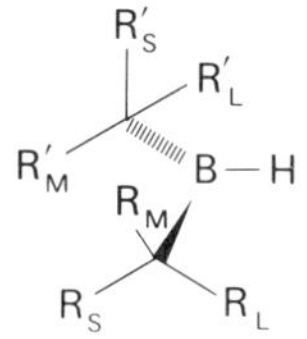 31B

Fig. 6-11. Conformational representation of di-3-pinanylborane.

(methyl) group of the alkene is positioned in the vicinity of the hydrogen at C-3′ rather than near the larger methylene, R_M (at C-4), as in **32B**.

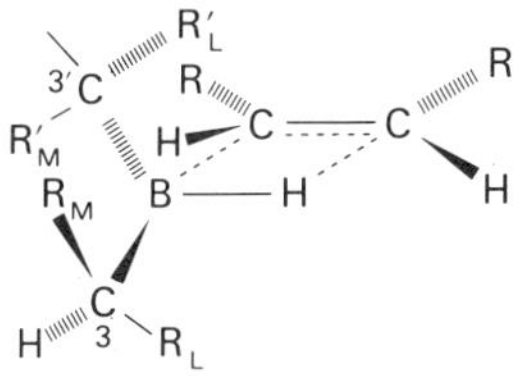

32A 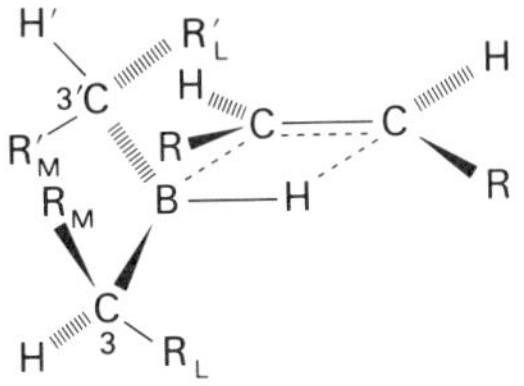 32B

To accommodate the results obtained with 2-substituted-1-alkenes (Table **6-3**), Brown and co-workers[12] proposed that slightly different steric interactions should be considered in arriving at stereochemical correlations using conformation **31**. For example, it was proposed that model **33A** is favored over **33B** by a modest factor because there is a small but significantly greater interaction of the ethyl group with R'_L in **33B** than with R_M in **33A**.

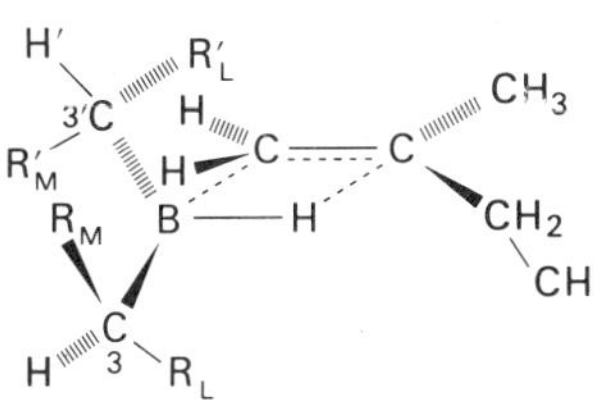

33A 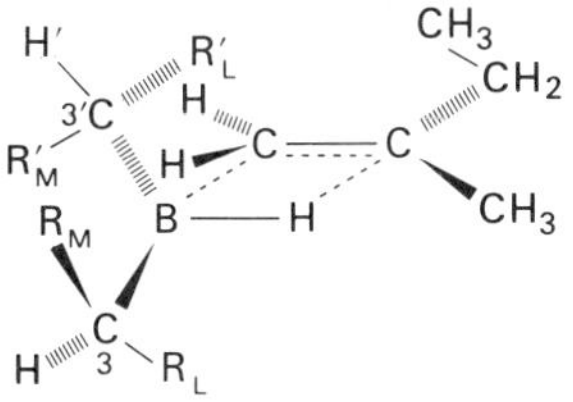 33B

Varma and Caspi[32] have used the conformation (**31**) proposed by Brown to generate a slightly different model which correlates hydroboration results with a different set of controlling interactions. In their view the di-3-pinanylborane (**31** from (+)-α-pinene) is oriented with the B—H bond along the intersection of two planes. Four quadrants (**34**) are thus defined. It is assumed that a *cis* alkene approaches so as to place both R groups in the

[32] K. R. Varma and E. Caspi, *Tetrahedron*, **24**, 6365 (1968).

upper right (UR) quadrant. This quadrant is claimed to be less crowded than the upper left (UL) quadrant into which three axial hydrogens of a pinanyl group protrude.* Because of the symmetry of conformation **31**, the

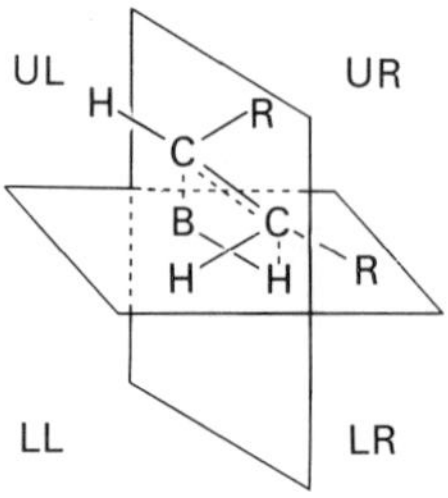

UL UR H R C B C H H R LL LR

34A

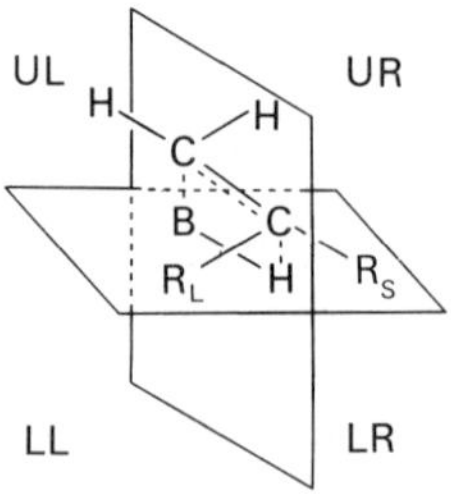

UL UR H H C B C R_L H R_S LL LR

34B

lower left (LL) quadrant has the same steric requirements as the upper right (UR) quadrant and is less crowded than the lower right (LR) quadrant. Therefore, viewing the transition state model as represented in **34A** correctly correlates the configuration of the predominant product from reductions of *cis* olefins by the reagent from (+)-α-pinene. Terminal alkenes, however, are envisaged as being better accommodated with the larger R group in the upper left (or lower right) quadrant where its crucial interactions are assumed to be with a C-2 axial proton versus a C-2′ equatorial methyl group in the upper right (or lower left) quadrant. Thus **34B** represents the preferred direction of attack for terminal alkenes. The quadrant model has also been used to attempt to rationalize the hydroboration of terminally deuterium labeled alkenes but the assumptions are too speculative to consider seriously. This model has the advantage of being more readily visualized than the others if it is only considered to be an empirical device for correlating known results.

Streitwieser and co-workers[17] modified Brown's model slightly in order to rationalize the stereochemistry of the hydroboration of *cis*-1-butene-1-*d* with P_2^*BH. Results with other terminal 1-deuterioalkenes (Table **6-4**) and unhindered *cis*-alkenes (Table **6-1**) are also accommodated by the modified model, but it does not rationalize the stereochemistry observed with unlabeled 2-substituted-1-alkenes (Table **6-3**). This model is also based on conformation **31** but invokes a π-complex representation rather than a four-center orientation. The interactions that mediate stereochemical control are presumed to be those of the alkyl group at C-2 of the alkene with H′ (R_S') versus R_M of the borane. For example, **35A** is favored over **35B** according to these criteria. This model considers any contribution to stereochemical control due to the isotopic disparity at the terminal carbon to be negligible. The data in Table **6-4** suggest, however, that this is probably not the case.

* Drawings **34A** and **34B** represent the reagent from (+)-α-pinene.

We have presented a brief picture of these models in order to demonstrate the extent to which different workers have been able to apply the same basic representation of the reagent (**31**) to various stereochemical results.

35A **35B**

A much more elaborate representation of the hydroboration process has been suggested by McKenna and co-workers.[16] Tetra-3-pinanyldiborane is assumed to be the actual hydroborating species. The model represents a kind of potential energy obstacle course imposed by steric interactions and orbital symmetry considerations. The alkene must maneuver toward a four-center transition state via several low energy valleys. The essential motions by which this is accomplished are represented in Fig. **6-12**.

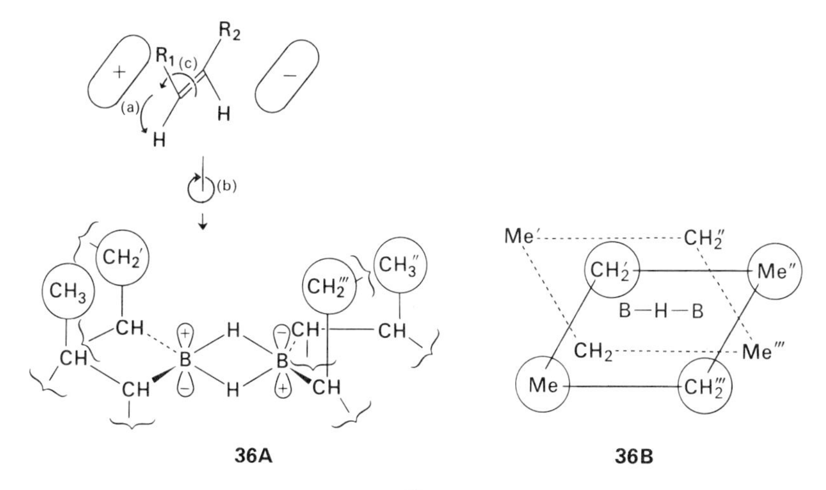

Fig. 6-12. The McKenna model[16] for reaction of (−)-tetra-3-pinanyldiborane with an acyclic *cis*-alkene. In the side view (**36A**) B—H bonds are in the plane of the paper. Motions of the alkene as it approaches a four-center transition state are (a) a dip, (b) a swing and (c) a rotation. Diagram **36B** is a top view of **36A**. In this representation B—H bonds lie in a plane that is perpendicular to the parallelograms defined by methylene and methyl groups of the four 3-pinanyl moieties. Groups that are part of the same bicyclic ring system are identically primed.

Mislow and his associates[33] have called attention to a potential hazard associated with the prediction of the absolute configuration of alcohols on the basis of models for hydroborations with di-3-pinanylborane. A freshly prepared THF solution of (+)-α-pinene and boron hydride (1:1 mole ratio) was allowed to react with benzonorbornadiene (**37**). After oxidation and acetylation there was obtained (+)-*exo*-2-benzonorbornenyl acetate (**39**) containing a 7.4% excess of the 1*R*,2*S* enantiomer. However, nearly the same enantiomeric excess of the 1*S*,2*R* enantiomer was obtained from a similar reaction sequence involving hydroboration of **37** with a 1:1 adduct of (+)-α-pinene and BH_3 that had been warmed and aged for 21 hr before using. Likewise, from *cis*-3-hexene there was prepared (−)-3-hexanol (20.6% e.e. *R* enantiomer) using fresh 1:1 adduct, whereas (+)-3-hexanol (10.7% e.e. *S* enantiomer) was obtained using the aged adduct.

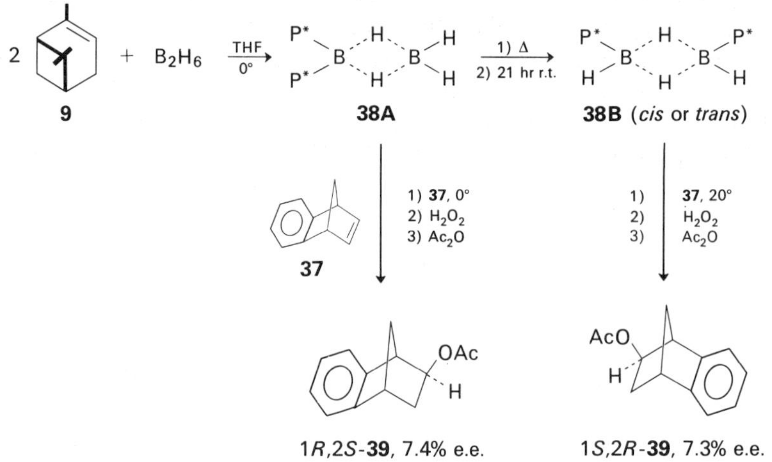

Fig. 6-13. Hydroboration of benzonorbornadiene (**37**) with 2:1 adduct of (+)-α-pinene and diborane.

It has been suggested[32b] that these results might be explained on the basis of a rearrangement of initially formed unsymmetrical adduct **38A** to *cis* and/or *trans* **38B**. These two different hydroborating species could produce enantiomeric products. Even though a 2:1 mole ratio of α-pinene to BH_3 has usually been employed (rather than the 1:1 ratio used in this work) so that tetra-3-pinanyldiborane or tri-3-pinanyldiborane are produced (rather than **38A** or **38B**), these observations are significant. The fact that the direction of asymmetric synthesis in the hydroboration of a given olefin with simple adducts of α-pinene and diborane depends upon the time elapsed between

[33] D. J. Sandman, K. Mislow, W. P. Giddings, J. Dirlam, and G. C. Hanson, *J. Am. Chem. Soc.*, **90**, 4877 (1968).

preparation of the adduct and its use provides dramatic evidence that one must be cautious when applying models to this and other complex reactions involving multiple chiral species as reactants.

6-2.6. Asymmetric Hydroboration Under the Control of Chiral Centers in the Olefin. There are numerous examples, particularly in cyclic systems, of hydroborations with diborane under the stereochemical control of asymmetric centers in the substrate. In almost all cases, the original asymmetric center is retained in the product. Hydroborations involving myrcene,[34a] α-pinene,[6] β-pinene,[6] 1-*p*-menthene,[13] 3-*p*-menthene,[34b] 2,3-dimethyl-3-*p*-tolylcyclopentene,[34c] and longifolene[34d] illustrate the stereochemistry to be expected.

Monohydroboration of (*R*)-(+)-limonene (**40**) with *bis*-(1,2-dimethylpropyl)-borane gives, after oxidative work-up, a mixture of (4*R*,8*R*)- and (4*R*,8*S*)-*p*-menth-1-ene-9-ol as expected.[34e] Hydroboration of **40** with excess diborane, however, gives what has been assumed to be[34f] a *bis*-borane adduct with one boron attached to a secondary position and the other to a terminal carbon. Thermal isomerization of this intermediate was presumed to convert it to a *bis*-borane with both boron atoms at terminal carbons. Oxidation produced 7,10-dihydroxy-*p*-menthane (**41**, $[\alpha]_D^{25}$ +20.1°). The net result of

40 **41**

this sequence of reactions is loss of the original asymmetric center at C-4 and generation of a new asymmetric center at C-8. Since the maximum rotation and configuration of the product are not known, it is not possible to determine the magnitude of the asymmetric bias or to speculate meaningfully concerning the stereochemical course of this sequence. In view of recent work on the hydroboration of dienes,[31c] cyclic intermediates may be involved. It is impossible to say whether asymmetric induction is intramolecular or intermolecular in this case.

The stereochemical course of the hydroboration-oxidation of the diastereomeric (1*R*)-isopulegols (**42**–**45**) has been studied in detail.[35] The

[34] (a) H. C. Brown, K. P. Singh, and B. J. Garner, *J. Organometal. Chem.*, **1**, 2 (1963). (b) F. Ascoli, A. M. Liquori, and B. Pispisa, *Chem. and Ind.*, 1579 (1964). (c) T. Irie, et al., *Tetrahedron Lett.*, 3619 (1965). (d) J. Lhomme and G. Ourisson, *Tetrahedron*, **24**, 3167 (1968). (e) B. A. Pawson, H. C. Cheung, S. Gurbaxani, and G. Saucy, *Chem. Commun.*, 1057 (1968). (f) H. G. Arlt, E. H. Sheers, and R. J. Chamberlain, *Chem. and Ind.*, 1409 (1961).

[35] K. H. Schulte-Elte and G. Ohloff, *Helv. Chim. Acta*, **50**, 153 (1967).

results are represented in Fig. **6-14**. The stereochemistry was better rationalized by assuming a *trans* addition involving an attack of external hydride on a borate intermediate (e.g., **46**) than by an intramolecular *cis* hydroboration

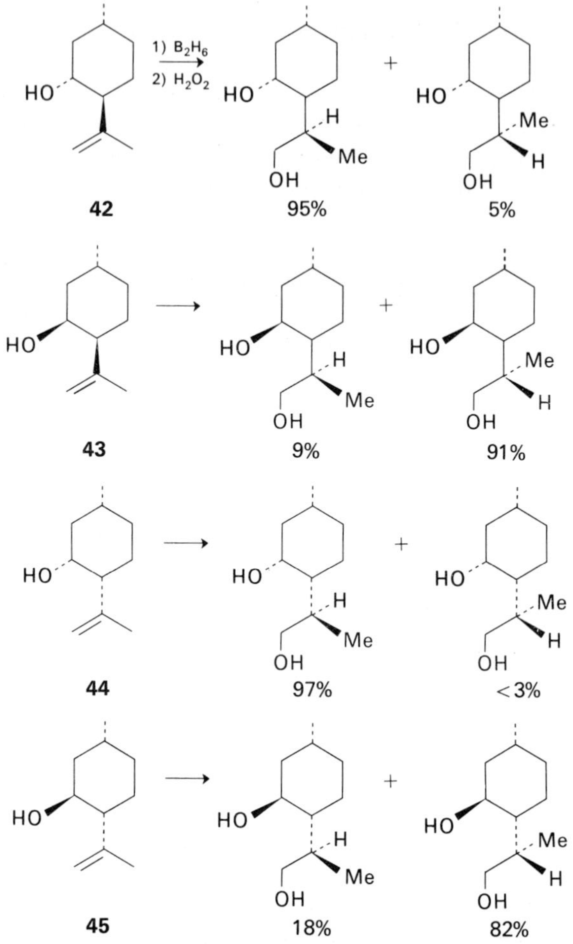

Fig. 6-14. Hydration of (1*R*)-isopulegols via hydroboration.

via an intermediate such as **47**. However, hydroboration of (1*R*,8*R*)-Δ^3-*p*-menthen-9-ol (**48**) most likely involves *cis* addition.[34]

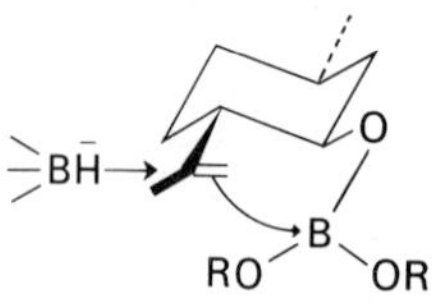

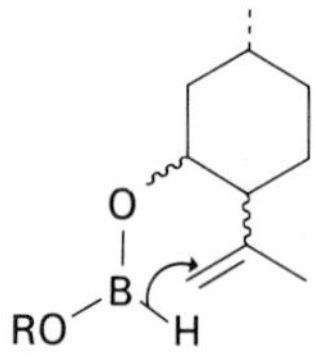

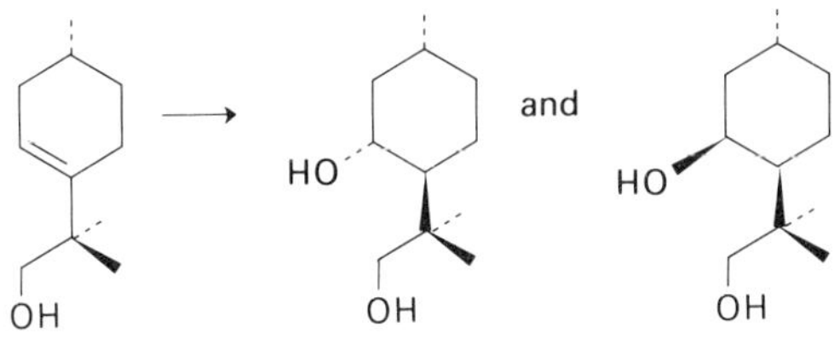

OH OH OH
HO HO
and
48 67% 33%

6-3. Asymmetric Cycloaddition Reactions

In this section we shall consider reactions in which a reagent adds across a carbon-carbon double bond in such a manner that a cyclic chiral product is formed either directly or in a subsequent cyclization step. Familiar reactions such as the formation of cyclopropyl compounds via carbene or carbenoid additions and Diels–Alder cycloadditions have been rather extensively studied, but other potential asymmetric cycloadditions involving, for example, dimerization, 1,3-dipolar addition, and nitrenes have not yet been reported.

6-3.1. The Synthesis of Chiral Cyclopropanes via Cycloaddition Reactions. Reaction of diphenyldiazomethane with either (−)-menthyl acrylate (**49**) or (−)-menthyl methacrylate (**50**) gives mixtures of diastereomeric esters which saponify to optically active cyclopropane carboxylic acids[36,37] **51** and **52** [2.2%e.e. in the former case (*R*)-(−)-**51**, R is H; and 10%e.e. in the latter case, (*R*)-(+)-**52**, R is Me]. Preliminary speculation[36] concerning the stereochemical course of these reactions was revised in view of the subsequently determined absolute configurations of the products.[37] The stereochemistry was rationalized on the basis of a Prelog-type model for the initial addition reaction followed by a "diazo exchange" during the cyclopropyl ring formation.[37] This model fixes the conformation of the substrate as shown in **49** and **50**. The C=CH_2 group replaces the C=O of the Prelog model for the asymmetric atrolactate synthesis (see Fig. **2-6**). In **49** and **50** the carbinol carbon of (−)-menthol is represented with the two smaller groups flanking the ester carbonyl. Preferential attack of the diphenyldiazomethane is visualized as taking place on the less hindered diastereotopic face of the double bond (from below the plane of the page). The pyrazoline (**53**) which is formed initially may decompose to

[36] (a) F. J. Impastato, L. Barash, and H. M. Walborsky, *J. Am. Chem. Soc.*, **81**, 1514 (1959). (b) H. M. Walborsky, L. Barash, A. E. Young, and F. J. Impastato, *J. Am. Chem. Soc.*, **83**, 2517 (1961).

[37] H. M. Walborsky and C. G. Pitt, *J. Am. Chem. Soc.*, **84**, 4831 (1962).

nitrogen and a diradical, or the ring may open by one of two ionic pathways. Only the ionic pathway, as represented by **54A** → **54B**, satisfactorily rationalizes the observed stereochemistry and this mechanism is preferred by Walborsky and co-workers. The reasons for choosing the *trans* coplanar conformation as the basis for the models (**49** and **50**) are not as persuasive as

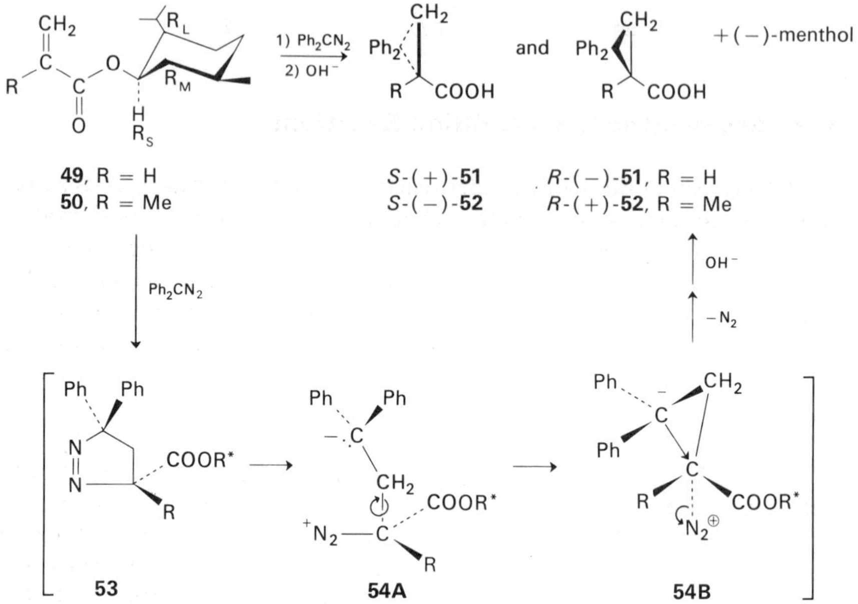

Fig. 6-15. Possible mechanism for the reaction of diphenyldiazomethane with chiral acrylates.

in the case of the oxygen analog (Fig. **2-6**) where this arrangement would appear to be more heavily favored by the two opposed carbonyl dipoles. A lower preference for the transoid arrangement may be one of the reasons for the lower asymmetric bias observed in these reactions.

Reaction of ethyl diazoacetate with (−)-menthyl α-isopropylacrylate (**55**) has been reported to give, after saponification, a 15% yield of (1*R*,2*R*)-(−)-*cis*-umbellularic acid, **56** (6% e.e.) and a 57% yield of (1*R*,2*S*)-(−)-*trans*-umbellularic acid, **57** (3% e.e.).[38] An analogous reaction of (−)-menthyl β,β-dimethylacrylate (**58**) gave (1*R*,3*R*)-(−)-*trans*-caronic acid, **59** (15.9% e.e.). Copper-catalyzed decomposition of dimethyldiazomethane in the presence of dimenthyl fumarate (**60**) produced an excess of (1*S*,3*S*)-(+)-*trans*-caronic acid (**61**). The configuration of the predominant isomer is predicted in each

[38] H. M. Walborsky, T. Sugita, M. Ohno, and Y. Inouye, *J. Am. Chem. Soc.*, **82**, 5255 (1960).

case by assuming attack of a carbene on the less hindered face of the double bond when the α,β-unsaturated ester is oriented in a manner analogous to a Prelog-type model[38] (*cf.* **49** and **50**).

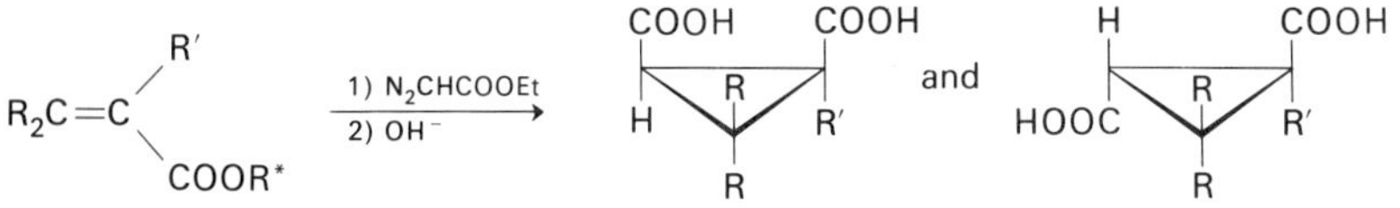

$R_2C{=}C(R')COOR^*$ —1) $N_2CHCOOEt$; 2) OH^-→ **56** and **57**

55, R = H, R′ = *i*Pr
58, R = Me, R′ = H
R* = (−)-menthyl

56

57, R = H, R′ = *i*Pr
59, R = Me, R′ = H

A reaction which is more complex than was at first supposed, is the base catalyzed condensation of ethyl acrylate (**62A**) and (−)-menthyl chloroacetate[39] (**63**) to yield a mixture of cyclopropyl esters which, after saponifica-

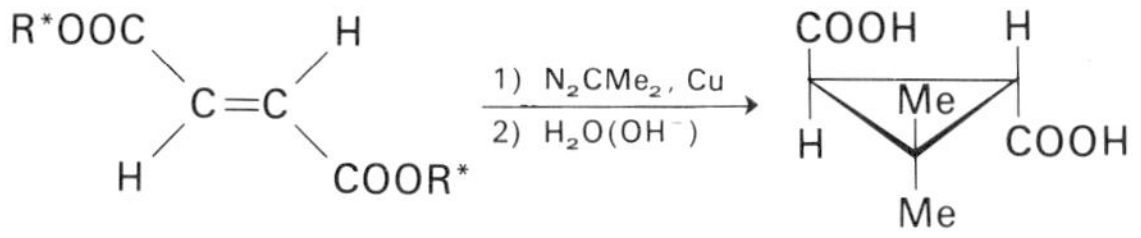

R*OOC–CH=CH–COOR* —1) N_2CMe_2, Cu; 2) $H_2O(OH^-)$→

60 **61**

tion, gives optically active *trans*-1,2-cyclopropanedicarboxylic acid (**64**). Only traces of the *cis*-acid (**65**) are found. Most intriguing is the fact that

$CH_2{=}CHCOOEt + ClCH_2COOR^*$ —*t*-BuO⁻→ mixed esters —OH⁻→

62A **63**

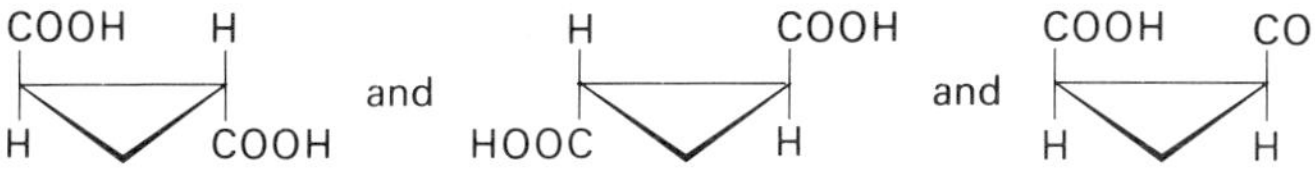

R,R-**64** and *S,S*-**64** and **65** (Trace)

the product from the reaction in toluene solvent was levorotatory (40–60% yields, 1.8–3.1% e.e.) while that from reaction in dimethylformamide (DMF) solvent was dextrorotatory (46–51% yield, 10.2–10.9% e.e.).*

On the basis of subsequent experiments with methyl acrylate (**62B**), McCoy[40a] proposed that the solvent effect was exercised primarily at the second step of the asymmetric synthesis (Fig. **6-16**) in determining the ratio

[39] Y. Inouye, S. Inamasu, M. Ōhno, T. Sugita, and H. M. Walborsky, *J. Am. Chem. Soc.*, **83**, 2962 (1961).

[40] (a) L. L. McCoy, *J. Org. Chem.*, **29**, 240 (1964). (b) *J. Am. Chem. Soc.*, **84**, 2246 (1962).

* This reversal of stereochemistry dramatizes the need for a thorough study of an asymmetric synthesis reaction before speculating about the factors that control the stereochemistry.

of *cis* and *trans* diesters but that this effect was masked by isomerization during saponification. By reducing the mixed esters with lithium aluminum hydride, acetylating, and analyzing by gas chromatography, McCoy showed that the ratio of *cis* and *trans* esters changed from 82:18 in toluene to 8:92 in DMF, in accord with earlier findings.[40b] He proposed that the variation in the optical activity of the product is the result of differences in the rates of hydrolysis and epimerization of the various ester intermediates. The *cis* (**66**) and *trans* (**67**) diesters are both chiral and the excess of one enantiomer over the other will be the same in each case because of their common origin,

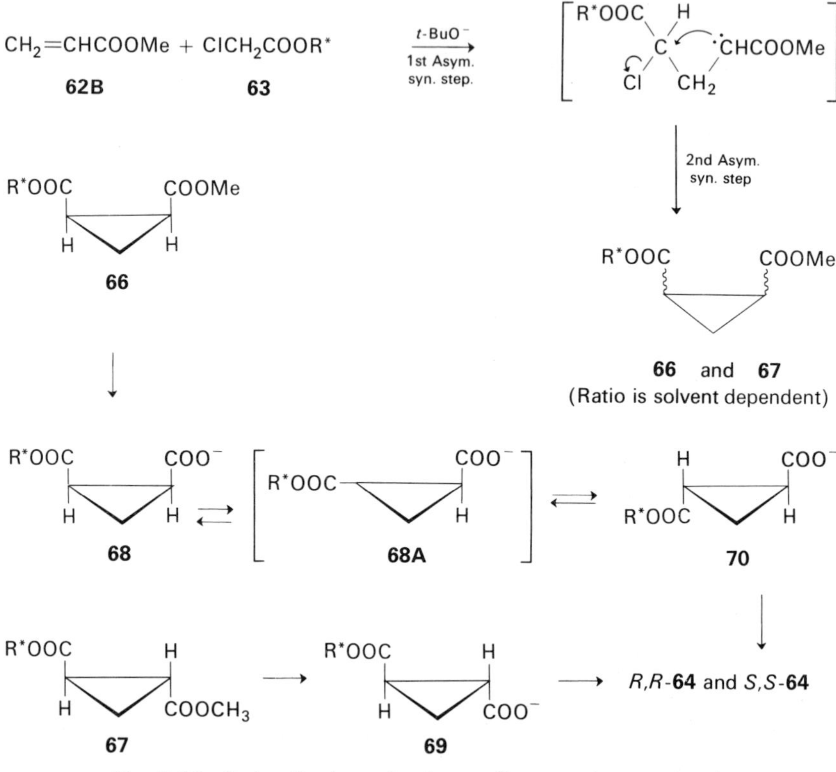

Fig. 6-16. Rationalization of solvent effect on the asymmetric synthesis of cyclopropane-1,2-dicarboxylic acids.[40] [R* = (−)-menthyl]. The reactions are shown for only one enantiomer of both the *cis* and *trans* series.

i.e., because asymmetry is induced at the carbon bearing the carbomenthoxy group during the first step of the synthesis. The saponification of these two diastereomers will proceed stepwise, the carbomethoxy group being hydrolyzed faster than the carbomenthoxy group. This is shown in Fig. **6-16** for only one enantiomer (**66**) of the *cis* pair and for one enantiomer (**67** with the same configuration *alpha* to the carbomenthoxy group) of the *trans* pair. The carboxylate ion resulting from rapid hydrolysis of the carbomethoxy group (shown in **68** and **69**) will retard epimerization at the asymmetric *alpha* carbon in contrast to the more facile epimerization via an ester anion (shown with localized charge in **68A**). So it is primarily the asymmetric center to which the carbomenthoxy group is attached which is subject to extensive isomerization (**68** ⇆ **70**) to give mainly the thermodynamically more stable *trans* half-ester (**70**). The stereochemical consequence of this is to produce from the *cis* diester **66** the enantiomer of the *trans* diacid which is obtained by direct saponification of the *trans* diester **67**. As a result, the rotation of the *trans* diacid (*R*,*R*-**64** and *S*,*S*-**64**) isolated after saponification will be dependent upon the amount of *cis* ester formed in the solvent-dependent second step of the asymmetric synthesis.

An unequivocal solvent effect on asymmetric synthesis has however been demonstrated in the closely related base-catalyzed reaction of (−)-menthyl chloropropionate (**71**) with methyl methacrylate[41] in which the intermediate mixed esters cannot isomerize. The reaction was carried out in varying ratios of DMF:benzene from pure DMF to pure benzene.

$CH_3CHClCOOR^*$ (**71**) + $CH_2{=}C(CH_3)COOMe$ (**72**) —NaH, DMF / benzene→ [cyclopropane with COOMe, CH_3 / CH_3, $COOR^*$ and cyclopropane with COOMe, $COOR^*$ / CH_3, CH_3] —1) OH^- 2) CH_2N_2→ **73** (COOMe, CH_3 / CH_3, COOMe) and **74** (COOMe, COOMe / CH_3, CH_3)

Following complete saponification of the intermediate mixed esters, the diacid mixture was treated with diazomethane and the resulting dimethyl esters were separated and analyzed by gas chromatography. More *cis* product was formed in benzene-rich solvent (almost entirely *cis* in pure benzene), as observed in the previous example.[40] More significant is the fact that the ratio of *R*,*R*:*S*,*S* product increased from 1:0.70 to 1:0.96 as the benzene:DMF solvent ratio increased from 0:10 to 9:1. In pure benzene the *R*,*R* enantiomer was produced in slight excess (*R*,*R*:*S*,*S* = 1.02). Plots of log *cis*:*trans* or log *trans*-*R*,*R*:*trans*-*S*,*S* versus the Kirkwood–Onsager

[41] (a) Y. Inouye, S. Inamasu, and M. Horiike, *Chem. and Ind.*, 1293 (1967). (b) Y. Inouye, S. Inamasu, M. Horiike, M. Ohno, and H. M. Walborsky, *Tetrahedron*, **24**, 2907 (1968).

reaction medium parameter (a measure of solvent polarity) were linear. This behavior was cited in support of a very detailed rationalization of the results in terms of differential solvation of the competing diastereomeric transition states.[41b] It has also been noted[41b] that the analogous reaction of (−)-menthyl chloroacetate and (−)-menthyl acrylate, where there can be no asymmetric isomerization of *cis*-diester during saponification, yields levorotatory *trans* acid in toluene ($[\alpha]_D$ − 8.0°) and dextrorotatory *trans* acid in DMF ($[\alpha]_D$ +5.7°).

Optically active 2-arylcyclopropane carboxylic acids have been synthesized by a mechanistically similar reaction between dimethyloxosulfonium or dimethylsulfonium methylide and *β*-arylacrylates.[42] In this case, the ester group of the acrylate is chiral, the asymmetric induction is quite long range (Fig. **6-17**) and the stereoselectivity is low. The fact that (−)-menthyl and (+)-bornyl cinnamates both gave the same enantiomer of acid **76** in excess is noteworthy. The order of small, medium, and large groups at the

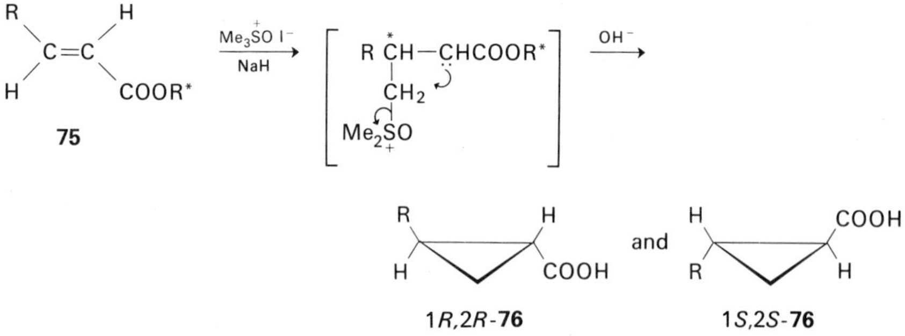

Fig. 6-17. Asymmetric synthesis of 2-arylcyclopropanecarboxylic acids from dimethyloxosulfonium methylide and chiral acrylates.[42] Examples studied were R = Ph (**75A**), *o*-MeOPh (**75B**), *p*-MeOPh (**75C**), and α-naphthyl (**75D**). Products **76A** [3–4% e.e. R,R-(−)-isomer] and **76B** were levorotatory whereas **76D** was dextrorotatory.

asymmetric carbinol carbon is reversed in going from (−)-menthyl to (+)-bornyl and one might therefore be tempted to propose that enantiomeric acids would be formed as they are in the Prelog atrolactic asymmetric synthesis (Sec. **1-4.4b**). This reverse behavior has been observed in the asymmetric Reformatsky reaction (Sec. **4-3**), the asymmetric Darzens condensation (Sec. **4-4**), and the asymmetric aldol-type condensations (Sec. **4-2**); possible rationalizations are briefly considered in Sec. **4-2**.

[42] H. Nozaki, H. Itô, D. Tunemoto, and K. Kondô, *Tetrahedron*, **22**, 441 (1966).

Higher asymmetric synthesis is observed when an optically active oxosulfonium methylide (**77**) is used to effect the cycloaddition.[43] Successive treatment of an oxosulfonium fluoroborate, derived from optically active methyl *p*-tolyl sulfoxide via the sulfoxamine, with sodium hydride in dimethylsulfoxide and an electrophilic alkene, gave a good yield of an optically active cyclopropane (Table **6-5**). The reaction was also adapted to the synthesis of several optically active epoxides by reaction of **77** with benzaldehyde, *p*-chlorobenzaldehyde, or acetophenone.*

Partial asymmetric syntheses of cyclopropanes have been accomplished using the Simmons–Smith reagent and the (−)-menthyl esters of several unsaturated carboxylic acids.[44] The *trans* olefin always gave the *trans*

TABLE 6-5

Asymmetric Synthesis of Cyclopropanes with a Chiral Oxosulfonium Methylide[a]

H(R')C=C(R'')H + $[^-CH_2—\overset{+}{S}(=O)(NMe_2)—p\text{-Tolyl}]$ (**77**) ⟶ *trans*-cyclopropane (H, R', R'', H) (**78**) + *p*-Tolyl-SONMe$_2$

No.	*trans*-Alkene		*trans*-1,2-Cyclopropane, **78**	
	R'	R''	% e.e.	config.
1	COOMe	Ph	30.4	(1*S*,2*S*)
2	COPh	Ph	35.3	(1*S*,2*S*)[b]
3	COPh	COPh	—	(1*S*,2*S*)[b]
4	COOMe	COOMe	15.2	(1*S*,2*S*)[c]

[a] Data from Reference 43.

[b] Tentative assignment based on configurations of products from other reactions involving **77**.

[c] Dimethyl maleate gave the same product, 17.8 % e.e. (1*S*,2*S*).

cyclopropane (Table **6-6**). The configuration of the predominant enantiomers, with the exception of the product from the cinnamate (No. 5), have been rationalized by assuming a model involving coordination of iodomethylzinc iodide to a twisted *cisoid* conformation of the substrate as in **79A** where the

[43] C. R. Johnson and C. W. Schroeck, *J. Am. Chem. Soc.*, **90**, 6852 (1968).

[44] (a) S. Sawada, K. Takehana, and Y. Inouye, *J. Org. Chem.*, **33**, 1767 (1968). (b) S. Sawada, J. Oda, and Y. Inouye, *J. Org. Chem.*, **33**, 2141 (1968).

* The asymmetric epoxidation of alkenes is discussed in Sec. **6-3.3**.

methylene attacks the double bond from below the plane of the page from the less hindered side of the (−)-menthyl group, as written.*

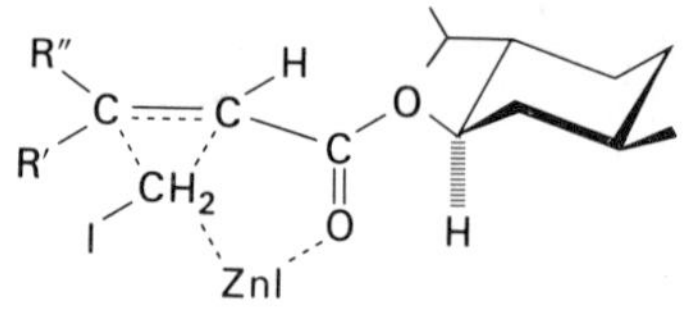

79A 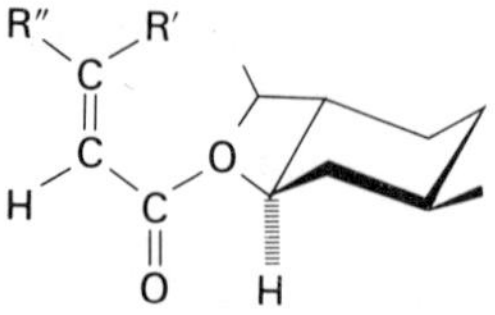 **79B**

The fact that (−)-menthyl cinnamate did not follow this pattern is rationalized[44a] by postulating that the cinnamate cannot achieve a *cisoid* conformation in the transition state such as **79A** without loss of resonance stabilization because the phenyl ring is crowded out of coplanarity with the olefinic bond. An alternative pathway for the cinnamate case is a *transoid*

TABLE 6-6

Asymmetric Simmons–Smith Syntheses with Chiral Substrates[44a]

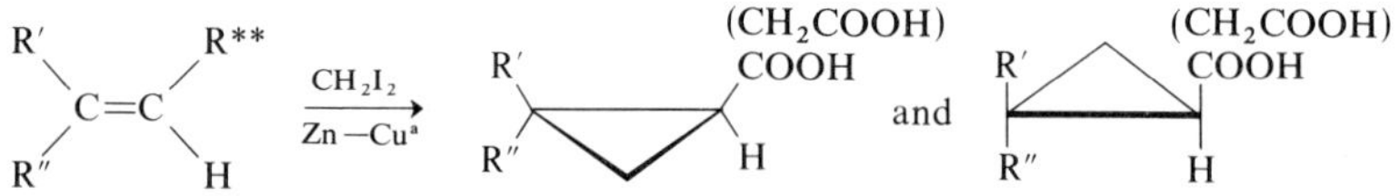

R′R″C=CHR** (**79**) $\xrightarrow[\text{Zn—Cu}^{a}]{CH_2I_2}$ **80A** and **80B**

No.	R′	R″	R**[b]	Cyclopropane config.	Stereoselectivity % e.e. or $[\alpha]_D$
1	H	Me	COOR*	1*S*,2*S* (**80A**)	2.8[c]
2	H	Me	CH_2COOR^*	1*R*,2*S*[d] (**80A**)	+1.77° [e,f]
3	Me	Me	COOR*	*S*[d] (**80A**)	+6.6.° [e,f]
4	H	COOR*	COOR*	1*S*,2*S* (**80A**)	6.4[g]
5	H	Ph	COOR*	1*R*,2*R* (**80B**)	9.3[h]
6	H	Ph	CH_2COOR^*	1*R*,2*S* (**80A**)	1.4

[a] Molar ratio of **79**:CH_2I_2:Zn—Cu = 1:2:4, ether solvent, 10–60 hr reflux.
[b] R** is the chiral substituent of which R* is (−)-menthyl.
[c] Based on $[\alpha]_D$ max 61° (neat) for acid (**80**).
[d] Configuration predicted on the basis of Brewster's rules.
[e] Methanol solvent.
[f] $[\alpha]_D$ max unknown.
[g] Based on $[\alpha]_D$ max 200° (neat) for dimethyl ester of **80**.
[h] Based on $[\alpha]_D$ max 311° (neat) for methyl ester of **80**.

* No particular conformation about the acyl-oxygen bond was suggested in Reference 44a; that shown in **79A** and **79B** is analogous to the Prelog atrolactate model (cf. Sec. **1-4.4b**, **68**).

Prelog-type model **79B** with one molecule of reagent coordinating with the carbonyl group and another attacking the double bond from the least hindered side (behind the plane of the page in **79B**). Coordination of the Simmons–Smith reagent with oxygen of the substrate is clearly indicated from the fact that both 2-cyclohexenol and 2-cyclohexenyl acetate give *cis*-2-bicyclo[4.1.0]heptanol (after LAH reduction in the latter case). The synthetic yield is highest in experiments 2 and 6, Table **6-6**, where the double bond is not conjugated with the carbonyl group, but the percent asymmetric synthesis is lower.

Achiral substrates have also been converted to optically active cyclopropanes via reaction with methylene iodide and zinc–copper couple in the presence of (−)-menthol[44b] (Table **6-7**). Apparently the reactive reagent is,

TABLE 6-7

Asymmetric Simmons–Smith Syntheses with Chiral Reagent[a]

R′(R″)C=C(R)(R‴) **81** —(CH_2I_2, Zn—Cu; (−)-Menthol)→ **82A** and **82B** (cyclopropanes bearing R′, R″, R, R‴)

No.	R′	R″	R‴	R	Predominant config.	Cyclopropane, **82B** % e.e. or $[\alpha]_D^{25}$
1	H	Me	H	COOMe	*R,R*	1.9[b]
2	H	Ph	H	COOMe	*R,R*	0.7[b]
3	H	COOMe	H	COOMe	*R,R*	3.4[b]
4	Me	COOMe	Me	COOMe	*R,R*	1.1[b]
5	H	Ph	H	Me	*R,R*	−3.2°[c]
6	H	Ph	H	Ph	*R*	−0.3°[c]
7	Ph	Ph	H	Me	*R*	0.3
8	Me	Me	H	COOH	*R*	−0.2°[c]
9	Me	Me	H	COMe	*R*	−0.7°[c]

[a] Data from Reference 44b. Reactions in (−)-menthol; Simmons–Smith reagent presumed to be chiral. Molar ratio of **81**:CH_2I_2:Zn—Cu:menthol = 0.5:1:2:0.3.

[b] In the case of esters the asymmetric synthesis is based on the rotation of the acid obtained by saponification of **82**.

[c] The $[\alpha]_D$ max of **82** or derivatives is unknown in these cases.

at least partially, iodomethylzinc (−)-menthoxide. It is seen, from those examples in Tables **6-6** and **6-7** which overlap, that products of opposite configurations are formed in excess from these different but related reactions.

Reaction of **83** or **84** with methylene iodide and a complex prepared from diethylzinc and *S*-(−)-leucine gave optically active products of undetermined enantiomeric purity.[45] Products **85** and **86** had low and opposite rotations: **85**, $[\alpha]_D^{25}$ +0.30° (c 80.6, $CHCl_3$); **86**, $[\alpha]_D^{25}$ −0.77° (c 38.2, $CHCl_3$). The nature of the complex, which was prepared by refluxing diethylzinc

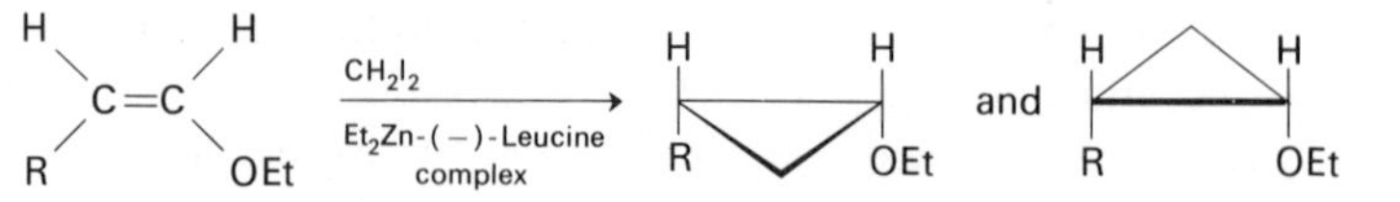

CH_2I_2; Et_2Zn-(−)-Leucine complex

83, R = Et
84, R = *i*Pr

S,R-**85**, R = Et
R,R-**86**, R = *i*Pr

and

R,S-**85**, R = Et
S,S-**86**, R = *i*Pr

and *S*-(−)-leucine in xylene, is unknown, but this reaction is probably mechanistically similar to a Simmons–Smith synthesis.

A chelate (**87**) prepared from *S*-(−)-α-methylbenzylamine, salicylaldehyde, and cupric ion has been used to effect asymmetric additions and insertions of diazoalkyl compounds.[46] When ethyl diazoacetate was decomposed

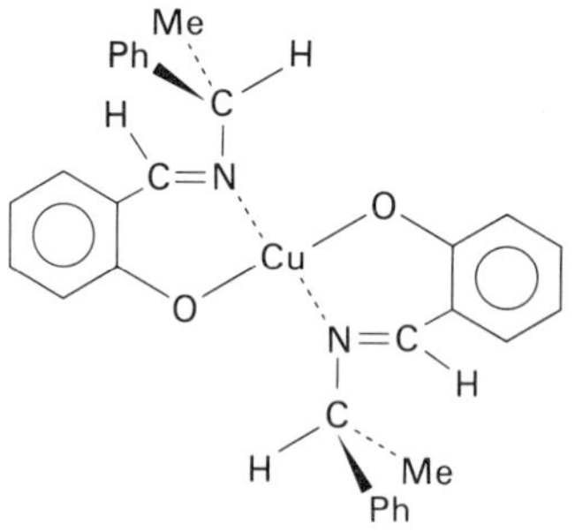

87

in styrene in the presence of **87**, excellent yields of a mixture of *cis* (**88**) and *trans* (**90**) ethyl 2-phenylcyclopropanecarboxylates were obtained. After separation (GLC), these diastereomeric esters were hydrolyzed to the corresponding optically active acids (**89** and **91**). The *trans* acid (**91**) contained a 6% excess of the (1*R*,2*R*)-(−)-enantiomer. The *cis* acid (**89**) had $[\alpha]_D^{25}$ −2.81 (c 50.5, $CHCl_3$), but its configuration and enantiomeric composition are unknown.

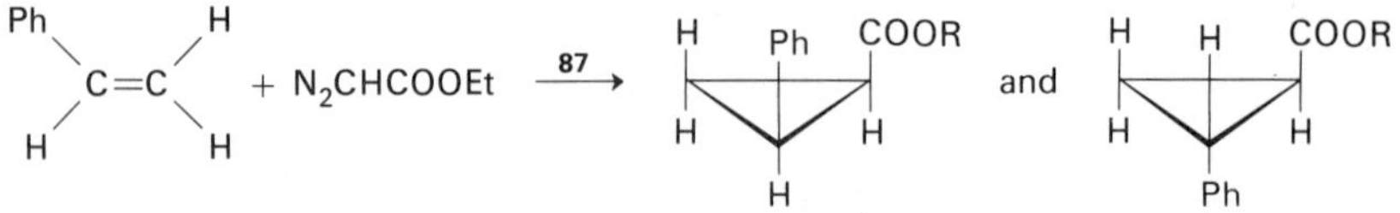

+ $N_2CHCOOEt$ —**87**→

88, R = Et
89, R = H

and

R,R-**90**, R = Et
R,R-**91**, R = H

[45] J. Furukawa, N. Kawabata, and J. Nishimura, *Tetrahedron Lett.*, 3495 (1968).
[46] H. Nozaki, H. Takaya, S. Moriuti, and R. Noyori, *Tetrahedron*, **24**, 3655 (1968).

Similarly, *trans*-1-phenylpropene (**92**) and diazomethane gave (1*R*,2*R*)-(−)-**93** (approximately 8% e.e.). Treatment of *cis, trans, trans*-1,5,9-cyclododecatriene (**94**) with diazomethane in the presence of **87** gave an optically

92 + CH_2N_2 $\xrightarrow{87}$ (1*R*,2*S*)-(−)-**93**

active cyclopropane (**95**) which was hydrogenated to levorotatory **96** ($[\alpha]_D$ − 1.18°). The latter was also obtained ($[\alpha]_D$ − 0.29°) from the analogous addition of diazomethane to *trans*-cyclododecene (**97**). Configurations are unknown. Asymmetric intramolecular additions of the same type were

94 $\xrightarrow[87]{CH_2N_2}$ **95** $\xrightarrow[[Ni/H_2]]{H_2}$ **96** $\xleftarrow[87]{CH_2N_2}$ **97**

accomplished with **98** and **100**.[46] It was estimated that the enantiomeric purity of (−)-**99** was at most 8%; neither the configuration of **99** nor the enantiomeric purity of **101** are known.

98 $\xrightarrow{87}$ **99**   **100** $\xrightarrow{87}$ **101**

An asymmetric ring enlargement of racemic 2-phenyloxetane (**102**) to optically active *cis* and *trans*-3-phenyltetrahydrofuran-2-carboxylates **103A** and **103B** has also been accomplished using this same chiral catalyst, **87**, and methyl diazoacetate. The specific rotations of compounds **103** and **104** were in the range of 4–6°, but enantiomeric purities and configurations are

RS-**102** + $N_2CHCOOMe$ $\xrightarrow{87}$ **103A**, R = Me; **104A**, R = H and **103B**, R = Me; **104B**, R = H unknown. A reactive chiral complex resulting from nucleophilic attack of diazoacetate on the chelate, followed by loss of nitrogen, was proposed as the carbenoid source in these reactions. Alternatively, a diazoacetate-copper chelate complex might react with olefin and simultaneously lose nitrogen.

Decomposition of 2-ethyl-1-diazobutane catalyzed by silver salts of chiral carboxylic acids, in the presence of (*S*)-α-methylbenzylamine and silver ion (but not in the absence of silver ion), or by thermolysis of the same compound in the presence of (*S*)-ethyl lactate or (*R*)-butyl tartrate gave optically active *trans*-1-ethyl-2-methylcyclopropane by intramolecular insertion of a carbenoid intermediate.[47] Products having very low specific

$$\underset{\mathbf{105}}{CH_3CH_2\underset{CHN_2}{\underset{|}{C}}HCH_2CH_3} \xrightarrow[\text{Catalyst}]{\text{Chiral}} \underset{\mathbf{106}\ (20\text{–}30\%\ \text{yield})}{\text{CH}_3,\ \text{H} / \text{H},\ \text{CH}_2\text{CH}_3\ \text{cyclopropane}}$$

rotations were obtained. The fact that optical activity was induced, even though of a low order, seems to implicate the chiral additive as a participant in diastereomeric transition states for the intramolecular insertion reaction since the intermediate carbene per se is achiral.

6-3.2. Diels–Alder Reactions. Optically active esters of fumaric acid react with 1,3-butadiene to yield diastereomeric esters of optically active 4-cyclohexene-1,2-dicarboxylic acid (Fig. **6-18**). Although Korolev and Mur,[48] who first studied this reaction, reported that di-(−)-menthyl fumarate (**107**) gave dextrorotatory diacid (**109**) after hydrolysis, Walborsky and co-workers found that levorotatory acid ($[\alpha]_D^{25}$ −8.65°, 5.4% e.e.) resulted when the diastereomeric ester precursors were completely saponified.[49] To circumvent this kinetic resolution, the mixture of esters (**108**) was reduced with lithium aluminum hydride. The predominant levorotatory diol was shown to be *R,R*-(−)-**110** (1.3% e.e.). Other examples of asymmetric Diels–Alder reactions have been studied as shown in Fig. **6-19**. In all but the last example the dienophile is chiral.

Careful studies of the reaction of butadiene with (−)-menthyl fumarate[49b] provide evidence that the reaction is sensitive to a variety of experimental parameters. Uncatalyzed reactions gave products varying between 0 and 3% enantiomeric excess when carried out in benzene at temperatures between 26 and 180°. The (1*R*,2*R*)-(−)-enantiomer of *trans*-1,2-dimethylol-4-cyclohexene (**110**) was the major isomer resulting from lithium aluminum

[47] W. Kirmse and H. Arold, *Angew. Chem.*, **80**, 534 (1968); *Internat. Edn.*, **7**, 539 (1968).

[48] A. Korolev and V. Mur, *Dokl. Akad. Nauk SSSR*, **59**, 251 (1948); *C.A.*, **42**, 6776 (1948).

[49] (a) H. M. Walborsky, L. Barash, and T. C. Davis, *J. Org. Chem.*, **26**, 4778 (1961). (b) H. M. Walborsky, L. Barash, and T. C. Davis, *Tetrahedron*, **19**, 2333 (1963).

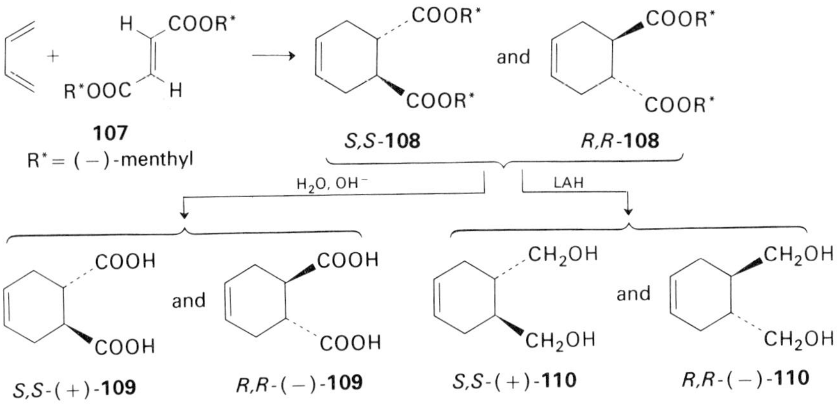

Fig. 6-18. Diels–Alder asymmetric synthesis (cf. Nos. 9 and 10, Table **6-8**).

hydride reduction of the intermediate epimeric esters. Asymmetric bias increased very slightly as the temperature increased, but the yield of Diels–Alder adduct decreased, apparently because of appreciable polymerization

TABLE 6-8

Asymmetric Catalyzed Diels–Alder Reaction of (−)-Dimenthyl Fumarate with 1,3-Butadiene[a]

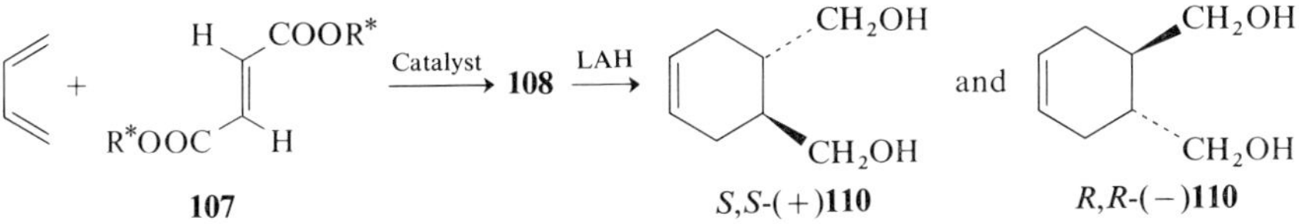

No.	Catalyst[a]	Solvent	Temp. °C	Predominant enantiomer of **110** config.	% e.e.
1	$TiCl_4$	Toluene—CH_2Cl_2	25	*S,S*-(+)	43
2	$TiCl_4$	Toluene	25	*S,S*-(+)	78
3	$SnCl_4$	Toluene	25	*S,S*-(+)	75
4	$AlCl_3$	Toluene	25	*S,S*-(+)	27
5	$AlCl_3$	CH_2Cl_2	−70	*S,S*-(+)	72
6	$AlCl_3$	CH_2Cl_2	−70	*S,S*-(+)	76
7	$AlCl_3$	Benzene	22	*S,S*-(+)	27
8	$AlCl_3$	Benzene	−70	*S,S*-(+)	59
9	None	Benzene	65	*R,R*-(−)	1.1
10	None	Benzene	25	Racemic	—

[a] Data from Reference 49b. One mole equivalent of catalyst used, except No. 6 where 0.5 mole equivalent used. R* = (−)-menthyl.

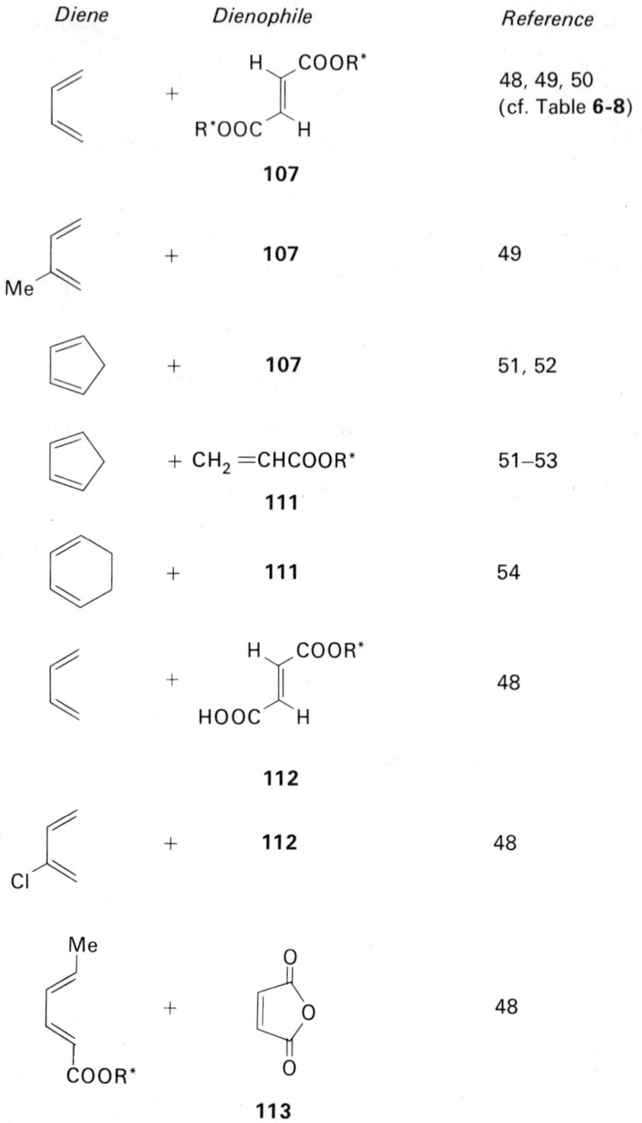

Diene		*Dienophile*	*Reference*
	+	H COOR* / R*OOC H **107**	48, 49, 50 (cf. Table **6-8**)
Me	+	**107**	49
	+	**107**	51, 52
	+	CH_2=CHCOOR* **111**	51–53
	+	**111**	54
	+	H COOR* / HOOC H **112**	48
Cl	+	**112**	48
Me / COOR*	+	O / O / O **113**	48

Fig. 6-19. Examples of asymmetric Diels–Alder reactions. R* = various chiral groups, e.g., (−)-menthyl, (−)-2-octyl, etc.

[50] A. Korolev, V. Mur, and V. G. Avakyan, *Zh. Obshch. Khim.*, **34**, 708 (1964); *Chem. Abst.*, **60**, 13153 (1964).

[51] J. Sauer and J. Kredel, *Angew. Chem.*, **77**, 1037 (1965); *Internat. Edn.*, **4**, 989 (1965).

[52] J. Sauer and J. Kredel, *Tetrahedron Lett.*, 6359 (1966).

[53] R. F. Farmer and J. Hamer, *J. Org. Chem.*, **31**, 2418 (1966).

[54] O. Červinka and O. Kříž, *Coll. Czech. Chem. Commun.*, **33**, 2342 (1968).

at higher temperatures. The effect of Lewis acids on the asymmetric cycloaddition reaction between di-(−)-menthyl fumarate and 1,3-butadiene changed the situation dramatically (Table **6-8**). Not only did the asymmetric bias increase greatly, but the chirality of product **110** from catalyzed reactions was the opposite of that from uncatalyzed reactions.

Similar studies with cyclopentadiene and di-(−)-menthyl fumarate or various chiral acrylates have supported the generality of higher enantiomeric yields in catalyzed reactions, but they have also revealed that product chirality is not always reversed (Tables **6-9** and **6-10**). The uncatalyzed reaction of di-(−)-menthyl fumarate (**107**) and cyclopentadiene in methylene chloride gave levorotatory **115**, whereas at a much lower temperature the same reaction catalyzed by $AlCl_3 \cdot OEt_2$ yielded dextrorotatory **115**.[52] This is analogous to Walborsky's observations for the reaction of the same fumarate with 1,3-butadiene.[49] However, the uncatalyzed reaction between **107** and cyclopentadiene also gave dextrorotatory **115** in acetone and dioxane,[52] whereas with 1,3-butadiene and **107** the chirality of the product from the uncatalyzed reaction was the same in various solvents (e.g., HOAc, CH_3CN, C_6H_6, and CCl_4).[49]

Chiral acrylates (Table **6-10**) and cyclopentadiene react to yield, after subsequent LAH reduction, two chiral products, **116** and **117**. The *endo* adduct **116** is formed in higher yield and higher enantiomeric purity in catalyzed reactions.* [55] Uncatalyzed reactions in triethylamine, dioxane,

TABLE 6-9

Asymmetric Diels–Alder Reaction of (−)-Dimethyl Fumarate with Cyclopentadiene[51,52]

H, COOR*, R*OOC, H —C_5H_6→ COOR*, COOR* —LAH→ CH_2OH, CH_2OH and CH_2OH, CH_2OH

107 R* = (−)-menthyl **114** 2*S*,3*S*-**115** 2*R*,3*R*-**115**

No.	Catalyst	Solvent	Temp. °C	Predominant enantiomer	
				rotation	% e.e.
1	None	Acetone	20	(+)	2.8
2	None	Dioxane	20	(+)	1.5
3	None	CH_2Cl_2	20	(−)	3.6
4	$AlCl_3 \cdot OEt_2$	CH_2Cl_2	−70	(+)	43

* It is possible that the Lewis acid forms a "sandwich" π-complex which lowers the energy of the transition state leading to *endo* product and thus accentuates the usual preference for *endo* isomer.

[55] J. Sauer and J. Kredel, *Tetrahedron Lett.*, 731 (1966).

TABLE 6-10

Asymmetric Diels–Alder Reactions of Chiral Acrylates and Cyclopentadiene[a,b]

$CH_2{=}CHCOOR^{*a}$ + cyclopentadiene ⟶ [diastereomeric esters] $\xrightarrow{LAH}$ (bicyclic, CH_2OH) and ($HOCH_2$, bicyclic)

111 2*R*-(+)-**116** 2*S*-(+)-**117**

Acrylate ester[a]	Catalyst	Solvent	Temp. °C	Predominant enantiomer of **116** %e.e.	Predominant enantiomer of **117** %e.e.
111A	None	Neat	25	(+)8	—
*R** = (−)-menthyl	None	CH_2Cl_2	0	(+)9	—
	None	CH_2Cl_2	35	(+)7	(+)3.1
	$AlCl_3 \cdot OEt_2$	CH_2Cl_2	0	(+)49	(−)36
	$AlCl_3 \cdot OEt_2$	CH_2Cl_2	−70	(+)67	—
	$BF_3 \cdot OEt_2$	CH_2Cl_2	0	(+)74	(−)43
	$BF_3 \cdot OEt_2$	CH_2Cl_2	−70	(+)82–85	—
	$SnCl_4$	Toluene	4–8	(+)41	—
	None	CH_2Cl_2	35	(+)4	(+)3
111B	$BF_3 \cdot OEt_2$	CH_2Cl_2	0	(+)23	(−)16
*R** = (−)-2-octyl	$BF_3 \cdot OEt_2$	CH_2Cl_2	−70	(+)27	—
111C	None	Neat	25	(−)3	—
*R** = (+)-2-octyl	$SnCl_4$	Toluene	4–8	(−)15	—
	$BF_3 \cdot OEt_2$	CH_2Cl_2	−70	(−)28	—
111D	None	CH_2Cl_2	35	(−)11	—
*R** = (+)-pinacolyl	$BF_3 \cdot OEt_2$	CH_2Cl_2	−70	(−)88	—
111E	None	Neat	25	(−)5	—
*R** = (+)-2-butyl	$SnCl_4$	Toluene	4–8	(−)24	—
111A *R** = (−)-menthyl	None	Neat	160	(−)6[c]	—

[a] R* represents the chiral group of the alcohol moieties of the various esters as follows: **111A**, R = (3*R*)-(−)-menthyl; **111B**, R = (2*R*)-(−)-2-octyl; **111C**, R = (2*S*)-(+)-2-octyl; **111D**, R = (2*S*)-(+)-3,3-dimethyl-2-butyl; **111E**, R = (2*S*)-(+)-2-butyl.

[b] Data from References 52, 53, and 54.

[c] Product from (−)-menthyl acrylate and 1,3-cyclohexadiene instead of 1,3-cyclopentadiene.[54]

methylene chloride, dimethoxyethane, acetone, and methanol gave products with low and fairly constant enantiomeric composition. The asymmetric bias in catalyzed reactions increased dramatically at lower temperatures, whereas temperature had little effect on the enantiomeric purities of products from uncatalyzed reactions. At −70° the reaction mixture became heterogeneous.

An interesting feature of the acrylate reactions is that the chirality of the predominant *exo* product (**117**) was inverted in going from uncatalyzed to catalyzed reactions, but the chirality of the *endo* product (**116**) did not change. One set of four diastereomeric transition states is involved in the uncatalyzed reaction and a different set defines the catalyzed reaction. The relative energies of the pair leading to enantiomeric *exo* products could be reversed when catalyst is present. However, one cannot exclude the possibility that there is appreciable thermodynamically controlled isomerization of the initially formed diastereomeric esters in the uncatalyzed (but not the catalyzed) reaction.

The effect of pressure on enantiomeric yield in Diels–Alder reactions has not been extensively studied. One report claims that the chirality of product **110** from the uncatalyzed reaction of 1,3-butadiene and di-(−)-menthyl fumarate in *m*-xylene can be reversed by increasing the pressure from 1 to 5000 atmospheres.[56]

An attempt has been made to rationalize the stereochemistry of the asymmetric Diels–Alder reaction of di-(−)-menthyl fumarate and 1,3-butadiene in terms of modified Prelog-type transition state models.[49] In the basic model (**118**), the diene and dienophile are allowed to approach in parallel planes, the approach of diene being from above (*re-re* face) to give 1*R*:2*R* product or from below (*si-si* face) to give 1*S*:2*S* product. All double bonds in the dienophile are frozen *transoid* in this model and the ester carbonyl is flanked by the small (hydrogen) and medium (—CH_2—) groups attached to the carbinol carbon of the (−)-menthyl group. It is then assumed that the predominant enantiomer is that one which would be predicted by approach of the diene from the side of the R_S groups (from above in **118**).

Attack on *re-re* face from above → COOR* / COOR* → LAH → CH_2OH / CH_2OH

118 *R,R*-**108** 1*R*,2*R*-(−)-**110**

Other models that predict the same result have been considered; still other models have been proposed for the catalyzed reactions. These are very speculative and many of the factors that contribute to the complexity of the situation have been discussed.[49b]

[56] B. S. El'yanov, E. I. Klabunovskii, M. G. Gonikberg, G. M. Parfenova, and L. F. Godunova, *Izv, Akad. Nauk SSSR, Ser. Khim.*, 1678 (1966); *Chem. Abst.*, **66**, 64939h (1967).

6-3.3. Asymmetric Epoxidation. The asymmetric epoxidation of a variety of monosubstituted ethylenes with (+)-mono-peroxycamphoric acid gives optically active epoxides slightly enriched in the levorotatory enantiomer (Table **6-11**). The asymmetric epoxidation of styrene has been

TABLE 6-11

Asymmetric Epoxidation of Alkenes with (+)-Monopercamphoric Acid[a]

$$RCH{=}CH_2 \;(\mathbf{119}) + \mathbf{1} \longrightarrow S\text{-}(-)\text{-}\mathbf{120} \text{ and } R\text{-}(+)\text{-}\mathbf{120} + \mathbf{4}$$

R	Solvent	α_D^{18} epoxide	Stereoselectivity[a] % e.e., *S*-(−)-**120**
CH_3	$CHCl_3$	−0.30°	2.0
$n\text{-}C_4H_9$	$CHCl_3$	−0.38°	2.5
$n\text{-}C_5H_{11}$	$CHCl_3$	−0.36°	2.5
$c\text{-}C_6H_{11}$	$CHCl_3$	—[b]	—[b]
$PhCH_2CH_2$	$CHCl_3$	−0.60°	2.1
$n\text{-}C_6H_{13}$	$CHCl_3$	—[b]	2.5
$n\text{-}C_6H_{13}$	Ether	—[b]	1.0
Ph	$CHCl_3$	(−)[c]	4.4
Ph	CH_2Cl_2	(−)[c]	3.5
Ph	Benzene	(−)[c]	2.5
Ph	Toluene	(−)[c]	2.4
Ph	CCl_4	(−)[c]	2.0
Ph	Ether	(−)[c]	2.0

[a] Data from Reference 1a and 1b. In Reference 1a, the configuration of the predominant levorotatory enantiomer is incorrectly reported to be *R*-(−) while in fact it is *S*-(−). This is given correctly in Reference 1b as established by E. Eliel and D. W. Delmonte, *J. Org. Chem.*, **21**, 596 (1956).

[b] No rotation was reported.

[c] The last six reactions gave levorotatory epoxides but no rotations were reported. The maximum rotation for styrene oxide is reported to be $[\alpha]_D$ + 34.2° (neat) [D. J. Pasto, C. C. Cumbo, and J. Fraster, *J. Am. Chem. Soc.*, **88**, 2194 (1966)] but the value used in this paper is not specified and appears to differ from this.

studied in various solvents. The asymmetric bias was highest in chloroform (4.4% e.e.) and lowest in ether or carbon tetrachloride (2.0% e.e.). These differences in stereoselectivity with solvent change must be due to preferential solvation of one of the competing transition states but are too small in terms of $\Delta\Delta G^{\ddagger}$ (maximum of *ca.* 30 cal/mole) to warrant attempted rationalization at this time.

The predominant enantiomer from asymmetric peroxidation of terminal olefins seems to be correctly predicted by a simple empirical model,[1] **121**.

1 **119** **121** *S*-**120**

There is good evidence for the mechanism of epoxidation[1,57] but the relative conformations of the established and the developing chiral centers in the transition state must be assumed, especially in view of the low stereoselectivities and the considerable distance between "inducing" and forming chiral centers.

The results of asymmetric epoxidations have been cited as evidence for the relative configurations of balfourodine (**123**) and isobalfourodine (**128**).[58] Epoxidation of the quinoline derivative **122** with (+)-monopercamphoric acid [or (*S*)-(+)-peroxyhydratropic acid or *S*-(−)-peroxy-*endo*-norbornane-2-carboxylic acid] gave (+)-balfourodine (**123**, 4–10% e.e.). The extension of model **121** for terminal olefins to a compound such as **122** is unwarranted and thus the configurations in Fig. **6-20** are relative but not necessarily absolute. However, assuming that the *S* configuration is induced at the chiral center of the intermediate epoxide, the configurations would be as shown. Epoxidation of the quinoline derivative **124** with (+)-monopercamphoric acid (**1**) gave two compounds, the furanoquinoline **126** and the pyranoquinoline **127**. Both must have arisen from the common intermediate epoxide **125**, the former via a displacement by the carbonyl group on the secondary carbon of the epoxide leading to inversion at the chiral center in **126**, and the latter via a displacement on the tertiary carbon of the epoxide leading to retention of configuration at the chiral center in **127**. Thus **126** and **127** must have "opposite" configurations as shown. Since **126** and **127** were converted, respectively, into (+)-balfourodine (**123**, 4.7% e.e.) and (−)-isobalfourodine (**128**, 9.3% e.e.) by processes which did not affect the chiral center, these two alkaloids must also have the "opposite" configurations.* This is the same conclusion reached by Rapoport and Holden[59] based on other chemical evidence.

[57] A. G. Davies, *Organic Peroxides* (London: Butterworth & Co., Ltd., 1961), p. 137.

[58] R. M. Bowman, J. F. Collins, and M. F. Grundon, *Chem. Commun.*, 1131 (1967).

[59] H. Rapoport and K. G. Holden, *J. Am. Chem. Soc.*, **81**, 3738 (1959); **82**, 4395 (1960).

* Bowman, Collins, and Grundon[58] reached this conclusion but apparently misinterpreted previous work in the literature and incorrectly concluded that their results were contrary to the findings of Rapoport and Holden.[59]

Asymmetric epoxidation of **129**, in which there is no quinolone oxygen available for subsequent ring opening, produced an optically active epoxide that was hydrolyzed to the (+)-orixine (2.4% e.e.) to which configuration **130** was tentatively assigned.[60]

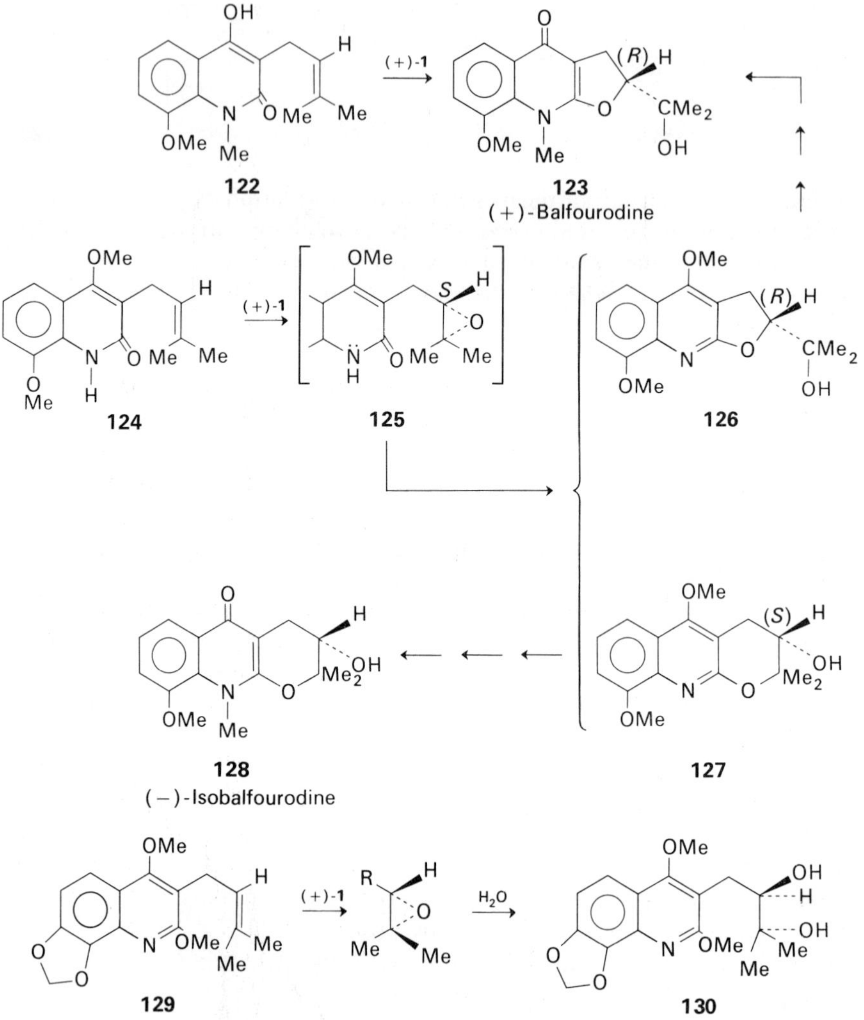

Fig. 6-20. Asymmetric synthesis of (+)-balfourodine (**127**) and (+)-isobalfourodine (**128**). These formulas represent relative configurations only since the absolute configurations are as yet unknown.

[60] R. M. Bowman and M. F. Grundon, *J. Chem. Soc.* (*C*), 2368 (1967). Formula XII in this paper is incorrect.

The previous examples of asymmetric epoxidation have involved chiral reagents and achiral alkenes. If the alkene is chiral, then achiral reagents may be used to accomplish an asymmetric epoxidation. For example, epoxidation of **131** and **133** with hydrogen peroxide gives optically active products after hydrolysis of intermediate diastereomeric epoxy cyanoesters (Fig. **6-21**).[61] The mechanism of the reaction is assumed to involve a Michael-type addition of a hydroperoxy anion or a pertungstate anion followed by displacement of hydroxide or tungstate as the epoxide is formed. Several solvents (methanol, ethanol, 1-propanol, 2-propanol, *t*-butyl alcohol, tetrahydrofuran, and pyridine) and various catalysts (Na_2WO_4, Na_3PO_4, Na_2CO_3) had only a minor effect on the asymmetric bias of this reaction. The isomeric composition of **132** is not known.

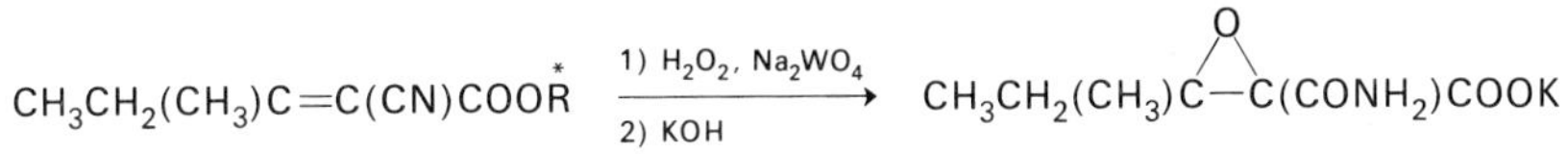

$CH_3CH_2(CH_3)C{=}C(CN)COOR^*$ → (1) H_2O_2, Na_2WO_4; 2) KOH) → $CH_3CH_2(CH_3)C{-}C(CONH_2)COOK$ (epoxide)

131, R* = (−)-menthyl
Unequal mixture of *cis-trans* diastereomers

132
Mixture of four diastereomers

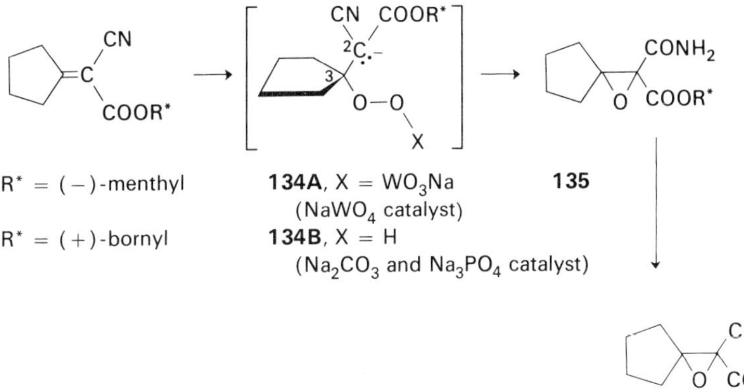

133A, R* = (−)-menthyl
133B, R* = (+)-bornyl

134A, X = WO_3Na ($NaWO_4$ catalyst)
134B, X = H (Na_2CO_3 and Na_3PO_4 catalyst)

135

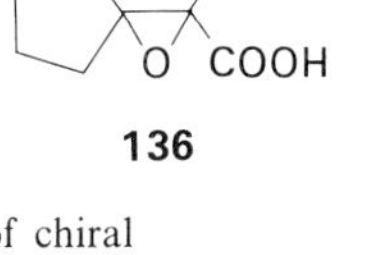

136

Fig. 6-21. Asymmetric epoxidation of unsaturated esters of chiral alcohols.

Although the (+)-bornyl and (−)-menthyl groups in **133A** and **133B** which are responsible for the asymmetric synthesis depicted in Fig. **6-21** have opposite chiral sequences of R_S, R_M, and R_L at the carbinol carbon (cf. Fig. **4-4**), they give the same enantiomer of **135** in excess. This result is in contrast to the behavior of these inducing groups in the atrolactate asymmetric synthesis (cf. Sec. **2-2.1**) but in accord with that observed in the

[61] M. Igarashi and H. Midorikawa, *Bull. Chem. Soc. Jap.*, **40**, 2624 (1967).

asymmetric aldol (Sec. **4-2**), Reformatsky (Sec. **4-3**), and Darzens (Sec. **4-4**) reactions. This indicates that asymmetric centers other than the α-carbon of the alcohol moiety of the ester play an important role in determining the stereoselectivity of the asymmetric epoxidations.

When the (−)-menthyl ester **135** was recrystallized to constant melting point and constant rotation, hydrolysis gave **136** with $[\alpha]_D^{25}$ −16.8°. Assuming that this represents enantiomerically pure product, the maximum asymmetric synthesis in these epoxidations was 14–15%.

The fact that **133A** gave optically active **136** when oxidized with peroxide and sodium tungstate but optically inactive **136** when the epoxidation was catalyzed with either sodium carbonate or sodium phosphate has been attributed to greater hindrance to rotation about the C-2 to C-3 bond in **134A** than in **134B**.

A related example of asymmetric epoxidation under the control of a chiral center in the substrate is the oxidation of linalool, **137**, with monoperphthalic acid to yield **138**. Hydrogenation of **138** followed by iodine catalyzed dehydration gave optically active **140**[62] in which the original chiral center in **137** has been destroyed but a new one established in the resulting mixture of epoxy olefins.

R—COOH; H_2/Pt; $-H_2O$

137 | **138** $[\alpha]_D$ + 3.63° | **139** $[\alpha]_D$ − 4.98° | **140** $[\alpha]_D$ −0.4°

6-4. Asymmetric Cyclization Reactions

Cyclization of acetal **141** with stannic chloride in benzene gives almost exclusively *trans* ring closure with the formation of two pairs of enantiomers **142A**–**142B** and **143A**–**143B** in which three new chiral centers have been created.[63] These diastereomeric pairs were separated and were converted to **144** in which the original inducing chiral centers as well as one of the three newly formed centers were removed. From **142** there was obtained 92% **144B** and 8% **144A**, while the reverse ratio was obtained from **143** (Fig. **6-22**).

[62] G. V. Pigulevskiĭ and G. V. Markina, *Dokl. Akad. Nauk SSSR*, **63**, 677 (1948); *Chem. Abst.*, **43**, 4628 (1949).

[63] W. S. Johnson, C. A. Harbert, and R. D. Stipanovic, *J. Am. Chem. Soc.*, **90**, 5279 (1968). (b) W. S. Johnson, *Accounts Chem. Res.*, **1**, 1 (1968).

The reaction therefore shows 84% asymmetric synthesis; in dry pentane or nitromethane, the percent asymmetric synthesis of **144B** via **142** was 72 and 48%, respectively. Aside from its intrinsic interest, this elegant asymmetric

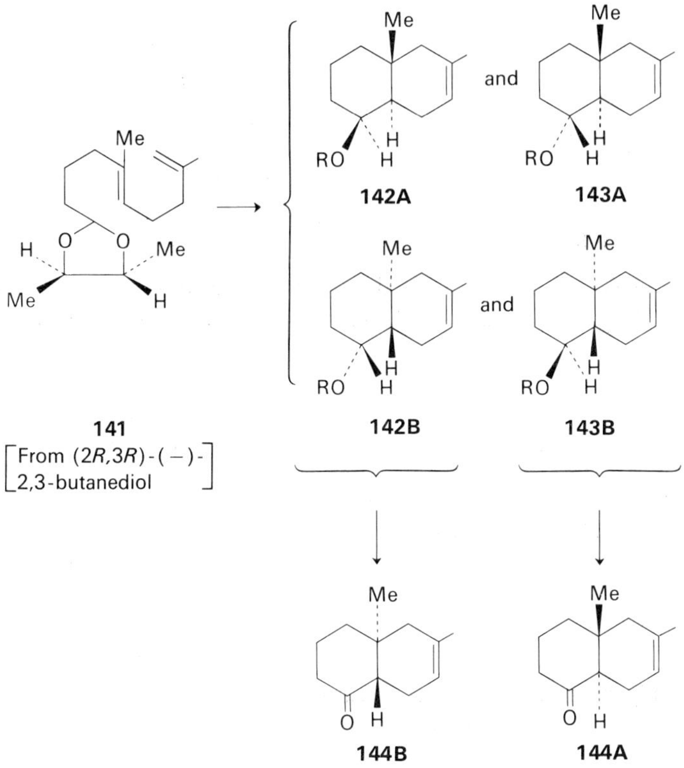

141
[From (2*R*,3*R*)-(−)-2,3-butanediol]

142A and **143A**

142B and **143B**

144B **144A**

Fig. 6-22. Asymmetric cyclization of optically active acetal **141**, $R = CH(CH_3)\overset{*}{C}H(OH)CH_3$ (asterisks mark both CH carbons: $\overset{*}{C}H(CH_3)\overset{*}{C}H(OH)CH_3$).

synthesis serves as a stereochemical model for biocyclizations,[63b] for example, the enzymatic cyclization of squalene (**145**) to give only one isomer of lanosterol (**146**) for which 128 stereoisomers are theoretically possible. Enzymatically this process has been shown to proceed via the initial formation of the 2,3-epoxide.[64]

In Sec. **6-3.1** (cf. Figs. **6-16** and **6-17**) some cycloaddition reactions that give chiral cyclopropanes were described. Chiral cyclopropanes can also be prepared via cyclizations that are not preceded by addition to an olefinic bond. For example, racemic 4-chloro-4-phenylbutanoic acid has been con-

[64] E. E. van Tamelen, J. D. Willett, and R. B. Clayton, *J. Am. Chem. Soc.*, **89**, 3371 (1967).

verted to chiral esters (**147**) with (−)-menthol and (+)-borneol and then cyclized with strong base (sodium or potassium *t*-butoxides or *t*-amyloxides) to esters of 2-phenylcyclopropanecarboxylic acid. Optically active *trans*-2-phenylcyclopropane carboxylic acid (**78**) was obtained after hydrolysis.[65]

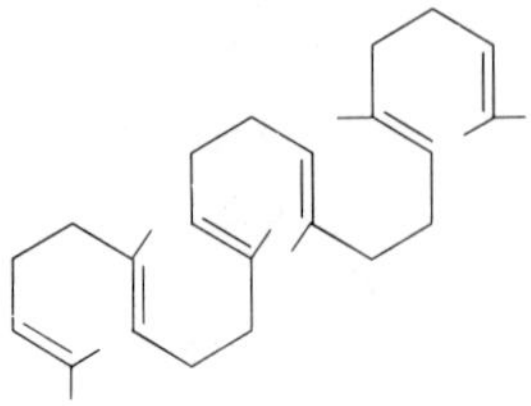

HO H

145 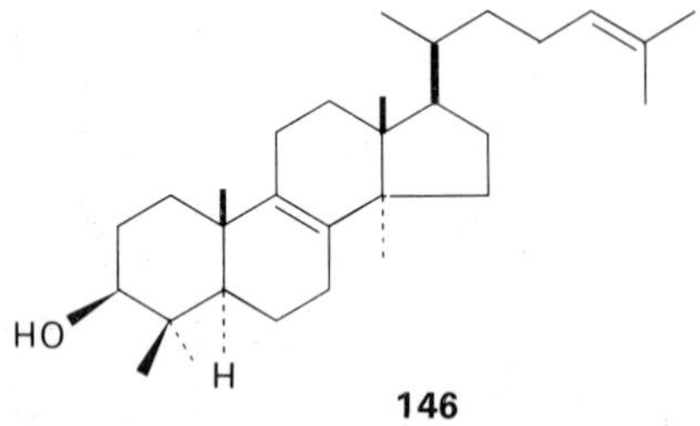 **146**

Since the absolute configuration and maximum rotations of **78** have now been established,[66] it can be calculated that the asymmetric synthesis (or kinetic

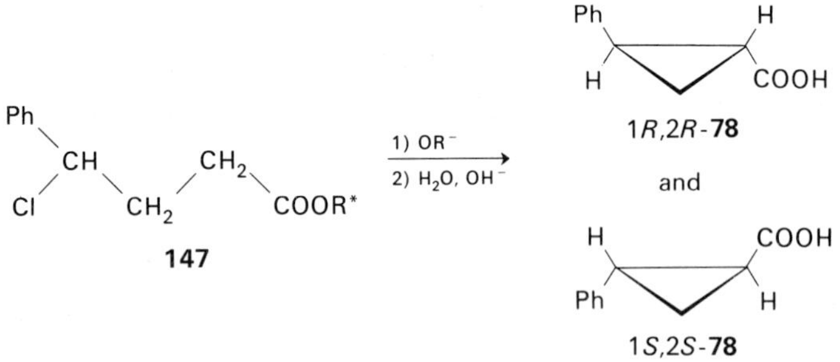

Ph CH Cl CH_2 CH_2 COOR* **147** 1) OR^- 2) H_2O, OH^-

Ph H H COOH 1*R*,2*R*-**78** and H COOH Ph H 1*S*,2*S*-**78**

Fig. 6-23. Asymmetric cyclopropane cyclizations, R* = (−)-menthyl or (+)-bornyl.

resolution) varies from 0.5 to 10%. Furthermore, the sign of rotation of **78** varied, depending upon the alkoxide employed. This represents a complex reaction in which the stereochemistry is not controlled solely by the configuration of the chiral moiety of the ester. In this particular case, the starting material is a racemate and the product has two chiral centers, one of which was in the original substrate, presumably in an inverted form. The *trans* compound appears to be produced more readily from one diastereomer than from the other, which either must remain unreacted or must be removed by some competing reaction. If a compound such as (−)-menthyl 4-methyl-4-chloropentanoate, could be cyclized to 2,2-dimethylcyclopropane carboxylic acid, an asymmetric synthesis without these complications could result.

[65] H. Nozaki, K. Kondô, O. Nakanisi, and K. Sisido, *Tetrahedron*, **19**, 1617 (1963).
[66] Y. Inouye, T. Sugita, and H. M. Walborsky, *Tetrahedron*, **20**, 1695 (1964).

Chiral esters (**148**) of diethylphosphonoacetic acid have been reported to react with racemic styrene oxide (**3**) in the presence of sodium hydride to give, after hydrolysis, *trans*-2-phenylcyclopropanecarboxylic acid (**78**) with low optical activity (3–5% e.e.).[65,67] Curiously, both the (+)[67] and (−)[65] enantiomers of **78** have been reported to be the predominant enantiomer when (−)-menthyl diethylphosphonoacetate was used. Esters from (−)-menthol and (+)-borneol (opposite chiral arrangements of R_S, R_M, and R_L at the carbinol center) gave (−)-**78** in excess in one study,[65] whereas esters from (−)-menthol and (−)-2-octanol (same chiral order of R_S, R_M, R_L) gave (+)-**78** in excess in another study.[67] An analogous reaction was

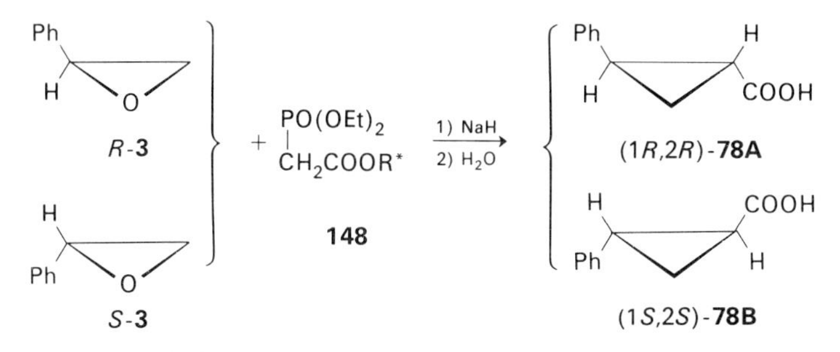

Fig. 6-24. Reaction of ylide from chiral esters of diethylphosphonoacetic acid with racemic styrene oxide, R* = (−)-menthyl, (+)-bornyl, and (−)-2-octyl.

done with chiral esters of diethylphosphonopropionic acid (**149**) in which the reagent is a mixture of diastereomers (Fig. **6-25**).[68] The ester group of the resulting cyclopropane derivative was converted to a methyl group, to give 1,1-dimethyl-2-phenylcyclopropane (**150**), thereby eliminating the chiral centers of the intermediate which were in the original reagent. Again the product showed only slight optical activity. Both the (−)-menthyl ester (**149A**) and the (+)-bornyl ester (**149B**) gave the same (*R*)-(−)-**150** in excess ($[\alpha]_D$ −0.75° and −0.28°, respectively). The maximum rotation of **150** was not determined but the configuration was assigned on the basis of the conversion of (*R*)-(+)-styrene oxide to (+)-**150** via this same reaction with assumed inversion of configuration.

If, as claimed, only the *trans* acid is formed and this exclusively by inversion at the chiral center of the styrene oxide, then the optically active product must arise by way of some kind of kinetic resolution, not by an asymmetric synthesis, since the starting material was racemic and the product incorporates the chiral center of the racemic substrate, albeit inverted.

[67] (a) I. Tömösközi, *Angew. Chem.*, **75**, 294 (1963); *Internat. Edn.*, **2**, 261 (1963). (b) I. Tömösközi, *Tetrahedron*, **19**, 1969 (1963).

[68] I. Tömösközi, *Tetrahedron*, **22**, 179 (1966).

Alternatively, it is possible that there is a competition between inversion (predominant) and retention (minor) pathways for the reaction. These competitive reactions could proceed at different rates for the enantiomers

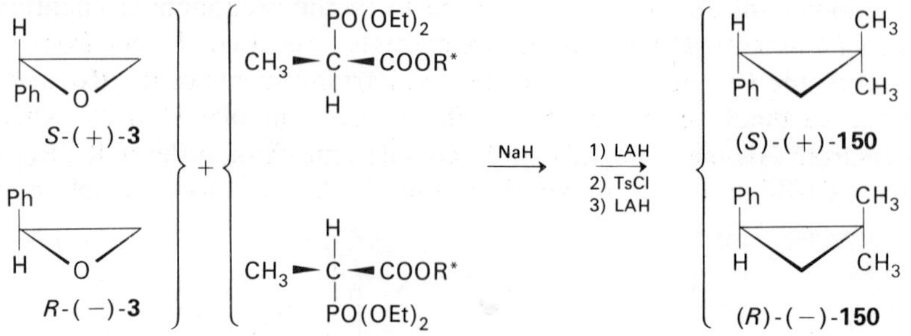

H Ph O S-(+)-3; Ph H O R-(−)-3 + CH_3–C(PO(OEt)$_2$)(H)–COOR* ; CH_3–C(H)(PO(OEt)$_2$)–COOR* $\xrightarrow{\text{NaH}}$ $\xrightarrow{\text{1) LAH, 2) TsCl, 3) LAH}}$ (S)-(+)-**150**; (R)-(−)-**150**

149A, R* = (−)-menthyl
149B, R* = (+)-bornyl

Fig. 6-25. Reaction of ylides of chiral esters of diethylphosphonopropionic acid with racemic styrene oxide.

of the racemic styrene oxide via diastereomeric transition states leading to unequal amounts of the cyclopropanes.

It is apparent that reaction of a symmetrical 1,1-di-substituted epoxide with a chiral reagent such as **148** would clarify the asymmetric synthesis aspects of this reaction considerably.

6-5. Asymmetric Oxymercuration and Alkoxymercuration

Mercuric salts add readily to alkenes in water or alcohol solvents to produce hydroxy or alkoxymercurials which can be demercurated to

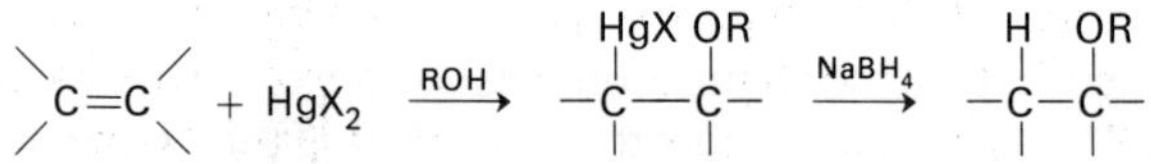

$$\text{>C=C<} + HgX_2 \xrightarrow{ROH} -\overset{HgX}{\underset{|}{C}}-\overset{OR}{\underset{|}{C}}- \xrightarrow{NaBH_4} -\overset{H}{\underset{|}{C}}-\overset{OR}{\underset{|}{C}}-$$

alcohols or ethers. This reaction has been reviewed.[69] Subsequent research has been concerned with the mechanism of the reaction, especially the stereochemistry of the addition of HgX and OR,[70] and demercuration

[69] (a) J. Chatt, *Chem. Rev.*, **48,** 7 (1951). (b) W. Kitching, *Organometal. Chem. Rev.*, **3,** 61 (1968).

[70] (a) T. G. Traylor, *J. Am. Chem. Soc.*, **86**, 244 (1964). (b) S. Wolfe and P. G. C. Campbell, *Can. J. Chem.*, **43**, 1184 (1965). (c) F. G. Bordwell and M. L. Douglass, *J. Am. Chem. Soc.*, **88**, 993 (1966). (d) F. R. Jensen and J. J. Miller, *Tetrahedron Lett.*, 4861 (1966).

procedures.[70c] Brown and co-workers[71] have developed reaction conditions which enable oxymercuration and demercuration to be accomplished easily and in excellent yield. This procedure frequently may be the method of choice for the stereoselective introduction of HO—H across a double bond. The stereochemistry of the addition is generally *trans* with unstrained olefins. Certain bicyclic olefins, however, undergo *cis* addition.[70] Cleavage of the carbon–mercury bond with sodium borohydride occurs predominantly, if not exclusively, with retention. Thus it is possible to convert norbornene to *exo-cis*-2-hydroxy-3-deuterionorbornane.[70c] Although there have been relatively few studies of asymmetric synthesis via oxymercuration, renewed interest in the reaction has stimulated some recent investigations along these lines.

The methoxymercuration of α,β-unsatured esters of optically active alcohols (**151**) followed by demercuration and hydrolysis of the ester (**152**)

$$\underset{\textbf{151}}{RR'C{=}CH(COOR^*)} \xrightarrow[HgX_2]{HOCH_3} \underset{\textbf{152}}{RR'C(OCH_3){-}CH(HgX)COOR^*} \xrightarrow[H_2O]{NaBH_4} \underset{\textbf{153}}{RR'C(OCH_3){-}CH_2COOH}$$

yields a potentially optically active β-methoxyacid (153). Asymmetric syntheses of this type have been achieved with a variety of chiral esters[72,73] (Table **6-12**) to give β-methoxy acids (up to 27% e.e.). This reaction sequence is an extension of much earlier work which suggested that asymmetric synthesis occurred in the methoxymercuration of (−)-menthyl and (+)-bornyl cinnamates.[74]

The acid catalyzed reaction of mercuric acetate with (−)-menthyl cinnamate (**151A**; R = H, R′ = Ph, R* = (−)-menthyl) in methanol gave what was apparently a mixture of isomers of the acetoxymercurial **152A** in 94% yield. Similarly two iodomercurials with opposite rotations, **155A** and **155B**, were obtained upon treatment of **152A** with sodium iodide, but only one chloromercurial (**156**) was isolated from reaction with sodium chloride. It could be converted into bromomercurial **154A** or iodomercurial **155B** according to the scheme shown on page 268.

[71] (a) H. C. Brown and P. Geoghegan, Jr., *J. Am. Chem. Soc.*, **89**, 1522 (1967). (b) H. C. Brown and W. J. Hammar, *ibid.*, **89**, 1524 (1967). (c) H. C. Brown, J. H. Kawakami, and S. Ikegami, *ibid.*, **89**, 1525 (1967).

[72] J. Oda, T. Nakagawa, and Y. Inouye, *Bull. Chem. Soc. Jap.*, **40**, 373 (1967).

[73] M. Kawana and S. Emoto, *Bull. Chem. Soc. Jap.*, **40**, 618 (1967).

[74] (a) L. T. Sandborn and C. S. Marvel, *J. Am. Chem. Soc.*, **48**, 1409 (1926). (b) E. Griffith and C. S. Marvel, *ibid.*, **53**, 789 (1931).

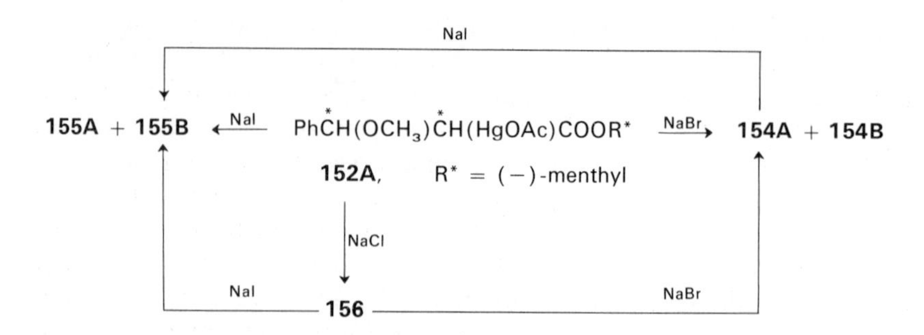

TABLE 6-12

Asymmetric Methoxymercurations[a]

$(R)(R')C{=}C(H)(COOR^*)$ **151** $\xrightarrow[\text{MeOH, }H^+]{Hg(OAc)_2}$ **152**

152 $\xrightarrow[\text{2) } H_2O]{\text{1) } NaBH_4 \text{ or } H_2S}$ $RR'C(OCH_3)CH_2COOH$ **153**

152 $\xrightarrow[\text{2) } LiAlH_4]{\text{1) } NaBH_4}$ $RR'C(OCH_3)CH_2CH_2OH$ **157**

No.	Catalyst	R	R′	R*	Stereoselectivity[b] **153** % e.e.	Stereoselectivity[b] **157** % e.e.	Configuration
1	HOAc	H	CH_3	(−)-Menthyl	12	—	*R*
2	BF_3	H	CH_3	(−)-Menthyl	15.5	—	*R*
3	HOAc	H	Ph	(−)-Menthyl	—	2.5	*S*
4	BF_3	H	Ph	(−)-Menthyl	—	4.9	*S*
5	BF_3	CH_3	Ph	(−)-Menthyl	—	2.3[c]	*S*
6[d]	HNO_3	H	Ph	(−)-Menthyl	19	—	*S*
7[e]	HNO_3	H	Ph	(−)-Menthyl	—	22	*S*
8[e]	HNO_3	H	Ph	(+)-Bornyl	4	—	*R*
9[e]	HNO_3	H	Ph	(+)-Bornyl	—	6	*R*
10[d]	HNO_3	H	Ph	**158**	19	—	*S*
11[e,f]	HNO_3	H	Ph	**158**	—	—	*S*
12[e]	HNO_3	H	Ph	**158**	—	20	*S*
13[e]	HNO_3	H	Ph	**159**	27	—	*S*
14[d]	HNO_3	H	Ph	**160**	22	—	*S*

[a] Experiments Nos. 1–5 are from Reference 72; the remainder from Reference 73.

[b] The maximum rotations, $[\alpha]_D$, used for calculations of the % e.e. are as follows: **153A** (R = H, R′ = CH_3), 11.6° (neat), P. A. Levene and R. E. Marker, *J. Biol. Chem.*, **102**, 297 (1933); **153B** (R = H, R′ = Ph), 67° (benzene), K. Balenović, B. Urbas, and A. Deljas, *Croat. Chem. Acta*, **31**, 153 (1959); **157B** (R = H, R′ = Ph), 120° (benzene), Balenovic, et al., *loc. cit.*, or 51.7° (neat)[72]; **157C** (R = CH_3, R′ = Ph), 24.2° ($CHCl_3$).[72]

[c] The authors[72] claim 2.3 % e.e. for a sample with $[\alpha]_D$ −8° (neat). This seems out of line with the reported[72] maximum $[\alpha]_D$ of 24.2° ($CHCl_3$); because of the different conditions (neat versus $CHCl_3$ solvent), these values cannot be compared.

[d] Reaction time, 1.5 hr.

[e] Reaction time, 5 hr.

[f] In this experiment the alcohol **157B** was obtained from the acid **153B** (from experiment No. 8, $[\alpha_D]$ − 13°) via treatment with CH_2N_2 and LAH.

These data can be rationalized in terms of a *trans*-methoxymercuration followed by conversion to the diastereomeric halomercurials and separation by fractional crystallization. If this is the correct interpretation, thc contemporary demercuration methods[70c,71–73] could be used to convert the pure halomercurials to enantiomerically pure β-methoxy acids (**153**) or alcohols (**157**).

Table **6-12** contains a summary of asymmetric methoxymercurations using (−)-menthyl, (+)-bornyl, diisopropylidene-D-glucosyl (**158**), dicyclohexylidene-D-glucosyl (**159**), and deoxyisopropylidene-D-xylosyl (**160**) α,β-

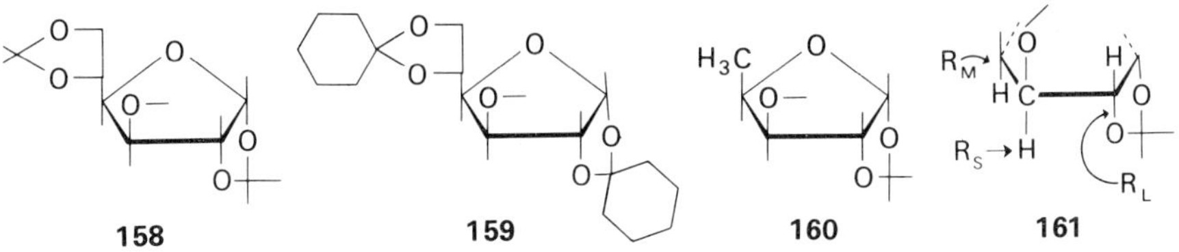

158 **159** **160** **161**

unsaturated esters. Unfortunately, interpretation of the quantitative aspect of these syntheses is very uncertain.* It was noted[73] that stereoselectivity decreased as the methoxymercuration reaction time increased. For example, when reaction No. 6 (Table **6-12**) was repeated with reaction times of 8 and 24 hr, the enantiomeric excess of (−)-**153** dropped to 10 and 3%, respectively. Reaction times of 48 and 72 hr gave the (+) enantiomer (1 and 4% e.e., respectively). Since the synthetic yield of **153** was fairly constant for reaction times from 1.5 to 72 hr, it seems that the kinetic product is not the thermodynamically more stable one and that equilibrium is not yet attained at 72 hr.

Cinnamates prepared from (−)-menthol and (+)-borneol (opposite chiral orders of R_S, R_M, and R_L) yielded enantiomeric products (Table **6-12**, Nos. 6 and 7 versus 8 and 9). This behavior parallels that observed in classical atrolactic acid asymmetric syntheses. The stereochemistry has been rationalized by assuming a Prelog-type model (**162**) in which the carbonyl group and the carbon-carbon double bond are *transoid* and coplanar, and R_S and R_M flank the carbonyl group. The configuration of the predominant product is then predicted on the assumption that the entering group approaches from the side of R_S. The resulting mercurinium ion (**163**) is then opened by S_N2 attack of the solvent to give the methoxymercurial **152** with inversion of configuration at the *beta* carbon atom. Both catalyzed and uncatalyzed reactions studied thus far give the same stereochemical result. In the reactions

* The importance of partial kinetic resolution due to incomplete hydrolysis of diastereomeric intermediates or decomposition to products, the importance of thermodynamic equilibration of epimeric products during extended reaction times, or the contribution of inadvertent concentration of one isomer during crystallization cannot be evaluated from the published data.

of esters of sugar derivatives (Table **6-12**, Nos. 10–14) the groups which play the role of R_S, R_M, and R_L have been assigned[73] as indicated in **161**. Since these derivatives have several oxygen atoms as well as multiple asymmetric centers, factors such as coordination with the reagent, hydrogen bonding

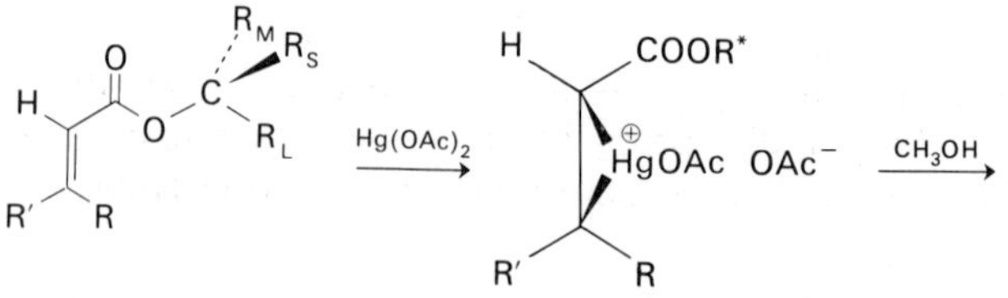

Fig. 6-26. Methoxymercuration of α,β-unsaturated chiral esters.

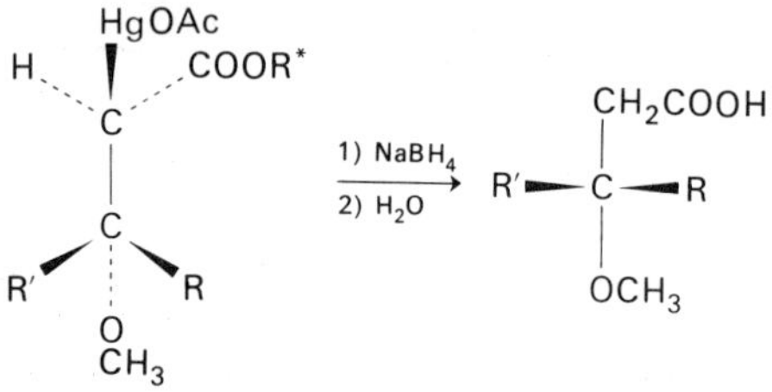

with solvent, and nonbonded interactions are undoubtedly significant in determining the actual nature of the asymmetric induction. In spite of these possible complications, this simple steric bulk model (**162**) seems to be applicable in the limited number of examples which have been reported. The fact that **160**, which has a methyl group at C-4 instead of an oxygen function as in **158** and **159**, gives a product with the same configuration and similar enantiomeric composition as **158** and **159**, suggests that coordination (at least by the oxygen substituent at C-4) is not critical. The same sugar derivatives have been used as inducing agents in atrolactate asymmetric syntheses (Table **2-5**), as chiral LAH complexes for ketone reductions (Sec. **5-3.5**), and in the asymmetric 1,4-addition of phenylmagnesium bromide to α,β-unsaturated esters (Table **6-13**). The same order of magnitude of asymmetric synthesis is observed in the atrolactate system but the asymmetric bias in the Grignard additions is generally higher.

The addition of mercuric salts to chiral α,β-unsaturated esters is an attractive system for additional study. It is possible that a Ugi-type linear free energy relationship (Sec. **1-4.4**) could be applied to this system. Asymmetric synthesis at the position *alpha* to the ester group should be demonstrable with a suitable α,β-unsaturated ester, e.g., **164** → **165**. There has been very little study of the use of other nucleophiles to "trap" the presumed mercurium ion intermediate. One recent paper does describe nitromercuration.[75] A modification of the schemes described could, in principle, lead to

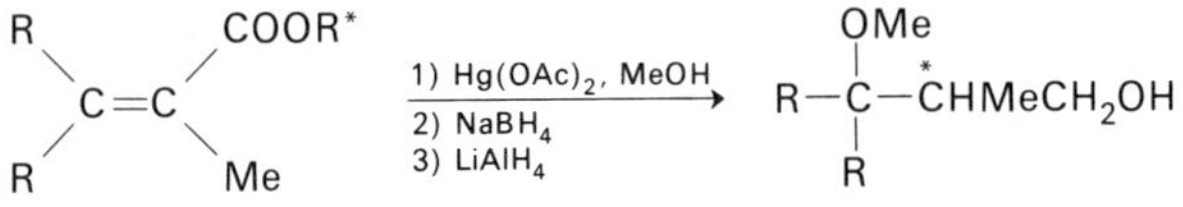

$$\underset{\mathbf{164}}{R_2C{=}C(COOR^*)Me} \xrightarrow[\substack{2)\ NaBH_4 \\ 3)\ LiAlH_4}]{1)\ Hg(OAc)_2,\ MeOH} \underset{\mathbf{165}}{R{-}C(OMe)(R){-}\overset{*}{C}HMeCH_2OH}$$

optically active β-aminoacids if the nitromercurial intermediates could be demercurated and the resulting nitroester could be reduced. Unfortunately attempts to accomplish demercuration of nitromercurials from simple alkenes have not been successful.[75]

The previous examples of asymmetric oxymercuration have all involved addition of achiral mercury salts to chiral alkenes. There is also evidence that chiral mercuric salts may add to alkenes in an asymmetric manner. The addition of (+)-mercuric lactate to cyclohexene in methanol gave a lactomercuri-2-methoxycyclohexane that appeared to be a 2:1 mixture of two substances.[76] These were separated, recrystallized, and allowed to react with sodium or potassium chloride. One gave an optically active chloromercuri-2-methoxycyclohexane, while the other was converted to an inactive isomer. In view of more recent evidence that oxymercuration of cyclohexene occurs in an exclusively *trans* manner,[77,78] these reactions can be formulated as shown in Fig. **6-27** and can be envisaged as follows. The reaction gives a

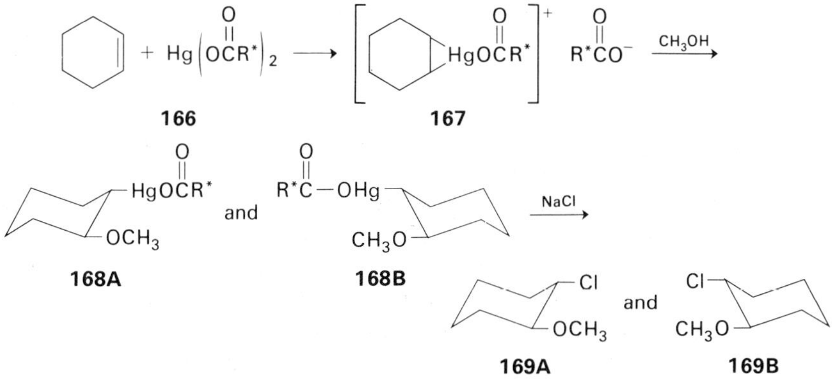

$$\underset{\mathbf{166}}{C_6H_{10} + Hg(OCOR^*)_2} \longrightarrow \underset{\mathbf{167}}{\left[C_6H_{10}HgOCOR^*\right]^+ \; R^*CO_2^-} \xrightarrow{CH_3OH} \underset{\mathbf{168A}}{\text{HgOCOR}^*/\text{OCH}_3} \text{ and } \underset{\mathbf{168B}}{R^*C(O){-}OHg/CH_3O} \xrightarrow{NaCl} \underset{\mathbf{169A}}{\text{Cl}/\text{OCH}_3} \text{ and } \underset{\mathbf{169B}}{\text{Cl}/\text{CH}_3\text{O}}$$

Fig. 6-27. Methoxymercuration of cyclohexene with chiral mercuric salts.

[75] G. B. Bachman and M. L. Whitehouse, *J. Org. Chem.*, **32**, 2303 (1967).
[76] J. Romeyn and G. F. Wright, *J. Am. Chem. Soc.*, **69**, 697 (1947).
[77] T. G. Traylor and A. W. Baker, *J. Am. Chem. Soc.*, **85**, 2746 (1963).
[78] S. Wolfe and P. G. C. Campbell, *Can. J. Chem.*, **43**, 1184 (1965).

cyclic lactomercurinium lactate (**167**) which opens by S_N2 solvent attack at either C-1 or C-2 to give an unequal mixture of enantiomeric *trans* isomers (**168A** and **168B**). The racemic compound of **168** crystallizes leaving an excess of one enantiomer. Separate treatment of the excess isomer with sodium chloride gives optically active *trans*-2-methoxy-1-chlorocyclohexane.

6-6. Miscellaneous Asymmetric Additions to Carbon-Carbon Double Bonds

6-6.1. Conjugate Additions to Chiral α,β-Unsaturated Esters. The addition of an achiral Grignard reagent to certain chiral α,β-unsaturated esters produces optically active β-substituted acids after saponification of the diastereomeric ester intermediates. Inouye and Walborsky[79]

$$\underset{\mathbf{170}}{(R)(H)C{=}C(H)(COOR^*)} \xrightarrow[2)\ H_2O(OH^-)]{1)\ R'MgX} \underset{\mathbf{171}}{RR'\overset{*}{C}H{-}CH_2COOH} + R^*OH$$

studied the addition of phenylmagnesium bromide to (−)-menthyl crotonate in ethyl ether at −8°. Uncatalyzed additions gave (*S*)-(+)-3-phenylbutanoic acid (5.4, 5.9, and 6.7% e.e.). Cadmium chloride did not appreciably affect the stereoselectivity [*S*-(+)-acid, 8.1% e.e.], but when the reaction was catalyzed by cuprous chloride, the predominant stereochemical course of the reaction was reversed and *R*-(−)-acid was produced (6.0, 6.3, and 10.2% e.e.).

Kawana and Emoto examined the reaction of phenylmagnesium bromide and benzylmagnesium chloride with chiral crotonates and cinnamates, respectively.[80] The reactions were carried out by dropwise addition of an ether solution of the ester (0.03 mole) to an ether solution of a Grignard reagent (0.075 mole) at −15 to −17°. The products were quantitatively saponified under conditions that did not promote any detectable racemization. Catalyzed reactions were carried out in exactly the same manner but with the addition of cuprous chloride (0.04 mole), in portions, beginning just before the ester addition. The experiments are summarized in Tables **6-13** and **6-14**.

[79] Y. Inouye and H. M. Walborsky, *J. Org. Chem.*, **27**, 2706 (1962).

[80] M. Kawana and S. Emoto, *Bull. Chem. Soc. Jap.*, **39**, 910 (1966). This paper incorrectly reports that J. Munch-Petersen and co-workers obtained a racemic 1,4-addition product upon reaction of benzylmagnesium chloride with (−)-*sec*-butyl cinnamate, whereas in fact racemic *sec*-butyl cinnamate was used.

TABLE 6-13

Asymmetric 1,4-Addition of Phenyl Grignard Reagent to Chiral Crotonates[a]

$$\underset{\textbf{170A}}{\text{CH}_3(\text{H})\text{C=C}(\text{H})\text{COOR*}} \xrightarrow[\text{2) H}_2\text{O(OH}^-\text{)} \quad \text{3) H}^+]{\text{1) PhMgBr}} \underset{S\textbf{-171A}}{\text{Ph}-\text{C}(\text{CH}_3)(\text{H})-\text{CH}_2\text{COOH}} \text{ and } \underset{R\textbf{-171A}}{\text{Ph}-\text{C}(\text{H})(\text{CH}_3)-\text{CH}_2\text{COOH}}$$

No.	R*	Catalyst	Synthetic yield %	Asymmetric synthesis[b] % e.e.	Configuration
1A	(−)-Menthyl	None	51	9	*S*
1B	(−)-Menthyl	None	52	6[c]	*S*
1C	(−)-Menthyl	Cu_2Cl_2	42	5	*R*
1D	(−)-Menthyl	Cu_2Cl_2	62	7[c]	*R*
1E	(−)-Menthyl	$CdCl_2$	50	8	*S*
2A	**158**[d,e]	None	12	32	*R*
2B	**158**[d,e]	Cu_2Cl_2	50	68	*R*
3A	**159**[d]	None	12	33	*R*
3B	**159**[d]	Cu_2Cl_2	58	74	*R*
4A	**160**[d]	None	32	16	*S*
4B	**160**[d]	Cu_2Cl_2	61	58	*R*

[a] Data from Reference 80 except runs 1B, 1D, and 1E from Reference 79.
[b] Based on $[\alpha]_D$ 57° for enantiomerically pure **171**.
[c] Average of three runs.
[d] Structural formulas for these groups are given in the text.
[e] Tetrahydrofuran solvent.

As seen in Table **6-13**, the addition of cuprous chloride to the reactions involving 1,4-addition of *phenyl* Grignard reagent to crotonate caused a significant change in the stereoselectivity, even reversing the chiral sense of the synthesis in two cases. However, in reactions involving *benzyl* Grignard reagent (Table **6-14**), the addition of cuprous chloride caused no significant change in the stereochemistry, which correlates with the negligible effect upon the synthetic yield of 1,4-addition product when cuprous chloride was added to reactions of benzyl Grignard reagent and achiral α,β-unsaturated esters.[81]

Two mechanisms, represented by **172** and **173**, have been considered for the 1,4-addition of Grignard reagents to α,β-unsaturated esters.[81] The first

[81] J. Munch-Petersen, P. Møller Jørgensen, and S. Refn, *Acta Chem. Scand.*, **13**, 1955 (1959). For a recent review of reactions of organomagnesium compounds with α,β-unsaturated carbonyl compounds, see J. Munch-Petersen, *Bull. Soc. Chim. Fr.*, **33**, 471 (1966). The role of copper in such reactions is discussed (with many references) by H. O. House, W. I. Respess, and G. M. Whitesides, *J. Org. Chem.*, **31**, 3128 (1966).

TABLE 6-14

Asymmetric 1,4-Addition of the Benzyl Grignard Reagent to Chiral Cinnamates[a,b]

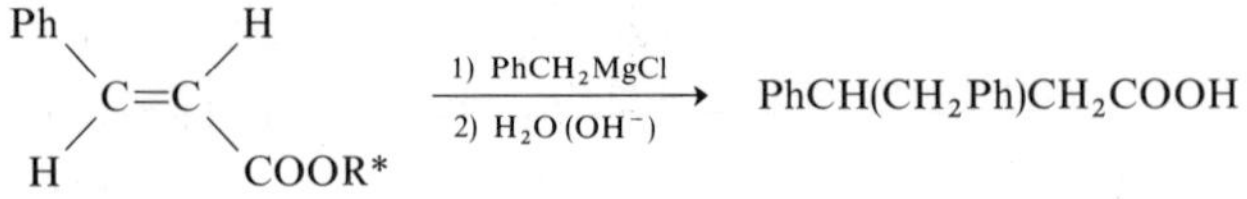

$$\underset{\textbf{170B}}{\text{Ph(H)C=C(H)COOR*}} \xrightarrow[\text{2) } H_2O\,(OH^-)]{\text{1) } PhCH_2MgCl} \underset{\textbf{171B}}{PhCH(CH_2Ph)CH_2COOH}$$

No.	R*	Catalyst	Synthetic yield %	Asymmetric synthesis[b] % e.e.	$[\alpha]_D$
5A	(−)-Menthyl	None	59	32	−19°
5B	(−)-Menthyl	Cu_2Cl_2	61	27	−16°
6A	**159**[c,d]	None	62	18	−11°
6B	**159**[c,d]	Cu_2Cl_2	58	17	−10°
7A	**158**[c]	None	49	18	−11°
7B	**158**[c]	Cu_2Cl_2	40	22	−13°
8A	**160**[c]	None	53	18	−11°
8B	**160**[c]	Cu_2Cl_2	56	17	−10°

[a] Data taken from Reference 80.
[b] The absolute configuration of **171B** has not been determined; $[\alpha]_D$ max −60° (benzene) was determined by resolution of partially active acid with quinine.
[c] Structural formulas for these groups are given in the text.
[d] THF solvent.

(**172**) represents a process in which two molecules of the reagent are involved per molecule of ester; the second represents a cyclic concerted process involving only one molecule of substrate and reagent. It is assumed that the dominant mechanism depends upon the specific Grignard and the relative amount of substrate and reagent. If the mechanism is as represented by **172**, then the stereochemical result of phenyl Grignard addition to (−)-menthyl crotonate will depend upon the conformation of the R* group in the transition state. If for this transition state one assumes model† **174**, in

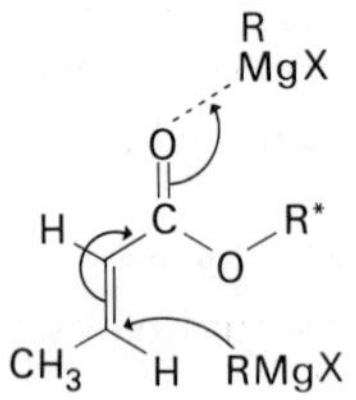

R MgX O C H R* O CH3 H RMgX

172

X Mg O R* R H C O CH3 H

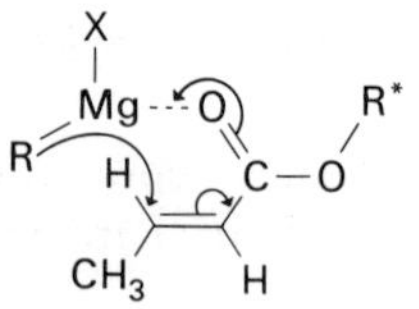

173

† Models **174** and **175** are modifications of those proposed in Reference 80 and are consistent with model **161** (Sec. 6-5) which was used to correlate the stereochemistry of the methoxy mercuration of chiral α,β-unsaturated esters.

which R_S, R_M, and R_L have the chiral distribution found in the (−)-menthyl group and the assumed conformation is analogous to that in the Prelog atrolactate asymmetric synthesis model, then *S*-(+)-**171A** will be the predicted

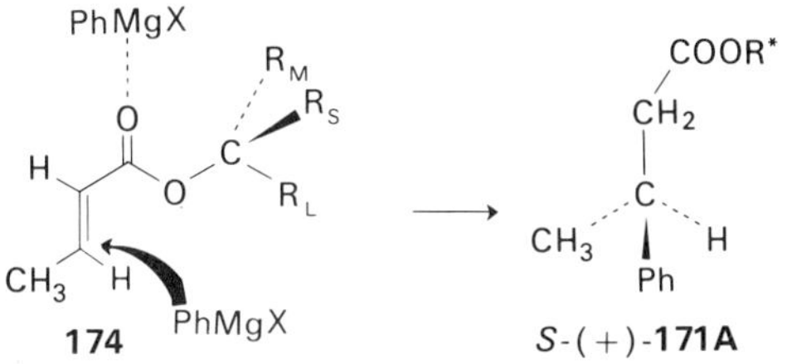

174 → *S*-(+)-**171A**

predominant isomer as observed. Alternatively when cuprous chloride is present, phenyl copper is presumed to be formed and to react with the substrate to give a complex with the double bond from the less hindered side so that subsequent nucleophilic attack by the phenyl Grignard reagent (or another phenyl copper) takes place from the opposite side, as in model **175** to give *R*-(−)-**171A**. Thus the stereochemical course is the reverse of that predicted by model **174**, as observed. Since neither the synthetic yields[81] nor

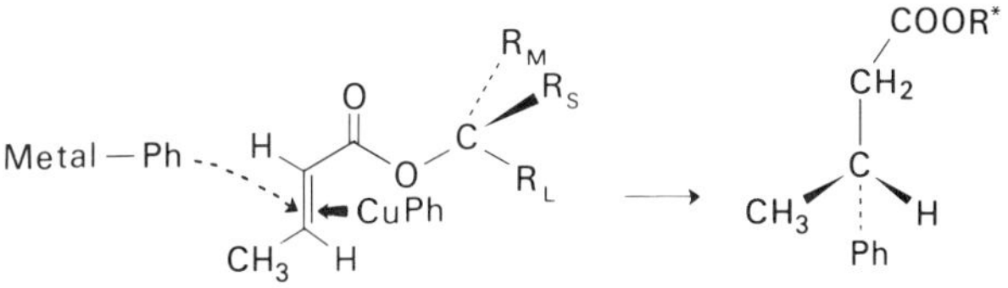

175 → *R*-(−)-**171A**

the stereoselectivities[80] (Table **6-14**) of the reactions of the benzyl Grignard reagent with cinnamate esters are significantly affected by the addition of cuprous chloride, it is concluded that benzyl copper intermediates are not involved under these circumstances.

The fact that the three chiral esters of **175** from the sugar derivatives **158**, **159**, and **160** give higher asymmetric induction than (−)-menthyl esters can be attributed in a general way to the greater conformational bias that should be possible due to chelation of the metalloorganic reagent with the several oxygen atoms. But the multiple possibilities for coordination of the metallic salts and/or metalloorganic reagent render any attempt to rationalize the results from these asymmetric syntheses much too speculative.

Reduction of the carbon-carbon double bond in ester **176** with sodium borohydride proceeds stereoselectively to yield (+)-lupinine (**179**, 10% e.e.) after reduction of the intermediate saturated ester **177** with lithium aluminum hydride.[82] Deuterium was introduced at C-10 when $NaBD_4$ was used.

[82] S. I. Goldberg and I. Ragade, *J. Org. Chem.*, **32**, 1046 (1967).

Protonation of the enolate intermediate gave only **177** and none of the C-1 epimer since no epilupinine was detected upon LAH reduction. Thus formation of the first chiral center at C-10 under the influence of the menthyl ester group was about 10% stereoselective but formation of the second chiral

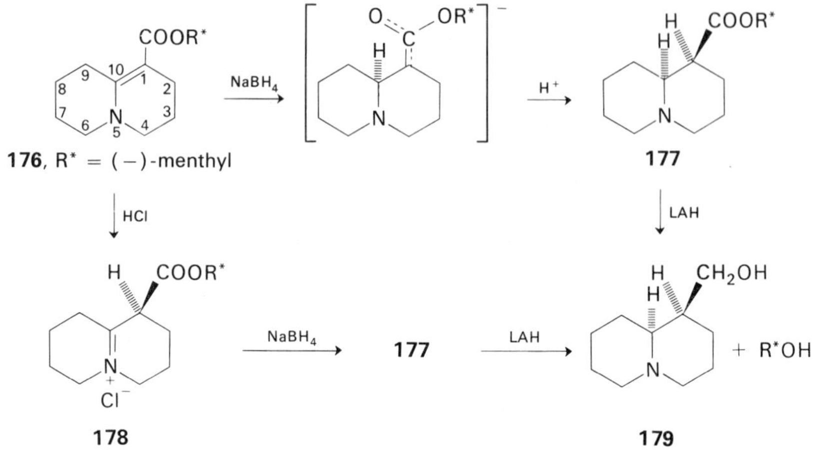

176, R* = (−)-menthyl

177

178

179

center at C-1 was completely stereoselective. Protonation of **176** at C-1 to give the immonium salt **178** was found to be slightly less stereoselective than hydride transfer to C-10 to produce **177**, but the direction of attack was the same in both instances. This was demonstrated by conversion of **178** to (+)-lupinine (6.8% e.e.) by successive reductions with sodium borohydride and lithium aluminum hydride. Again, no epilupinine was found (see Sec. **5-3.4** for other examples of immonium ion asymmetric reductions).

6-6.2. Uncatalyzed Additions of Chiral Alcohols and Amines to Ketenes. Ketenes react rapidly with alcohols and 1° or 2° amines to form esters or amides. Two types of asymmetric ketene additions have been studied; uncatalyzed addition of a chiral alcohol or amine and addition of an achiral alcohol in the presence of an optically active 3° amine catalyst.

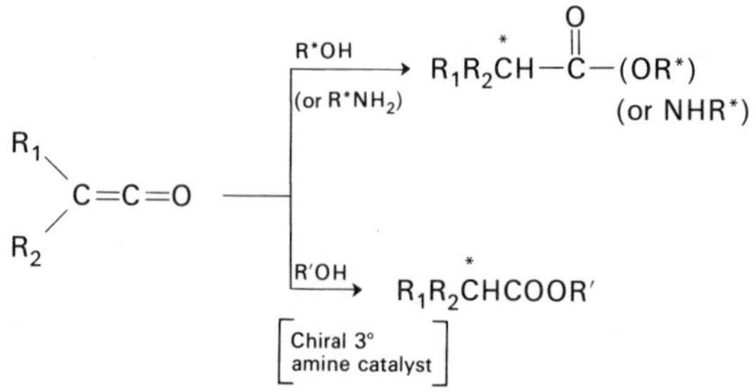

$$R_1R_2C{=}C{=}O \xrightarrow[\text{(or } R^*NH_2)]{R^*OH} R_1R_2\overset{*}{C}H{-}\overset{O}{\overset{\|}{C}}{-}(OR^*) \text{ (or } NHR^*)$$

$$R_1R_2C{=}C{=}O \xrightarrow[\text{[Chiral 3° amine catalyst]}]{R'OH} R_1R_2\overset{*}{C}HCOOR'$$

Some examples of the first type of asymmetric synthesis are discussed in this section while other similar, uncatalyzed reactions relating to amino acid syntheses are treated in Chapter 7.

Weiss[83] treated phenyl-*p*-tolylketene with (−)-menthol and obtained levorotatory menthyl phenyl-*p*-tolylacetate which he claimed was almost entirely one diastereomer. Later work showed that methylethylketene under the same conditions gave a 50:50 mixture of diastereomers[84] and that Weiss probably had also obtained, within experimental limits, equal amounts of the two possible diastereomers. Four reviews[85] propagated the erroneous conclusion that an asymmetric synthesis had been accomplished by claiming that Weiss obtained optically active phenyl-*p*-tolylacetic acid. While it was true that he claimed to have isolated mainly one diastereomeric ester, Weiss actually reported that the acid produced by saponification of this ester was racemic! As McKenzie has pointed out,[84a] even if the ester had been diastereomerically pure, it probably would have epimerized under the saponification conditions[86] to give racemic acid.

A very detailed and important study of the reaction of phenylmethylketene (**180**) with (*S*)-(−)-α-methylbenzylamine to give the diastereomeric amides *S,S*-**181** and *R,S*-**181** (Fig. **6-28**) has produced some extremely interesting

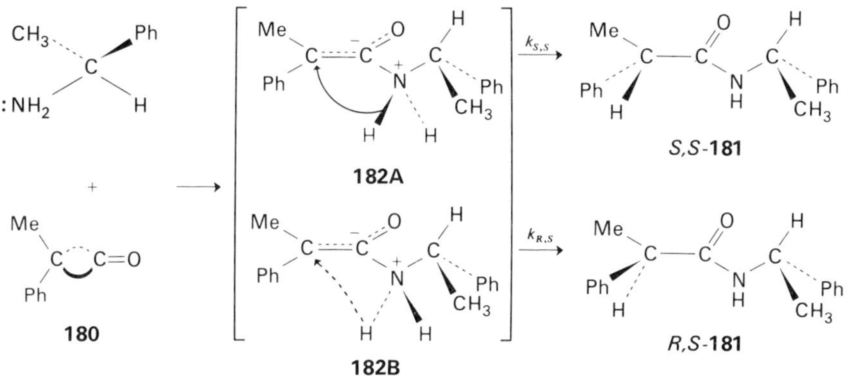

Fig. 6-28. Asymmetric reaction of a ketene with chiral amine.

[83] R. Weiss, *Monatsh. Chem.*, **40**, 391 (1919).

[84] (a) A. McKenzie and E. W. Christie, *J. Chem. Soc.*, 1070 (1934). (b) P. D. Ritchie, *Asymmetric Synthesis and Asymmetric Induction* (London: Oxford Univ. Press, 1933), p. 25.

[85] (a) R. Robinson, *Ann. Reports, Chem. Soc. London*, **17**, 74 (1920). (b) G. Wittig, *Stereochemie* (Leipzig: Akademische Verlagsgesellschaft M.B.H., 1930), p. 40. (c) S. Goldschmidt, *Stereochemie*, Vol. 4 of *Hand und Jahrbuch der Chemische Physik* (Leipzig: Akademische Verlagsgesellschaft M.B.H., 1933), p. 29. (d) F. Ebel in *Stereochemie*, ed. K. Freudenberg (Leipzig and Vienna: Franz Deuticke, 1933), p. 582.

[86] A. McKenzie and S. T. Widdows, *J. Chem. Soc.*, **107**, 702 (1915).

data and interpretations.[87] The stereochemically determining step must be the addition of the hydrogen to the α-carbon of the ketene moiety and the asymmetric bias must be determined by the chirality of the nitrogen substituent. There are several mechanistic possibilities. If the reaction is in essence a four-centered attack on the carbon-carbon double bond, then the stereochemistry is determined by the difference in energy of activation for approach of the reagent from the *re* versus the *si* face of the double bond; i.e., from either the left or the right, as shown in **182C** and **182D**. But if the attack is stepwise

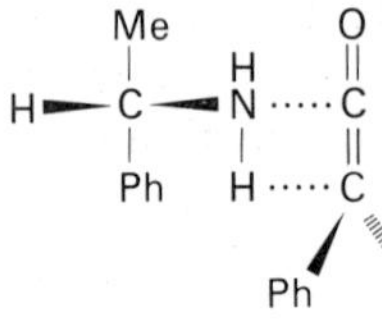

182C 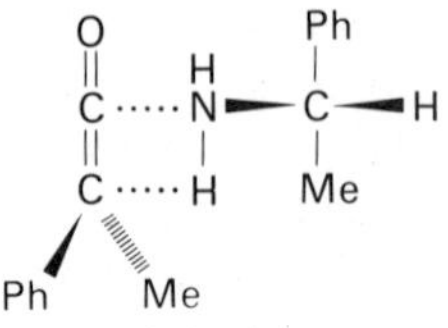 **182D**

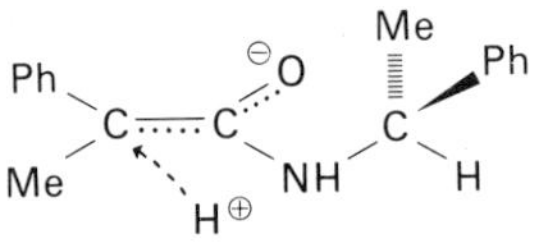

182E 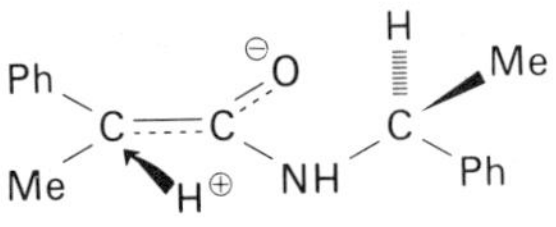 **182F**

to given an intermediate such as shown in **182A** and **182B**, in which the anion is stabilized by resonance, then the initial reagent approach is not the stereochemically determining step but it is the subsequent hydrogen transfer either as shown in Fig. **6-28** or by transfer of the proton of the ion pair as shown in **182E** and **182F**. Which product will be favored must depend upon the conformation of the chiral group attached to nitrogen and, in **182A**–**182B** or **182E**–**182F**, it also depends upon whether the phenyl or methyl group of the ketene moiety is *cis* to the oxygen in the stabilized anion. There are three conformations in **182** in which the groups on the chiral center are staggered with respect to the carbonyl; one of these has the more bulky phenyl group in front, one in back, and one in the plane of the page. There are also three conformations in which the groups on the chiral center are eclipsed with respect to the carbonyl; again one of these has the phenyl group on the back side, one on the front, and one staggered. Furthermore there are two opposing ways to view the hydrogen transfer in the concerted (**182A** and **182B**) or ion pair (**182E** and **182F**) processes; one can consider the hydrogen will approach from the less hindered side but one must also consider the possibility that crowding by the larger group will assist the hydrogen from the more bulky side to be transferred more readily. Pracejus

[87] (a) H. Pracejus, *Ann. Chem.*, **634**, 23 (1960). (b) H. Pracejus and A. Tille, *Chem. Ber.*, **96**, 854 (1963). Interpretations presented earlier[87a] are modified in this paper.

chose, as a working hypothesis, model **182E** for the favored conformation of the low temperature process. For the high temperature process he proposed the model represented by **182F** in which the chiral group has been rotated 120° so that the methyl and hydrogen are staggered with respect to the carbonyl group. The rationale for choosing one conformation over another and especially for postulating different favored conformations at different temperatures is obscure.

As shown in Fig. **6-29**, a plot of log $k_{S,S}/k_{R,S}$ (log Q, where Q is the diastereomer ratio) versus $1/T$ was linear with a positive slope below about 25° in toluene solvent, but it deviated from ideal behavior above this temperature. The reaction in methylcyclohexane, *p*-xylene, and mesitylene solvents showed similar behavior over the portion of this temperature range that was accessible within the melting point limits of these solvents. At about −105° in toluene a 60% excess of *S,S*-**181** was obtained. The percent asymmetric synthesis dropped to nearly 0 at +100°. Above this temperature a slight excess of *R,S*-**181** was obtained. Diethyl ether, dipropyl ether, ethyl acetate, and chloroform behaved much like the hydrocarbon solvents, but log Q versus $1/T$ plots had less positive slopes. Such a plot for dipropyl ether, which

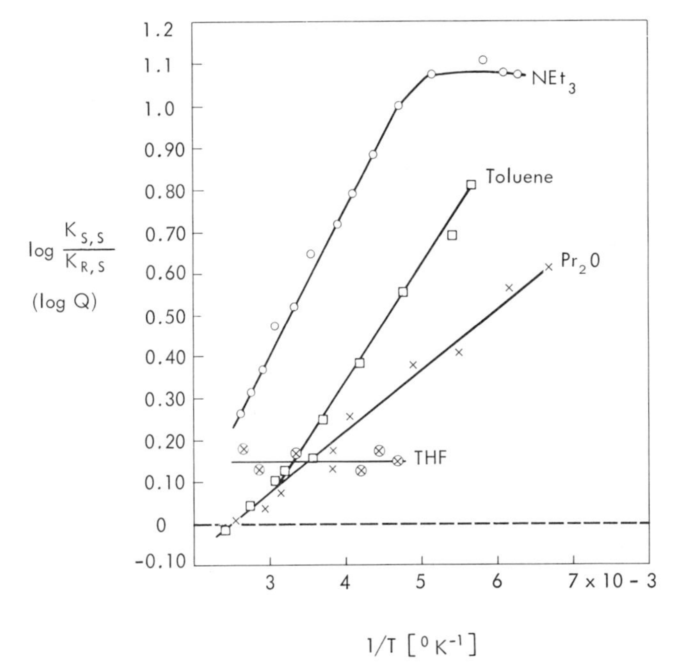

Fig. 6-29. Plots of log diastereomer ratio (log Q = log $k_{S,S}/k_{R,S}$) versus $1/T$ for reaction of phenyl methyl ketene with S-(−)-α-methylbenzylamine in various solvents.

gives the most positive slope of the three oxygenated solvents mentioned above, is shown in Fig. **6-29**. In tetrahydrofuran and N,N-dimethylformamide, little sensitivity of the asymmetric bias to reaction temperature was found. Triethylamine gave the most positive slope of any solvent and produced the highest asymmetric bias (84–85% e.e.) below −65°, the temperature at which the plot of log Q versus $1/T$ becomes "nonideal."

Studies[87] show that there is a regular decrease in $\Delta\Delta H^{\ddagger}$ as the solvating ability or basicity of the solvent increases. When $\Delta\Delta H^{\ddagger}$ in a series of solvents is plotted against the H—Cl stretching frequency in the same solvents (a measure of solvent basicity), an almost linear relationship is observed. These studies also reveal that $\Delta\Delta S^{\ddagger}$ has the same sign as $\Delta\Delta H^{\ddagger}$ in the lower basicity solvents and that it decreases nearly in proportion to the decrease in $\Delta\Delta H^{\ddagger}$, eventually changing sign in the most basic solvents. The approximate relationship between $\Delta\Delta H^{\ddagger}$ and $\Delta\Delta S^{\ddagger}$ can be expressed by the equation $\Delta\Delta H^{\ddagger} = 150 + \beta'\Delta\Delta S^{\ddagger}$. The parameter β' has the dimensions of temperature and is related to that temperature at which log Q versus $1/T$ plots should intersect (about 20°). Since at this temperature the stereospecificity in all solvents (except triethylamine) is practically the same, β' has been called the *isospecificity temperature*. A hypothesis concerning the origin of the opposing effect of $\Delta\Delta H^{\ddagger}$ and $\Delta\Delta S^{\ddagger}$ has been suggested.[88]

Ketene **183** also reacts with *S*-(−)-α-methylbenzylamine to give an amide enriched in one diastereomer of **184**,[88] but the stereoselectivities are not as high as those with phenylmethylketene (**180**). Plots of log Q versus $1/T$ for

183 =C=O + PhC*H(NH$_2$)CH$_3$ ⟶ **184** H, *–CONHC*HCH$_3$Ph

a variety of solvents showed that the percent asymmetric synthesis is a much more complex function of temperature and solvent characteristics in these cases. Perhaps atropisomeric forms of an intermediate analogous to **182**, due to rotation of the aryl group of the ketene portion, introduce an additional asymmetric influence which is a complex function of solvent and temperature.

The asymmetric addition of achiral alcohols to ketenes may also be accomplished if an optically active base is used to catalyze the reaction.[88] For example, it has been observed that the addition of methanol to phenylmethylketene in the presence of brucine yields optically active ethyl 2-phenyl-

[88] (a) H. Pracejus, *Ann.*, **634**, 9, (1960). (b) H. Pracejus and H. Maetje, *J. prakt. Chem.*, **24**, 195 (1964). (c) H. Pracejus, *Tetrahedron Lett.*, 3809 (1966). (d) H. Pracejus, *Fortsch. Chem. Forsch.*, **8**, 493–553 (1967).

propionate.[88a] The direction and extent of asymmetric synthesis depends strongly on the temperature, approximately a 25% excess of *S*-(+) ester being produced at −110°, whereas a 10% excess of *R*-(−) ester was produced at +80°.

Similarly other alcohol additions to ketenes **180** or **183** catalyzed by chiral bases such as strychnine, brucine, quinine, dihydroquinine, cinchonidine, quinidine, and cinchonine (often as the acetyl derivative) frequently gave enantiomeric products when carried out at widely different temperatures.[88b] Pracejus postulates[88b,88d] that there is an antagonism of enthalpy and entropy effects on the stability of the various diastereomeric transition states which can be envisaged for the reaction. The mechanism is thought to involve a slow formation of an ion pair **185**, comparable to **182E** and **182F** within which there is rapid transfer of a proton to the incipient asymmetric center of the ester.

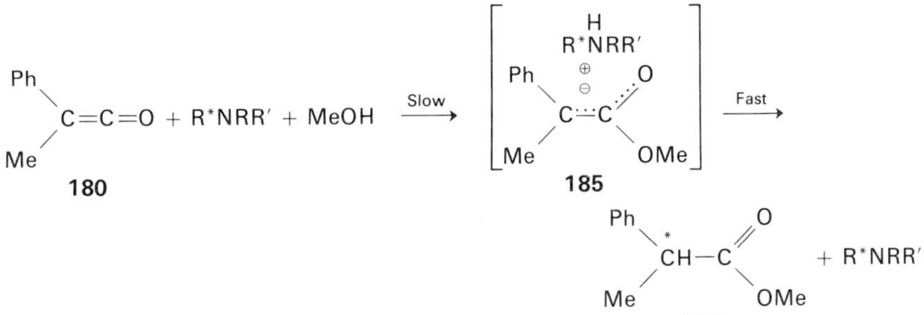

Ph(Me)C=C=O + R*NRR′ + MeOH —Slow→ [Ph(Me)C⊖⋯C(=O)OMe with H–R*NRR′⊕] —Fast→ Ph(Me)C*H–C(=O)OMe + R*NRR′

180 **185** **186**

The cation may transfer its proton to one face of **185** giving the *R*-(−) enantiomer of **186** or to the opposite face giving *S*-(+)-**186**.[88d] The conformation that is favored at high temperatures may not be the one favored at low temperatures because of entropy factors.[88d] With bases having a bridgehead nitrogen, the reaction is extremely sensitive to size differences between the groups around nitrogen. This is reflected in the high percent asymmetric synthesis achieved with **180** and benzoylquinine at −110° [76% e.e., *R*-(−)-**186**]. Even more dramatic is the observation that with the chiral bicyclic amine 1*S*,4*R*-**187** as catalyst there was obtained a slight excess of *S*-(+)-**186** (0.13% e.e.).[88c] Apparently even the small steric difference between CD_2 and CH_2 is sufficient to cause a detectable energy difference between diastereomeric conformations of **185**.

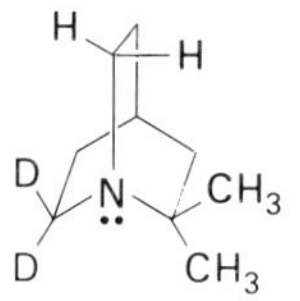

H H D D N CH₃ CH₃

187

It is possible that the asymmetric reaction[89] of the acid chloride **188** with trichloroacetaldehyde in the presence of brucine to give the β-lactone **189** proceeds via a ketene intermediate,[89] although one cannot exclude an aldol-type addition followed by lactone formation. It is clear however that the optically active base is intimately involved in the mechanism. The configuration and enantiomeric purity of **189** were not determined.

$$\text{188} + CCl_3CHO \xrightarrow{\text{Brucine}} \text{189},\ [\alpha]_D\ -70.6°$$

188 **189**, $[\alpha]_D$ −70.6°

6-6.3. Reduction of Enynols to Optically Active Allenes with Chiral Lithium Aluminum Hydride Complexes. Enynols have been reduced to allenes by a chiral reagent made by allowing LAH to react with a limited amount of a chiral alcohol such as (−)-menthol (R*OH) and various sugar derivatives. This and related reactions are discussed in Sec. **9-2.2** and summarized in Table **9-1** where the reduction of carbonyl compounds by the same reagent is considered.

6-6.4. Hydroxylation of Chiral α,β-Unsaturated Esters. Some years ago McKenzie and Wren[90] reported the asymmetric synthesis of tartaric acid by hydroxylation of chiral esters or half-esters of fumaric acid with potassium permanganate. Half-esters gave tartaric acid solutions having higher rotations than those from the hydroxylation of diesters, but in both cases the observed rotations were small. Since the rotations were taken on tartaric acid or potassium tartrate solutions of unknown concentrations, the percent asymmetric synthesis cannot be determined. However, making reasonable assumptions concerning percent yields and

$$\text{190} + MnO_4^- \longrightarrow [\text{191}] \xrightarrow[H^+]{HOH} \text{tartaric acid} + R^*OH$$

190A, R,R* = (−)-bornyl
190B, R,R* = (−)-menthyl
190C, R = H, R* = (−)-bornyl
190D, R = H, R* = (−)-menthyl

191

[89] D. Borrmann and R. Wegler, *Chem. Ber.*, **99**, 1245 (1966).
[90] A. McKenzie and H. Wren, *J. Chem. Soc.*, **91**, 1215 (1907).

concentrations, one can crudely estimate stereoselectivities in the range of 2–20%. Levorotatory product was obtained from (−)-menthyl and (−)-bornyl diesters, whereas (+)- and (−)-bornyl hydrogen fumarates gave dextro- and levorotatory products, respectively. Formation of active rather than *meso* tartaric acid is consistent with the mechanism shown (**191**).[91] There are certain *formal* similarities between the formation of the nonisolable intermediate **191** and the corresponding Diels–Alder adduct (Sec. **6-3.2**), but more data are required before meaningful stereochemical comparisons between the two reactions can be made.

6-6.5. Addition of Ammonia and Amines. Some asymmetric amino acid syntheses involve the addition of a chiral amine to the carbon-carbon double bond of an achiral α,β-unsaturated acid[92,93] or the addition of an achiral amine to a chiral derivative of an α,β-unsaturated acid.[94,95] These are discussed (Sec. **7-3**) along with other asymmetric syntheses of amino acids.

A related reaction is that of ammonia with D-*arabo*-3,4,5,6-tetraacetoxy-1-nitro-1-hexene (**192**). During the reaction, an acetyl group migrates from oxygen to nitrogen. Rearrangement may take place from C-3 as shown in **193** or from C-4 via an analogous six-membered intermediate not illustrated. Only one diastereomer, 2-acetamido-1,2-dideoxy-1-nitro-D-mannitol (**194**), was found originally[96] but it has been subsequently shown[97] that the product was a 6:1 mixture of **194** and its C-2 epimer. The configuration of **194** at C-2 was found by converting the *aci*-nitro salt to D-mannosamine hydrochloride (**195**) via a Nef reaction and subsequent acetylation and acidification[96] (Fig. **6-30**). Recent advances in the synthesis of starting materials[98] and interest in the biological role of amino sugars[99] should promote related asymmetric syntheses of other amino sugars from unsaturated nitro sugars.

The addition of ammonia to D-arabo-3,4,5,6-tetraacetoxy-1,1-bis-(ethanesulfonyl)-1-hexene (**196**) gives an 84% yield of the D-glucosamine derivative

[91] K. B. Wiberg and K. A. Saegebarth, *J. Am. Chem. Soc.*, **79**, 2822 (1957).

[92] A. P. Terent'ev, R. A. Gracheva, L. F. Titova, and T. F. Dedenko, *Dokl. Akad. Nauk SSSR.*, **154**, 1406 (1964); *Chem. Abst.*, **60**, 12099 (1964).

[93] A. P. Terent'ev, R. A. Gracheva, and T. F. Dedenko, *Dokl. Akad. Nauk SSSR.*, **163**, 386 (1965); *Chem. Abst.*, **63**, 11344 (1965).

[94] K. Harada and K. Matsumoto, *J. Org. Chem.*, **31**, 2985 (1966).

[95] Y. Liwschitz and A. Singerman, *J. Chem. Soc.* (C), 1200 (1966).

[96] A. N. O'Neill, *Can. J. Chem.*, **37**, 1747 (1959).

[97] J. C. Sowden and M. L. Oftedahl, *J. Am. Chem. Soc.*, **82**, 2303 (1960).

[98] W. A. Szarek, D. G. Lance, and R. L. Beach, *Chem. Commun.*, 356 (1968).

[99] D-Mannosamine occurs naturally as a component of N-acetylneuraminic acid [D. G. Comb and S. Roseman, *J. Am. Chem. Soc.*, **80**, 497, 3166 (1958)]. Other amino sugars are structural units of antibiotics [K. L. Rinehart, Jr., *The Neomycins and Related Antibiotics* (New York: John Wiley & Sons, Inc., 1964)] and bacterial polysaccharides [N. Sharon and R. W. Jeanloz, *J. Biol. Chem.*, **235**, 1 (1960)].

197 (none of the D-mannosamine epimer was found).[100] Thus even though the chiral portion of the molecule is identical to that present in **192**, an

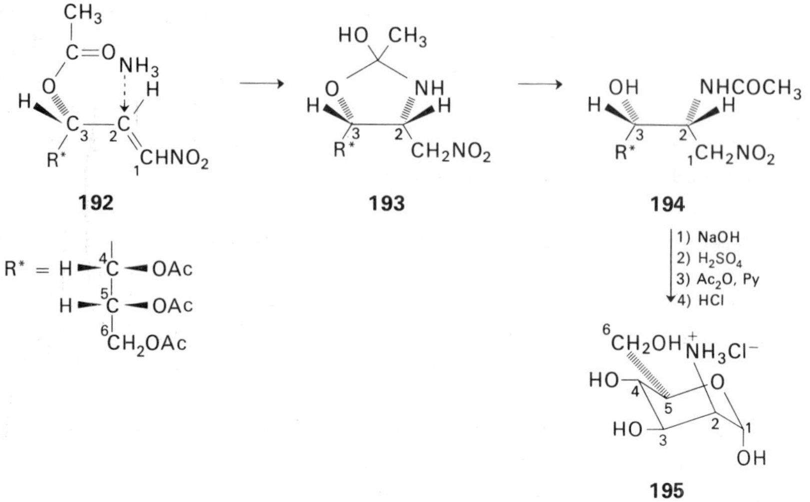

Fig. 6-30. Asymmetric synthesis of D-mannosamine (**195**).[96,97]

opposite chirality is induced at C-2 in the ammonia adduct. Additional studies of this type of asymmetric synthesis may be stimulated by a more

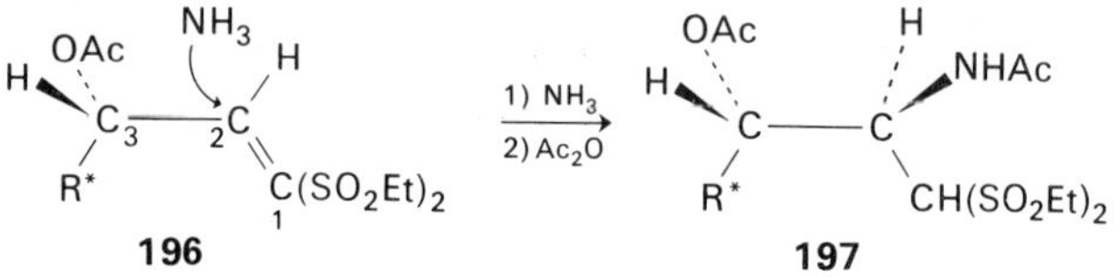

recently developed synthesis of 1,1-bis-(alkylsulfonyl)-1-alkenes.[101]

6-6.6. Asymmetric Addition of Bromine. Addition of bromine to chiral alkenes, such as **198**, takes place in a *trans* manner to produce optically active dibromo alcohols (**199**) which upon oxidation give optically active dibromo ketones (**200**). The diastereomers of **199A** and **199B** have been separated by fractional crystallization and individually oxidized to enantiomerically pure ketones.[102] Bromination of (−)-**198A** at −15° gave the highest stereoselectivity (41% e.e.), whereas at 60° the sign of rotation of the product was reversed (5% e.e.). Convincing but not conclusive evidence was

[100] D. L. MacDonald and H. O. L. Fischer, *J. Am. Chem. Soc.*, **74**, 2087 (1952).
[101] M. L. Oftedahl, J. W. Baker, and M. W. Dietrich, *J. Org. Chem.*, **30**, 296 (1965).
[102] J. Kenyon and S. M. Partridge, *J. Chem. Soc.*, 1313 (1936).

presented to show that there was no thermodynamic equilibration of isomers during the higher temperature reactions.

Bromination of (+)-**198B** at 15° gave a nearly quantitative yield of **199B** which was separated into a (+)-diastereomer (m.p. 114–115°, 60%) and a (−)-diastereomer (m.p. 93–94°, 40%).[103] Bromination of (−)-**198B** gave

$$\underset{\mathbf{198}}{\text{RCH=CH}\overset{*}{\text{C}}\text{HOHR}'} \xrightarrow{Br_2} \underset{\mathbf{199}}{\text{R}\overset{*}{\text{C}}\text{HBr}\overset{*}{\text{C}}\text{HBr}\overset{*}{\text{C}}\text{HOHR}'} \xrightarrow{CrO_3} \underset{\mathbf{200}}{\text{R}\overset{*}{\text{C}}\text{HBr}\overset{*}{\text{C}}\text{HBrCOR}'}$$

	R	*R'*
A	Ph	Me
B	Ph	Et
C	H	Et
D	Me	Me
E	H	Ph

200A (R = Ph, R' = Me)

Temp.	*% e.e.*
−15°	41(−)
−6°	34(−)
+18°	19(−)
+60°	5(+)
+75°	7(+)

Fig. 6-31. Asymmetric additions to chiral alkenes.

the same mixture of diastereomers with the relative yields reversed. Chromic anhydride oxidation of (−)-**199B** produced (−)-**200B** in about 90% yield, while (+)-**199B** gave (+)-**200B**, an observation consistent with the assumption that addition of bromine to **198B** is *trans* and that there is no isomerization during the reaction sequence. The enantiomeric excess of **200B** obtained from (+)-**198B** (without separation of the diastereomers of **199B**) increased as the temperature at which the bromination was conducted decreased (19% e.e. at 22° versus 32% e.e. at −10°).

$$(+)\text{-}\mathbf{198B} \xrightarrow{Br_2} \begin{cases} 40\%\ (+)\text{-}\mathbf{199B} \xrightarrow{CrO_3} (+)\text{-}\mathbf{200B} \xleftarrow{CrO_3} (+)\text{-}\mathbf{199B}\ 60\% \\ 60\%\ (-)\text{-}\mathbf{199B} \xrightarrow{CrO_3} (-)\text{-}\mathbf{200B} \xleftarrow{CrO_3} (-)\text{-}\mathbf{199B}\ 40\% \end{cases} \longleftarrow (-)\text{-}\mathbf{198B}$$

Bromination of chiral alkene **198C** (Fig. **6-31**) followed by oxidation with CrO_3 gave optically active **200C**.[104] This result shows that asymmetric bromination resulting in the formation of a single new asymmetric carbon in the ketone can be observed in certain cases. Bromination of alkene **198E**, however, leads to inactive ketone.[104] In this case there is a phenyl group *alpha* to the carbonyl and the chances for racemization via an enol form during acid catalyzed oxidation are greatly enhanced.

[103] M. P. Balfe, J. Kenyon, and D. Y. Waddan, *J. Chem. Soc.*, 1366 (1954).
[104] S. M. Partridge, Thesis, London (1936); quoted in Reference 103.

Brominations of menthyl cinnamate,[105] dicinnamyl tartrate, and salts from cinnamic acid and several amines[106] have been reported. In the first two cases, dibromocinnamic acid and dibromocinnamyl alcohol could not be obtained pure due to dehydrobromination during hydrolysis of the dibromoesters. Cinnamates of (+)-dihydrocinchonine and 2-aminoglucose were converted to optically active dibromocinnamic acid [2–16% excess of (+)-enantiomer]; due to the fact that several crystallizations were involved, one cannot be certain that these represent true asymmetric syntheses.

6-6.7. Attempted Prévost-Type Asymmetric Synthesis with a Chiral Electrophile. A mixture of bromine and the silver salt of optically active 2-ethylhexanoate (**201**) in carbon tetrachloride reacts with styrene to yield optically active 2-bromo-1-phenylethyl 2-ethylhexanoate (**203**). Saponification of the ester produces inactive 1-phenyl-1,2-ethanediol (**205**) and partially racemized 2-ethylhexanoic acid (**204**).[107] It was concluded that

$$\underset{\mathbf{201}}{\mathrm{BuEt\overset{*}{C}HCOOAg}} \xrightarrow{Br_2} \left[\mathrm{BuEt\overset{*}{C}H\overset{O}{\overset{\|}{C}}OBr}\right] \xrightarrow{PhCH=CH_2} \underset{\mathbf{202}}{\left[\mathrm{H_2C\overset{Br}{\diagup\diagdown}CHPh}\right]^+} \;\; [\mathrm{BuEt\overset{*}{C}HCOO}]^- \longrightarrow$$

$$\underset{\mathbf{203}}{\mathrm{BuEt\overset{*}{C}H\overset{O}{\overset{\|}{C}}-O-\overset{CH_2Br}{\overset{|}{C}}HPh}} \xrightarrow{OH^-} \underset{\mathbf{204}}{\mathrm{BuEt\overset{*}{C}HCOOH}} + \underset{\mathbf{205}}{\mathrm{PhCH(OH)CH_2OH}}$$

the isolated ester **203** was an equal mixture of the two possible epimers, that no racemization of the alcohol moiety occurred during saponification, and that the formation of racemic glycol implied no stereoselectivity in the formation of bromonium ion (**202**).

More definitive experiments of this type can be envisaged using chiral Simonini complexes[108] from reaction of the silver salt of an optically active acid and iodine (preferably from acids having an aryl group on the asymmetric carbon). Hydroxylation via either a net *cis* or *trans* mechanism could be accomplished by modifying the reaction conditions.[109]

6-6.8. Asymmetric Electrophilic Addition to an Enamine. The enamine formed from cyclohexanone (**206**) and (−)-isobornylamine (**207**) was allowed to react successively with isopropyl Grignard reagent (to

[105] J. B. Cohen and C. E. Whiteley, *J. Chem. Soc.*, **79**, 1305 (1901).
[106] H. Erlenmeyer, *Helv. Chim. Acta*, **13**, 731 (1930).
[107] D. C. Abbott and C. L. Arcus, *J. Chem. Soc.*, 1515 (1952).
[108] C. V. Wilson, *Organic Reactions*, **9**, 332 (1957).
[109] H. O. House, *Modern Synthetic Reactions* (New York: W. A. Benjamin, Inc., 1965), pp. 141–143.

form the N-bromomagnesium salt) and methyl iodide.[110] After hydrolysis, optically active 2-methylcyclohexanone [72% e.e., (*S*)-(+)-**208**] was obtained.

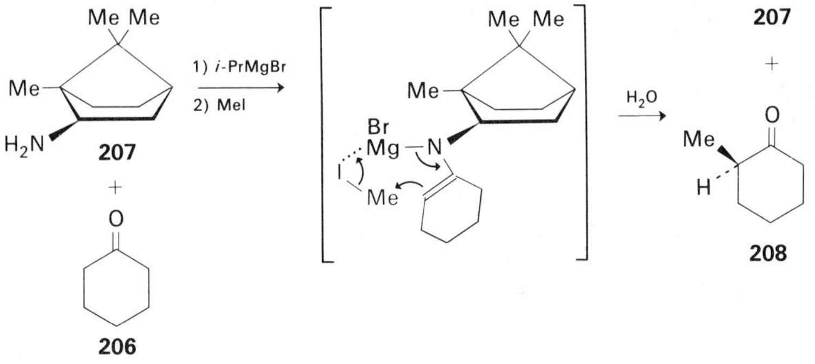

Me Me; Me; H_2N; **207**; + **206**; 1) *i*-PrMgBr 2) MeI; Br Mg N Me; H_2O; **207** + Me H **208**

6-6.9. Asymmetric Addition of Olefins to Dienophiles. Olefins **209A** and **209B** undergo a thermal addition reaction with maleic anhydride to yield optically active **211** (Fig. **6-32**).[111] Although olefins **209A**

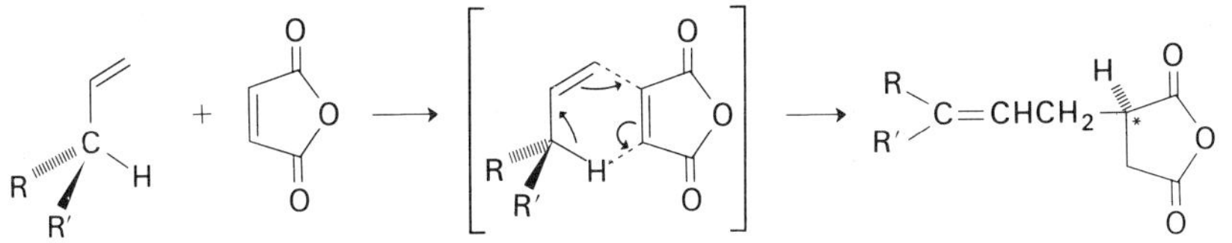

R-(−)-**209A**, R = Ph, R′ = Me
R-(−)-**209B**, R = Me, R′ = *i*-hexyl **210**

R, R′ C=CHCH$_2$ H O O O

211A, $[\alpha]_D^{20}$ −14.5°
211B, $[\alpha]_D^{20}$ −1.90°

Fig. 6-32. Asymmetric addition of chiral olefins to maleic anhydride.

and **209B** both have the *R*-(−) configuration, they in fact have opposite arrangements of R_S, R_M, and R_L about the chiral center if one considers *i*-hexyl larger than vinyl and vinyl larger than methyl. Although the enantiomeric purities of **211A** and **211B** are not known, analysis of optical rotatory dispersion (ORD) spectra of these compounds indicates that they have opposite configurations and suggests that **211A** is *R* and **211B** is *S*. Thus both olefins react in the same stereochemical sense. The reaction is believed to proceed via a cyclic transition state (e.g., **210**) in which there is concerted bond reorganization. A study of models suggests that only when the bulkier groups (phenyl or *i*-hexyl) are oriented away from the dienophile is the correct stereochemistry predicted.[111]

[110] D. Mea-Jacheet and A. Horeau, *Bull. Soc. Chim. Fr.*, 4571 (1968).
[111] R. K. Hill and M. Rabinovitz, *J. Am. Chem. Soc.*, **86**, 965 (1964).

6-7. Asymmetric Catalytic Hydrogenation

6-7.1. Introduction. The asymmetric additions of hydrogen to carbon-carbon double bonds may be divided into two main classes. The alkene may contain a chiral center which mediates the hydrogenation process so that one diastereotopic face of the alkene is preferentially saturated; alternatively, the hydrogen may be transferred from a catalyst that is chiral, with the result that enantiotopic faces of an achiral alkene are selectively saturated via diastereomeric transition states which incorporate the catalyst.

The most important advances in asymmetric hydrogenations have made use of soluble chiral catalysts; these are discussed first. Subsequently, selected examples of hetereogeneous-asymmetric hydrogenation are considered but no attempt is made to realize a comprehensive coverage of this vast area.

6-7.2. Asymmetric Homogeneous Hydrogenation with Chiral Rhodium Complexes. Although the activation of molecular hydrogen by transition metal complexes had been known for some time,[112–114] preliminary accounts of the use of soluble rhodium catalysts,[115–117] especially complexes of Rh(I) and triphenylphosphine,[117] so captivated the imagination of organic chemists that reports[118–120] on the utility of this system appeared virtually simultaneously with the first detailed study[121] of the structure of the catalyst and the mechanism of its catalytic action. Interest in $Rh(Ph_3P)_3Cl$, which is commercially available, and related soluble catalysts has continued unabated.[122] It was inevitable that contemporary interest in soluble rhodium-phosphine reduction catalysts, on the one hand, and in optically

[112] S. W. Weller and G. A. Mills, *Adv. in Catalysis*, **8**, 163 (1956).

[113] J. Halpern, *Adv. in Catalysis*, **11**, 301 (1959).

[114] M. F. Sloan, A. S. Matlack, and D. S. Breslow, *J. Amer. Chem. Soc.*, **85**, 4014 (1963) and references therein.

[115] R. D. Gillard, J. A. Osborn, P. B. Stockwell, and G. Wilkinson, *Proc. Chem. Soc.*, 284 (1964).

[116] J. A. Osborn, G. Wilkinson, and J. F. Young, *Chem. Commun.*, 17 (1965).

[117] J. F. Young, J. A. Osborn, F. H. Jardine, and G. Wilkinson, *Chem. Commun.*, 131 (1965).

[118] C. Djerassi and J. Gutzwiller, *J. Am. Chem. Soc.*, **88**, 4537 (1966).

[119] A. J. Birch and K. A. M. Walker, *J. Chem. Soc.*, (*C*), 1894 (1966).

[120] A. J. Birch and K. A. M. Walker, *Tetrahedron Lett.*, 4939 (1966).

[121] J. A. Osborn, F. H. Jardine, J. F. Young, and G. Wilkinson, *J. Chem. Soc.*, (*A*), 1711 (1966).

[122] (a) G. Wilkinson, *Bull. Soc. Chim. Fr.*, 5055 (1968). (b) A. S. Hussey and Y. Takeuchi, *J. Am. Chem. Soc.*, **91**, 672 (1969). (c) H. van Bekkum, F. van Rantwijk, and T. van de Putte, *Tetrahedron Lett.*, 1 (1969). (d) Y. Chevallier, R. Stern, and L. Sajus, *Tetrahedron Lett.*, 1197 (1969). (e) L. Horner, H. Büthe, and H. Siegel, *Tetrahedron Lett.*, 4023 (1968). (f) C. O'Connor and G. Wilkinson, *Tetrahedron Lett.*, 1375 (1969).

active *tert*-phosphines,[123,124] on the other hand, would lead to studies of soluble rhodium complexes of chiral phosphines as asymmetric hydrogenation catalysts.

Asymmetric hydrogenation of α-phenylacrylic acid (**212A**) and itaconic acid (**212B**) was achieved[125] with a catalyst prepared from trichlorotris-[(−)-methylpropylphenylphosphine]-rhodium. The exact nature of the catalytically active species is somewhat uncertain, although it is probable

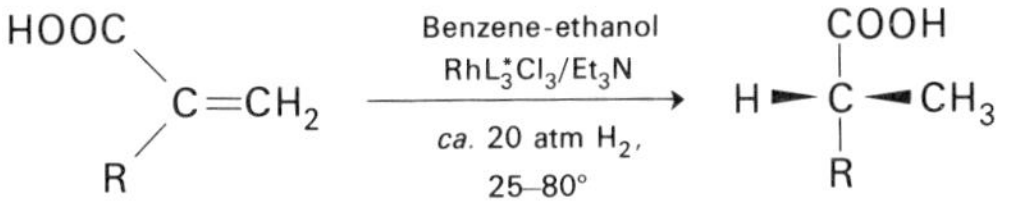

$$\text{HOOC(R)C{=}CH}_2 \xrightarrow[\textit{ca. } 20\text{ atm H}_2,\ 25\text{–}80^\circ]{\text{Benzene-ethanol, RhL}^*_3\text{Cl}_3/\text{Et}_3\text{N}} \text{H}\!-\!\text{C(COOH)(R)}\!-\!\text{CH}_3$$

212A, R = Ph
212B, R = CH_2COOH

213A, 15% e.e., *S*-(+)
213B, 3% e.e. (configuration unreported)

Fig. 6-33. Asymmetric homogeneous hydrogenation of α-phenylacrylic and itaconic acids. The chiral ligand L* is (*R*)-methylpropylphenylphosphine (69% e.e.).[125]

that Rh(II) or Rh(I) complexes are involved. No free rhodium is deposited during the hydrogenation in accord with expectation based on the recognized ability of π-acceptor ligands (such as phosphines) to stabilize the metal toward complete reduction.

In another study[126] two substituted styrenes (**214A** and **214B**) were asymmetrically hydrogenated with a Rh(I) catalyst prepared *in situ* from a Rh(I) μ-complex of 1,5-hexadiene and (*S*)-(+)-methylpropylphenylphosphine. In this case the structure of the catalyst is more certain. The mechanism of its action probably parallels that for $Rh(Ph_3P)_3Cl$ and can be summarized

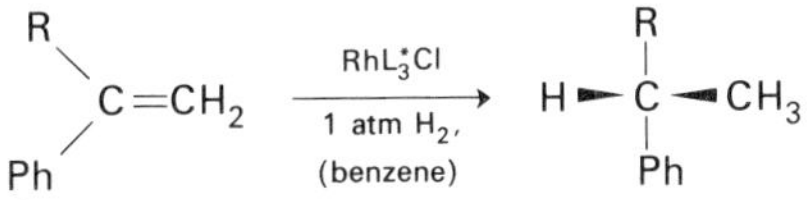

$$\text{R(Ph)C{=}CH}_2 \xrightarrow[\text{1 atm H}_2,\ \text{(benzene)}]{\text{RhL}^*_3\text{Cl}} \text{H}\!-\!\text{C(R)(Ph)}\!-\!\text{CH}_3$$

214A, R = Et
214B, R = OMe

215A, 7–8% e.e., *S*-(+)
215B, 3–4% e.e., *R*-(+)

Fig. 6-34. Asymmetric homogeneous hydrogenation of α-substituted styrenes; L* is (*S*)-methylpropylphenylphosphine.[126]

[123] O. Korpiun, R. A. Lewis, J. Chickos, and K. Mislow, *J. Am. Chem. Soc.*, **90**, 4842 (1968).
[124] T. L. Emmick and R. L. Letsinger, *J. Am. Chem. Soc.*, **90**, 3459 (1968).
[125] W. S. Knowles and M. J. Sabacky, *Chem. Commun.*, 1445 (1968).
[126] L. Horner, H. Siegel, and H. Büthe, *Angew. Chem.*, **80**, 1034 (1968); *Internatl. Edn.*, **7**, 942 (1968).

as follows. The square planar rhodium complex, **216**, dissociates one of the triphenylphosphine ligands and undergoes oxidative addition of hydrogen to give a coordinatively unsaturated Rh(III) intermediate which is shown in Fig. **6-35** as the solvent saturated species **217**. By solvent-alkene interchange, **217** yields a new complex (**218**) which transfers hydrogen from rhodium to carbon, thus producing the saturated compound (**219**) and regenerating the square planar complex (**220**) which can repeat the hydrogenation cycle via **217** and **218**.

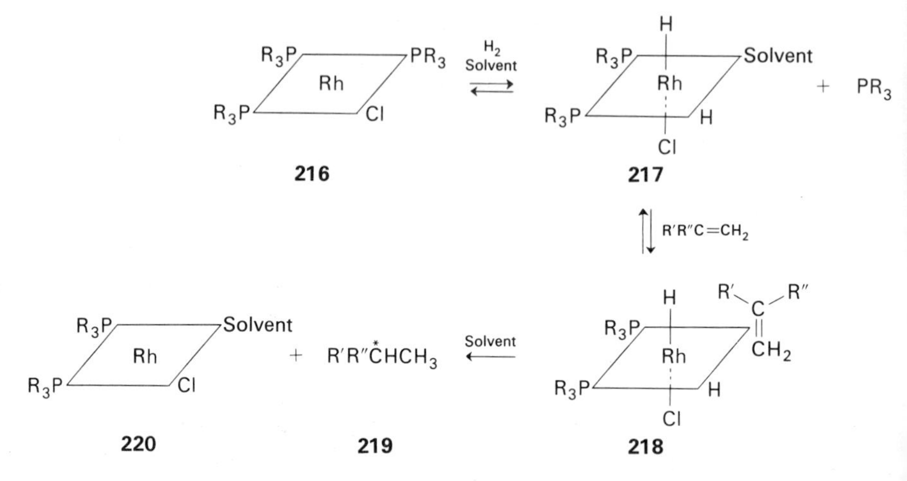

Fig. 6-35. Mechanism of homogeneous hydrogenation.

In one experiment, (Table **6-15**, No. 5) in which the chirality of the catalyst is centered at a carbon on the group attached to phosphorus and not at phosphorus itself, the asymmetric induction was negligible. It would be a mistake to construe this result to mean that all similar catalysts would have low stereoselectivity. The asymmetric inducing ability of the primary active amyl group has usually been low. Furthermore, it is possible that the combined effect of two chiral ligands would counteract rather than reinforce each other. Further studies with chiral phosphines are required in order to clarify this point.

Rhodium (I) complexes containing pyridine and chiral amides have been used to catalyze the asymmetric hydrogenation of some α,β-unsaturated esters (Table **6-15**, Nos. 6–8).[127] A catalyst prepared by reduction of $(py)_3RhCl_3$ with sodium borohydride in the presence of (*R*)-(+)-PhMe$\overset{*}{C}$HNHCHO catalyzed the reduction of (*E*)-methyl β-methylcinnamate (**221**) to optically active methyl 3-phenylbutanoate [54% e.e., *S*-(+)]. This is the highest percent asymmetric synthesis yet reported for an

[127] P. Abley and F. J. McQuillin, *Chem. Commun.*, 477 (1969).

TABLE 6-15

Asymmetric Hydrogenation with Soluble Rhodium Catalysts

$$\mathrm{RR'C{=}CHR''} \xrightarrow[\mathrm{H_2}]{\text{Soluble chiral catalyst}} \mathrm{RR'\overset{*}{C}H{-}CH_2R''}$$

No.	Alkene			Catalyst	Product		Ref.
	R	R′	R″		% e.e.		
1	Ph	COOH	H	$\{[\mathrm{Me{-}\ddot{P}(\mathit{n}Pr){-}Ph}]_3\mathrm{RhCl}\}$[a]	15[b]	*S*-(+)	125
2	CH_2COOH	COOH	H		3[c]	—	125
3	Ph	Et	H	$\{[\mathrm{Ph{-}\ddot{P}(\mathit{n}Pr){-}Me}]_3\mathrm{RhCl}\}$[d]	7–8	*S*-(+)	126
4	Ph	OMe	H		3–4	*R*-(+)	126
5	Ph	COOH	H	$[\mathrm{PhP(CH_2\overset{*}{C}HMeEt)_2}]_3\mathrm{RhCl}$[e]	1[c]	—	125
6	Ph	CH_3	COOMe	$\mathrm{(py)_2Rh[(-)\text{-}PhMe\overset{*}{C}HNHCHO]Cl_2(BH_4)}$[g]	47	*R*-(−)	127
7	Ph	CH_3	COOMe	$\mathrm{(py)_2Rh[(+)\text{-}PhMe\overset{*}{C}HNHCHO]Cl_2(BH_4)}$[f]	54	*S*-(+)	127
8	Ph	CH_3	COOMe	$\mathrm{(py)_2Rh[(-)\text{-}Me\overset{*}{C}H(OH)CONMe_2]Cl_2(BH_4)}$[h]	16	*R*-(−)	127

[a] The structure of the catalyst is uncertain. It was prepared *in situ* from $\mathrm{Rh(PhMe\text{-}\mathit{n}\text{-}PrP)_3Cl_3}$ by treatment with 300–400 psi hydrogen and 3.5 moles of triethylamine per mole of Rh(III) complex in an ethanol-benzene solution of the alkene at 20–80°. The phosphine contained a 69 % e.e. of the (*R*)-(−)-enantiomer.

[b] Not corrected for % e.e. of phosphine.

[c] Configuration not reported.

[d] The catalyst was formed *in situ* from $\mathrm{[Rh(1,5\text{-}hexadiene)Cl]_2}$ and (*S*)-(+)-methylphenyl-*n*-propylphosphine in benzene; hydrogenations were carried out at 1 atmosphere of hydrogen and room temperature. The % e.e. of the phosphine was not reported.

[e] Neither the configuration nor the enantiomeric purity were reported.

[f] Catalyst was prepared by reduction of $\mathrm{(py)_3RhCl_3}$ with $NaBH_4$ in (*R*)-(+)-PhMeCHNHCHO solvent having $[\alpha]_D$ +180°. Py represents pyridine.

[g] Catalyst was prepared as in Footnote f but with (*S*)-(−)-amide $[\alpha]_D$ −172°.

[h] Catalyst was prepared from (*S*)-(−)-amide ($[\alpha]_D$ not reported).

asymmetric homogeneous hydrogenation, but it is reasonable to suppose that even higher asymmetric bias can be achieved with catalyst systems of this type.

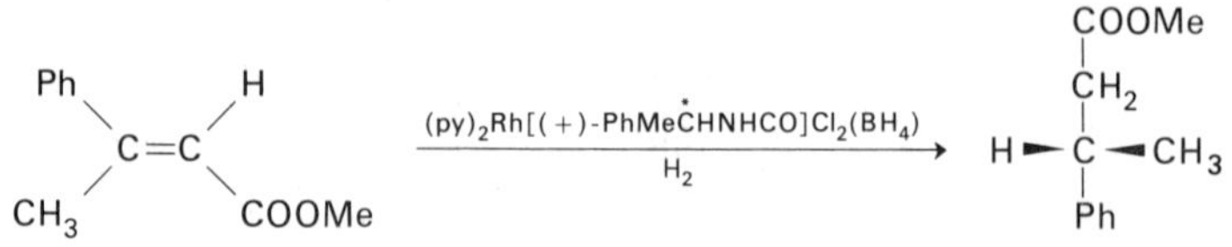

$$\underset{\textbf{221}}{Ph(CH_3)C{=}CH(COOMe)} \xrightarrow[H_2]{(py)_2Rh[(+)\text{-}PhMe\dot{C}HNHCO]Cl_2(BH_4)} \underset{\textbf{222},\ 54\%\ e.e.}{H{-}C(CH_2COOMe)(Ph){-}CH_3}$$

6-7.3. Asymmetric Heterogeneous Hydrogenation.

Several reviews have discussed asymmetric heterogeneous hydrogenations,[128] and in Secs. **7-1.1** and **7-1.2** a number of such reactions leading to amino acids are described in which the catalyst is deposited on a chiral carrier such as powdered silk or quartz. In general, the asymmetric bias that has been achieved by such reductions has been rather low. Such a catalyst has been made by depositing palladium on a silica gel that has been pretreated with an alkaloid and then all the alkaloid leached out.[129] Using this catalyst,

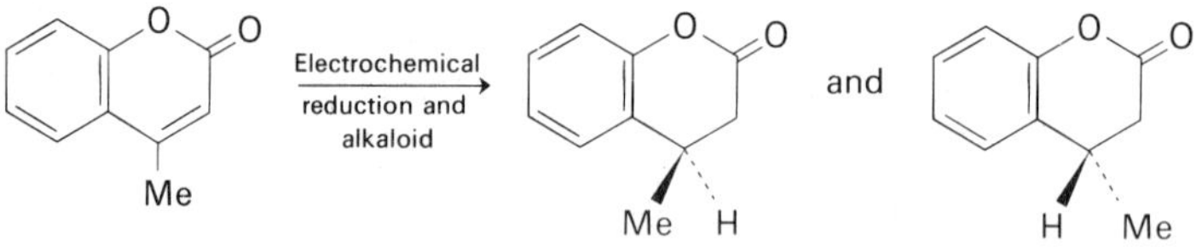

223 $\xrightarrow{\text{Electrochemical reduction and alkaloid}}$ *R*-(+)-**224** and *S*-(−)-**224**

Alkaloid	*% e.e.*
Narcotine	17 *S*
Codeine	13 *R*
Brucine	0 *RS*
Emetine	13 *R*
Apocodeine	1 *S*
Sparteine	19 *R*
Yohimbine	13 *R*

Fig. 6-36. Asymmetric electrochemical reduction of 4-methylcoumarin (**223**) in the presence of alkaloids.[130]

[128] (a) E. I. Klabunovskii, *Asymmetric Synthesis* (Moscow: Gosundart Nauchtekl, Izdatel Khim., 1960). German translation by G. Rudakoff (Berlin: Deutscher Verlag der Wissenschaften, 1963). (b) E. I. Klabunovskii, *Stereospecific Catalysis* (Moscow: Nauka, 1968). (c) L. Velluz, J. Valls, and J. Mathieu, *Angew. Chem.*, **79**, 774 (1967); *Internatl. Edn.*, **6**, 778 (1967). (d) H. Pracejus, *Fortsch. Chem. Forsch.*, **8**, 493 (1967), pp. 531–535. (e) D. R. Boyd and M. A. McKervey, *Quart. Rev.*, **22**, 95 (1968), pp. 114–115. (f) J. Mathieu and J. Weill-Raynal, *Bull. Soc. Chim. Fr.*, 1211 (1968), pp. 1212–1213. (g) S. Mitsui, *Yuki Gosei Kagaku Kyokai Schi.*, **22**, 989 (1964). (h) J. D. Morrison, *Survey of Progress in Chemistry*, Vol. **3**, ed., A. F. Scott (New York: Academic Press, 1966), pp. 172–174.

[129] R. E. Padgett and R. L. Beamer, *J. Pharm. Sci.*, **53**, 689 (1964).

[130] R. N. Gourley, J. Grimshaw, and P. G. Millar, *Chem. Commun.*, 1278 (1967).

α-methylcinnamic acid was reduced to optically active 2-methyl-3-phenylpropanoic acid (1.7–3.3 % e.e.)

A formally similar asymmetric electrochemical reduction[130] (at a mercury cathode in aqueous methanol, pH 5–6) of 4-methylcoumarin (**223**) to 3,4-dihydro-4-methylcoumarin (**224**) has been reported in the presence of various alkaloids (Fig. **6-36**). The observation of optically active products appears to implicate the alkaloid as an intimate part of the hydrogen transfer transition state.

Table **6-16** lists a few selected examples of reductions of chiral unsaturated substrates with the formation of a mixture of diastereomeric products. The stereoselectivity in such cases may be quite high. These particular examples are of interest since, in each case, the inducing chiral center can be removed by hydrolysis (No. 1) or rendered achiral either by oxidation (Nos. 2–5) or by further hydrogenation (No. 6), thus demonstrating clearly the nature of the asymmetric hydrogenation process. It is important to note, however, that these examples differ only in this respect from the reduction of innumerable optically active alkaloids, steroids, or other natural products where an asymmetric synthesis in the Marckwald sense is usually difficult to demonstrate (cf. Sec. **1-2**).

TABLE 6-16

Selected Asymmetric Hydrogenations of Chiral Compounds

Substrate		Product	Ref.
(Ph)(CH_3)C=C(H)(COOR*)	⟶	$Ph\overset{*}{C}H(CH_3)CH_2COOR^*$	131, 132
(Ph)(H)C=C(CH_3)($\overset{*}{C}HOHCH_3$)	⟶	$PhCH_2\overset{*}{C}H(CH_3)\overset{*}{C}H(OH)CH_3$	133, 134
2-furyl–$\overset{*}{C}HOHCH_3$	$\xrightarrow{Ni}$	2-tetrahydrofuryl (C*, H)–$\overset{*}{C}HOHCH_3$	136

[131] G. Vavon and B. Jakubowicz, *Compt. rend.*, **196**, 1614 (1933).

[132] V. Prelog and H. Scherrer, *Helv. Chim. Acta*, **42**, 2227 (1959).

[133] C. L. Arcus, L. A. Cort, T. J. Howard, and Le Ba Loc, *J. Chem. Soc.*, 1195 (1960).

[134] C. L. Arcus and T. J. Howard, *J. Chem. Soc.*, 670 (1961).

[135] C. L. Arcus, J. M. J. Page, and J. A. Reid, *J. Chem. Soc.*,1213 (1963).

[136] (a) D. J. Duveen, *C. R. Acad. Sci., Paris*, **206**, 1974 (1938). (b) D. J. Duveen and J. Kenyon, *Bull. Soc. Chim. Fr.*, [5] **7**, 165 (1940).

TABLE 6-16—continued

Substrate	Product	Ref.
(2-pyridyl)–$CH_2\overset{*}{C}HOHCH_3$	$\xrightarrow{PtO_2}$ (2-piperidyl, N–H, *C–H)–$CH_2\overset{*}{C}HOHCH_3$	137
(2-pyridyl)–$\overset{*}{C}HOHC_2H_5$	$\xrightarrow{PtO_2}$ (2-piperidyl, N–H, *C–H)–$\overset{*}{C}HOHC_2H_5$	138
–O–C(=O)–$\overset{*}{C}H$(phenyl)(cyclohexyl)	$\longrightarrow$ *–O–C(=O)–$\overset{*}{C}H$(phenyl)(cyclohexyl)	139

1969 Addenda

Sec. 6-2.3. Hydroboration of norbornene with "di-3-pinanylborane" was reported to give (1*R*,2*R*)-(+)-exonorborneol, 95% e.e. (cf. results in Table **6-1**). Curiously, oxidation of norborneol with chromium trioxide gave *R*-(−)-norcamphor having approximately 70% of the optical rotation of that obtained from (+)-norborneol.[140] A similar sequence has already been reported[24] (Fig. **6-9**) to give norcamphor with one-third the optical rotation reported in this work.

Sec. 6-3. The *trans* enamine formed from (*R*)-(+)-N-methyl-α-methylbenzylamine and propionaldehyde was treated with sulfene (from $MeSO_2Cl$ and Et_3N) to give a (2 + 2) cycloaddition product. The chiral inducing amino moiety was removed by quaterization and Hofmann elimination to give (*S*)-(+)-4-methylthiete-1,1-dioxide, 6% e.e.[141]

Four optically active diastereomeric 1-ethylidene-2-methyl-3-cyanocyclobutanes of unknown enantiomeric purity were produced when (*R*)-(−)-1,3-dimethylallene was allowed to react in a (2 + 2) cycloaddition with

[137] G. Fodor and G. A. Cooke, *Tetrahedron Suppl.* **8**, *Part 1*, 113 (1966).
[138] G. Fodor and E. Bauerschmidt, *J. Heterocyclic Chem.*, **5**, 205 (1968).
[139] R. Bousset, *Bull. Soc. Chim. Fr.*, [5] **5**, 479 (1938).
[140] R. N. McDonald and R. N. Steppel, *J. Am. Chem. Soc.*, **91**, 782 (1969).
[141] L. A. Paquette and J. P. Freeman, *J. Am. Chem. Soc.*, **91**, 7548 (1969).

acrylonitrile.[142] Ozonolysis converted these adducts to chiral *cis*- and *trans*-2-methyl-3-cyanocyclobutanones whose configurations at C-2 were assigned *R* on the basis of their *CD* curves.

Sec. 6-3.1. Attempted Simmons–Smith asymmetric syntheses on (−)-menthyl crotonate and cinnamate, using a modified reagent, failed to give any of the desired cyclopropane derivative.[143]

A thorough study of the stereochemistry of the methylene iodide–zinc–copper couple of cyclic allylic alcohols has been reported.[144]

The decomposition of ethyl diazoacetate in presence of styrene and [(−)-tribornylphosphite] copper (I) chloride gave (1*S*,2*R*)-*trans*- and (1*R*,2*S*)-*trans*-1-carbethoxy-2-phenylcyclopropane, 3.2 and 2.6% e.e. respectively.[145]

Sec. 6-3.3. Further details concerning the preparation of chiral epoxidizing reagents including the new (−)-*cis*-permyrtanic acid have been described.[146]

Styrene and four different nonterminal alkenes were oxidized with chiral peracids to give chiral epoxides (7.5% e.e. maximum, cf. Table **6-10**). A stereocorrelation model was proposed for such asymmetric peroxidations.[147]

The relative configurations of balfourodine and isobalfourodine, which had been based in part on asymmetric epoxidations[58] as given in Fig. **6-20**, have now been confirmed with more definitive experiments.[148]

The epoxidation of eight racemic, acyclic, allyl alcohols with *p*-nitroperbenzoic acid gave *erythro–threo* mixtures of the diastereomeric epoxy alcohols in a ratio from 9:1 to 50:1. These epoxidations differ from the others discussed in the last part of Sec. **6-3.3** in that the reagent is achiral, rather than chiral, and the asymmetric induction is under the control of the chiral center in the substrate.[149]

Sec. 6-6. The synthesis of branched chain sugars via stereoselective sodium borohydride reduction of an α,β-unsaturated nitro derivative has been applied to ketoses.[150]

Sec. 6-6.2. Pracejus and Kohl[151] have extended their studies on the asymmetric addition of methanol to phenylmethylketene in the presence of

[142] J. E. Baldwin and U. V. Roy, *Chem. Commun.*, 1225 (1969).
[143] S. Sawada and Y. Inouye, *Bull. Chem. Soc. Jap.*, **42**, 2669 (1969).
[144] C. D. Poulter, E. C. Friedrich, and S. Winstein, *J. Am. Chem. Soc.*, **91**, 6892 (1969).
[145] W. R. Moser, *J. Am. Chem. Soc.*, **91**, 1135 (1969).
[146] J. F. Collins and M. A. McKervey, *J. Org. Chem.*, **34**, 4172 (1969).
[147] F. Montanari, I. Morre, and G. Torre, *Chem. Commun.*, 135 (1969).
[148] J. F. Collins and M. F. Grundon, *Chem. Commun.*, 1078 (1969).
[149] J. L. Pierre, P. Chautemps, and P. Arnaud, *Bull. Soc. Chim. Fr.*, 1317 (1969).
[150] G. J. Lourens, *Tetrahedron Lett.*, 3733 (1969).
[151] H. Pracejus and G. Kohl, *Ann. Chem.*, **722**, 1 (1969).

sixteen additional chiral basic catalysts over a temperature range of +60 to −100°. Based on these new results, the previous model for this reaction (cf. Formula **185**) has been modified.

Sec. 6-6.8. The addition of acrylonitrile or methyl acrylate to the chiral enamine obtained from several L-proline esters and cyclohexanone, followed by removal of the chiral inducing moiety, gave (*S*)-2-cyano- and (*S*)-2-carbomethoxy-cyclohexanone (15 to 59% e.e.).[152] The enamines obtained by condensation of L-proline esters and amides with various α,α-disubstituted acetaldehydes were treated with methyl vinyl ketone to give, after hydrolysis and ring closure, a variety of chiral 4,4-disubstituted cyclohexenones (14 to 49% e.e.).[153]

A special example of a (2 + 2) cycloaddition of sulfene to an enamine has been described.[141]

Sec. 6-7.3. An additional appropriate example of asymmetric catalytic hydrogenation (cf. Table **6-16**) is the reduction of (−)-menthyl 2-furan-carboxylate to give, after hydrolysis, tetrahydrofuran-2-carboxylic acid. The asymmetric induction was 24% using 10% Pd–C in ethanol but 0% using Raney nickel.[154]

[152] S. Yamada, K. Hiroi, and K. Achiwa, *Tetrahedron Lett.*, 4233 (1969).

[153] S. Yamada and G. Otani, *Tetrahedron Lett.*, 4237 (1969).

[154] E. I. Klabunovskii, L. F. Godunova, R. A. Karakhanov, E. Y. Beilinson, T. G. Rybina, and I. A. Rubtsov, *Izv. Akad. Nauk SSSR*, 298 (1969); *Chem. Abst.*, **70**, 114356 (1969).

7

Asymmetric Synthesis of Amino Acids and Amino Acid Derivatives*

7-1. Hydrogenation of Carbon-Carbon Double Bonds

7-1.1. Achiral Substrates and Chiral Catalysts. Hydrogenation of 4-benzylidene-2-methyl-5-oxazolone (**1**) over a palladium on silk fibroin† catalyst yields (*R*)-(+)-phenylalanine.[1-4] The enantiomeric excess achieved is a function of the origin of the fibroin and its chemical pretreatment.

[1] T. Isoda, A. Ichikawa, and T. Shimamoto, *Rikagaku Kenkyusho Hokoku*, **34**, 134 (1958); *Chem. Abst.*, **54**, 287f (1960).

[2] S. Akabori, S. Sakurai, Y. Izumi, and Y. Fujii, *Nature*, **178**, 323 (1956).

[3] S. Akabori, Y. Izumi, Y. Fujii, and S. Sakurai, *Nippon Kagaku Zasshi*, **77**, 1374 (1956); *Chem. Abst.*, **53**, 5149b (1959).

[4] S. Akabori, Y. Izumi, and Y. Fujii, *Nippon Kagaku Zasshi*, **78**, 886 (1957); *Chem Abst.*, **54**, 9889e (1960).

* Difficulty was encountered in evaluating some experiments in this area. First, the optical rotation of an amino acid is very dependent on pH and, to a lesser extent, on concentration; in some cases authors have reported the specific rotations of products under conditions for which no comparison data for enantiomerically pure materials could be found. Unless the author estimated the enantiomeric composition, there was no alternative but to report specific rotations and the conditions under which they were obtained. Second, solid diastereomeric or enantiomeric products were usually obtained, and in some cases it was impossible to determine from literature data if partial resolution might have occurred during crystallization.

† Fibroin is the water insoluble structural protein from silk.

Values between approximately 30 and 70% e.e. have been reported. The effectiveness of catalysts prepared from three kinds of silk fibroin has

1) H_2 [Pd-silk fibroin] 2) H_2O

1 → $R(+)$, 30–70% e.e.

been studied.[4] Palladium on fibroin from the silk of wild silkworms gave phenylalanine having a higher enantiomeric purity than that resulting from hydrogenation over palladium on fibroin from the silk of cultured silkworms.* However, a catalyst prepared by depositing palladium on acetylated cultured silk fibroin gave phenylalanine [approximately 70% e.e., *R*-(+)] with much higher stereoselectivity than that obtained with unacetylated fibroin from wild silkworms.

A similar asymmetric hydrogenation is that of α-acetamidocinnamic acid (**2A**) to phenylalanine over palladium on poly-L-leucine.[5] A 92% yield of phenylalanine containing approximately a 5% excess of the (*S*)-(−)-enantiomer was obtained. Hydrogenation of **2A** and **2B** in the presence of

1) H_2[Pd-poly-L-leucine] 2) H_2O

2A, R = Me
2B, R = Ph

S-(−), 5% e.e. and *R*-(+)

Raney nickel in an alkaline glucose solution gave phenylalanine with a maximum 10% excess of one enantiomer.[6] The enantiomeric excess and configuration of the predominant enantiomer were found to depend upon the method of preparation of the catalyst, the amount of solvent, the order of mixing reagents, the temperature, and the amount of Raney nickel. A Japanese patent has been issued for the preparation of (*S*)-(−)-phenylalanine by the hydrogenation of **2** with Pd-C in the presence of small amounts of (*S*)-methionine.[7] The generally low percent asymmetric synthesis observed for hydrogenations over metals adsorbed on dissymmetric supports may be

[5] R. L. Beamer, C. S. Fickling, and J. H. Ewing, *J. Pharm. Sci.*, **56,** 1029 (1967).
[6] M. Nakazaki, *Nippon Kagaku Zasshi*, **75**, 831 (1954); *Chem. Abst.*, **49**, 13937f (1955).
[7] S. Senoh, S. Ouchi, and K. Tsunoda, Jap. Pat. 13,307 (1963); *Chem. Abst.*, **60**, 3092h (1964).

* Fibroin from cultured silkworms has a higher glycine and lower alanine content than that from wild silkworms.

due in part to the nonuniform dissymmetric environment of the catalytic sites because of metal clumps and a certain amount of random adsorption of the chiral agent on the catalyst surface. Several highly intuitive schemes have been devised to rationalize asymmetric syntheses of this type,[5,8] but the evidence for their validity is rather fragmentary.

7-1.2. Chiral Substrates. Hydrogenation of the (*S*)-diketopiperazine, **3A**, over platinum oxide catalyst resulted in saturation of the benzylidene double bond. It was reported that hydrolytic work-up produced a dipeptide (**4A**). Although there was no direct evidence regarding the amounts of the two possible diastereomers of **4A** that were produced, in one experiment hydrolytic cleavage of the crude dipeptide yielded phenylalanine that contained a 16% excess of the *S*-(−) enantiomer.[9] Similar treatment of *R*-**3A** produced an excess of (*R*)-(+)-phenylalanine. However, since the yield of phenylalanine from crude dipeptide was only about 55% (as the hydrochloride), one cannot be certain that the enantiomeric composition of the phenylalanine reflects the dipeptide diastereomer ratio. Furthermore, recrystallization of the crude dipeptide resulted in enrichment in one of the diastereomers of **4A**, so that cleavage gave much purer (*S*)-(−)-phenylalanine (84–90% e.e.).[9,10] Therefore, 16% asymmetric synthesis is a maximum (and

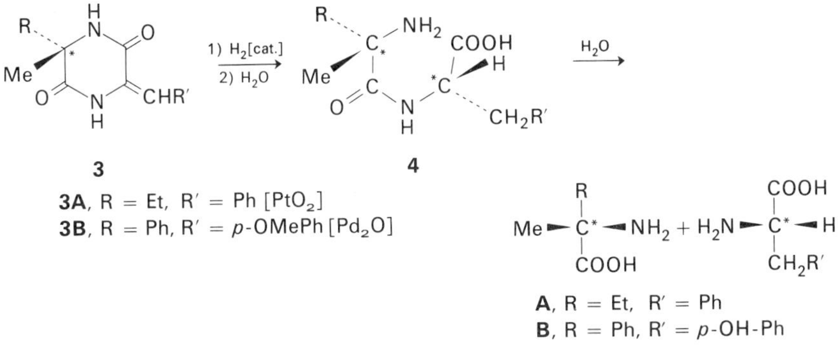

1) H_2[cat.] 2) H_2O → H_2O →

3

4

3A, R = Et, R′ = Ph [PtO_2]
3B, R = Ph, R′ = *p*-OMePh [Pd_2O]

A, R = Et, R′ = Ph
B, R = Ph, R′ = *p*-OH-Ph

Fig. 7-1. Asymmetric hydrogenation of diketopiperazines.[9–11]

highly suspect) value for this system. Starting with **3B**, Maeda[11] adapted this procedure to the synthesis of tyrosine having a 35% excess of the (−)-enantiomer. This probably does not reflect the true percent asymmetric

[8] R. L. Beamer, J. D. Smith, J. Andrako, and W. H. Hartung, *J. Org. Chem.*, **25**, 798 (1960).

[9] S. Akabori, T. Ikenaka, and K. Matsumoto, *Nippon Kagaku Zasshi*, **73**, 112 (1952); *Chem. Abst.*, **47**, 9938g (1953).

[10] S. Akabori, T. Ikenaka, and K. Matsumoto, *Proc. Jap. Acad.*, **27**, 7 (1951); *Chem. Abst.*, **47**, 379h (1953).

[11] G. Maeda, *Nippon Kagaku Zasshi.*, **77**, 1011 (1956); *Chem. Abst.*, **53**, 5147h (1959).

synthesis since the intermediate dipeptide was purified by recrystallization, and furthermore partial racemization may have occurred during the ether cleavage step.[11]

Essentially complete stereoselectivity has been achieved in the hydrogenation[12] of **5A**, which was prepared from (+)-1,2-diphenylethanolamine and dimethyl acetylenedicarboxylate. The product **5B** was hydrogenolyzed in nearly quantitative yield to optically active (*S*)-(+)-methyl aspartate which was at least 98% enantiomerically pure! Apparently the axial phenyl

+ MeOOCC≡CCOOMe ⟶

(+)-Isomer resolved with L-glutamic acid

5A

Ni (Raney) H_2

$Pd(OH)_2$ H_2

5B

S-(+), at least 98% e.e.

group in **5A** prevents adsorption and hydrogenation from the axial face of the exocyclic double bond. The yield of **5B** from **5A** was essentially quantitative.

Chiral α-benzoylaminocinnamates (**6**) have also been hydrogenated to diastereomeric amino acid precursors (*S*-**7** and *R*-**7**). In this case more reliable estimates of the percent asymmetric synthesis are possible, since in some experiments both diastereomers have been isolated in high yield and in a reasonably pure state.[13] From **6A**, for example, there was obtained a 95% yield of a mixture of *R*-**7A** and *S*-**7A** which was collected as two crops of crystals having different melting ranges (32% *R*-**7A**, m.p. 168–170°; 68% *S*-**7A**, m.p. 127–135°). After recrystallization from ethyl acetate, the higher melting crop (m.p. 175–176°, $[\alpha]_D$ −9°) was transesterified with methanol and hydrolyzed to (*R*)-(+)-*p*-aminophenylalanine ($[\alpha]_D$ +45°, 96% e.e.). The lower melting crop was crystallized from ether-ligroin (m.p. 130–136°, $[\alpha]_D$ −46.5°) and similarly hydrolyzed to (*S*)-(−)-*p*-amino phenylalanine ($[\alpha]_D$ − 42°, 90% e.e.). From these data it can be estimated that there was approximately 30% asymmetric synthesis.

[12] J. P. Vigneron, H. Kagan, and A. Horeau, *Tetrahedron Lett.*, 5681 (1968).

[13] (a) A. Pedrazzoli, *Chimia* (Switz.), **10**, 260 (1956). (b) A. Pedrazzoli, *Helv. Chim. Acta*, **40**, 80 (1957).

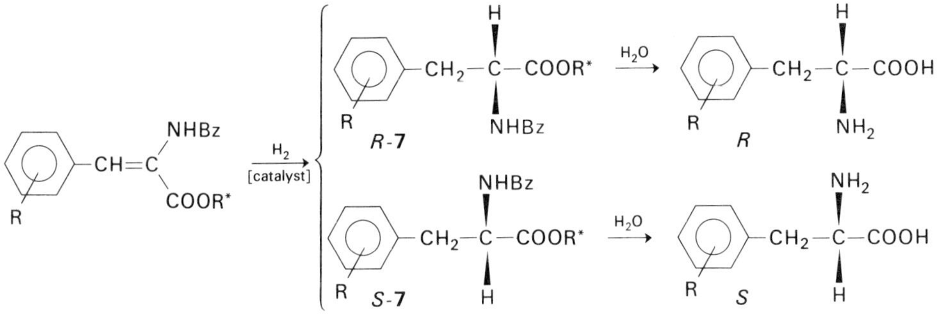

$$\text{R-C}_6\text{H}_4\text{-CH=C(NHBz)COOR}^* \xrightarrow[\text{[catalyst]}]{H_2} \begin{cases} \text{R-C}_6\text{H}_4\text{-CH}_2\text{-C(H)(NHBz)-COOR}^* \ (R\text{-}7) \xrightarrow{H_2O} \text{R-C}_6\text{H}_4\text{-CH}_2\text{-C(H)(NH}_2\text{)-COOH} \ (R) \\ \text{R-C}_6\text{H}_4\text{-CH}_2\text{-C(NHBz)(H)-COOR}^* \ (S\text{-}7) \xrightarrow{H_2O} \text{R-C}_6\text{H}_4\text{-CH}_2\text{-C(NH}_2\text{)(H)-COOH} \ (S) \end{cases}$$

6A, R = *p*-NO_2 (NH_2 in products)
R* = (−)-menthyl
catalyst = Pd-Al_2O_3
6B, R = *p*-NO_2 (NH_2 in products)
R* = (−)-bornyl
catalyst = Pd-Al_2O_3
6C, R = 3,4-methylenedioxy
R* = (−)-menthyl
catalyst = Pd-C

Chirality at new center

	% *S*	% *R*
A	65	35
B	48	52
C	49	51

Fig. 7-2. Asymmetric hydrogenation of chiral esters of α-benzoylaminocinnamates.[13,14]

The *p*-nitrocinnamate from (−)-borneol (**6B**) was hydrogenated under the same conditions to a mixture of diastereomeric esters of *p*-aminophenylalanine having only a slight excess of the *R*-(+) enantiomer (approximately 4% e.e.).* Yamada[14] et al. hydrogenated **6C** and obtained a 95% yield of a mixture of two diastereomers. Fractional crystallization (89% recovery) yielded a 2% excess of one of them, but hydrolysis of the unfractionated mixture produced optically inactive 3,4-methylenedioxyphenylalanine.

Sheehan and Chandler[15] studied the hydrogenation of several chiral α,β-unsaturated amides and observed 6–39% asymmetric synthesis. Hydrogenation of *S*-**8A** with Raney nickel in methanol followed by hydrolysis yielded (*R*)-(−)valine (**9A**, 90% yield, 39% e.e.). Similarly (*S*)-α-methylbenzyl-α′-benzamido-β′,β′-dimethylacrylamide (*S*-**8B**) gave (*R*)-(−)-valine (9**B**, 10% e.e.) and the corresponding (*R*)-α-methylbenzylamide (*R*-**8B**) gave (*S*)-(+)-valine (18% e.e.). Amide **8C** from *S*-amine and α-benzamidocinnamic acid was used to synthesize (*R*)-(+)-phenylalanine (**9C**, 6% e.e.).

[14] S. Yamada, T. Fujii, and T. Shiori, *Chem. Pharm. Bull.* (Tokyo), **10**, 680 (1962); *Chem. Abst.*, **58**, 11463e (1963).

[15] J. C. Sheehan and R. Chandler, *J. Am. Chem. Soc.*, **83**, 4795 (1961).

* Apparently there is no significant amount of racemization associated with the hydrogenation and hydrolysis reactions, since *p*-aminophenylalanine from hydrogenation of (−)-*p*-nitrophenylalanine had $[\alpha]_D$ −47° and that from hydrogenation and hydrolysis of the ethyl ester had $[\alpha]_D$ −46°.

It is interesting that there is a two-fold decrease in percent enantiomeric excess in going from an α-acetamido to an α-benzamido group even though the acyl group is at a position rather remote from both the developing and

$$R'R''C{=}C(NHCOR''')(CONH\overset{*}{C}HMePh) \xrightarrow[2)\ H_2O]{1)\ H_2\ [Ni]} R'R''CH{-}CH(NH_2)COOH$$

8 → **9**

A, R′ = R″ = R‴ = Me — Valine, 39% e.e.
B, R′ = R″ = Me, R‴ = Ph — Valine, 18% e.e.
C, R′ = H, R″ = R‴ = Ph — Phenylalanine, 6% e.e.

inducing asymmetric centers. It would be intriguing to examine the effect of alkyl groups larger than methyl and aryl groups larger than phenyl.

A modified Prelog-type model (**10**) nicely accomodates the results that have thus far been reported for catalytic hydrogenations of carbon-carbon double bonds in α-acylamido unsaturated esters or amides. The essential features of such a model are (a) a *transoid* relationship of the carbonyl and alkene moieties, (b) the carbonyl located between the small and medium groups on the asymmetric carbon, and (c) the hydrogen approaching the double bond from the least hindered side (from below the plane of the page in **10** as illustrated for the reduction of **8A**). It is impossible to apply more than empirical justification to such a model, since there is little evidence concerning the conformational preferences of the substrates involved. Furthermore, the utility of such evidence would be questionable even if it were available, since the transition states for these reactions involve adsorbed species.

8A ⟶ [AcNH–C(=CMe₂)–C(=O)–NH–C(S)(Me)(H)Ph] (**10**) ⟶ AcNH–C(R)(H)(C(=O)NH–C(S)(Me)(H)Ph)–C(H)Me₂ (**11**)

7-2. Reduction of Carbon-Nitrogen Double Bonds

Experiments involving reduction of an imine bond to produce an amino acid parallel those discussed in Sec. **7-1**. There are few examples of chiral catalysts being used, but the reduction of a C=N bond in a chiral amino acid precursor has been one of the most thoroughly studied asymmetric amino acid syntheses.

7-2.1. Achiral Substrates and Chiral Catalysts. The work of Nakamura[16] represents one of the earliest attempts to reduce asymmetrically a C=N bond using a dissymmetric hydrogenation catalyst. Reduction of acetophenone oxime over platinum black containing a small amount of ethyl menthoxyacetate or tartaric acid gave optically active α-methylbenzylamine (15–18% e.e.). Hexahydrofluorenone oxime was also reduced to an optically active amine under similar conditions. Later Akabori and co-workers[2,3] hydrogenated benzil dioxime over palladium on silk fibroin to obtain optically active 1,2-diphenyl-1,2-diaminoethane. The specific rotation of the product showed an enormous variation with concentration ($[\alpha]_D$ +8.75°, 800 mg/100 ml ether; $[\alpha]_D$ +431°, 30 mg/100 ml ether). Its configuration and enantiomeric purity are unknown. Oxime acetates **12** and **13** have also been hydrogenated over palladium on silk fibroin catalysts to give amino acids having a 7–30% excess of one enantiomer. The results of these experiments are summarized in Table **7-1**. These reactions are similar to those described in Sec. **7-1.1**.

$$EtOOC\overset{\overset{\displaystyle N-OAc}{\|}}{C}CH_2CH_2COOEt \qquad PhCH_2\overset{\overset{\displaystyle N-OAc}{\|}}{C}COOEt$$

12 **13**

TABLE 7-1

Hydrogenation of Oxime Acetates over Palladium on Silk Fibroin

Oxime acetate	Product	$[\alpha]_D$	Configuration	Approximate % e.e.[a]	Ref.
12	Glutamic acid	2.25[b]	*S*	7	2,3
12	Glutamic acid	4.5[c]	*S*	14	4
12	Glutamic acid	—	*S*	15	1
13	Phenylalanine	9.25[d]	*R*	30	2,3

[a] Different values for the same compound may reflect differences in origin or pretreatment of the fibroin (see Sec. **7-1.1**).
[b] 7% in 2*N* HCl.
[c] 500 mg/18 ml 2*N* HCl.
[d] 9% in H_2O.

7-2.2. Achiral Substrates and Chiral Reagents. There have been numerous studies of asymmetric syntheses which bear a formal resemblance to pyridoxal-dependent transaminations. In the enzyme reaction an α-keto acid is converted to an amino acid in the presence of a transaminase

[16] Y. Nakamura, *J. Chem. Soc. Jap.*, **61**, 1051 (1940); *Chem. Abst.*, **37**, 377 (1943).

and the ubiquitous coenzyme, pyridoxal-5′-phosphate (vitamin B_6), at the expense of another amino acid as the nitrogen source.* Model reactions with other pyridoxal derivatives are believed to involve similar mechanistic features. A pathway for a pyridoxal dependent transamination is shown in Fig. 7-3.

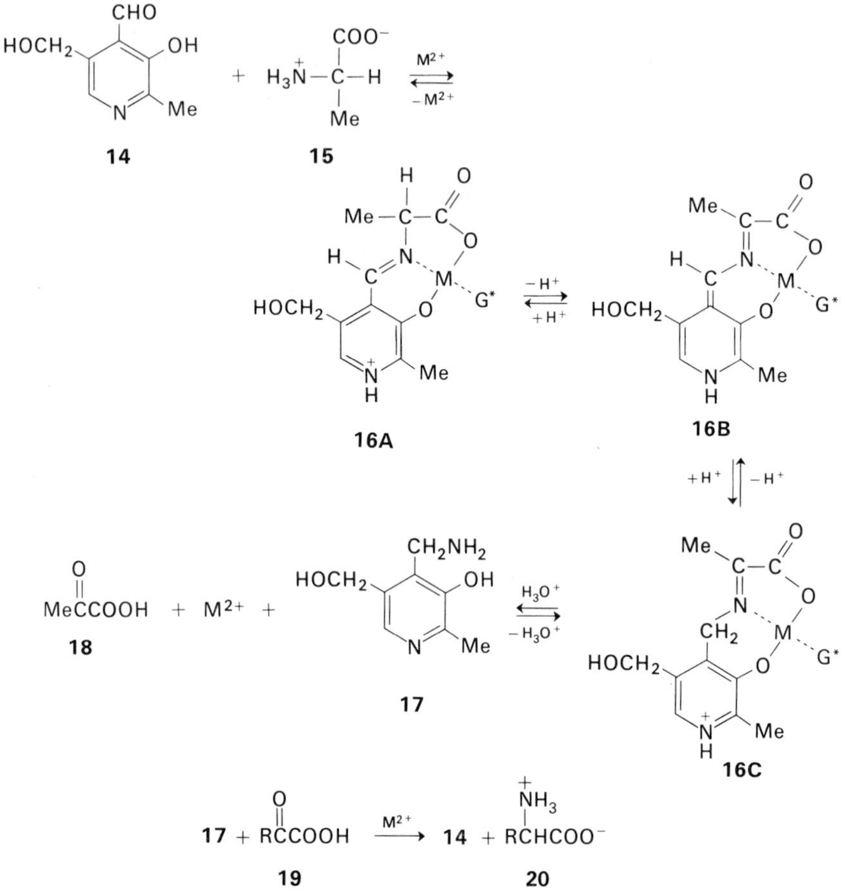

$$\mathbf{17} + \underset{\mathbf{19}}{RCOCOOH} \xrightarrow{M^{2+}} \mathbf{14} + \underset{\mathbf{20}}{RCH(\overset{+}{N}H_3)COO^-}$$

Fig. 7-3. The role of pyridoxal in transamination. The chiral group G* might be an amino acid moiety or an enzyme.

* For a more complete account of the transamination reaction, the reader should consult detailed reviews such as the series of papers on vitamin B_6 appearing in *Vitamins and Hormones*, **22**, 359–927 (1964) and references therein; T. C. Bruice and S. J. Benkovic, Chapter 8 in Vol. II of *Bioorganic Mechanisms* (New York: W. A. Benjamin, Inc., 1966); B. M. Guirard and E. E. Snell, Chapter 5 in Vol. 15 of *Comprehensive Biochemistry*, ed., M. Florkin and E. H. Stotz (New York: Elsevier Pub. Co., 1964). E. E. Snell, A. E. Braunstein, E. S. Severin, and Yu. M. Torchinsky, eds., *Pyridoxal Catalysis* (New York: Interscience Pub., Inc., 1968).

Pyridoxal (**14**) reacts with an amino acid, e.g., alanine (**15**), and a metal ion* to form a metal complex (**16A**). Imine chelate **16A** is in tautomeric equilibrium with **16C** by way of **16B**. Hydrolysis of **16C** yields pyridoxamine (**17**) and an α-keto acid (**18**) related to the original amino acid (**15**). The pyridoxamine (**17**) can react with any other keto acid (**19**) present to convert it, via the reverse of this process, to pyridoxal and a new amino acid (**20**).

A nonenzymatic transamination was studied by Longenecker and Snell[17] who showed that pyridoxal promotes the transamination of 2-ketoglutaric acid to glutamic acid in the presence of alanine or phenylalanine. Duplicate experiments with alanine confirmed that the reaction consistently produced a glutamic acid (5–7% e.e.) with the same configuration as the alanine. (*S*)-Phenylalanine behaved similarly, yielding (*S*)-glutamic acid (5–10% e.e.).

$$\underset{\text{2-Ketoglutaric acid}}{HOOCCH_2CH_2\overset{O}{\overset{\|}{C}}COOH} \xrightarrow[\text{Pyridoxal (14)},\ Cu^{2+},\ 100^\circ]{(S)\text{-Alanine}} \underset{(S)\text{-Glutamic acid}}{HOOCCH_2CH_2\overset{NH_2}{\underset{H}{C}}COOH}$$

Kinetic studies have confirmed that enzymatic transaminations which are pyridoxal dependent proceed by a mechanism much like that proposed for nonenzymatic transamination.[18] There is evidence[19] that pyridoxal (as the 5′-phosphate) is bound to some enzymes by imine formation with the ε-amino group of a lysine residue. Transamination must therefore precede the steps represented in Fig. **7-3**, but the resulting imine intermediates are thought to *resemble* **16A**–**16C**.

The hydrogenation of chiral imines (**21**) formed from α-keto acids and optically active amines also leads to derivatives of optically active α-amino acids (**22**). Any optically active amine, including an amino acid or a peptide,

$$\underset{\mathbf{21}}{R^*N{=}C(COOH)R} \longrightarrow \underset{\mathbf{22}}{R^*NH\overset{*}{C}H(R)COOH} \longrightarrow \underset{\mathbf{23}}{H_2N\overset{*}{C}H(R)COOH}$$

can be used, but it is desirable for R* to be a group that can be quantitatively removed from the product **22** without affecting the configurational integrity

[17] J. B. Longenecker and E. E. Snell, *Proc. Natl. Acad. Sci. U.S.*, **42** [5], 221 (1956).

[18] G. G. Hammes and P. Fasella, *J. Am. Chem. Soc.*, **84**, 4644 (1962).

[19] (a) W. T. Jenkins and I. W. Sizer, *J. Biol. Chem.*, **234**, 1179 (1959). (b) E. E. Snell and W. T. Jenkins, *J. Cell. Comp. Physiol.*, **54**, *Suppl.* 1, 161 (1959).

* Metal ions are required for nonenzymatic pyridoxal catalyzed transaminations, but not for most enzymatic transaminations where the protein exerts an equivalent influence.

of the new asymmetric center. If R* is benzylic, it can be cleaved by hydrogenolysis to give **23**. Alternatively, the percent asymmetric synthesis can be obtained by determining the diastereomeric composition of **22**. The usual precautions must be taken to ensure that the determination is quantitative. This has not always been done for the experiments reported in Table **7-2** and

TABLE 7-2

Asymmetric Synthesis of Amino Acids by Reductive Amination[a,b]

$$\underset{\mathbf{19}}{R-\overset{O}{\overset{\|}{C}}-COOH} + R^*NH_2 \longrightarrow [\mathbf{21} \longrightarrow \mathbf{22}] \longrightarrow \underset{\mathbf{23}}{H_2N-\overset{*}{\underset{R}{\underset{|}{C}}}H-COOH} + \overset{*}{R}H$$

No.	*R*	R*NH₂[c]	Major product (**22** or **23**)	Asymmetric synthesis (DNP)[d] % e.e.	Ref.
1	Me	(*S*)-(+)-Arginine	(*S*)-(+)-Allooctopine	70	20,21
2	Me	(*R*)-(+)-MBA	(*R*)-(−)-Alanine	70	23
3	Me	(*S*)-(−)-MBA	(*S*)-(+)-Alanine	81	23
4[e]	Me	(*S*)-(−)-MBA	(*S*)-(+)-Alanine	74	23
5	Me	(*S*)-(−)-MBA	(*S*)-(+)-Alanine	(67)63	27
6	Me	(*R*)-(+)-MBA	(*R*)-(−)-Alanine	(65)77	27
7	Me	(*R*)-(−)-PhGly	(*R*)-(−)-Alanine	(56)41	26
8[f]	Me	(*R*)-(−)-PhGly	(*R*)-(−)-Alanine	(54)53	26
9[g]	Me	(*R*)-(−)-PhGly	(*R*)-(−)-Alanine	(47)43	26
10[f,h]	Me	(*S*)-(+)-PhGly	(*S*)-(+)-Alanine	(64)41	25,26
11	Me	(*S*)-(−)-EBA	(*S*)-(+)-Alanine	(52)52	26
12	Me	(*R*)-(+)-NpEtA	(*R*)-(−)-Alanine	(80)83	29a
13[i]	Me	(*R*)-$PhCH_2CH(CH_3)CONHNH_2$	(*S*)-(+)-Alanine	8	30
14[j]	Me	(*S*)-(−)-MBA	(*S*)-(−)-Alanylgylcine	54	24
15[j]	Me	(*R*)-(+)-MBA	(*R*)-(+)-Alanylgylcine	58	24
16[k,f]	Me	N-amino-2-hydroxymethylindoline	(*S*)-(−)-Alanylgylcine	82	32
17[f]	Me[l]	(*S*)-N-Aminoanabasine	(*S*)-(+)-Alanine	40	31
18	Et	(*S*)-(−)-MBA	(*S*)-(+)-2-Aminobutyric acid	63[m]	23
19	Et	(*S*)-(−)-MBA	(*S*)-(+)-2-Aminobutyric acid	(39)37	27
20[n]	Et	(*R*)-(+)-MBA	(*R*)-(−)-2-Aminobutyric acid	43	23
21[h]	Et	(*R*)-(+)-MBA	(*R*)-(−)-2-Aminobutyric acid	13	23
22	Et	(*R*)-(+)-MBA	(*R*)-(−)-2-Aminobutyric acid	(37)38	27
23	Et	(*S*)-(−)-EBA	(*S*)-(+)-2-Aminobutyric acid	(33)36	27

TABLE 7-2—continued

No.	*R*	R*NH$_2$[c]	Major product (**22** or **23**)	Asymmetric synthesis (DNP)[d] % e.e.	Ref.
24	Et	(*R*)-(+)-EBA	(*R*)-(−)-2-Aminobutyric acid	(36)39	27
25	Et	(*R*)-(−)-PhGly	(*R*)-(−)-2-Aminobutyric acid	(30)32	27
26	Et	(*R*)-(−)-PhGly	(*R*)-(−)-2-Aminobutyric acid	(42)39	27
27[f,h]	Et	(*S*)-(+)-PhGly	(*S*)-(+)-2-Aminobutyric acid	(44)36	27
28[f,k]	Et	N-Amino-2-hydroxymethylindoline	(*S*)-(+)-2-Aminobutyric acid	98	32
29	$PhCH_2$	(*S*)-(−)-MBA	(*S*)-(−)-Phenylalanine	12	23
30	$PhCH_2$	(*S*)-(−)-MBA	(*S*)-(−)-Phenylalanine	(14)12	27
31	$PhCH_2$	(*R*)-(+)-MBA	(*R*)-(+)-Phenylalanine	13	23
32	$PhCH_2$	(*R*)-(+)-MBA	(*R*)-(+)-Phenylalanine	(14)13	27
33	$PhCH_2$	(*S*)-(−)-EBA	(*S*)-(−)-Phenylalanine	(10)5	27
34	$PhCH_2$	(*R*)-(+)-EBA	(*R*)-(+)-Phenylalanine	(11)5	27
35[i]	$PhCH_2$	(*R*)-$PhCH_2CH(CH_3)CONHNH_2$	(*S*)-(−)-Phenylalanine	5	30
36	Ph	(*S*)-(−)-MBA	(*S*)-(+)-Phenylglycine	73	28
37	Ph	(*S*)-(−)-MBA	(*S*)-(+)-Phenylglycine	(30)28	27
38	Ph	(*R*)-(+)-MBA	(*R*)-(−)-Phenylglycine	(31)29	27
39	Ph	(*S*)-(−)-EBA	(*S*)-(+)-Phenylglycine	(24)25	27
40	Ph	(*R*)-(+)-EBA	(*R*)-(−)-Phenylglycine	(26)24	27
41	iPr	(*R*)-(+)-MBA	(*R*)-(−)-Valine	28	23
42	$HOOCCH_2$	(*S*)-(+)-PhGly	(*S*)-(+)-Aspartic Acid	(58)[o]	26
43[f,h]	$HOOCCH_2$	(*R*)-(−)-PhGly	(*R*)-(−)-Aspartic Acid	(47)[o]	26
44	$HOOCCH_2$	(*R*)-(−)-PhGly	(*R*)-(−)-Aspartic Acid	(45)[o]	26
45	$HOOCCH_2$	(*S*)-(−)-MBA	(*S*)-(+)-Alanine	(69)74	29b
46	$HOOCCH_2$	(*S*)-(−)-EBA	(*S*)-(+)-Alanine	(52)42	29b
47	$HOOCCH_2$	(*R*)-(+)-NpEtA	(*R*)-(−)-Alanine	(73)74	29b
48	$HOOCCH_2CH_2$	(*S*)-(−)-MBA	(*S*)-(+)-Glutamic Acid	35	24
49	$HOOCCH_2CH_2$	(*S*)-(−)-MBA	(*S*)-(+)-Glutamic Acid	(12)13	27
50	$HOOCCH_2CH_2$	(*S*)-(−)-EBA	(*S*)-(+)-Glutamic Acid	(6)2	27
51[f,h]	$HOOCCH_2CH_2$	(*R*)-(−)-PhGly	(*R*)-(−)-Glutamic Acid	(48)35	26
52[f,m]	$HOOCCH_2CH_2$	(*R*)-(−)-PhGly	(*R*)-(−)-Glutamic Acid	(49)56	25,26

[a] Unless otherwise noted, reactions were carried out by successive hydrogenation (10% Pd-C) and hydrogenolysis ($Pd(OH)_2$—C) with a mole ratio of keto acid to amine of 1:2.

[b] See Table **7-4** for summary of solvent effects on some of these reactions.

[c] MBA = α-methylbenzylamine, PhGly = phenylglycine, EBA = α-ethylbenzylamine, and NpEtA = 1-naphthylethylamine.

[d] Values in parentheses are based on data for the 2,4-dinitrophenyl (DNP) derivative prepared after hydrogenolysis and separated by column chromatography on Celite.

[e] Benzyl ester.

the original publications should be consulted for their quantitative reliability.

(*S*)-(+)-Arginine, **24**, (as the carbonate) was allowed to react with an excess of pyruvic acid (Table **7-2**, No. 1);[20] the reaction mixture was then treated with hydrogen over platinum oxide to yield a product identified as (+)-octopine, an amino acid found in octopus muscle, scallops, and a few other marine animals. This was repeated[21] (PdO_2 catalyst, with both arginine carbonate, and hydrochloride) and the product identified as (+)-isooctopine (**26**) [now called *S*-(+)-allooctopine[22]], an epimer of (+)-octopine. If one assumes

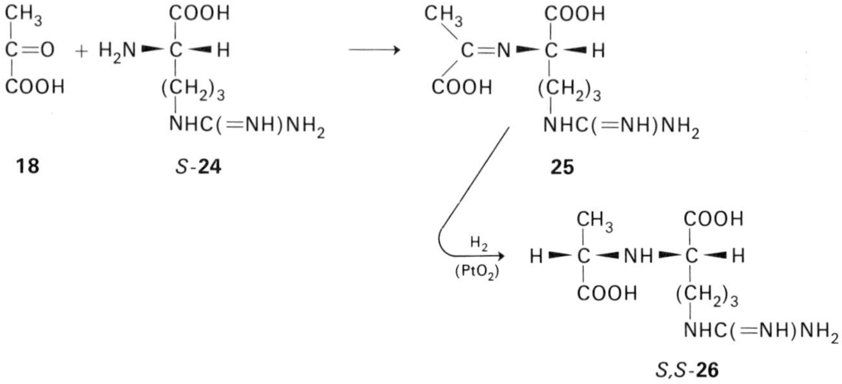

Fig. 7-4. Reductive amination of pyruvic acid by (*S*)-arginine.

that the (+)-allooctopine (*S*-alanine moiety) has been completely separated from epimeric (+)-octopine (*R*-alanine moiety), then the reported 85% yield represents a 70% asymmetric synthesis.

[20] F. Knoop and C. Martius, *Z. Physiol. Chem.*, **258**, 238 (1939).

[21] R. M. Herbst and E. A. Swart, *J. Org. Chem.*, **11**, 368 (1946).

[22] J. P. Greenstein and M. Winitz, *Chemistry of the Amino Acids* (New York: John Wiley & Sons, Inc., 1961), Vol. 1, p. 205.

---

[f] The mole ratio of keto acid to amine was 1:1.

[g] The mole ratio of keto acid to amine was 1:3.

[h] 10% Pd-C only.

[i] Pt_2O only.

[j] N-Pyruvyl-glycine.

[k] Al/Hg reduction.

[l] The ethyl ester of pyruvic acid was used; reduction with Zn/HCl.

[m] Reference 23 claims a hydrochloride with 91.4% e.e., whereas the correct value from the rotations reported appears to be 83%. The higher value for the purified hydrochloride is apparently due to enantiomeric fractionation during its isolation by crystallization.

[n] $Pd(OH)_2$—C only.

[o] The product was a mixture of aspartic acid and alanine; isolated only as DNP derivative, cf. also Table **7-4**.

The most extensive investigations of asymmetric hydrogenolytic transaminations have been carried out by Hiskey and Northrop[23,24] and Harada and co-workers.[25–27] Asymmetric syntheses of alanine, 2-aminobutyric acid, phenylalanine, phenylglycine, valine, aspartic acid, and glutamic acid have been studied using α-methylbenzylamine, α-ethylbenzylamine, and phenylglycine as chiral aminating agents. The two groups have used different work-up and isolation procedures, and their results, while generally comparable, do vary somewhat. In most cases Hiskey and Northrop formed the imine *in situ*, hydrogenated with 10% palladium on carbon (Pd—C) and then cleaved the original benzylic chiral group, R*, by hydrogenolysis with palladium hydroxide on carbon, a particularly effective catalyst for this purpose. The amino acid was then isolated by crystallization. However, Harada and co-workers found that this procedure sometimes caused fractionation of the chiral amino acid and racemate. Therefore, after hydrogenolysis, 2,4-dinitrofluorobenzene was added to form the 2,4-dinitrophenyl (DNP) derivative of the amino acid, which was isolated by chromatography, presumably without fractionation, and the optical rotation determined. In Table **7-2**, percent enantiomeric excess values obtained by the latter procedure are listed in parentheses along with those calculated from data obtained after isolation by the procedure of Hiskey and Northrop.

It has also been demonstrated[23,24] that isolation and purification of the intermediate amino acid derivative (**22**) from α-keto acid and chiral amine results in amino acid of higher enantiomeric purity than that obtained by the standard procedure in which **22** is hydrogenolyzed without isolation. This is not unexpected since **22** is a mixture of two diastereomers that can be readily separated in some cases.

One conclusion that can be drawn from the data in Table **7-2** is that the use of a 10% Pd—C catalyst to saturate the imine bond gives a higher percent asymmetric synthesis than either PtO_2 or $Pd(OH)_2$ catalysts. The superior stereoselectivity of hydrogenations over Pd catalysts has been noted by other workers.[8]

In order to emphasize some of the data on which other conclusions can be based, percent asymmetric synthesis values for a number of reactions under similar conditions are given in Table **7-3** as a function of the structures of chiral amine and α-keto acid. For the three amines listed, it is seen that an *S* amine yields *S* amino acid, and an *R* amine yields *R* amino acid. This correlation can be symbolized: *S*-**21** → *S*-**23**, as shown at the top of Table **7-3**, for the examples studied.

[23] R. G. Hiskey and R. C. Northrop, *J. Am. Chem. Soc.*, **83**, 4798 (1961).
[24] R. G. Hiskey and R. C. Northrop, *J. Am. Chem. Soc.*, **87**, 1753 (1965).
[25] K. Harada, *Nature*, **212**, 1571 (1966).
[26] K. Harada, *J. Org. Chem.*, **32**, 1790 (1967).
[27] K. Harada and K. Matsumoto, *J. Org. Chem.*, **32**, 1794 (1967).

TABLE 7-3

Hydrogenolytic Transamination—Percent Asymmetric Synthesis as a Function of Keto Acid and Amine Structure[a]

$$\mathrm{H\!-\!\underset{R''}{\overset{R'}{C}}\!-\!NH_2} + \mathrm{O{=}\underset{R}{\overset{COOH}{C}}} \longrightarrow \left[\mathrm{H\!-\!\underset{R''}{\overset{R'}{C}}\!-\!N{=}\underset{R}{\overset{COOH}{C}}}\right] \longrightarrow \mathrm{H_2N\!-\!\underset{R}{\overset{COOH}{C}}\!-\!H}$$

19 **21** **23**

No.	Keto acid R =	(S)-α-Methylbenzylamine R′ = Me, R″ = Ph % e.e.	(S)-α-Ethylbenzylamine R′ = Et, R″ = Ph % e.e.	(S)-Phenylglycine R′ = Ph, R″ = COOH % e.e.
1	Me	70, 81, 74, 63, 77, (65), (67)	39, 52, (51), (52)	41, (56)
2	Et	63, 37, 38, (37), (39)	36, 39, (33), (36)	32, (30)
3	Ph	73, 28, 29, (30), (31)	24, 25, (24), (26)	—
4	*i*-Pr	28	—	—
5	CH_2Ph	12, 12, 13, 13, (14), (14)	4, 5, (10), (11)	—
6	CH_2CH_2COOH	35, 13, (12)	2, (6)	35, (48)[b]
7	CH_2COOH	74, (69)	42, (52)	(47), (58)[b]

[a] Values in parentheses are based on data for 2,4-dinitrophenyl derivatives and are presumably more accurate indications of the degree of asymmetric bias. In other cases the amino acid was isolated. Unless otherwise noted, reactions involved sequential hydrogenation and hydrogenolysis with Pd—C and $Pd(OH)_2$, respectively, and a 1:2 mole ratio of keto acid to amine. These data are condensed from Table **7-2** where the references are given.

[b] 1:1 mole ratio of keto acid to amine; 10% Pd—C for both hydrogenation and hydrogenolysis.

The percent asymmetric synthesis varies with the nature of R in the keto acid. The general order appears to be Me > CH_2COOH > Et > Ph > CH_2Ph, CH_2CH_2COOH when either α-methyl or α-ethylbenzylamine are used. The relative position of isopropyl is uncertain (one experiment), but it appears to be comparable to phenyl. With phenylglycine the approximate order is Me, CH_2COOH > CH_2CH_2COOH > Et. For a given keto acid, α-methylbenzylamine induces a higher percent asymmetric synthesis than α-ethylbenzylamine. The effectiveness of phenylglycine relative to the other two amines does not fit a consistent pattern.

These general trends have prompted a great deal of detailed speculation concerning the nature of the transition states involved, and a number of topolographic models have been proposed.[26–28] Several factors make such models little more than interesting intellectual exercises at the present time. (a) There has not been a very extensive study of the effect of various chiral groups. (b) There is insufficient evidence concerning the geometrical

[28] A. Kanai and S. Mitsui, *Nippon Kagaku Zasshi*, **87**, 183 (1966); *Chem. Abst.*, **65**, 16835 (1966).

isomerism and conformational preferences of similar imines in solution. (c) Asymmetric induction is 1,3 and comparison with the more common 1,2- or 1,4- (Cram or Prelog) type, for which many examples are available, is hazardous. (d) Finally, these are catalytic reactions, and the nature of adsorbed species is not known.

Models for hydrogenolytic transaminations with phenylglycine will be considered first. These reactions have been carried out in a basic medium so the imine (**21**) should be primarily in the form of its dianion. The repulsion of the carboxylate groups in the two portions of the imine may introduce a factor not present in the case of the imines from α-alkylbenzylamines. Therefore one must consider that asymmetric hydrogenolytic transaminations with phenylglycine may not follow the same steric course as those with α-alkylbenzylamines. Harada[26,29] has proposed models similar to *E*-**27** and *Z*-**27**. The model for the *Z* isomer is constructed with phenyl and hydrogen flanking the carboxylate group of the imino acid portion.* In this conformation the carboxylate groups are as far apart as is possible for the isomer having a *Z*-arrangement of groups around the double bond. Saturation of the C=N bond in either *E*-**27** or *Z*-**27** with hydrogen from below the plane of the paper predicts the observed stereochemistry. Harada has attached considerable significance to the fact that these models do not show a serious interaction of R on the keto acid with the chiral center of the amine. This is, in his view, quite compatible with the observation that the percent asymmetric synthesis does not appear to be very sensitive to the nature of R when phenylglycine is the transaminating agent (Table **7-3**).

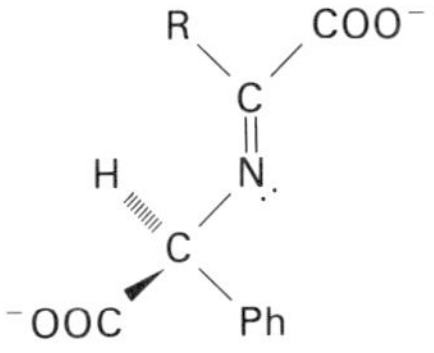

R, COO⁻, C, N, H, C, ⁻OOC, Ph

E-**27**

R, COO⁻, C, N, H, C, ⁻OOC, Ph

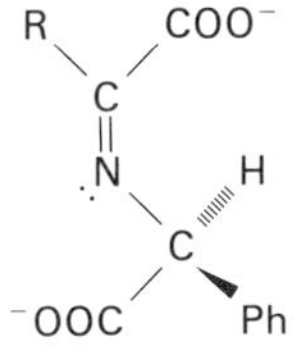

Z-**27**

Kanai and Mitsui have proposed **28A**, with hydrogen attacking from the less hindered side (below the plane of the page), as a model for hydrogenolytic transaminations with optically active (*S*)-α-methylbenzylamine.[28] Harada[27] has argued against this model on the grounds that it does not predict the sensitivity of percent asymmetric synthesis to the steric bulk of R that is observed when α-alkylbenzylamines are used (Table **7-3**). He favors a

[29] (a) K. Harada and K. Matsumoto, *J. Org. Chem.*, **33**, 4467 (1968). (b) K. Matsumoto and K. Harada, *ibid.*, **33**, 4526 (1968).

* Harada[26] actually proposed a conformation for *Z*-**27** having hydrogen on the chiral center in the plane of the imino acid COO⁻ group, but the conformation shown here is equally consistent with the results.

competition of *E*-**28B** and *Z*-**28C** with the latter becoming increasingly important for larger R groups, and he also notes that when R′ is varied from methyl to ethyl, model **28B–28C** predicts that the difference between the energy required for approach of the hydrogen from below versus above the plane of the page would *decrease* in accord with the observed decrease in percent asymmetric synthesis on going from α-methyl to α-ethylbenzylamine. More elaborate and even more speculative representations which suggest the role of the catalyst have also been proposed.[27,29]

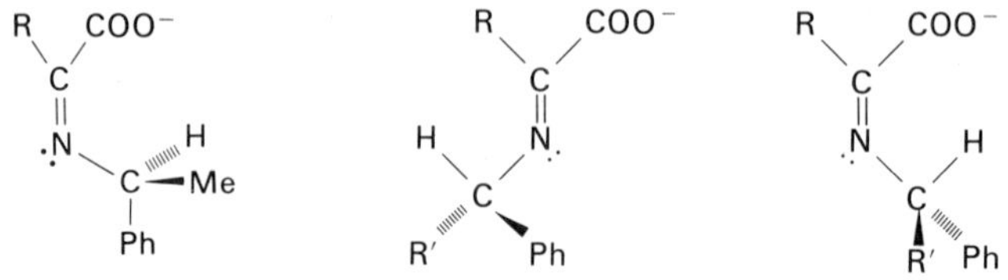

Z-**28A** *E*-**28B** *Z*-**28C**

The detailed mechanism of these reactions is certainly complex and the favored conformation of adsorbed species depends not only upon the steric and electronic properties of groups in the substrate but also upon the reaction medium. Rather extreme solvent effects have been noted for some palladium catalyzed hydrogenolytic transaminations.[29] For example, Table **7-4** lists the results obtained in systematic studies involving several α-keto acids and several benzylic amines in various solvents. Harada[29a] has interpreted these effects in terms of less chelated character for the adsorbed intermediate imine in polar solvents and therefore lower asymmetric bias. In solvents of low polarity a more rigid chelated species is envisaged as the predominate adsorbed intermediate. The importance of solvent effects is dramatized by Nos. 39 and 40 in Table **7-4** in which enantiomeric amines induce the predominate formation of the same (*R*)-(−)-glutamic acid. Similarly (*S*)-(−)-α-methylbenzylamine induces the formation of (*S*)-(+)-glutamic acid in isopropanol-triethylamine solvent (No. 33, 14% e.e.) but the formation of (*R*)-(−)-glutamic acid (25% e.e.) in water-pyridine solvent (No. 38).

Hydrogenation of the imino lactone **29**, prepared from racemic *erythro*-1,2-diphenylethanolamine via the sequence of reactions shown below,[12] gave only the diastereomer **30** (as a *dl* pair), indicating that hydrogenation was completely stereoselective. A similar stereoselective hydrogenation of an exocyclic carbon-carbon double bond in another derivative of (+)-1,2-diphenylethanolamine (**5A**) has been described in Sec. **7-1.2**. The high stereoselectivity in the case of **29** indicates essentially complete shielding of one face of the C=N bond so that hydrogen transfer occurs only from the axial direction. Compound **30** may be considered to be a derivative of phenylglycine and, if *R,S* or *S,R* ethanolamine had been used, the phenylglycine moiety would have been formed in only one configuration; *i.e.*, complete asymmetric synthesis would have been achieved.

HO NH$_2$ PhCHCHPh (*dl-erythro*) $\xrightarrow{Ph_3CCl}$ HO NHCPh$_3$ PhCHCHPh $\xrightarrow{PhCOCOCl}$ NHCPh$_3$ PhCHCHPh OC(=O)—C(=O)—Ph $\xrightarrow{1)\ HCl;\ 2)\ AgNO_3}$ **29** $\xrightarrow{H_2(Pd \cdot C),\ EtOAc}$ **30**

A few amino compounds other than simple amines have also been employed as transaminating agents. Akabori[30] used the hydrazide of (*R*)-2-methyl-3-phenylpropanoic acid (**31**) to synthesize (*S*)-(+)-alanine (approximately 8% e.e.) and (*S*)-(−)-phenylalanine (approximately 5% e.e.). The low percent asymmetric synthesis might reflect the relatively large distance between the original and developing chiral centers or the fact that methyl and benzyl may not exert a large steric bulk difference, or both (Table **7-2**, Nos. 13 and 35).

31 CONHNH$_2$ / CH$_3$—C—CH$_2$Ph / H $\xrightarrow{R'COCOOH}$ CONHN=C(COOH)R' / CH$_3$—C—CH$_2$Ph / H $\xrightarrow{1)\ PtO_2;\ 2)\ P,\ HI}$ H$_2$N—C(COOH)(R')—H (*S*-**23**)

R′ = Me or Benzyl

A hydrazone intermediate (**33**) was isolated (95% recrystallized yield) from the reaction of ethyl pyruvate and (*S*)-N-aminoanabasine (**32**). Reduction of this hydrazone with zinc and ethanolic hydrogen chloride gave, after hydrolysis and purification by adsorption chromatography, (*S*)-(+)-alanine (10% yield, 40% e.e.) and recovered anabasine (90% of the original optical activity retained).[31] It would be desirable to have data for similar reactions in which the pyridine moiety was replaced by an alkyl or aryl group.

[30] S. Akabori and S. Sakurai, *Nippon Kagaku Zasshi*, **78**, 1629 (1957); *Chem. Abst.*, **53**, 21687b (1959).

[31] A. N. Kost, R. S. Sagitullin, and M. A. Yurovskaja, *Chem. and Ind.*, 1496 (1966).

TABLE 7-4

Solvent Effects on Hydrogenolytic Transaminations[a]

$$R{-}\overset{O}{\overset{\|}{C}}{-}COOR' + R^*{-}NH_2 \longrightarrow (R)(R'OOC)C{=}N{-}R^* \xrightarrow[Pd]{H_2} R{-}\overset{*}{C}H(NH_2)COOH$$

No.	R	R′	R*NH₂[a]	Solvent[b]	Product amino acid	% e.e.[c]
1	Me	H	S-(−)-MBA	THF	S-(+)-Alanine	62 (66)
2				EtOH		63 (67)
3				H_2O/py		28 (40)
4				H_2O/NaOH		26 (33)
5	Me	H	S-(−)-EBA	EtOH		52 (52)
6				EtOH/H_2O/NaOH		28 (37)
7				1:4 MeOH/H_2O/NaOH		9 (31)
8	Me	H	R-(+)-NpEtA	EtOH	R-(−)-Alanine	80 (83)
9				EtOH/H_2O/NaOH		74 (73)
10	Me	Benzyl	S-(−)-MBA	Hexane[d]	S-(+)-Alanine	55 (72)
11				EtOAc		44 (60)
12				*i*-PrOH		34 (46)
13				DMF		— (50)
14				MeOH		10 (38)
15				Dioxane/H_2O		— (29)
16				2:1 MeOH/H_2O		— (35)
17				1:2 MeOH/H_2O		— (39)
18				1:4 MeOH/H_2O		14 (29)
19				H_2O[d]		21 (41)

TABLE 7-4—continued

20	Me	Benzyl	*R*-(+)-MBA[e]	Dioxane	R-(−)-Alanine	29 (58)
21	Me	Benzyl	*R*-(+)-EBA[e]	Hexane[d]	R-(−)-Alanine	39 (74)
22			R-(+)-NpEtA	Hexane[d]	R-(−)-Alanine	61 (86)
23				EtOAc		47 (78)
24	Ph	H	S-(−)-MBA	EtOH	S-(+)-Phenylglycine	28 (30)
25				H_2O/NaOH		23 (24)
26A	CH_2COOH	H	S-(−)-MBA	EtOH/H_2O/NaOH	S-(+)-Alanine[f]	50 (58)
26B					R-(+)-Aspartic	44 (45)
27			S-(−)-MBA	EtOH	S-(+)-Alanine	74 (69)
28A			S-(−)-EBA	EtOH/H_2O/NaOH	S-(+)-Alanine[f]	38 (37)
28B					R-(+)-Aspartic	27 (25)
29			S-(−)EBA	EtOH	S-(+)-Alanine	42 (52)
30			R-(+)-NpEtA	EtOH/H_2O/NaOH	R-(−)-Alanine	74 (73)
31A			S-(+)-PhGly	H_2O/NaOH	S-(+)-Alanine[f]	56 (60)
31B					S-(−)-Aspartic	48 (53)
32A			R-(−)-PhGly	H_2O/NaOH	R-(−)-Alanine[f]	54 (62)
32B					R-(+)-Aspartic	53 (53)
33	CH_2CH_2COOH	H	S-(−)-MBA	*i*-PrOH/Et_3N	S-(+)-Glutamic	12 (14)
34				EtOH		13 (12)
35				MeOH/Et_3N		6 (5)
36				Dioxane/H_2O/Et_3N	R-(−)-Glutamic	19 (19)
37				MeOH/H_2O/NaOH	R-(−)-Glutamic	26 (27)
38	CH_2CH_2COOH	H	S-(−)-MBA	H_2O/py	R-(−)-Glutamic	27 (27)
39			R-(+)-EBA	EtOH	R-(−)-Glutamic	2 (6)
40			S-(−)-EBA	H_2O/MeOH/NaOH	R-(−)-Glutamic	25 (30)

[a] Numbers 27–32B from Reference 29b; all others from Reference 29a. Abbreviations for chiral amines are the same as those in Table **7-2**.
[b] THF = tetrahydrofuran, EtOH = ethanol, py = pyridine, MeOH = methanol, EtOAc = ethyl acetate, *i*-PrOH = isopropanol, DMF = N,N-dimethylformamide.
[c] Values in parentheses are based on rotations for dinitrophenyl derivatives and are presumably more reliable.
[d] The solvent phase was not homogeneous.
[e] Both enantiomers of amine were of nearly the same, but not identical, enantiomeric purities.
[f] The intermediate imine partially decarboxylated, thus giving two amino acid products.[29]

Results from the use of N-aminoconiine or some other chiral 1-amino-2-substituted piperidine, for example, would be of interest.

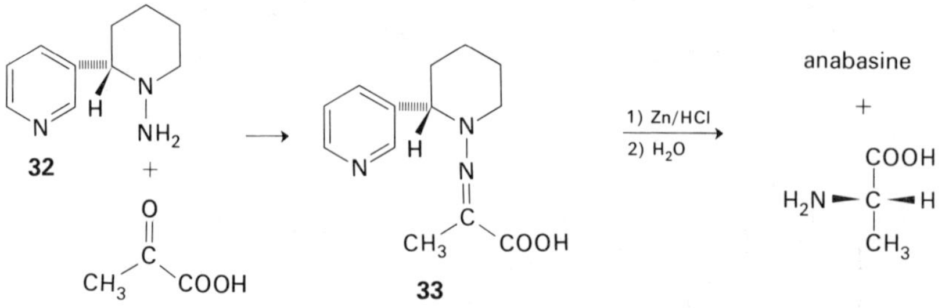

32 + $CH_3COCOOH$ → 33 —1) Zn/HCl, 2) H_2O→ anabasine + H_2N–C(COOH)(H)–CH_3

Similar, but much more stereoselective reductions have been reported in preliminary form.[32] Using N-amino-2-hydroxymethylindoline (**34**) of unknown configuration, (*S*)-(+)-alanine (82% e.e.) and (*S*)-(+)-α-aminobutyric acid (98% e.e.) were synthesized via aluminum amalgam reduction of intermediate **35** (Table **7-2**, Nos. 16 and 28). On the basis of the stereochemical result and the high percent asymmetric synthesis, one can speculate that the configuration at C-2 of the indoline is *R* (as shown in Fig. **7-5**) assuming that hydrogen saturates the imine double bond of **35** from the less hindered side (above the plane of the page).

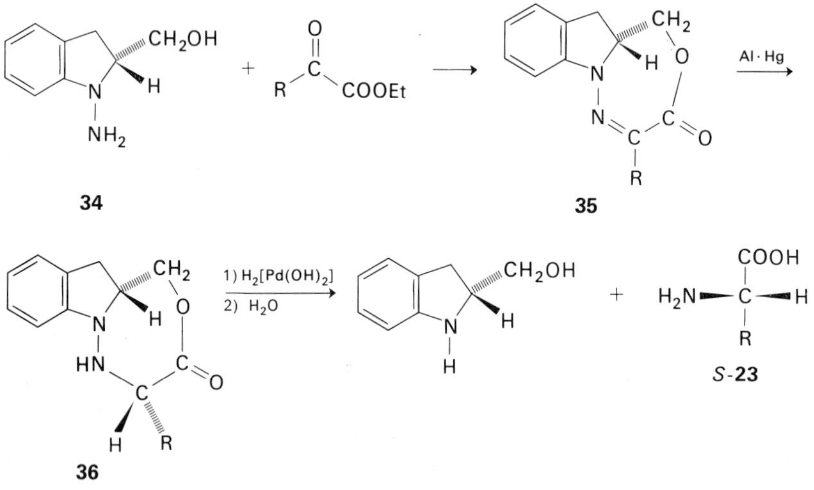

34 + RCOCOOEt → 35 —Al·Hg→ 36 —1) $H_2[Pd(OH)_2]$, 2) H_2O→ 2-hydroxymethylindoline + H_2N–C(COOH)(H)–R (*S*-**23**)

Fig. 7-5. Hydrogenolytic transamination with N-amino-2-hydroxymethylindoline.

[32] R. J. McCaully, *Diss. Abst.*, **26**, 1919 (1965); *Chem. Abst.*, **64**, 3677 (1966) (with E. J. Corey, Harvard University).

It is interesting to compare this result with that observed for the catalytic hydrogenation of **29**. In both **29** and **35** the imine is prevented from undergoing *syn-anti* isomerism, and this fact, plus the rigidity of the heterocyclic system, leads to an enhanced extent of asymmetric synthesis. Similar conformational bias is almost certainly associated with enzymatic transaminations.

All the asymmetric syntheses described so far have had the inducing chiral center attached to the nitrogen of the imine bond. Syntheses in which this center is located elsewhere have also been investigated. For example, Harada and co-workers[27,33] and Hiskey and Northrop[24] have studied hydrogenations of oximes and imines formed from menthyl pyruvate or chiral amides of pyruvic acid. In addition, there have been a few studies of substrates with multiple chiral groups, one attached to the imine nitrogen and another to the carbonyl group of a keto acid moiety. In such cases with multiple chiral groups one encounters the intriguing prospect of the chiral forces either reinforcing or canceling each other.

Table **7-5** summarizes the stereochemistry of hydrogenations of oximes and N-benzyl imines from amides of pyruvic acid and chiral amines. Table **7-6** lists similar data for oximes and N-benzyl imines of menthyl pyruvate, menthyl 2-ketobutyrate and menthyl phenylglyoxylate. It is seen from Nos. 4–7 in Table **7-5** that there is a reversal of the stereochemical course in the asymmetric hydrogenation of **37** when the methyl group (R″) on the chiral α-methylbenzyl moiety of the amide center is replaced with ethyl. Thus, although a Prelog-type model (**40**) will rationalize the results when R_M = ethyl and R_L = Ph, it is incompatible with the results when R_M = Me and R_L = Ph. Harada[27] has attempted to explain this reversal of stereochemistry by the use of different and specific models, but there seems to be no rational reason for the shift from one model to the other to accommodate the replacement of methyl with ethyl and one is forced to the conclusion that much more

$[H_2]$ → $R'C(=NR)C(=O)NHC(H)(R_L)(R_M)$ $\xrightarrow[(-RH)]{H_2\ [catalyst]}$ $R'C(H)(NH_2)COOH$ + $H_2NC(H)(R_L)(R_M)$

R-**40** *R*-**23**

data will be required before the topology of the transition state in this reaction is elucidated.

In contrast, the reduction of all the (−)-menthyl α-imino esters reported in Table **7-6** are analogous to the stereochemistry observed in the reduction

[33] K. Matsumoto and K. Harada, *J. Org. Chem.*, **31**, 1956 (1966).

TABLE 7-5

Asymmetric Hydrogenation of Chiral Amides of α-Imino and α-Oximino Acids[24,27]

$$\underset{\textbf{37}}{R{-}N{=}C(R')(\overset{*}{C}ONHCHR''R''')} \xrightarrow[(-RH)]{H_2,\ catalyst} \underset{\textbf{38}}{R'\overset{*}{C}H(NH_2)CONHCHR''R'''} \xrightarrow{H_2O} \textbf{23}$$

No.	R	R′	R″	R‴	Original chiral center	Catalyst[a]	New chiral center	Asymmetric synthesis[b] % e.e.
1	Benzyl	Me	Me	COO*i*Bu	*S*	A	*R*	34
2	Benzyl	Me	*i*Pr	COO*i*Bu	*S*	A	*S*	32
3	Benzyl	Me	*i*Pr	COO*i*Bu	*R*	A	*R*	32
4	OH	Ph	Me	Ph	*S*	B	*R*	5.5
5	OH	Ph	Me	Ph	*R*	B	*S*	8.8
6	OH	Ph	Et	Ph	*S*	B	*S*	4.9
7	OH	Ph	Et	Ph	*R*	B	*R*	10

[a] A = 10% Pd—C; B = $Pd(OH)_2$.
[b] For the first reaction listed, the percent asymmetric synthesis value is based upon analysis of **38** after isolation by chromatography;[24] for other reactions the DNP derivative of the amino acid (**23**) was isolated after hydrolysis of **38**.[27]

TABLE 7-6

Asymmetric Hydrogenation of Chiral Esters of α-Imino or α-Oximino Acids[33]

$$\underset{\textbf{39}}{R{-}N{=}C(COO(-)menthyl)(R')} \xrightarrow[(-RH)]{H_2,\ catalyst} \xrightarrow{H_2O} \underset{\textbf{23}}{R'{-}CH(NH_2)COOH} + (-)\text{-menthol}$$

No.	R′	R	Catalyst[a]	New chiral center	Asymmetric synthesis[b] % e.e.
1	Me	Benzyl	A	*RS*-(±)	0 (0)
2	Me	Benzyl	B	*R*-(−)	13.6 (16.3)
3	Et	Benzyl	A	*RS*-(±)	0 (0)
4	Et	Benzyl	B	*R*-(+)	0 (7.6)
5	Me	OH	A	*R*-(−)	11.5 (24.6)
6	Me	OH	B	*R*-(−)	11.0 (24.5)
7	Et	OH	A	*R*-(+)	10.3 (20.5)
8	Et	OH	B	*R*-(+)	9.4 (20.7)
9	Ph	OH	A	*R*-(−)	24.9 (49.1)
10	Ph	OH	B	*R*-(−)	22.7 (44.2)

[a] A = 10% Pd—C; B = $Pd(OH)_2$.
[b] First value listed is for isolated amino acid; value in parentheses is based on the rotation of the DNP derivative which was isolated by chromatography.

of the corresponding (−)-menthyl α-keto esters (compare with data of Table **2-4**). These results are accommodated by a Prelog-type model such as **40** in which an ester replaces the amide group.

The data in Table **7-7** reveal the combined asymmetric inductive effects of the chiral (−)-menthyl ester group and the chiral α-alkybenzylimino

TABLE 7-7

Asymmetric Hydrogenation of Imines Having Two Chiral Centers

$$\underset{\mathbf{41}}{\begin{matrix}\text{COO}(-)\text{-Menthyl}\\ |\\ \text{C=O}\\ |\\ \text{CH}_3\end{matrix}} \xrightarrow{\text{R*NH}_2} \underset{\mathbf{42}}{\begin{matrix}\text{COO}(-)\text{-Menthyl}\\ |\\ \text{C=NR*}\\ |\\ \text{CH}_3\end{matrix}} \xrightarrow[\text{2) Hydrolysis}]{\text{1) H}_2\text{, Pd(OH)}_2} \underset{\mathbf{13}}{\begin{matrix}\text{COOH}\\ |\\ \text{*CHNH}_2\\ |\\ \text{CH}_3\end{matrix}} + (-)\text{-Menthol}$$

No.	R*NH$_2$[a]	Configuration of alanine	% e.e.[b]	Ref.
1	*S*-(−)-MBA	*S*-(+)	16 (19)	27
2	*R*-(+)-MBA	*R*-(−)	56 (60)	27
3	*S*-(−)-MBA[c]	*S*-(+)	18 (24)	33
4	*R*-(+)-MBA[c]	*R*-(−)	64 (66)	33
5	*S*-(−)-EBA	*S*-(+)	12 (15)	27
6	*R*-(+)-EBA	*R*-(−)	34 (36)	27

[a] MBA = α-methylbenzylamine; EBA = α-ethylbenzylamine.
[b] Values in parentheses are for DNP derivative; all values have been corrected for racemization during hydrolysis.
[c] 5 % Pd—C catalyst.

group. As seen in Table **7-6**, in each case where the (−)-menthyl ester group alone was the inducing moiety, (*R*)-alanine was produced in excess. In Table **7-2** (Nos. 5 and 6), it was seen that the reduction of α-methylbenzyliminopyruvic acid gave about 65 % asymmetric synthesis and that the imine made from (*R*)-α-methylbenzylamine leads to the formation of (*R*)-alanine. One might conclude from these observations that a (−)-menthyl ester group and an (*R*)-α-methylbenzylimino group in the same molecule would have additive "double asymmetric inductive" effects while a (−)-menthyl ester group and an (*S*)-α-methylbenzylimino group in the same molecule would work in opposition to each other. The data in Table **7-7** show that the α-alkylbenzylimino group controls the asymmetric synthesis since (*R*)-amine induces formation of (*R*)-alanine while (*S*)-amine induces formation of (*S*)-alanine in the presence of a (−)-menthyl ester group. The (*S*)-amine moiety and (−)-menthyl ester moiety act in opposition to each other leading to reduced stereoselectivity, but additive effects of (*R*)-amine and (−)-menthyl moieties are not observed.

Hiskey and Northrop[24] have also investigated multiple chiral influences. As part of a study of dipeptide asymmetric synthesis via hydrogenolytic transamination, they obtained the data shown in Table **7-8**. From these

TABLE 7-8

Asymmetric Hydrogenation of Chiral Imino Amides[24]

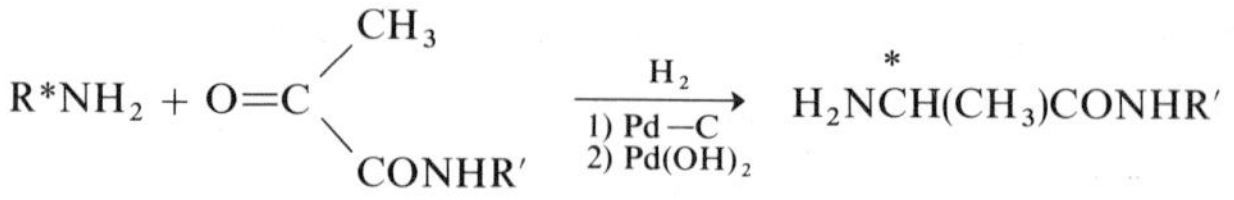

$$R^*NH_2 + O{=}C(CH_3)(CONHR') \xrightarrow[\text{1) Pd—C} \;\; \text{2) Pd(OH)}_2]{H_2} H_2N\overset{*}{C}H(CH_3)CONHR'$$

43

No.	R*—NH$_2$[a]	R′NH	Major product	Asymmetric synthesis % e.e.
1	*S*-MBA	Glycyl	*S*-Alanylglycine	54
2	*R*-MBA	Glycyl	*R*-Alanylglycine	58
3	*R*-MBA	(*S*)-Alanyl	*R,S*-Alanylalanine	48
4	*S*-MBA	(*S*)-Alanyl	*S,S*-Alanylalanine	28
5	$PhCH_2NH_2$	(*S*)-Alanyl	*R,S*-Alanylalanine	34

[a] MBA = α-methylbenzylamine.

data it can be concluded that an (*S*)-alanyl center in the amide portion (No. 5) should oppose the chiral influence of an (*S*)-α-methylbenzyl center in the imine moiety (No. 1). That it does is apparent from the result of experiment No. 4. When an (*S*)-alanyl amide center is present with an (*R*)-α-methylbenzyl imine center (No. 3), the percent asymmetric synthesis is greater than that observed for chiral amide alone, but it is less than that found for chiral imine alone. Again *additivity is not observed.*

Kanai and Mitsui[28] have reported additional evidence regarding opposing chiral influences. Hydrogenation of **44**, in which (*S*)-α-methylbenzylamine forms the chiral moiety of both the imine and amide portions, gave a very low percent enantiomeric excess of (*R*)-(−)-phenylglycine. However, hydro-

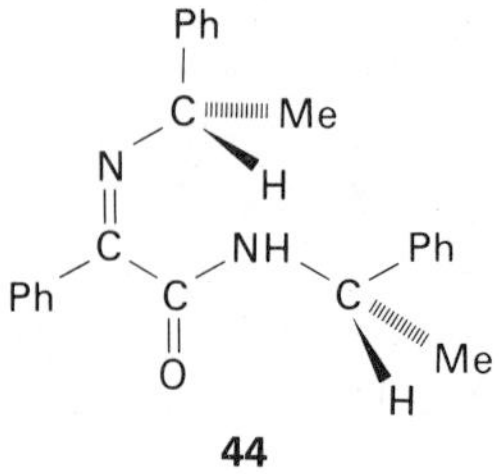

Ph, C, Me, N, H, C, NH, Ph, Ph, C, C, Me, O, H

44

genations of the imino acid of **44**, lacking the chiral amide moiety, gives up to 73% e.e. of (*S*)-(+)-phenylglycine (cf. Table **7-2**, Nos. 36–38). Thus the two chiral substituents in **44** must exercise opposing chiral effects during the asymmetric hydrogenation.

7-3. Nucleophilic Additions to Carbon-Carbon Double Bonds

7-3.1. Addition of Chiral Amines to the Carbon-Carbon Double Bond of an Achiral Substrate.* Terent'ev and co-workers[34] reported that the reaction between maleic acid and two equivalents of (*S*)-α-methylbenzylamine gave a mixture of readily separable diastereomeric N-α-methylbenzylaspartic acids. One isomer, obtained in 43% yield, was subjected to hydrogenolysis to give *R*-(−)-aspartic acid (88% e.e., 100% yield). This cannot be considered a verified asymmetric synthesis because the intermediate diastereomers were separated. In fact, Liwschitz and Singerman[35] have described a comparable experiment where the intermediate (74% yield) was hydrogenolyzed to racemic aspartic acid (67% yield). In this case crystallization of the product could have concentrated the racemate and thus the question of the extent of asymmetric synthesis is still unsettled, although it cannot be particularly high.

Treatment of crotonic acid with (*S*)-α-methylbenzylamine also gave a readily separable mixture of diastereomers. In one case, however, the intermediate diastereomers were hydrogenolyzed without prior crystallization to give (+)-β-aminobutyric acid (10% e.e.).[36] Harada and Matsumoto[37] have studied this type of reaction, taking care to avoid fractionation of intermediates or products. When diethyl maleate (**45**) was allowed to react with excess optically active α-methylbenzylamine and the resulting product was subjected to hydrogenolysis and hydrolysis, optically active aspartic acid was obtained (13.7–15.4% e.e.). Similar treatment of diethyl fumarate (**47**) gave optically active product (4.6% e.e.). In both cases the use of (*S*)-α-methylbenzylamine resulted in an excess of (*S*)-aspartic acid while (*R*)-amine gave an excess of (*R*)-amino acid.

A detailed study of the intermediates involved in these reactions revealed that both N-alkyl succinimides (**51**) and N-alkyl aspartic acid diamides (**50**)

[34] A. P. Terent'ev, R. A. Gracheva, L. F. Titova, and T. F. Dedenko, *Dokl. Akad. Nauk SSSR*, **154**, 1406 (1964); English translation, 206 (1964); *Chem. Abst.*, **60**, 12099a (1964).

[35] Y. Liwschitz and A. Singerman, *J. Chem. Soc.*, (*C*), 1200 (1966).

[36] A. P. Terent'ev, R. A. Gracheva, and T. F. Dedenko, *Dokl. Akad. Nauk SSSR*, **163**, 386 (1965); English translation, 674 (1965); *Chem. Abst.*, **63**, 11344b (1965).

[37] K. Harada and K. Matsumoto, *J. Org. Chem.*, **31**, 2985 (1966).

* Because of the intermediates presumed to be involved these reactions might also be classified under the heading for Sec. **7-3.3.**

were formed and that these underwent hydrogenolysis to enantiomeric products (Fig. **7-6**). More of intermediate **50** was produced from fumarate than from maleate, which is consistent with the lower percent asymmetric synthesis observed in the former case.

Phthalimido *t*-butyl ketene, **52**, reacts with (*R*)-(+)-α-methylbenzylamine (**53A**) to give a mixture of diastereomers of N-phthaloyl-"*t*"-leucine-α-methylbenzylamide in which the *R*,*R* amide (**54**) is in excess by 1.4 to 4.7% over a 130° temperature range (Table **7-9**).[38] With the isopropyl ester

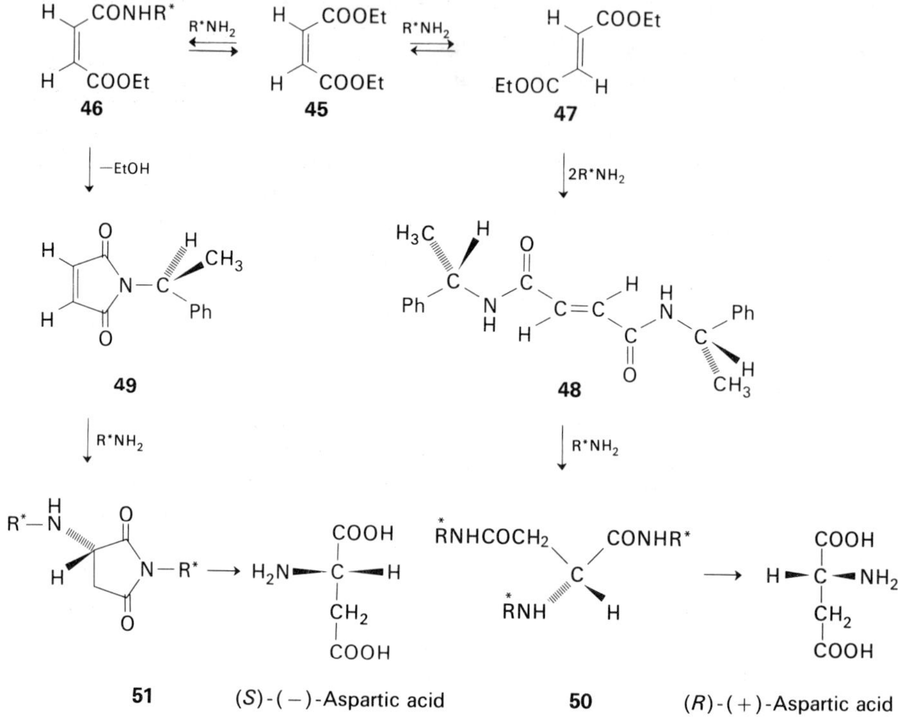

Fig. 7-6. Reaction of diethyl fumarate and diethyl maleate with (*S*)-α-methylbenzylamine (R^*NH_2). Both esters yield an excess of (*S*)-aspartic acid; fumarate gave a lower % e.e., presumably due to the greater contribution of **48** and **50**.

of (*R*)-alanine (**53B**), however, much higher percents asymmetric synthesis are observed, ranging between 44 and 69%, the highest value being a 69.5% excess of *R*,*R*-dipeptide ester (**54B**) at −62°.

The steric result was rationalized in terms of a model (**55**) in which a proton shift from nitrogen to carbon, which presumably determines the configuration at the incipient asymmetric center, is assumed to occur more

[38] S. Winter and H. Pracejus, *Chem. Ber.*, **99**, 151 (1966).

readily from the side of the smallest group attached to the inducing chiral center of the amine moiety.*

TABLE 7-9

Reaction of Chiral Amines with *t*-Butyl Phthalimido Ketene[38]

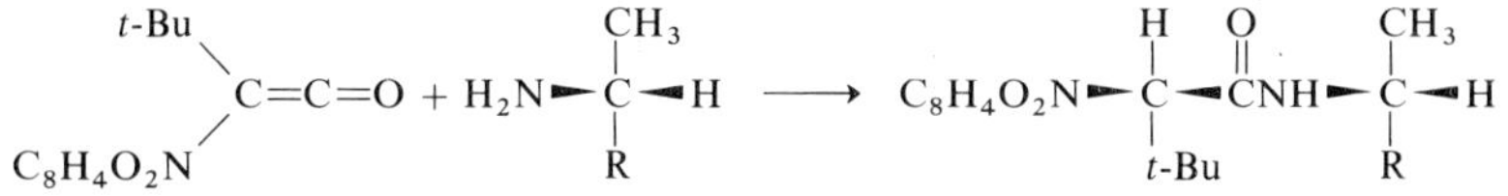

$$(t\text{-Bu})(C_8H_4O_2N)C{=}C{=}O + H_2N\text{—}C(CH_3)(H)(R) \longrightarrow C_8H_4O_2N\text{—}C(H)(t\text{-Bu})\text{—}C(=O)NH\text{—}C(CH_3)(H)(R)$$

52 $C_8H_4O_2N$=Phthalimido (*R*)-**53A**, R = Ph (*R*)-**53B**, R = COO-*i*Pr (*R,R*)—**54** Predominant diastereomer

R-53A		R-53B	
Temp.°	% e.e. (*R,R*)**54A**	Temp.°	% e.e. (*R,R*)**54B**
40.4	3.1	45	49
20.0	4.7	24	53
0.9	1.9	5	55
−49	4.1	3	44
−62	2.1	−20	55
−90	1.4	−40	63
		−62	69
		−95	63
		−110	63

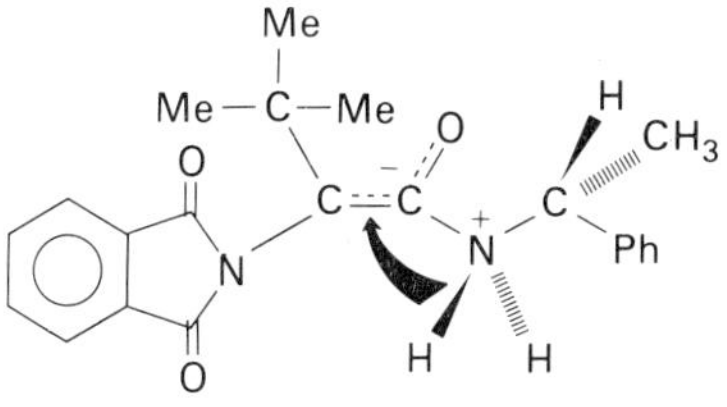

Me
Me—C—Me
O O
C==C
N N
H CH3
C
Ph
H H
O

55

* The model embodies the assumption that **55** is an intermediate, that the favored conformation of the chiral center with respect to the carbonyl group is as shown, that *t*-Bu is larger than phthalimido [as is suggested by the relative conformational free energies of these groups, H. Booth and G. C. Gidley, *Tetrahedron Lett.*, 1449 (1964)], that the best orientation for the presumably larger *t*-Bu group would be "*trans*" to the chiral alkylammonium moiety, and that hydrogen is transferred from the least hindered side. The analogy to the situation discussed in Sec **6-6.2** (Fig. **6-28**) is apparent.

7-3.2. Addition of Achiral Nucleophiles to the Carbon-Carbon Double Bond of a Chiral Substrate. Liwschitz and Singerman[35] prepared (*R*)-N-benzyl-aspartic acid, **58** (60% e.e.) by the addition of benzyl amine to (*R*)-N-α-methylbenzylmaleamic acid (**56**). Since the intermediate half amide, **57**, was isolated and recrystallized (which may have caused partial fractionation of diastereomers) and since the yield was only about 54%, the exact percent of asymmetric synthesis is uncertain but the results above (Sec. **7-3.1**) suggest that it was quite significant.

$\xrightarrow{RNH_2}$ $\xrightarrow{HCl}$

56 $R = CH_2Ph$ — **57** (54% yield) — (*R*)-**58**, 60% e.e. (62% yield)

A similar reaction of (*R*,*R*)-diamide **59** with benzylamine was also studied.[37] The intermediate diamide of N-benzylaspartic acid was hydrolyzed (presumably completely) and the N-benzyl group was cleaved by hydrogenolysis to give (*S*)-aspartic acid (5.6–7.6% e.e.) which was isolated as the DNP derivative by column chromatography. It is interesting that the

1) $PhCH_2NH_2$ 2) H_2O 3) H_2, Catalysis 4) DNP-F

59 — (*S*)-**58** R = 2,4-Dinitrophenyl

(*R*,*R*)-diamide **59** gave the (*S*)-amino acid whereas the (*R*)-half amide **56** gave (*R*)-amino acid when treated similarly with benzylamine. This may be a result of the different directing influence for the diamide versus the half amide or of the difference in maleamide versus fumaramide or the two reactions may differ mechanistically. There is, in addition to these complexities and those described in Sec. **7-3.1** (Fig. **7-6**), the possibility of amide interchange with the benzylamine.

7-3.3. Addition of Chiral Amine to the Carbon-Carbon Double Bond of a Chiral Substrate.* Chiral diamide **59** has also been treated with chiral amine having the same configuration as that used to form

* See also examples discussed in Sec. **7-3.1**.

the amide.[37] Following hydrolysis and hydrogenolysis of the resulting diastereomers of the N-α-methylbenzyl aspartic acid diamide, the product was isolated both as the free amino acid (approximately 11 % e.e.) and as the DNP derivative (approximately 15% e.e.). It was shown that crystallization of the free aspartic acid resulted in concentration of the racemate; therefore, values based on the rotation of the DNP derivative are more reliable. In this case, (*R*,*R*)-diamide plus (*R*)-amine gave (*S*)-aspartic acid and (*S*,*S*)-diamide plus (*S*)-amine gave (*R*)-aspartic acid. The stereochemical relationships for this and several related reactions that have been discussed in Sec. **7-3** are summarized in Table **7-10**. Harada and Matsumoto have speculated concerning models consistent with the limited data available.[37]

TABLE 7-10

Configurational Correlations for the Addition of Amines to Fumaric Acid Derivatives[37]

$$\text{R-C(=O)-CH=CH-C(=O)-R}' \xrightarrow[\text{2) Hydrolysis; 3) Hydrogenolysis}]{\text{1) R}''\text{NH}_2} \text{HOOCCH}_2\text{CH(NH}_2\text{)COOH}$$

No.	R[a]	R′[a]	R″NH$_2$	Predominant enantiomer of aspartic acid	% e.e.
1	OEt	OEt	*S*-MBA	*S*	4.6
2	OEt	OEt	*R*-MBA	*R*	4.5
3	*R*-MBA	*R*-MBA	Benzylamine	*S*	5.6
4	*S*-MBA	*S*-MBA	Benzylamine	*R*	7.6
5	*R*-MBA	*R*-MBA	*R*-MBA	*S*	15.2
6	*S*-MBA	*S*-MBA	*S*-MBA	*R*	15.3

[a] When R is derived from α-methylbenzylamine, the symbols *R*-MBA and *S*-MBA are used for *R* and *S* amines.

7-3.4. Addition of Achiral Nucleophiles to the Carbon-Carbon Double Bond of an Achiral Substrate Under the Influence of a Chiral Catalyst. Just as many of the experiments described in Sec. **7-2.2** were formally similar to enzymatic transaminations, those reactions described in the preceding three sections are reminiscent of aminations catalyzed by aspartase* (also called aspartic ammonia lyase).

* Aspartase is found in plants and microorganisms but not in man. Investigation of the quaternary structure suggests that it is tetrameric with a subunit molecular weight of about 45,000.[39]

[39] V. R. Williams and D. J. Lartigue, *J. Biol. Chem.*, **242**, 2973 (1967).

Aspartase promotes a stereospecific, reversible addition of NH_4^+ to fumarate, with a *trans* addition of D and ND_3 resulting in the formation of (2*S*,3*R*)-3-

$$\text{H(}^-\text{OOC)C=C(COO}^-\text{)H} \underset{\text{Aspartase}}{\overset{\overset{+}{N}D_4,\ D_2O}{\rightleftarrows}} \text{D}_3\overset{+}{N}-\text{C(H)(COO}^-\text{)}-\text{C(D)(H)(COO}^-\text{)}$$

60

deuterio-aspartic acid (**60**) when the reaction is carried out in the presence of D_2O.[40] Those reactions cited in Sec. **7-3.2** come closest to imitating the symmetry characteristics of the aspartase catalyzed reaction. It may be that aspartase functions in a manner that is formally similar, for example, by binding the substrate to a chiral prosthetic group as a labile ester, thioester, or amide.

Dehydrochlorination of N-phthaloyl derivatives of amino acid chlorides gives ketenes* (**61**) to which alcohols can add stereoselectively (up to 33% e.e.) in the presence of alkaloid catalysts.[41,42] The resulting N-phthaloylamino acid esters (**62**) were optically active (Table **7-11**).

TABLE 7-11

Alkaloid Catalyzed Additions of Achiral Alcohols to N-Phthaloyl Alkyl Ketenes

$$C_8H_4O_2N-C(R)=C=O \xrightarrow[\text{Alkaloid}]{R'OH} C_8H_4O_2N-CH(R)-C(=O)-OR'$$

61 → **62**

No.	Ketene R =	Alcohol R′ =	Alkaloid	Temp. °	Product Configuration	% e.e.
1	*t*-Bu[a]	*n*-Bu	Brucine	20	*S*-**62**	≅8
2	*t*-Bu[a]	Ph	Brucine	20	*S*-**62**	≅2[b]
3	CH_2Ph[c]	Me	Brucine	22	*S*-**62**	7
4	CH_2Ph[c]	Me	Brucine	−96	*S*-**62**	22
5	CH_2Ph[c]	Me	Acetylquinine	23	*RS*-**62**	0
6	CH_2Ph[c]	Me	Acetylquinine	96	*R*-**62**	33

[a] Ketene **61**, R = *t*-Bu, was prepared *in situ*; cf. Reference 42.
[b] See Reference 41.
[c] Ketene **61**, R = CH_2Ph, was not isolated; cf. Reference 41.

[40] (a) O. Gawron and T. P. Fondy, *J. Am. Chem. Soc.*, **81**, 6333 (1959). (b) H. J. Bright, R. E. Lundin, and L. L. Ingraham, *Biochemistry*, **3**, 1224 (1964).

[41] G. Pracejus (*nee* Schneider), *Ann.*, **621**, 42 (1959).

[42] S. Winter and H. Pracejus, *Chem. Ber.*, **99**, 151 (1966).

* Ketene **61** R = *t*-Bu is isolable; for R = CH_3, Ph, and $CH(CH_3)_2$ there is evidence for ketenes *in situ*, but they have not been isolated.[42]

7-4. Miscellaneous Asymmetric Amino Acid Syntheses

Several asymmetric syntheses of amino acids, which are not conveniently classified in terms of simple symmetry characteristics, are considered here.

7-4.1. Asymmetric Strecker Amino Acid Synthesis. The asymmetric cyanohydrin reaction (Sec. **4-1**) has been extended to the Strecker synthesis.[43] For example, by treating acetaldehyde with (*S*)-(−)-α-methylbenzylamine in an aqueous methanol solution of sodium cyanide at room temperature for 5 days, there is obtained a product (**66**) which, on hydrolysis to **63**, and hydrogenolysis gives (*S*)-(+)-alanine (68% e.e., 17% yield). After crystallization, the product contained a 90% excess of *S*-(+)-alanine. It has been tentatively suggested that this asymmetric synthesis might involve the

$$CH_3CHO + H_2N\text{—}C(Ph)(Me)\text{—}H \xrightarrow{CN^-} CH_3CH(CN)NH\text{—}C(Ph)(Me)\text{—}H \xrightarrow{H_3O^+} CH_3CH(COOH)NH\text{—}C(Ph)(Me)\text{—}H$$

S,S-**66** (major)
R,S-**66** (minor)

63

reaction of the chiral amine at different rates with equilibrating enantiomeric cyanohydrin intermediates.[44] This interpretation is in accord with the observation that racemic acetaldehyde cyanohydrin reacts with an equimolar amount of (*S*)-(−)-α-methylbenzylamine to give, after the usual hydrolytic-hydrogenolytic work-up, (*S*)-(+)-alanine (86–99% e.e. after one recrystallization).[44] However, it seems more probable that the amine reacts with the free aldehyde, which is in equilibrium with the cyanohydrin to give an imine (**64**) or immonium ion (**65**), the latter as an oriented ion pair with cyanide ion. Since both of these proposed intermediates are chiral, preferential attack of cyanide from one of the diastereotopic faces would be expected to give unequal amounts of *R,S*- and *S,S*-**66** (Fig. **7-7**) which, in turn, would ultimately lead to optically active alanine via **63** after hydrolytic-hydrogenolytic work-up.

7-4.2. Asymmetric Oxazolone Peptide Synthesis. The reaction of racemic 5-oxazolones with *S* amino acid esters in the presence of triethylamine gives optically active peptides (Fig. **7-8**).[45] Although this reaction does not strictly qualify as an asymmetric synthesis because both chiral

[43] K. Harada, *Nature*, **200**, 1201 (1963).
[44] K. Harada and S. W. Fox, *Naturwiss.*, **51**, 06 (1964).
[45] F. Weygand, W. Steglich, and X. Barocio de la Lama, *Tetrahedron, Suppl.* No. 8, 9 (1966).

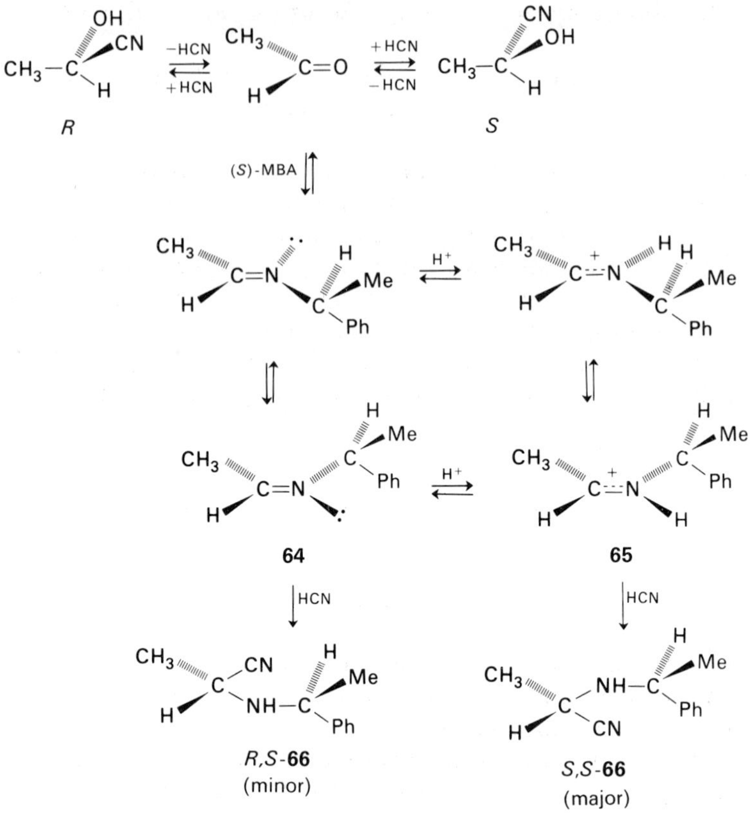

Fig. 7-7. Asymmetric Strecker synthesis of alanine; (*S*)-MBA is (*S*)-α-methylbenzylamine.

centers of the peptide are already present in the starting materials, it accomplishes the same end since more than 50 % of the racemic oxazolone can be incorporated into the peptide as one enantiomer. It has been established that the racemization of the oxazolone is fast compared to the aminolysis reaction. Thus racemic 4-isopropyl-2-phenyloxazolone (*RS*-**67**, R = Ph) reacts with (*S*)-valine to give an 83 % yield of (*R*,*S*)-benzoylvalylvaline methyl ester (66.8 % e.e., Table **7-12**, No. 2). This is possible because the *R* and *S* forms of the oxazolone **67** are equilibrated in the presence of base, presumably via the achiral anion **68**. The formation of the major product, *R*,*S*-**70**, is determined by the relative rates ($k_{R,S}$ versus $k_{S,S}$) of the competing reactions as shown in Fig. **7-8**. Since the reagent, *S*-**69**, is chiral, the relative rates of attack on the enantiomeric oxazolones *S*-**67** and *R*-**67** must be different.* This is an

* An alternative mechanism consistent with these observations involves attack of the chiral reagent **69** at different rates on the enantiotopic faces of a symmetrical intermediate such as **68**.

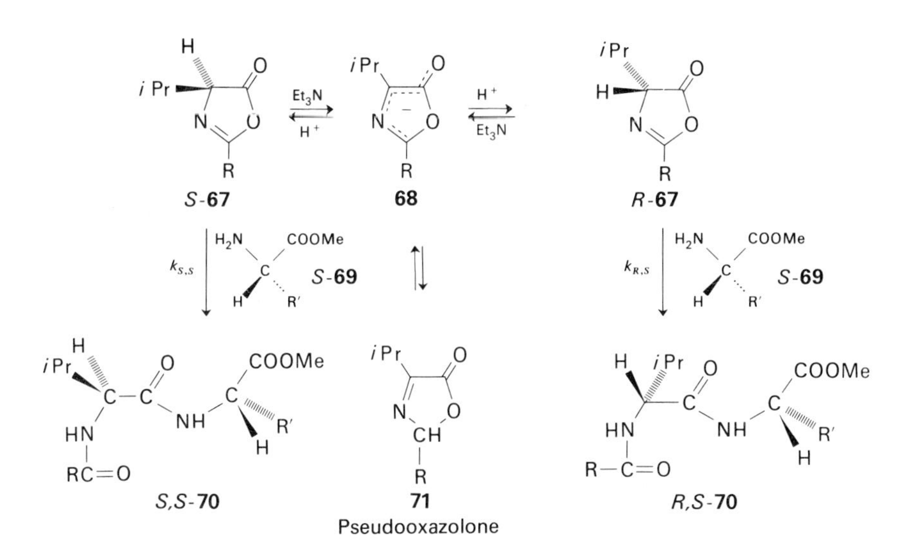

Fig. 7-8. Reaction of *S* amino acid esters with racemic 2-substituted 4-isopropyl 5-oxazolones.[45]

example of a kinetic resolution of stereochemically labile enantiomers (Secs. **1-4.2** and **1-4.3**).

As evident from the data in Table **7-12**, when *R* is a phenyl, methyl, or carbobenzoxy-protected alkylamino group, reaction with an *S* amino acid ester produces an excess of diastereomer **70** having the *R* configuration in the terminal valine residue. However, it has been reported[46] that the reaction of an *S* amino acid ester with racemic 2-trifluoromethyl-4-alkyl-pseudooxazol-5-ones (**71**, R = CF_3) yields an excess of the diastereomer having the *S* configuration at the N-terminal residue.

The reaction types summarized in Fig. **7-8** and Table **7-12** have been analyzed using Ugi's linear free enthalpy method[45] (see Sec. **1-4.4b**). For the reaction of racemic **67** (with R = Ph) and (*S*)-alanine methyl ester (**69**, $R' = CH_3$), the ratio of diastereomeric products, Q, was readily determined by N-trifluoroacetylation and analytical GLPC to be 0.653 (*S,S*-**70**:*R,S*-**70**). The chirality parameter calculated from the ligand constant of the groups attached on the chiral center of alanine methyl ester ($\lambda_H = 0.0$, $\lambda_{Me} = 1.0$, and $\lambda_{COOMe} = 0.88$) is 0.106. With an antisymmetric factor of -1, the τ value is calculated to be 1.75. A τ value of the same magnitude (1.65) was calculated from the reaction of **67** (R = Ph) with (*S*)-valine methyl ester ($Q = 0.192$, $\lambda_{iPr} = 1.27$). By using average values of τ (1.70) and Q (0.422), obtained from

[46] W. Steglich, D. Meyer, X. Barocio de la Lama, H. Tanner, and F. Weygand, European Reptile Symposium, Nordwijk, Holland, September, 1966; as reported by H. Pracejus, *Fortsch. Chem. Forsch.*, **8**, 493–553 (1967).

TABLE 7-12

Reaction of Racemic 2-Substituted 4-Isopropyl-5-oxazolones with *S* Amino Acid Methyl Esters[a]

No.	2-Substituted oxazolone, **67** R[a]	Amino acid ester **69** R′	Diastereomer excess of *R,S*-**70** %[b]
1	Ph	Me (Alanine)	21
2	Ph	*i*Pr (Valine)	67
3	Ph	*i*Bu (Leucine)	57
4	Ph	CH_2Ph (Phenylalanine)	48
5	Me	*i*Pr (Valine)	21
6	CH_2NHCBZ[c]	*i*Pr (Valine)	35
7	CH_2NHCBZ[c]	CH_2Ph (Phenylalanine)	22
8	CH(CH₂Ph)NHCOCH(NHCBZ[c])*i*Pr	*i*Pr (Valine)	42

[a] See Reference 45 and Fig. **7-8**.
[b] Determined by N-trifluoroacetylation and analysis by GLC.
[c] CBZ represents the carbobenzoxy group.

the reaction of **67** (R = Ph) with (*S*)-leucine and (*S*)-phenylalanine methyl esters, ligand constants were calculated for isobutyl ($\lambda = 1.20$) and benzyl ($\lambda = 1.15$). These are reasonable values relative to those for other groups.

As a critical test, values thus obtained were used to predict the diastereomer ratio for a new reaction of the same type, but with **67** having R = CH_2NHCBZ rather than R = Ph. A τ value for the new system ($\tau = 0.721$) was obtained using Q (0.486) for the reaction of **67** (R = CH_2NHCBZ) with (*S*)-valine methyl ester. Using this τ value and the ligand constant that had been calculated for benzyl, the percent excess of *R,S* diastereomer expected from reaction of **67** (R = CH_2NHCBZ) with (*S*)-phenylalanine methyl ester was calculated to be 22%, the value observed.

Interesting four-component asymmetric syntheses have been investigated by Ugi and co-workers as models for potential asymmetric peptide synthesis reactions.[47] An aldehyde, isonitrile, chiral primary amine, and carboxylic

$$\underset{\mathbf{72}}{RR'CH(H)C{=}O} + \underset{\mathbf{73}}{R''N{\equiv}C:} + \underset{\mathbf{74}}{R^*NH_2} + \underset{\mathbf{75}}{R'''COOH} \longrightarrow \underset{\mathbf{76}}{R''NH{-}C({=}O){-}\overset{*}{C}H(CHRR'){-}N(R^*)COR'''}$$

[47] (a) I. Ugi, K. Offermann, H. Herlinger, and D. Marquarding, *Ann.*, **709**, 1 (1967). (b) *ibid.*, **709**, 11 (1967) and references therein. See especially I. Ugi, Chapter 1 in *New Methods in Preparative Organic Chemistry*, ed., W. Foerst (New York: Academic Press, 1968), Vol. 4, for a general review of four-component reactions.

acid react to produce excellent yields of adduct **76** which can be hydrolyzed to an amino acid. As a result of over 250 carefully detailed experiments the essential characteristics of this complex reaction have been elucidated. Some selected results are summarized in Table **7-13**; Fig. **7-9** outlines the probable mechanism.

TABLE 7-13

Four-Component Asymmetric Syntheses[47]

$$\underset{\mathbf{72}}{R-\overset{R'}{\overset{|}{C}}HCHO} + \underset{\mathbf{73}}{t\text{-}BuN{\equiv}C} + \underset{\mathbf{74}}{Ph\overset{*}{C}H(NH_2)CH_3} + \underset{\mathbf{75}}{PhCOOH} \longrightarrow$$

$$\underset{\mathbf{76}}{t\text{-}Bu-NH-\overset{\overset{O}{\|}}{C}-\overset{CHRR'}{\overset{|}{\overset{*}{C}H}}-\underset{\underset{COPh}{|}}{N}\overset{*}{C}H(CH_3)Ph}$$

R	R′	Conc.[a] M/Kg	Solvent	Temp. °	Amine, **74** Configuration	New chiral center, **76** Configuration	Yield[b] %
Me	Me	1.6	EtOH	−70	*R*	*S*	72
		0.29	MeOH	0	*R*	*R*	72
		0.05	MeOH	−40	*S*	*S*	77
		1.0	MeOH	−80	*S*	*R*	75
Et[c]	Me	1.9	EtOH	−70	*R*	*S*	72
		0.38	MeOH	+10	*R*	*R*	73
		0.38	MeOH	+10	*S*	*S*	73
		2.1	EtOH	−70	*S*	*R*	68
Et	H	1.0	MeOH	−70	*S*	*R*	70–75[d]
		1.0	MeOH	−70	*R*	*S*	
		0.1	MeOH	0	*S*	*S*	
		0.1	MeOH	0	*R*	*R*	

[a] Mole/kg solvent; all components in any one example have the same concentration.
[b] Percent yield (unnormalized) of the diastereomer having the chirality indicated.
[c] *S*-(+)-Aldehyde was used.
[d] Exact values were not reported.

The following generalities are evident from the data in Table **7-13**. At low temperatures and high concentrations of reactants, (*R*)-α-methylbenzylamine (**74**) promotes the formation of **76** having the *S*-configuration at the new asymmetric center; but at higher temperatures and lower concentrations, the induced configuration is *R*. In all cases the asymmetric bias is high* under optimum conditions. The chiral inducing influence of the asymmetric center in the amine predominates over that in the aldehyde. The last result is particularly dramatic when one considers the mechanism shown in Fig. **7-9**, since the amine induction is seen to be 1:3 while an α-asymmetric center in an aldehyde exerts a 1:2 influence on the developing chiral center.

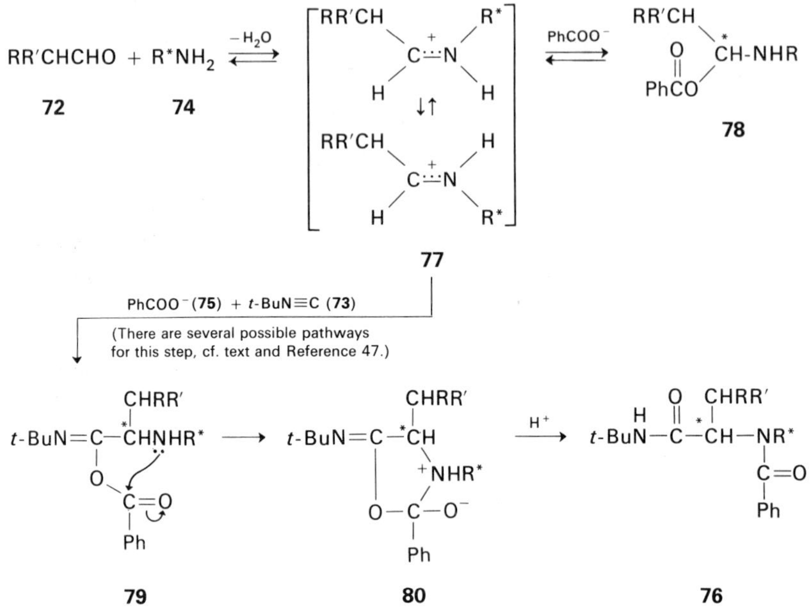

RR′CHCHO + R^*NH_2 $\xrightleftharpoons{-H_2O}$ [77] $\xrightleftharpoons{PhCOO^-}$ 78

72 **74** **77** **78**

$PhCOO^-$ (**75**) + *t*-BuN≡C (**73**)

(There are several possible pathways for this step, cf. text and Reference 47.)

79 ⟶ **80** $\xrightarrow{H^+}$ **76**

Fig. 7-9. "Four-component" asymmetric peptide synthesis.

The mechanism is believed to involve the formation of an immonium ion (**77**) which may react reversibly with the acid anion to form **78**. In the presence of isonitrile, the so-called α-adduct **79** is formed, but there is no good evidence concerning the details of this step. Presumably there could be a two-step reaction (Eq. 1) via a nitrilium ion (**81**) or a three-center reaction involving either an insertion of the nitrile into the C—O bond of **78** (Eq. 2) or addition of an oriented ion pair to the nitrile (Eq. 3). Ugi has suggested[47] that in dilute solution the immonium ion reacts predominantly via the two-step pathway (Eq. 1) to give **79** with the *R* configuration at the new chiral center, when (*R*)-α-methylbenzylamine is used; but as the concentration of reactants is increased, other processes (Eq. 2) or (Eq. 3) predominate, yielding product having the *S* configuration at the new chiral center.

* As can be seen from Table **7-13**, under optimum conditions about a 70 per cent yield of one epimer of **76** was isolated.

$$R^*\text{—}\overset{+}{NH}\text{⋯}C\langle \; + \; :C{\equiv}N\text{—}R'' \longrightarrow \left[R^*\text{—}NH\text{—}\underset{|}{\overset{|}{C}}\text{—}C{\equiv}\overset{+}{N}R''\right] \xrightarrow{PhCOO^-} \mathbf{79} \qquad (1)$$

77 **81**

$$\mathbf{77} \rightleftarrows \mathbf{78} \longrightarrow \left[R^*\text{—}NH\text{—}\overset{|}{C}\text{—}O\text{—}\overset{O}{\overset{\|}{C}}\text{—}Ph \quad :C{\equiv}NR''\right] \longrightarrow \mathbf{79} \qquad (2)$$

$$\mathbf{77} \rightleftarrows \left[R^*\text{—}NH\overset{+}{\text{⋯}}C\langle \quad O\overset{-}{\text{⋯}C\text{⋯}}O \quad (\text{—}Ph)\right] + :C{\equiv}N\text{—}R'' \longrightarrow \mathbf{79} \qquad (3)$$

7-4.3. Chiral Metal Complex Catalyzed Asymmetric Condensations. Amino acids have also been synthesized using chiral metal complexes as templates.[48] Treatment of (−)-glycinato-*bis*-(ethylenediamine)-cobalt(III) iodide (**80**) with acetaldehyde in a carbonate solution gives, after hydrolytic work-up and precipitation of cobalt as the sulfide, glycine and an optically active mixture of allothreonine (16% yield) and threonine (**81**, 56% yield, *ca.* 8% e.e.). An asymmetric condensation apparently takes place at the prochiral carbon of glycine under the influence of the dissymmetric environment of the chiral complex (**80**, asymmetric at cobalt). The stereoselectivity was lower with the corresponding (−)-propylenediamine complex.

$$[\overset{*}{Co}(H_2NCH_2CH_2NH_2)_2(NH_2CH_2COO)]^{2+}I_2^- + CH_3CHO \xrightarrow{Na_2CO_3}$$

80

$$CH_3\overset{*}{C}H(OH)\overset{*}{C}H(NH_2)COOH$$

81

In a similar study,[49] a chiral cobalt(III) complex (**82A**), formed from α-amino-α-methylmalonic acid and the chiral tetradentate ligand, (*S*,*S*)-α,α′-dimethyltriethylenetetramine, was decarboxylated to a complex (**83**) containing mainly (*S*)-alanine ligands. It was thought that hydrolysis of the complex to give *S*-alanine (14% e.e.) caused partial racemization. Complex **82B**, a structural isomer of **82A**, may be an important intermediate in this reaction. Diastereomers of **82** (epimeric at the incipient asymmetric carbon of alanine) may form at different rates and/or be decarboxylated at different

[48] M. Murakami and K. Takahashi, *Bull. Chem. Soc. Jap.*, **32**, 308 (1959).
[49] R. G. Asperger and Chui Fan Liu, *Inorg. Chem.*, **6**, 796 (1967).

rates. The overall transformation may be described in terms of selective decarboxylation of one of the prochiral carboxylate groups of the malonate under the influence of the chiral cobalt complex ion.

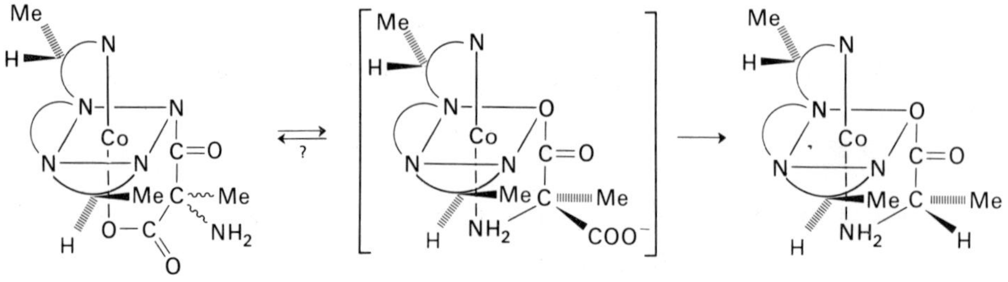

82A

82B

83

8

Asymmetric Synthesis at Hetero Atoms

8-1. Asymmetric Synthesis at Sulfur

8-1.1. Introduction. Suitably substituted tricoordinate sulfur compounds such as sulfinates, sulfoxides, and sulfilimines, as well as sulfonium salts, can be obtained in optically active forms.[1,2] The stereochemistry of sulfinates and sulfoxides, in particular, has been studied rather extensively. Early resolutions of sulfoxides,[1,2] which required a functional handle elsewhere in the molecule,[3] so that resolving agents such as alkaloids or camphorsulfonic acid could be used, have been complemented by a more general resolution method[4] involving a chiral platinum(II)-α-methylbenzylamine complex. Furthermore, direct stereoselective syntheses[5] (detailed in Sec. **8-1.4**) now play an important role in such investigations.

[1] R. L. Shriner, R. Adams, and C. S. Marvel, *Organic Chemistry*, ed., H. Gilman (New York: John Wiley & Sons, Inc., 1943), Vol. I, p. 419ff.

[2] C. C. Price and S. Oae, *Sulfur Bonding* (New York: The Ronald Press, 1962), pp. 129–132, 149–150.

[3] P. W. B. Harrison, J. Kenyon, and H. Phillips, *J. Chem. Soc.*, 2079 (1926).

[4] A. C. Cope and E. A. Caress, *J. Am. Chem. Soc.*, **88**, 1711 (1966).

[5] (a) K. K. Andersen, *Tetrahedron Lett.*, 93 (1962). (b) K. K. Andersen, W. Gaffield, N. E. Papanikolaou, J. W. Foley, and R. I. Perkins, *J. Am. Chem. Soc.*, **86**, 5637 (1964). (c) K. K.

A major interest in the stereochemistry of sulfoxides stems from the fact that compounds containing chiral sulfinyl groups can be isolated from natural sources. For example, one of the first optically active natural products to be isolated, whose chirality was due to a hetero atom, was sulphoraphen[6] ($CH_3\overset{*}{S}OCH{=}CHCH_2CH_2NCS$) whose sulfide analog is a principal constituent of radish. Enzymatic degradation of mustard oil thioglucosides yields other chiral sulfinyl isothiocyanates[7] that have been the object of ORD[8] and X-ray crystallographic[9] studies. Related investigations have been carried out with sulfinyl amino acids such as S-methylcysteine S-oxide.[10]

In the following section, we shall discuss the asymmetric synthesis of compounds containing a dissymmetric sulfur atom. Examples of asymmetric induction at another center due to chiral sulfur are given in Sec. **9-3**.

8-1.2. Asymmetric Oxidation of Achiral Sulfur Compounds with Chiral Reagents. The reaction of (+)-monopercamphoric acid (**1**) with an unsymmetrical sulfide produces a chiral sulfoxide. This kind of reaction has been the subject of extensive study and considerable controversy. This oxidation (a general representation of which is shown in Fig. **8-1**) involves attack by the chiral peracid upon one or the other of the enantiotopic electron pairs of the prochiral sulfur atom (or vice versa) to give enantiomeric sulfoxides (**3** and **4**) which are usually produced in measurably unequal amounts.

A number of other chiral peracids have also been used, and the structures of these are compared with that of (+)-monopercamphoric acid in Fig. **8-2**. Table **8-1** summarizes the results of oxidations with these peracids, the first twenty-eight entries being examples that involve (+)-monopercamphoric acid. The percents enantiomeric excess are usually low. Since sulfoxides generally have rather large rotations, this is not as serious a handicap as might be presumed. Thus the determination of a few tenths of a percent enantiomeric excess can be determined with confidence in some cases. However, the small observed rotations may be responsible for such discrepancies

Andersen, *J. Org. Chem.*, **29**, 1953 (1964). (d) K. Mislow, T. Simmons, J. T. Melillo, and A. L. Ternay, Jr., *J. Am. Chem. Soc.*, **86**, 1452 (1964). (e) K. Mislow, M. M. Green, P. Laur, J. T. Melillo, T. Simmons, and A. L. Ternay, Jr., *J. Amer. Chem. Soc.*, **87**, 1958 (1965). (f) M. Axelrod, P. Bickart, J. Jacobus, M. M. Green, and K. Mislow, *J. Am. Chem. Soc.*, **90**, 4835 (1968).

[6] H. Schmid and P. Karrer, *Helv. Chim. Acta*, **31**, 1017 (1948).

[7] B. W. Christensen and A. Kjaer, *Acta Chem. Scand.*, **17**, 846 (1963) and references therein.

[8] W. Klyne and J. Day, *Acta Chem. Scand.*, **14**, 215 (1960).

[9] K. K. Cheung, A. Kjaer, and G. A. Sim, *Chem. Commun.*, 100 (1965).

[10] (a) R. Hine and D. Rodgers, *Chem. and Ind.*, 1428 (1956). (b) J. F. Carson and F. F. Wong, *J. Org. Chem.*, **26**, 4997 (1961). (c) W. Gaffield, F. F. Wong, and J. F. Carson, *J. Org. Chem.*, **30**, 951 (1965).

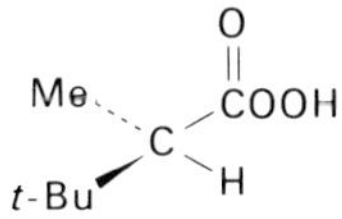

2R-**5** 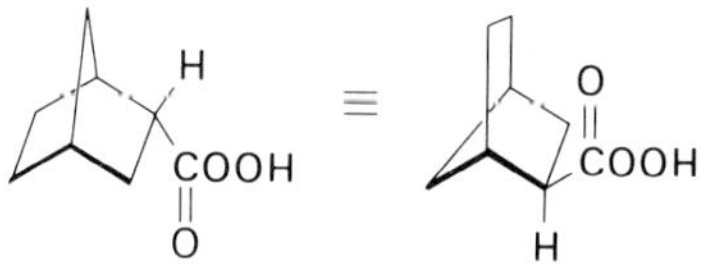 2*R*-**6** 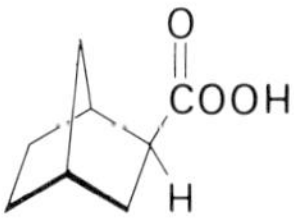 2*S*-**7**

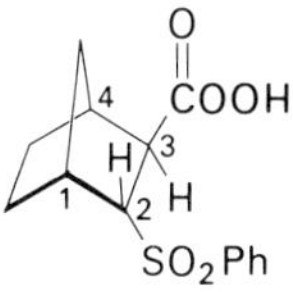

3*S*-**15** 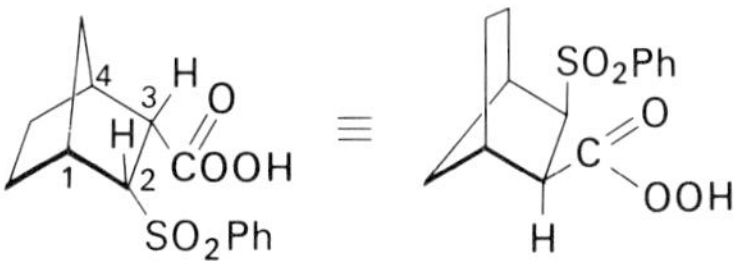 3*R*-**16**

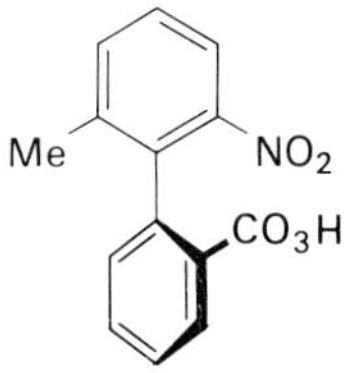

23 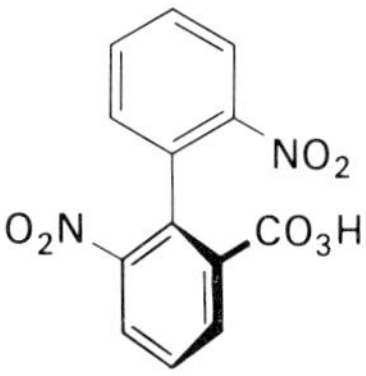 **24**

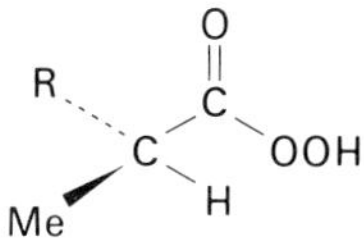

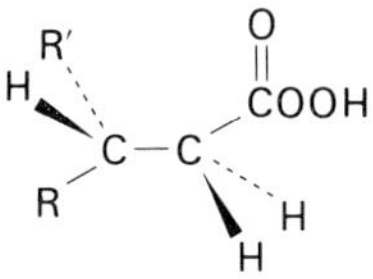

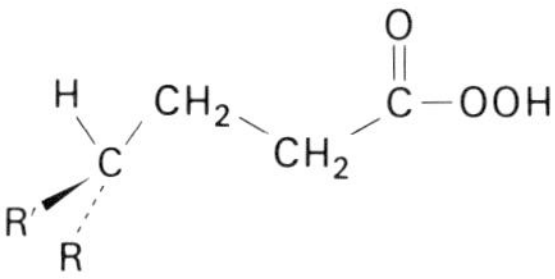

R		
2*S*-**8**, Ph	3*R*-**17**, R′ = Me, R = Ph	4*S*-**20**, R = COOH, R′ = Me
2*S*-**9**, α-Naph	3*S*-**18**, R′ = COOH, R = Me	4*R*-**21**, R = COOH, R′ = $PhCH_2OCONH$
2*S*-**10**, Et	3*S*-**19**, R′ = COOH, R = *p*-$MeC_6H_4SO_2NH_2$	4*S*-**22**, R = COOMe, R′ = N-Phthalimido
2*S*-**11**, Cyclohexyl		
2*R*-**12**, Phenoxy		
2*S*-**13**, *p*-NO_2Ph		
2*S*-**14**, N-Phthalimido		

Fig. 8-2. Chiral peracids used to oxidize sulfides.

TABLE 8-1

Asymmetric Oxidation of Sulfides with Chiral Peracids[a]

$$R''{-}S{-}R''' + R^{*}CO_3H \longrightarrow R''{-}\overset{O}{\underset{}{\overset{\|}{S}}}{}^{*}{-}R''' + R^{*}CO_2H$$

No.	Peracid	R″	R‴	Sulfoxide % e.e. ([α]$_D$)	Sulfoxide Config.	Ref.
1	1*S*-**1**	*o*-C_6H_4COOMe	Me	4.3	*R*	11
2	”	”	Et	1.2	*R*	11
3	”	”	*i*-Pr	0.4	*S*	11
4	”	”	*t*-Bu	2.4	*S*	11
5[b]	”	”	Me	(2.2°, MeOH)	*R*	12
6[b]	”	”	Et	(2.5°, MeOH)	*R*	12
7	1*S*-**1**	Ph	Me	3.8	*R*	13
8	”	”	*n*-Pr	(4.9°, 4.2°)[c]	*R*	14
9	”	”	*n*-Bu	(4.9°, 4.4°)[c]	*R*	14
10	”	”	CH_2Ph	(5.4°, 4.8°)[c]	*R*	14
11	”	”	*i*-Bu	(6.0°, 5.3°)[c]	*R*	14
12	”	”	*i*-Pr	0	—	14
13	”	”	*s*-Bu	0	—	14
14	”	”	*p*-Cl-C_6H_4	0	—	14
15	”	”	*t*-Bu	1.3	*S*	14, 13
16	1*S*-**1**	CH_2Ph	Me	0.9, 1.0[c]	*S*	15
17	”	”	Et	3.0, 2.5[c]	*S*	14, 15
18	”	”	*n*-Bu	1.7, 4.4[c]	*S*	14, 15
19	”	”	*i*-Pr	20, 4.5[c]	*R*	14, 15
20	”	”	*t*-Bu	4.3, 4.3[c]	*R*	14, 15
21	1*S*-**1**	*n*-Bu	*n*-Pr	0	—	14
22	1*S*-**1**	Adamantyl	Me	(−2.9°)[d]	*R*	16
23	1*S*-**1**	PhthNCH$_2$[e]	Me	(4.1°)[f]	—[g]	17
24	1*S*-**1**	CH_2Ph	Pr	(−3.0°, 0.7°)	—[g]	14
25	”	”	*i*-Bu	(−3.1°, 0°)[c]	—[g]	14
26	”	”	*s*-Bu	(0.7, 3.8)[c]	—[g]	14
27	”	”	*i*-Pr	0.1	*S*	11
28	”	”	*t*-Bu	1.6	*S*	11
29	2*S*-**11**	*o*-C_6H_4COOMe	Me	—	*R*	18
30	2*R*-**11**[h]	”	Me	—	*S*	18
31	2*S*-**8**	”	Me	2.7	*R*	11
32	”	”	Et	1.0	*R*	11
33	2*S*-**8**	Ph	Me	1.0	*R*	13
34	”	”	*t*-Bu	0.6	*S*	13
35	2*R*-**5**	Ph	Me	4.7	*S*	13
36	”	”	*t*-Bu	0.9	*R*	13
37	2*S*-**9**	Ph	Me	2.9	*R*	13
38	”	”	*t*-Bu	1.2	*S*	13
39	2*S*-**10**	Ph	Me	0.8	*R*	13
40	”	Ph	*t*-Bu	0.3	*S*	13
41	2*S*-**11**	Ph	Me	2.0	*R*	13
42	”	”	*t*-Bu	0.8	*S*	13
43	2*R*-**12**	Ph	Me	0.8	*R*	13

TABLE 8-1—continued

No.	Peracid	R″	R‴	Sulfoxide		Ref.
				% e.e. ($[\alpha]_D$)	Config.	
44	2*R*-**12**	Ph	*t*-Bu	0	—	13
45	2*S*-**13**	Ph	Me	2.3	*R*	13
46	,,	,,	*t*-Bu	0	—	13
47	2*R*-**6**	Ph	Me	2.9	*S*	13
48	,,	,,	*t*-Bu	2.4	*R*	13
49	2*S*-**7**	Ph	Me	2.1	*R*	13
50	,,	,,	*t*-Bu	0	—	13
51	3*R*-**16**	Ph	Me	5.4	*R*	13
52	,,	,,	*t*-Bu	10.3	*S*	13
53	3*S*-**15**	Ph	Me	2.7	*R*	13
54	,,	,,	*t*-Bu	5.2	*S*	13
55	2*S*-**14**	$PhthNCH_2$[e]	Me	(+2.5°)[f]	—[g]	17
56	3*R*-**17**	Ph	Me	2.3	*S*	13
57	,,	,,	*t*-Bu	1.2	*S*	13
58	3*S*-**18**	$PhthNCH_2$[e]	Me	0	—	17
59	3S-**19**	$PhthNCH_2$[e]	Me	0	—	17
60	4*S*-**20**	$PhthNCH_2$[e]	Me	(+2°)[f]	—[g]	17
61	4*R*-**20**[h]	$PhthNCH_2$[e]	Me	(−2.3°)[f]	—[g]	17
62	4R-**21**	$PhthNCH_2$[e]	Me	(+3.7°)[f]	—[g]	17
63	4*S*-**22**	$PhthNCH_2$[e]	Me	(+3.3°)[f]	—[g]	17
64	(?)-**23**[g]	Ph	Me	1.4	*R*	13
65	(?)-**23**	,,	*t*-Bu	3.2	*R*	13
66	(?)-**24**[g]	Ph	Me	1.8	*S*	13
67	(?)-**24**	,,	*t*-Bu	0.6	*S*	13

[a] $CHCl_3$ solvent, 0 to −10°, unless otherwise annotated.
[b] Et_2O solvent, 0°.
[c] The first value is the % e.e. (or specific rotation) based on optical data in chloroform and the second value is derived from optical data in ethanol. The disagreement is probably a reflection of the error in the very small observed rotations.[15]
[d] Ethanol solvent.
[e] PhthN = N-Phthalimido.
[f] Methanol-water, 4:1, C = 2.
[g] Configuration unknown.
[h] The enantiomer of the acid shown in Fig. **8-2**.

[11] A. Maccioni, F. Montanari, M. Secci, and M. Tramontini, *Tetrahedron Lett.*, 607 (1961).
[12] K. Balenović, N. Bregant, and D. Francetić, *Tetrahedron Lett.*, [6], 20 (1960).
[13] U. Folli, D. Iarossi, F. Montanari, and G. Torre, *J. Chem. Soc.*, (*C*), 1317 (1968).
[14] A. Mayr, F. Montanari, and M. Tramontini, *Gazz. Chim. Ital.*, **90**, 739 (1960); *Chem. Abst.*, **55**, 16460b (1960).
[15] K. Mislow, M. Green, and M. Raban, *J. Am. Chem. Soc.*, **87**, 2761 (1965).
[16] D. R. Rayner, A. J. Gordon, and K. Mislow, *J. Am. Chem. Soc.*, **90**, 4854 (1968).
[17] K. Balenović, I. Bregovec, D. Francetić, I. Monković, and V. Tomašić, *Chem. and Ind.*, 469 (1961).
[18] A. Maccioni, *Boll. Sci. Fac. Chim. Ind. Bologna*, **23**, 41 (1965); *Chem. Abst.*, **63**, 8239f (1965).

as the fourfold difference in percent enantiomeric excess noted[15] when rotations in two different solvents were used to compute the results from the same asymmetric oxidation (Table **8-1**, No. 19).

It can be seen from the results with (+)-monopercamphoric acid that when the sulfide ligands are of similar steric dimensions at the point of

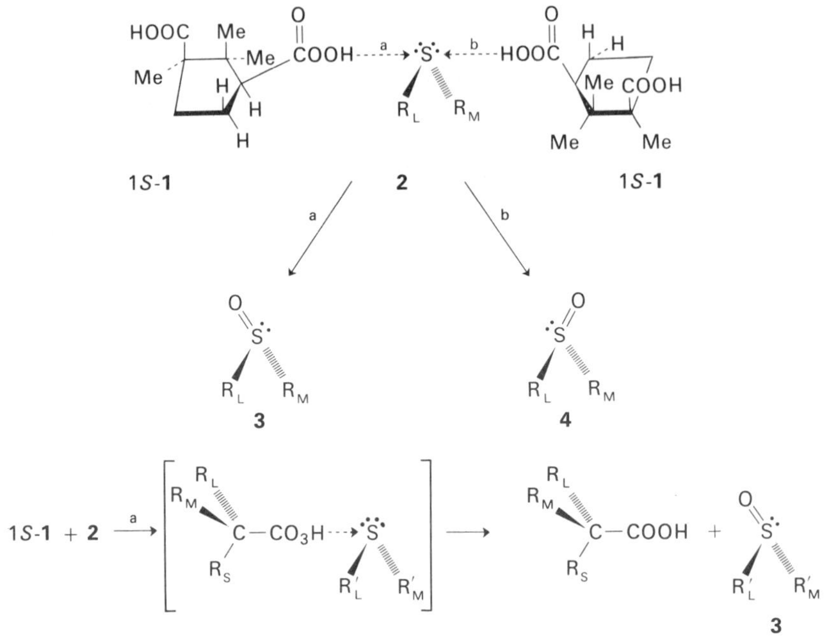

Fig. 8-1. Oxidation of an unsymmetrical sulfide with (+)-monopercamphoric acid. The groups R_L and R_M have decreasing steric requirements. The formula in brackets represents the Montanari proposed model for the favored mode of attack.

attachment to sulfur, there is low or negligible asymmetric bias during the oxidation (e.g., Table **8-1**, Nos. 3, 12–14, 21, and 27). A more surprising observation is that the structure of the chiral peracid has relatively little influence on the *magnitude* of the asymmetric bias; in fact, if any trend can be discerned, it is the reverse of what might be anticipated. For example, peracids **8** and **14**, which have the chiral center α to the peroxycarboxylate group, are less effective than peracids **17** and **22**, which may be considered to be the β and γ analogs of **8** and **14**, respectively.

Montanari[11,14] proposed a transition state model, based on the representation in Fig. **8-1**, for predicting the configuration of the sulfoxide resulting from asymmetric peroxidations with (+)-monopercamphoric acid. According to this model, pathway (a) is preferred since the approach of the reagent to the sulfide is such that the smaller group (CH_2) attached to the chiral center, rather than the larger group ($-CMe_2$), opposes the larger group of the substrate sulfide. Mislow and co-workers[19] have rightfully taken vigorous exception to these transition state models, deeming them to be "as fanciful as they are successful." It was stated[19] that to presume the model embodies the rationale of its success is "extravagant conjecture." The model was also shown to fail for *n*-alkyl benzyl sulfoxides[15] (Table **8-1**, Nos. 16–18, and presumably also then for 24). Montanari countered this criticism[13,20] by showing that, under carefully controlled conditions, when methyl phenyl sulfide and *t*-butyl phenyl sulfide were oxidized by fifteen different chiral peracids (the absolute configurations of which were known in all but two cases), they gave chiral sulfoxides with configurations in accord with the model in nine cases (7/15, 33/34, 35/36, 37/38, 39/40, 41/42, 47/48, 51/52, and 53/54). This correspondence of the model with the experimental results is based on the assumption that although methyl is smaller than phenyl, *t*-butyl is larger than phenyl. In three other examples (43/44, 45/46, and 49/50) the percent enantiomeric excess of the sulfoxide from the asymmetric oxidation of the *t*-butyl phenyl sulfide member of the pair was essentially zero. In one example where the absolute configuration of the reagent was known (56/57) and two others where it was unknown (64/65 and 66/67), the sulfoxides had the same instead of the opposite configurations in a manner clearly deviating from the pattern of the other examples. It was contended[13] that "the constant trend of correct results descending from our working hypothesis may hardly be attributed to a constant trend of casual events" but, on the other hand, that this "line of reasoning does not pretend to give any detailed or rigorous interpretation of the transition state in the oxidation of sulfides to sulfoxides by peracids." It is relevant to consider the development of this controversy,[11,14,19,20] since its essence is not unique to the asymmetric oxidation of sulfides. The appropriateness of transition state models has been a matter of general concern throughout this book and opinions on the matter have been expressed here (see, for example, Secs. **1-4.4b**, **2-2**, **3-3**, **5-2.3a**, and **6-2.5**) and elsewhere.[21] It is pertinent to point out that, in asymmetric reductions of alkyl phenyl ketones, *t*-butyl usually acts as though it is smaller than phenyl (Tables **5-7** and **5-12**), whereas in sulfide oxidations Montanari concludes that it behaves as though it is larger

[19] K. Mislow, M. M. Green, P. Laur, J. T. Melillo, T. Simmons, and A. L. Ternay, Jr., *J. Amer. Chem. Soc.*, **87**, 1958 (1965).

[20] F. Montanari, *Tetrahedron Lett.*, 3367 (1965).

[21] (a) J. D. Morrison, *Survey of Progress in Chemistry*, ed., A. F. Scott (New York: Academic Press, 1966), Vol. 3, pp. 147–182. (b) D. R. Boyd and M. A. McKervey, *Quart. Revs.*, **22**, 95 (1968).

than phenyl. Also Montanari must assume that isopropyl is larger than phenyl in order to accommodate his results to the model. This is the reverse of the experience in asymmetric reductions (Table **5-7**) and open-chain systems where the Cram rule applies (Table **3-1**). Thus it is apparent that the steric requirements in the asymmetric oxidations of sulfides do not parallel those in these other systems.

Mislow has emphasized the following salient points with respect to the asymmetric oxidation of achiral sulfides with chiral peracids.[19] There are three bonds separating the electrophilic oxygen and the nearest asymmetric carbon in (+)-monopercamphoric acid (and peracids **5-16**, Fig. **8-2**). Thus a detailed model such as **25** may appear superficially analogous to a Prelog-type model **26**. However, in **26** the new chiral center is created at the third atom from the established chiral center, while in **25** it is the fourth atom from the established chiral center. By analogy with Cram's rule the peracid conformation chosen for **25** places the carbonyl oxygen between the small and

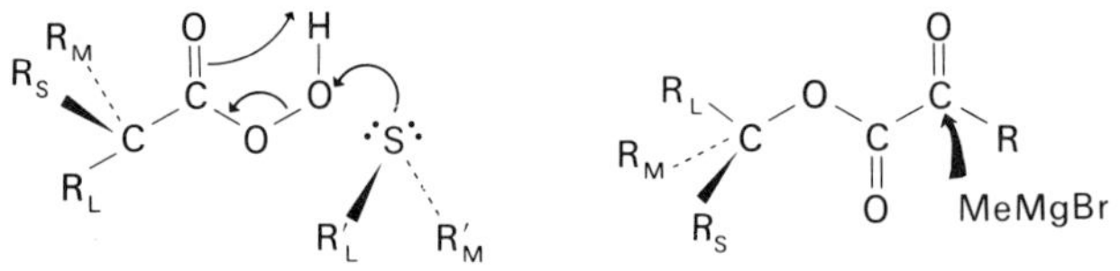

medium size groups at the asymmetric carbon. As we have seen, however, gross steric effects do not parallel those in the Cram system; therefore, the conformational effects are probably quite different. There is no independent evidence concerning the peracid conformation at the α-carbon in the ground state,* let alone in the transition state. Furthermore, the sulfide may rotate about the incipient S—O bond which defines its locus of attack so that the groups R'_L and R'_M may sweep through the volume of a cone, and any rotameric conformation thus generated leads to sulfoxide of the same configuration. In terms of physical reality, Mislow pointed out that there is no basis for a distinction between the cones of attack which describe the approach of the peroxy acid toward one or the other enantiotopic electron pairs of the sulfide. Indeed, considering that all but two of the asymmetric oxidations reported in Table **8-1** are less than 6% while two-thirds are under 3% (which corresponds to a difference in energies of activations in the competing diastereomeric transition states of only 30 cal/mole at 25°), the steric and/or electronic differences which constitute the basis of these

* The O—O bond in hydrogen peroxide can be envisaged as the spine of a book opened approximately 94° with the two hydrogens located, respectively, at the opposite corners of the opposing pages. There is no evidence that the O—O bond in acyl hydroperoxides is also skewed in the same way and there is no evidence concerning the favored conformation with respect to the carbonyl group and α-carbon.

asymmetric oxidations are slight and may be very subtle. Nevertheless there are regularities in these observed asymmetric syntheses and these may prove valuable as a basis for empirical correlation models for reactions run under identical conditions.

Some other limitations should be emphasized. Table **8-2** shows the effect of solvent on the asymmetric synthesis. The Montanari empirical model

TABLE 8-2[13]

Solvent Effects on Asymmetric Oxidation of Aryl Alkyl Sulfides at 0°

No.	Solvent	1*S*-(+)-Percamphoric acid (**1**) % e.e., Config. of sulfoxide: PhSO-Me	PhSO-*t*-Bu	2*S*-(+)-Perhydratropic acid (**8**) % e.e., Config. of sulfoxide: PhSO-Me	PhSO-*t*-Bu
1	CH_3CN	1.2 (*R*)	0.4 (*R*)	—	—
2	$(CH_3)_2CHOH$	1.7 (*R*)	0.4 (*R*)	0.6 (*R*)	0.2 (*R*)
3	$CHCl_3$	3.8 (*R*)	1.3 (*S*)	1.0 (*R*)	0.6 (*S*)
4	$(C_2H_5)_2O$	4.5 (*R*)	1.5 (*R*)	1.9 (*R*)	0
5	C_6H_6	3.6 (*R*)	1.4 (*R*)	—	—
6	CCl_4	1.6 (*R*)	0.7(*S*)	1.3 (*R*)	0.5 (*S*)
7	Dioxane	3.1 (*S*)	0.2 (*S*)	—	—

(Fig. **8-1**) predicts that the configurations will be reversed in going from phenyl methyl to phenyl *t*-butyl sulfide since the *t*-butyl group acts as though it is larger than phenyl according to this model. With the two peracids shown, the configuration should be *R* for phenyl methyl sulfoxide and *S* for phenyl *t*-butyl sulfoxide. As can be seen from the table, these predictions are fulfilled only in chloroform and carbon tetrachloride solvents. Considering the distance between the inducing and new chiral centers and the low order of asymmetric synthesis involved, mediation of the asymmetric influence by the solvent is not unexpected. Studies of the mechanism of epoxidation[22] and sulfide oxidation with hydrogen peroxide[23] also suggest the intimate involvement of solvent in the transition state. It is apparent that more strongly coordinating solvents yield results that are not in accord with the model, and it must therefore be further qualified as to reaction conditions.

The effect of temperature (Table **8-3**) has also been examined. At −50° the oxidations were more stereoselective by factors of from about two to four.

[22] R. C. Ewins, H. B. Henbest, and M. A. McKervey, *Chem. Commun.*, 1085 (1967). H. B. Henbest, *Chem. Soc. Special Publ.*, No. 19, 83 (1965); Sec. **6-3.3.**

[23] M. A. P. Dankleff, R. Curci, J. O. Edwards, and H. Y. Pyun, *J. Am. Chem. Soc.*, **90**, 3209 (1968).

TABLE 8-3[13]

Effect of Temperature on the Asymmetric Synthesis of Aryl Alkyl Sulfoxides

Peracid	Sulfoxide, % e.e., and configuration			
	PhSO-Me		PhSO-*t*-Bu	
	0°	−50°	0°	−50°
1*S*-**1**	3.8 (*R*)	6.5 (*R*)	1.3 (*S*)	4.4 (*S*)
2*S*-**8**	1.0 (*R*)	2.7 (*R*)	0.6 (*S*)	2.5 (*S*)
2*S*-**9**	2.9 (*R*)	—	1.2 (*S*)	—
2*R*-**9**	—	4.7 (*S*)	—	2.8 (*R*)
2*S*-**10**	0.8 (*R*)	3.4 (*R*)	0.3 (*S*)	1.3 (*S*)
2*R*-**5**	4.7 (*S*)	11.8 (*S*)	0.9 (*R*)	5.8 (*R*)
2*S*-**11**	2.0 (*R*)	7.2 (*R*)	0.8 (*S*)	2.2 (*S*)

The synthetic utility of the asymmetric peroxidation of sulfides to give chiral sulfoxides is limited by the low asymmetric bias. In most cases the reaction of a Grignard reagent with an optically active sulfinate[5] (Sec. **8-1.4**) is a superior method. In at least one case,[16] however, the sulfinate procedure was abandoned because of difficulty with the Grignard reagent. Oxidation of 1-adamantyl methyl sulfide with (+)-percamphoric acid (Table **8-1**, No. 22) gave a 44% yield of the *R*-sulfoxide (% e.e. unknown). The asymmetric oxidation method is also applicable to cyclic sulfides, whereas the sulfinate procedure is not.

Microbiological oxidation can give high stereoselectivity with certain sulfides.[24] The entries in Table **8-4** are such that preferential attack on the electron pair to the right as written will give **29**; with the exception of Nos. 4 and 10, this gives the chiral sulfoxide with the *R* configurational designation. In general the higher stereoselectivities are observed when R_A is *p*-tolyl. It is also very interesting that the stereoselectivity increases with increased branching of the R_B alkyl group. It is revealing that the stereoselectivity should drop sharply from the oxidation of *p*-tolyl 4-methylbenzyl sulfide (No. 2, 88–100%) and *p*-tolyl benzyl sulfide (No. 5, 56–81%) to that of phenyl benzyl sulfide (No. 10, 5.4–27%). The series of reactions represented by Nos. 5, 10, and 11 are further illustrated by **30**. This again shows the great sensitivity of this enzyme system to changes in substrate structure at positions remote from the sulfur; such sensitivity would not be anticipated from experience with chemical asymmetric syntheses.

[24] (a) R. M. Dodson, M. Newman, and H. M. Tsuchiya, *J. Org. Chem.*, **27**, 2707 (1962). (b) B. J. Auret, D. R. Boyd, and H. B. Henbest, *Chem. Commun.*, 66 (1966). (c) B. J. Auret, D. R. Boyd, S. Ross, and H. B. Henbest, *J. Chem. Soc.* (C), 2371 (1968).

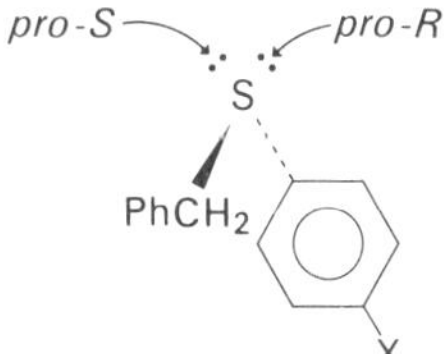

X =	Average ratio *pro-R*:*pro-S* oxidation
H	43:57 [14% e.e., *S*(−)]
Me	86:14 [72% e.e., *R*(+)]
t-Bu	56:44 [12% e.e., *R*(+)]

30

Any further oxidation of chiral sulfoxide to the sulfone by the chiral enzyme system would be expected to proceed asymmetrically and would introduce a serious complication to the interpretation of these results. This

TABLE 8-4

Oxidation of Sulfides with *Aspergillus niger*[a]

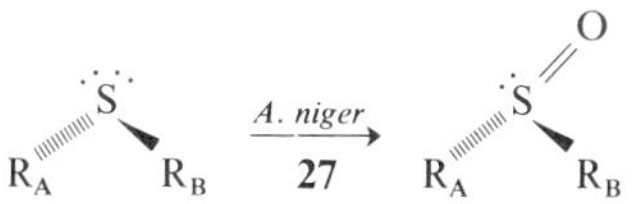

$$R_A\text{-S-}R_B \ (\mathbf{28}) \xrightarrow[\mathbf{27}]{A.\ niger} R_A\text{-S(O)-}R_B \ (\mathbf{29})$$

	Sulfide, R_A-S-R_B		Sulfoxide[b] **29**		
No.	R_A	R_B	$[\alpha]_D$[b]	Method A[b,c] % e.e.	Method B[b,d] % e.e.
1	*p*-Tolyl	*t*-Bu	(+)	94, 91	100, 98
2	*p*-Tolyl	$CH_2C_6H_4$-*p*-Me	(+)	88, 88[e]	94, 100
3	$PhCH_2$	*t*-Bu	(−)	77, 71[e]	91, 91
4	$CH_2C_6H_4$-*p*-Me	*t*-Bu	(−)[b]	—	78, 88
5	*p*-Tolyl	$CH_2C_6H_5$	(+)	82, 71[e]	81, 56
6	*p*-Tolyl	*i*-Pr	(+)	—	71, 69
7	*p*-Tolyl	Me	(+)	38, 32, 87[e]	—[f]
8	*p*-Tolyl	*n*-Bu	(+)	—	34, 33, 30
9	$CH_2C_6H_5$	Me	(+)	46, 25[e]	—[f]
10	$CH_2C_6H_5$	Ph	(−)[b]	5, 4,[e] 18[g]	27, 25
11	$CH_2C_6H_4$-*p*-*t*-Bu	$CH_2C_6H_5$	(+)	—[f]	13, 12
12	α-Naphthyl	CH_3	(−)	$[\alpha]_D$ −3.20° ($CHCl_3$)[g]	—

[a] Reference 24c, unless otherwise annotated.

[b] Percent enantiomeric excess of configuration **29**, *based on rotation of once recrystallized sulfoxide*. All sulfoxides have the configuration represented by formula **29** which is *R* except for Nos. 4 and 10 which are *S*.

[c] Growing fungus method, 4 days at 30°.

[d] Acetone powder method; fungus extracted with cold acetone before addition of sulfide for 6 day reaction at 28–30°.

[e] Sulfone also obtained (see text).

[f] No sulfoxide isolated.

[g] Reference 24a.

has been shown to occur[25] by subjecting racemic sulfoxides to the action of *Aspergillus niger* (**27**). Recovered sulfoxides from partial oxidation in this manner were optically active (Table **8-5**) in four of the six cases studied.*

TABLE 8-5

Comparison of Asymmetric Synthesis and Kinetic Resolution of Sulfoxides with *Aspergillus niger*[25]

R_A–S(:)–R_B (**28**) —*A. niger*, **27**→ R_A–S(O)–R_B (**29**) —*A. niger*, **27**→ R_A–S(O)(O)–R_B (**31**)

No.[a]	R_A	R_B	Asymmetric synthesis: % e.e. **29**; configuration / % yield sulfoxide / % yield sulfone (**31**)	Kinetic resolution: % e.e. **29**; configuration / % yield sulfoxide / % yield sulfone (**31**)
3	$PhCH_2$	*t*-Bu	77, 71; *S*-(−) 24, 61 15, 10	0 70 5
5	*p*-Tolyl	CH_2Ph	82, 71; *R*-(+) 19, 26 4, 7	7, 7; *R*-(+) 32, 26 32, 21
7	*p*-Tolyl	Me	38, 32, 87; *R*-(+) 38, 48, 7 0, 5, 19	30; *R*-(+) 59 8
9	$PhCH_2$	Me	46, 25; *R*-(+) 18, 17 18, 10	0 35 20
10	$PhCH_2$	Ph	5, 4; *S*-(−) 9, 35 3, 10	5; *R*-(+)[b] 43 14
13	**32**		29, 14; *R*-(+) 38, 16 28, 1	16; *R*-(+) 32 29[c]

[a] Numbers correspond to those used in Table **8-4**.
[b] The enantiomer of **29**.
[c] The sulfone contained a 24% e.e. of the (+) enantiomer (**34**).

[25] B. J. Auret, D. R. Boyd, and H. B. Henbest, *J. Chem. Soc.* (*C*), 2374 (1968).
[26] (a) M. Kobayashi and A. Yabe, *Bull Chem. Soc. Jap.*, **40**, 224 (1967). (b) U. Folli, D. Iarossi, and F. Montanari, *J. Chem. Soc.* (*C*), 1372 (1968).
[27] K. Balenović and N. Bregant, *Chem. and Ind.*, 1577 (1964).

* Similarly it has been shown that partial oxidation of racemic sulfoxides with chiral peracids[26] or partial reduction with cysteine[27] leads to a low order of asymmetric kinetic destruction of the racemate.

Partial reaction of *p*-tolyl methyl sulfoxide with *A. niger* leads to residual sulfoxide appreciably enriched in the *R*-(+)-enantiomer (30% e.e.). However, this partial kinetic resolution is not entirely due to preferential oxidation of the *S*-(−)-sulfoxide to sulfone since the yield of sulfone is only 8%. Apparently other reactions, of an undetermined nature, are also preferentially destroying the *S*-(−)-sulfoxide in this case. In two cases (Nos. 5 and 7) the effect of preferential destruction of one enantiomer of the racemic sulfoxide should be to increase the amount of the enantiomer formed preferentially by asymmetric synthesis. In accord with this conclusion is the observation that a higher enantiomeric excess of **29** results when the asymmetric oxidation of the sulfide is carried out under conditions that give higher yields of sulfone (e.g., 87% e.e. versus 38% e.e. in No. 7). In No. 10 it is clear that kinetic resolution would tend to lower the enantiomeric excess of sulfoxide due to asymmetric synthesis.

The substrate **32** for experiment No. 13 (Table **8-5**) is a racemate and the enzymatic process involves kinetic resolution at both stages of oxidation. Because of the C_2 symmetry axis of the starting material, oxidation at either face of a given enantiomer of the sulfide (**32**) yields the same enantiomer of sulfoxide (**33**); i.e., the sulfur faces are equivalent not enantiotopic. Incubation of racemic sulfide with *A. niger*, however, results in the formation of optically active sulfoxide since the oxidative enzyme system of the microorganism can distinguish between the enantiomers of **32** and preferentially oxidizes one of them to sulfoxide. A 14% e.e. of (+)-**33** is obtained when the

±**32** —*A. niger*→ (1*S*, 6*S*)-(+)-**33** (14% e.e.)

sulfoxide yield is 16% and the sulfone yield is 1%. When racemic sulfoxide (**33**) is treated with *A. niger*, the recovered sulfoxide (32% yield) contains

(±)-**33** —*A. niger*→ (+)-**33** (16% e.e.), 32% yield and (−)-**34** (24% e.e.), 29% yield

a 16% excess of the (+) enantiomer. The kinetic relationships may be summarized by noting that $k_1 > k_2$ and $k_4 > k_3$.

$$(+)\text{-}\mathbf{32} \xrightarrow{k_1} (+)\text{-}\mathbf{33} \xrightarrow{k_3} (+)\text{-}\mathbf{34}$$

$$(-)\text{-}\mathbf{32} \xrightarrow{k_2} (-)\text{-}\mathbf{33} \xrightarrow{k_4} (-)\text{-}\mathbf{34}$$

When the sulfone yield increases from 1 to 28% during oxidation of the sulfide, the enantiomeric excess of (+)-**33** increases by 15% due to preferential consumption of the (−)-**33** in the second stage of the oxidation.

There are other isolated examples of the stereoselective oxidation of sulfides by microorganisms. Biotin (**35**) is oxidized exclusively to a levorotatory sulfoxide (**36**) by a growing culture of *A. niger*,[28] in contrast to *in vitro* oxidation by peroxide which produces an excess of dextrorotatory sulfoxide (stereoselective chemical oxidation of chiral sulfides is discussed in Sec. **8-1.3**).

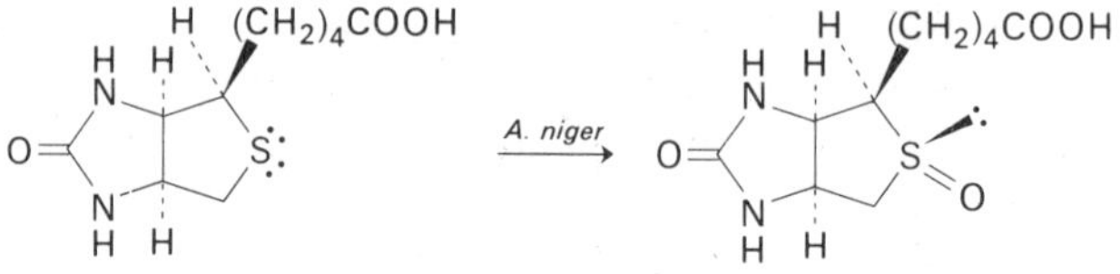

H $(CH_2)_4COOH$ — *A. niger* →

35 (−)-**36**

The epimeric 17-methylthioandrostenones, 17α-**37** and 17β-**37**, have been oxidized by incubation with cultures of *Rhizopus stolonifer* A.T.C.C. 6227-6. Only one 11-α-hydroxysulfoxide was isolated from the oxidation of the 17α or 17β epimer,[24a,29] but the 11-deoxysulfides isolated were found in the epimer ratio of 7.5:1. Similarly *Calonectria decora* oxidizes **38** to a single 17β-hydroxysulfoxide[30] which when reacylated gives a single 17β-acetoxy-7α-methylsulfinylandrost-4-en-3-one. Oxidation of **38** with monoperphthalic acid[30,31] (MPA) gave two 17β-acetoxy-7α-methylsulfinylandrost-4-en-3-ones. The one formed in lesser amount was identical with that from microbial oxidation of **38**. Both isomers from MPA oxidation gave the same sulfone, thus proving that they were epimeric at sulfur.

An analogy has been drawn between enzymatic reactions and an asymmetric sulfide oxidation involving an optically active additive that is not

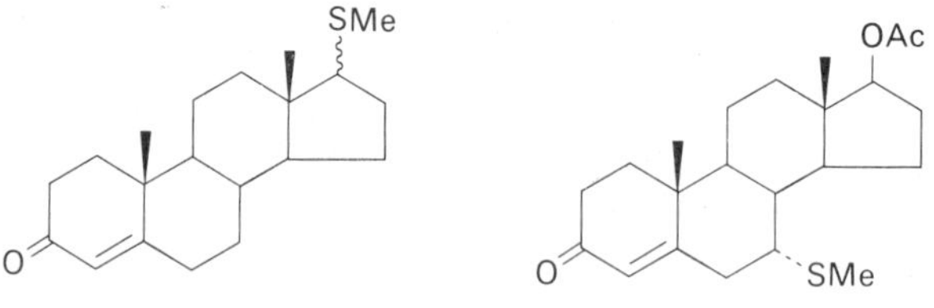

SMe OAc SMe

37 **38**

[28] L. D. Wright, E. L. Cresson, J. Valiant, D. E. Wolf, and K. Folkers, *J. Amer. Chem. Soc.*, **76**, 4163 (1954).

[29] R. M. Dodson and P. B. Sollman, U.S. Pat. 2,999,101 (1961); *Chem. Abst.*, **56**, 2488 (1962).

[30] C. E. Homlund, K. J. Sax, B. E. Nielsen, R. E. Hartman, R. H. Evans, Jr., and R. H. Blank, *J. Org. Chem.*, **27**, 1468 (1962).

[31] R. E. Schaub and M. J. Weiss, *J. Org. Chem.*, **27**, 2221 (1962).

incorporated into the final product.[32] The oxidation of benzyl methyl sulfide by iodine in the presence of a (*S*)-(+)-2-phenylsuccinate buffer to optically active benzyl methyl sulfoxide [6.4% e.e., *R*-(+)] serves as a chemical model for such an enzymatic process. The mechanism is believed to involve equilibrium formation of a racemic sulfonium salt (**39**) which undergoes nucleophilic attack by the chiral succinate (**40**) to give epimeric sulfonium zwitterions (**41**). Intramolecular displacement to give a 6.4% excess of the *R*-(+) sulfoxide can be due to preferential conversion of one enantiomer of **39** to **41**, preferential conversion of one epimer of **41** to sulfoxide, or both.

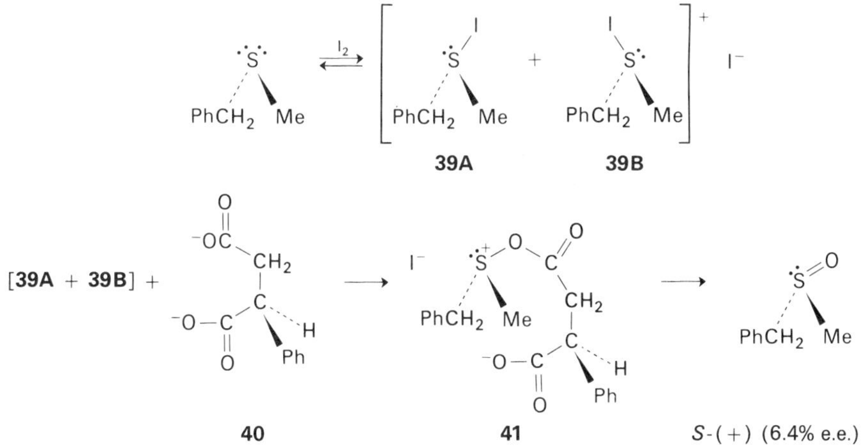

Fig. 8-3. Oxidation of benzyl methyl sulfide in the presence of (*S*)-(+)-2-phenylsuccinate. The structure of the diastereomeric intermediates **41** (epimeric at sulfur) is uncertain since either carboxylate group might be attached to sulfur.

Asymmetric oxidation has been used to obtain optically active sulfur compounds other than monosulfoxides; a number of disulfides have been asymmetrically oxidized,[33–35] as have a few sulfinates[35,36] and sulfenamides[35] (Table **8-6**). The configurations as well as the minimum percents enantiomeric excess were determined for two products,[35] No. 1 (1.5% e.e., *R*) and No. 2 (2.2% e.e., *R*). Reaction of the initial oxidation product with benzyl Grignard reagent gave benzyl *p*-tolyl sulfoxide. This Grignard reaction proceeds stereoselectively with inversion of configuration (Sec. **8-1.4**), thus

[32] T. Higuchi, I. H. Pitman, and K.-H. Gensch, *J. Amer. Chem. Soc.*, **88**, 5676 (1966).
[33] W. E. Savige and A. Fava, *Chem. Commun.*, 417 (1965).
[34] J. L. Kice and G. B. Large, *J. Amer. Chem. Soc.*, **90**, 4069 (1968).
[35] L. Sagramora, P. Koch, A. Garbesi, and A. Fava, *Chem. Commun.*, 985 (1967).
[36] E. Ciuffarin, M. Isola, and H. Fava, *J. Amer. Chem. Soc.*, **90**, 3594 (1968).

permitting assignment of configurations* and enantiomeric excesses to these products.

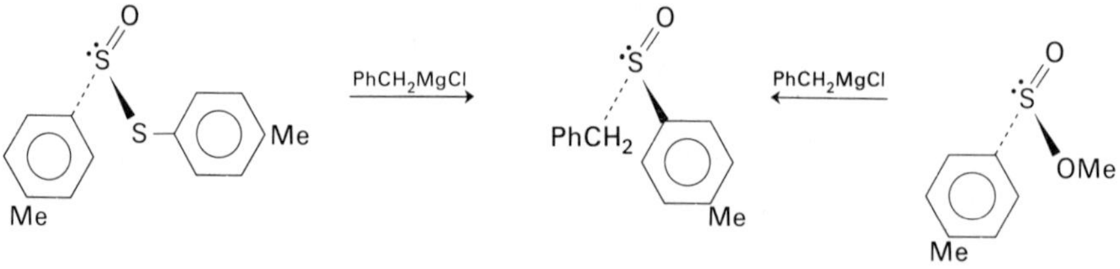

$\xrightarrow{PhCH_2MgCl}$ $\xleftarrow{PhCH_2MgCl}$

R-(+) *S*-(−) *R*-(+)

TABLE 8-6

Asymmetric Oxidation of Disulfides, Sulfenates, and Sulfenamides with (1*S*)-(+)-Monopercamphoric Acid[a]

No.	Sulfur compound	Product	Rotation,[b] $[\alpha]_D^{25}$ ($CHCl_3$)	Ref.
1	*p*-Tolyl-S-S-*p*-tolyl	*p*-Tolyl-$\overset{O}{\overset{\|}{S}}$-S-*p*-tolyl	+4.4°, 7.6°	35
2	*p*-Tolyl-S-OMe	*p*-Tolyl-$\overset{O}{\overset{\|}{S}}$-OMe	+2.33°	35
3	Ph-S-S-Ph	Ph-$\overset{O}{\overset{\|}{S}}$-S-Ph	+8.5–14.0° [c]	34
4	Ar-S-S-Ar[d]	Ar-$\overset{O}{\overset{\|}{S}}$-S-Ar[d]	*ca.* +10° [e]	33
5	Dodecyl-S-S-dodecyl	Dodecyl-$\overset{O}{\overset{\|}{S}}$-S-dodecyl	0°	33
6	$PhCH_2$-S-S-CH_2Ph	$PhCH_2$-$\overset{O}{\overset{\|}{S}}$-S-$CH_2Ph$	0°	33
7	*p*-Tolyl-S-$OCHPh_2$	*p*-Tolyl-$\overset{O}{\overset{\|}{S}}$-$OCHPh_2$	+6.5° [f]	36
8	*p*-Tolyl-S—N	*p*-Tolyl-$\overset{O}{\overset{\|}{S}}$—N		

[a] Reaction in $CHCl_3$ solvent at below 10°.
[b] Rotation $[\alpha]_{436}^{25}$ ($CHCl_3$) unless otherwise noted.
[c] The range of rotations ($[\alpha]_{436}^{25}$, dioxane) is due to racemization during work-up.
[d] Ar = Ph, *p*-ClPh, and *β*-naphthyl.
[e] The exact rotation was not reported.
[f] Rotation, $[\alpha]_{436}^{25}$ (HOAc).
* See footnote in Sec. **8-1.4**, p. 362.

An interesting application of asymmetric oxidation at sulfur is its use to determine which of two diastereomeric disulfoxides is the *dl* and which the *meso* form. Such an example is shown in the oxidation of **42A** to **42B**.[37] Disulfoxide **42B** exists in α (higher melting) and β (lower melting) forms. Oxidation of **42A** (or racemic monosulfoxide) with monopercamphoric acid

$$\underset{\textbf{42A}}{Me{-}S{-}CH_2{-}S{-}Me} \xrightarrow[\text{acid}]{\text{Monopercamphoric}} \underset{\textbf{42B}}{Me{-}\overset{\overset{O}{\|}}{S}{-}CH_2{-}\overset{\overset{O}{\|}}{S}{-}Me}$$

gives a mixture of diastereomers which can be separated into the levorotatory α-form and inactive β-form. Therefore the β- (lower melting) form is the *meso* modification. This conclusion is confirmed by the nmr spectra which show the methylene hydrogens of the β-form as an AB quartet while the corresponding hydrogens in the α-form give a singlet. Only the *meso* form has diastereotopic methylene hydrogens capable of chemical shift non-equivalence; those in the *dl* isomer are equivalent in an achiral environment.

8-1.3. Asymmetric Oxidation of Chiral Sulfides with Achiral Reagents. There are numerous examples of the oxidation of a sulfide to a sulfoxide so that one of a pair of diastereomers (epimeric at sulfur) is produced in excess. These oxidations are stereochemically analogous to the addition reactions at a carbonyl group in a chiral aldehyde or ketone (situations where the Cram rule pertains, Secs. **3-3** and **3-4**). As in the case of aldehydes and ketones, in order to fulfill the conditions for an asymmetric synthesis, the substrate should be optically active. However, if a method for analyzing and identifying the diastereomeric products is available, a study can be carried out equally as well using a racemic substrate. A general reaction of this type is illustrated by the conversion of **43** to **44A** and **44B** which have opposite configurations at sulfur.

$$\underset{\textbf{43}}{RR'R''C{-}\ddot{S}{-}Me} \xrightarrow{PhCO_3H} \underset{\textbf{44A}}{RR'R''C{-}S(O){-}Me} \text{ and } \underset{\textbf{44B}}{RR'R''C{-}S(O){-}Me}$$

Some of the interest in reactions of this kind stems from the fact that sulfoxides with multiple chiral centers comprise a rather large and important

[37] R. Louw and H. Nieuwenhuyse, *Chem. Commun.*, 1561 (1968).

class of natural products* ranging from penicillin derivatives[38a] to a metabolic product of blowflies.[38b] They have also been key compounds in pioneering stereochemical studies. For example, (*S*)-(+)-methyl-L-cysteine sulfoxide, which can be isolated from onions,[10b] was used in the first X-ray determination of the absolute configuration of a compound asymmetric at sulfur,[10a] and has been the subject of more recent ORD studies.[10c]

There have been no systematic studies of the stereochemistry of the oxidation of chiral acyclic sulfides and only a few such studies of cyclic sulfide oxidation. Thus, although there are numerous isolated examples of this kind of asymmetric synthesis, especially in cyclic systems, very little is known about the factors which mediate asymmetric induction. The following treatment of reactions of this type is, therefore, based on selected examples which are chosen to illustrate the scope of the phenomenon and is not intended to be exhaustive.

8-1.3a. *Acyclic Sulfides.* Oxidation of (−)-methionine (**45**) with hydrogen peroxide, iodine, or iodate produces unequal amounts of diastereomeric sulfoxides.[39] The percent enantiomeric excess of the *S,S* epimer (**46A**)

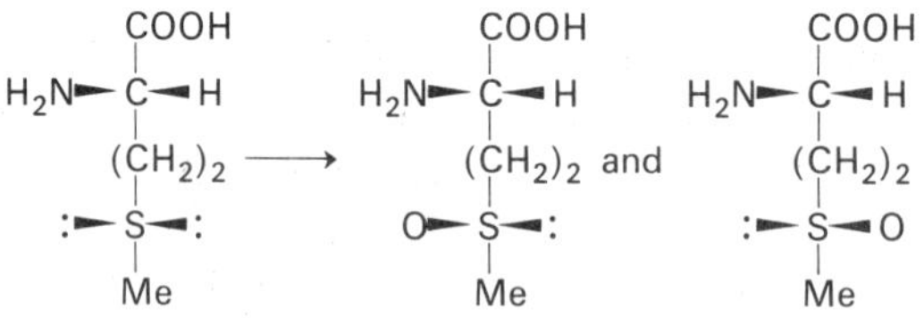

COOH | H_2N—C—H | $(CH_2)_2$ | :—S—: | Me → COOH | H_2N—C—H | $(CH_2)_2$ | O—S—: | Me and COOH | H_2N—C—H | $(CH_2)_2$ | :—S—O | Me

S-**45** *S,S*-(+)-**46A** *S,R*-(−)-**46B**

varies from about 1% (H_2O_2) to approximately 30% (I_2 at pH 7), depending upon the oxidizing agent and the conditions. Similar oxidations have been carried out on methionine derivatives.[40a] The configurations of the products are now known to be represented by (+)-**46A** and (−)-**46B**.[40b]

Various S-alkyl cysteines have also been oxidized to mixtures of diastereomeric sulfoxides with hydrogen peroxide.[10b] In these cases the asymmetric induction is 1,3 rather than 1,4 as in oxidations of methionine derivatives. It is difficult to be sure which cysteine diastereomer predominates in any

[38] (a) J. D. Cocker, S. Eardley, G. I. Gregory, M. E. Hall, and A. G. Long, *J. Chem. Soc.* (*C*), 1142 (1966). (b) F. Lucas and L. Levenbook, *Biochem. J.*, **100**, 473 (1966).

[39] T. F. Lavine, *J. Biol. Chem.*, **169**, 477 (1947).

[40] (a) B. Iselin, *Helv. Chim. Acta*, **44**, 61 (1961). (b) B. W. Christensen and A. Kjaer, *Chem. Commun.*, 225 (1965).

* See E. A. Barnsley, *Tetrahedron*, **24**, 3747 (1968), for a number of examples and leading references. Also, A. I. Virtoner, *Angew. Chem. Internat. Edn.*, **1**, 299 (1962).

given case. In one investigation[41] the oxidation mixture from S-methyl-L-cysteine was evaporated to dryness and the residue was found to be levorotatory ($[\alpha]_D$ $-16°$). The presence of both isomers was qualitatively confirmed by chromatography, but by-products or unreacted starting material could have affected the rotation of the crude mixture. Another study[42] of the hydrogen peroxide oxidation of S-(*cis*-1-propenyl)-L-cysteine suggests that about 67% of a dextrorotatory epimer ($[\alpha]_D$ max $+118.5°$) and 33% of a *levo* epimer ($[\alpha]_D$ max $-106.6°$) are produced.

Quantitative estimation of the diastereomer ratios in such reactions is subject to more than the usual amount of experimental error. The products are frequently solids or are isolated as solid derivatives; there is the possibility of partial oxidation to sulfones with concommitant kinetic resolution as well as the danger of epimerization at sulfur.[43] In the methione oxidations,[40] for example, while there was negligible oxidation to sulfone and probably no epimerization at sulfur, the product was crystallized and, in fact, it was demonstrated that recrystallization led to enrichment in the (+)-diastereomer.

Each member of the series of *ortho*-thiomethylbenzoates (**47**) shown in Table **8-7** was subjected to oxidation with perbenzoic acid in chloroform at $-5°$.[44] After saponification of the resulting sulfoxide esters, the optically

TABLE 8-7

Asymmetric Oxidation of *ortho*-Thiomethylbenzoates[44]

$$o\text{-}(MeS)C_6H_4C(=O)OR^* \xrightarrow[2)\ H_2O,\ OH^-]{1)\ PhCO_3H} o\text{-}(Me\overset{*}{S}=O)C_6H_4C(=O)OH + R^*OH$$

47 → **48**

No.	R*OH	Predominant enantiomer of **48**, % e.e.
47A	(3*R*)-(−)Menthol	6%, *S*-(−)
47B	(*R*)-Mesitylmethylcarbinol	20%, *S*-(−)
47C	(*S*)-Methyl-α-naphthylcarbinol	9%, *R*-(+)

[41] R. L. M. Synge and J. C. Wood, *Biochem. J.*, **64**, 252 (1956).

[42] J. F. Carson and L. E. Boggs, *J. Org. Chem.*, **31**, 2862 (1966).

[43] (a) J. Jacobus and K. Mislow, *J. Am. Chem. Soc.*, **89**, 5228 (1967). (b) K. Mislow, *Rec. Chem. Prog.*, **28**, 217 (1967).

[44] A. Maccioni, F. Montanari, M. Secci, and M. Tramontini, *Tetrahedron Lett.*, 607 (1961).

active sulfoxide acid (**48**) was obtained. The assignment of configuration for this acid must be considered provisional since it is based upon results obtained from asymmetric peroxidations using (+)-monopercamphoric acid and achiral esters (**47**; R* = Me, *i*-Pr, and *t*-Bu; see Table **8-1**, Nos. 1–6, for which an empirical configurational correlation model was developed[44] (Sec. **8-1.2**), which may not be reliable for this system.

It can be seen that for the limited number of examples studied the configuration of the sulfoxide can be correlated with that of the carbinol carbon (this is true whether or not the assignment of configuration is correct). Thus **47** prepared from an *R* carbinol leads to (−)-**48** (presumably *S*) while that from an *S* carbinol is oxidized to (+)-**48**. The asymmetric bias is appreciable even though five bonds separate the inducing chiral center from sulfur. The model (**49**) proposed[44] to rationalize the results orients the carbonyl group in the less hindered region between R_S and R_M with the largest group R_L shielding one of the diastereotopic electron pairs on sulfur; attack on the *pro-R* electron pair by peroxide is thus favored (from above in **49**).

This model can only be considered a correlation device for the results in Table **8-7**. Other conformational models such as **50** would also rationalize the results just as satisfactorily.

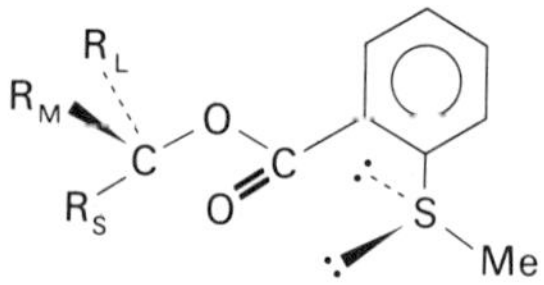

49 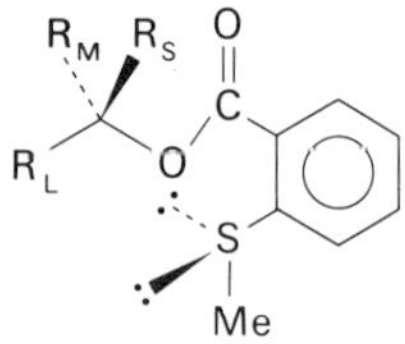 **50**

As pointed out in the previous section and emphasized in the following discussion, empirical conformational rules a priori cannot be extended from one type of reaction to another with any confidence. Perhaps the example which best justifies this admonition is the oxidation of (*R*)-2-octyl phenyl sulfide (**51**) with *t*-butyl hydroperoxide. Unequal amounts of the epimeric sulfoxides *R,R*-**52** and *R,S*-**52** are produced. In the initial investigation[45] of this reaction it was assumed that the *R,R* epimer would

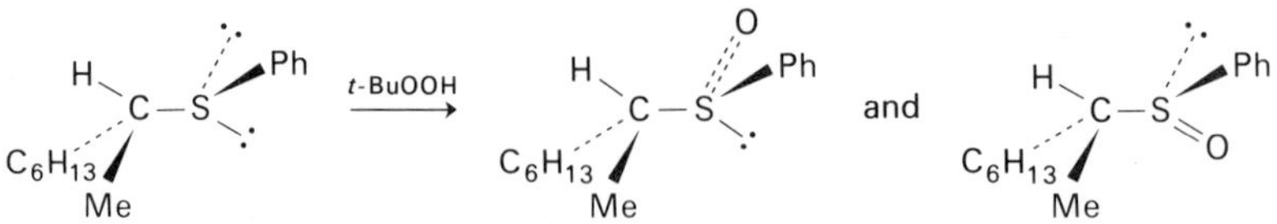

R-**51** —*t*-BuOOH→ *R,R*-**52** and *R,S*-**52**

[45] D. J. Cram and S. H. Pine, *J. Am. Chem. Soc.*, **85**, 1096 (1963).

predominate since it would result from oxidation involving the less hindered electron pair when the sulfide was viewed in what was assumed to be the most stable conformation (**51**). This prediction was based upon an extrapolation of Cram's empirical rules resulting from extensive studies of the addition reactions of chiral aldehydes and ketones (Chapters 3 and 4). Experimentally it was found that a levorotatory sulfoxide was formed in 24% excess over its dextrorotatory epimer.* Therefore, using the extrapolated rule of asymmetric induction described above, the *R,R* configuration was assigned to the levorotatory sulfoxide. An equilibration study was assumed to support this assignment since, when either the *dextro* or *levo* epimer was treated with potassium *t*-butoxide in dimethyl sulfoxide, the resulting equilibrium mixture contained an excess of the *levo* isomer. It was thought that *R,R*-**52**, which in the conformation shown has the two smallest groups staggered between the two larger groups on the adjacent atoms, ought to be thermodynamically more stable than *R,S*-**52**.

As reasonable as these arguments appear, they lead to an erroneous conclusion as demonstrated by Mislow, et al.[19] Using a highly reliable method of stereospecific synthesis (Sec. **8-1.4**), it was shown that both (*R*)-2-octyl *p*-tolyl (*R*)-sulfoxide and (*S*)-2-octyl *p*-tolyl (*R*)-sulfoxide are dextrorotatory, and that their ORD curves are essentially mirror images for those of (−)-**52**. Therefore, contrary to the earlier conclusions,[45] the predominant diastereomer from the oxidation of *R*-**51** is *R,S*-(−)-**52**. Furthermore, it was confirmed[19] that the thermodynamically more stable diastereomer was also *R,S*-(−)-**52**. It is clear that extrapolation of the conformational rules developed for addition reactions of chiral aldehydes and ketones to the oxidation of chiral sulfides is not justified. Mislow's point[19] is well taken that, even if the prediction had been correct, the extrapolation would be equally objectionable.

Another example pertinent to the discussion of 1,2-asymmetric induction in the oxidation of chiral sulfides is the perbenzoic acid oxidation[46] of the *erythro* and *threo* diastereomers corresponding to structure **53**. Racemic sulfides were used; thus the sulfoxides were optically inactive. From the *erythro* sulfide there was isolated a 34% excess of a higher melting diastereomer (m.p. 184° versus 132° for the other diastereomer); whereas

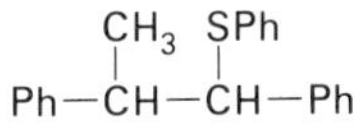

$Ph-CH(CH_3)-CH(SPh)-Ph$ **53**

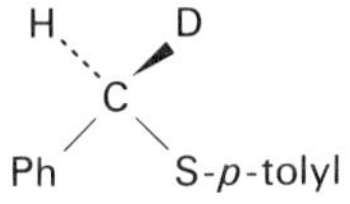

H, D, Ph, S-*p*-tolyl on C **54**

$Ph-CH(CH_3)-S-CH_2COOH$ **55**

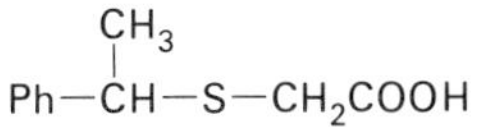

[46] C. A. Kingsbury and D. J. Cram, *J. Am. Chem. Soc.*, **82**, 1810 (1960).

* Epimers of this type usually have opposite rotations, since the contribution of the sulfoxide center to the rotatory power is normally dominant.

oxidation of the *threo* sulfide gave a 50% excess of a lower melting (75° versus 155°) diastereomer. Provisional assignments of configuration at sulfur were made on the basis of inspection of models and consideration of reactivity and ultraviolet spectra. In view of the failure of such models in the case of **51**, these assignments should be checked.

The periodate oxidation of the chiral sulfide **54**, which was optically active by virtue of deuterium substitution,[47] gave a 50:50 mixture of the enantiomeric products ("racemic" at sulfur), indicating no measurable preferential directing influence by the deuterium labeled asymmetric carbon. Finally, hydrogen peroxide oxidation of (+)-**55** gave a dextrorotatory sulfoxide ($[\alpha]_D$ +180°) as an initial precipitate after acidification of the reaction mixture.[48] A second crop of crystals had a lower rotation ($[\alpha]_D$ +111°).

TABLE 8-8

Oxidation of 4-Substituted Thianes[a]

56 $\xrightarrow{\text{Oxidizing reagent}}$ *cis*-**57** + *trans*-**57**

No.	Oxidizing agent	Solvent	Temp.	*cis*:*trans* Ratio R = *t*-Bu	Me	*p*-Cl-C_6H_4
1	$PhICl_2$	Pyridine-H_2O	−40°	—	—	5:95[b]
2	$PhICl_2$	Pyridine-H_2O	20°	—	—	10:90[b]
3	Ozone	CCl_4, CH_2Cl_2	−40°	10:90	15:85	16:84
4	DABCO-$2Br_2$[c]	HOAc-H_2O	20°	—	—	14:86[b]
5	PhIO	Pyridine-H_2O	20°	—	—	17:83[b]
6	CrO_3	Pyridine	25°	27:73	—	—
7	*t*-BuOOH	MeOH	50°	27:73	32:68	—
8	H_2O_2	AcOH	25°	35:65	—	—
9	*t*-BuOOH	Benzene	50°	36:64	35:64	35:65
10	*m*-Cl-$C_6H_4CO_3H$	CH_2Cl_2	0°	36:64	30:70	33:67
11	H_2O_2	Acetone	25°	37:63	40:60	30:70
12	PhIO	Benzene	80°	46:54	49:51	51:49
13	HNO_3	Ac_2O	0°	67:33	—	—
14	$NaIO_4$	H_2O	0°	75:25	72:28	76:24
15	N_2O_4[d]	None	0°	81:19	76:24	81:19
16	*t*-BuOCl	MeOH	−70°	100:0	89:11	98:2

[a] Reference 49a unless otherwise noted.
[b] Reference 49b.
[c] Bromine complex of 1,4-diazabicyclo[2.2.2]octane.
[d] This reagent causes equilibration of the resulting sulfoxides.

[47] E. G. Miller, D. R. Rayner, H. T. Thomas, and K. Mislow, *J. Am. Chem. Soc.*, **90**, 4861 (1968).

[48] B. Holmberg, *Arkiv Kemi*, **13A**, No. 15, 8 (1939); *Chem. Abst.*, **33**, 8586 (1939).

Recrystallization of the higher rotating material gave a product having $[\alpha]_D$ +187° which may have been enantiomerically pure.

8-1.3b. *Cyclic Sulfides.* Studies of the factors which control stereoselective synthesis in the oxidation of *cyclic* sulfides have added materially to our understanding of this process. The results indicate that the nature of the oxidizing agent and, to a lesser extent in most cases, the reaction conditions are important variables; these must be held constant if meaningful conclusions concerning the effect of structural variations in the substrate are to be obtained. However, as was true for chiral acyclic sulfides, many studies have not been designed to quantitatively evaluate diastereomer ratios and one must exercise caution in the interpretation of existing data. Several reactions have been selected to illustrate some of the important generalities which have emerged from studies with various types of cyclic sulfides.

Two key investigations in this area involve the stereochemistry of 4-substituted thiane oxidations with a variety of reagents.[49a,b] The results are summarized in Table **8-8**. The data are presented so as to reflect a general increase in the percentage of axial oxidation as one moves down the tabular listing of reactions. Several trends are readily apparent. Peroxy reagents (Nos. 7–11) give a rather similar product ratio with all three of the thianes studied; equatorial oxidation is preferred. Iodobenzene dichloride and *t*-butyl hypochlorite give dramatically opposed results, equatorial oxidation being heavily favored with the former and axial with the latter. The substituent in the 4-position appears to exert a negligible effect upon the stereoselectivity.*

Contrary to intuitive expectation on the basis of an extrapolation from cyclohexyl systems, the thermodynamically more stable conformation for **57** is that in which the S—O bond is axial.[49,50] Therefore axial oxidation of **56**

[49] (a) C. R. Johnson and D. McCants, Jr., *J. Am. Chem. Soc.*, **87**, 1109 (1965). (b) G. Barbieri, M. Cinguini, S. Colonna, and F. Montanari, *J. Chem. Soc.* (*C*), 659 (1968).

[50] R. S. McEwen, G. A. Sim, and C. R. Johnson, *Chem. Commun.*, 885 (1967).

* The three thianes listed in Table **8-8** (R = *t*-Bu, Me, and *p*-ClC_6H_4) have been the most carefully studied, but in only one of these cases (R = *p*-Cl-C_6H_4) is there a significant dipole interaction between the S—O and R groups. In the case of 4-substituted cyclohexanones, dipole effects (when R = OMe, for example) change the stereoselectivity of hydride reduction appreciably (see Reference 51, Sec. **3-4.2**) from that expected solely on the basis of the conformational free energy of R. The dipole interaction tends to favor equatorial hydride attack.** Similar effects might be important in the transition state for thiane oxidation also, but this possibility has not been experimentally tested. The oxidation of 4-chlorothiane [J. C. Martin and J. J. Uebel, *J. Am. Chem. Soc.*, **86**, 2936 (1964)] with periodate gives a mixture of *cis* and *trans* sulfoxides in which *cis* predominates, but accurate *cis*:*trans* ratios have not been reported. Furthermore, both the *cis* and *trans* sulfoxides prefer an *S*—O axial conformation, at the expense of a C—Cl equatorial orientation in the case of the *trans* compound.

** See however Reference 74, p. 132, for another interpretation.

might be described as being "product development controlled," whereas with a "large" reagent equatorial oxidation would correspond to "steric approach control."[49a] However, as emphasized in Sec. **3-4**, the product development control concept does not appear to be valid for addition reaction of cyclic ketones, for which it was originally proposed. It also appears from the meager data available that it should not be applied to cyclic sulfide oxidations. In the sulfide oxidations there is not even a superficial relationship between the "size" of a reagent and its preference for axial or equatorial oxygenation. Undoubtedly, vastly different mechanisms are involved in contrast to the general similarity of most carbonyl addition mechanisms, particularly most hydride reductions. Clearly it is possible to obtain significantly larger amounts of the *more* thermodynamically stable isomer than is present at equilibrium (Table **8-9**). For these and other

TABLE 8-9

***cis-trans* Equilibration of 4-Substituted Thianes, 57**

No.	**57**, R =	Equilibration method	% *cis*-**57**
1a	*t*-Bu	HCl-Dioxane, 25°	90
1b		Decalin, 190°	80
1c		N_2O_4	81
2a	Me	HCl-Dioxane, 25°	80
2b		Decalin, 190°	65
2c		N_2O_4	76
3	*p*-Cl-C_6H_4	HCl-Dioxane, 25°	80
4	Cl	HCl-Dioxane, 25°	65

reasons (Sec. **3-4**), it is preferable to rationalize the stereochemistry of cyclic sulfide oxidations in terms of the effect of substrate structure on the intrinsic axial or equatorial preference of a given reagent using transition state "steric strain" arguments. The intrinsic preference can be considered to be that reflected by the data in Table **8-8** for unhindered thianes. Unfortunately, there has been no attempt to systematically vary the nature and the position of substituents on a thiane ring so that conclusions analogous to those developed in Sec. **3-4** could be drawn. However, some interesting experiments with oxathians have revealed the kind of behavior that might be expected with at least some of the reagents listed in Table **8-8**.

Periodate oxidation of **58** gives more than 90% of the equatorial sulfoxide (**59A**).[51] Ozone and hydrogen peroxide behave similarly. Nuclear magnetic

[51] (a) K. W. Buck, A. B. Foster, A. R. Perry, and J. M. Webber, *Chem. Commun.*, 433 (1965). (b) K. W. Buck, T. A. Hamor, and D. J. Watkin, *Chem. Commun.*, 759 (1966). (c) K. W. Buck, A. B. Foster, W. D. Pardoe, M. H. Qadir, and J. M. Webber, *Chem. Commun.*, 759 (1966). (d) A. B. Foster, Q. H. Hasan, D. R. Hawkins, and J. M. Webber, *Chem. Commun.*, 1084 (1968).

resonance[51c] and X-ray studies[51b] support the assignment of conformation **59A** both in solution and in crystalline form. The stereochemistry is in

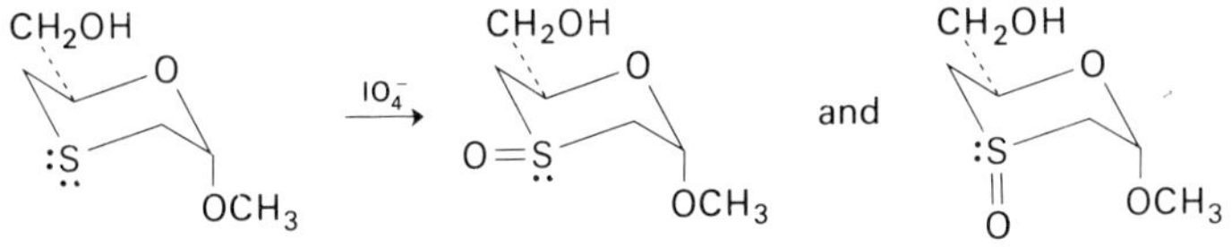

58 $\xrightarrow{IO_4^-}$ **59A** and **59B**

contrast to that observed[49a] for periodate and hydrogen peroxide oxidations of 4-substituted thianes (Table **8-8**) where axial oxidation predominated. Ozone gave preferential equatorial oxidation in both systems however. In the case of **58** there is the possibility of steric and dipole interactions in the transition state that are not present in the thiane examples. Particularly evident is the steric interaction due to the axial methoxy group which could hinder the formation of axial sulfoxide (**59B**). Periodate oxidation of **60**, in which the methoxy group is equatorial, gives more than 90% of the axial sulfoxide (**61A**).[51d,52] Likewise, periodate oxidation of **62** gives almost

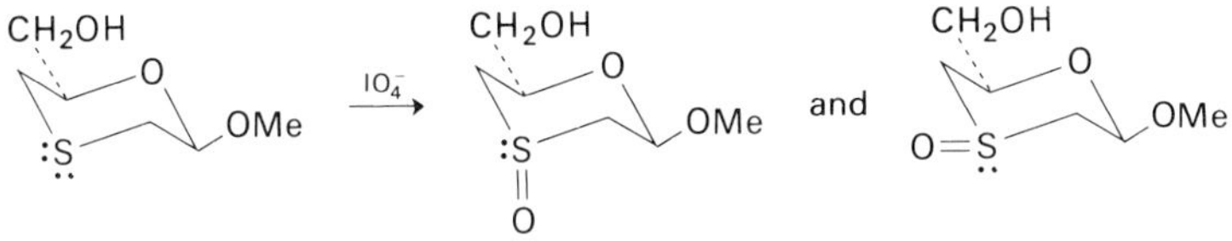

60 $\xrightarrow{IO_4^-}$ **61A** and **61B**

exclusively axial product and **63** gives mainly axial sulfoxide. Thus the configuration at the anomeric carbon appears to exert a dominant influence upon the stereochemistry of the oxidation with periodate. It is tempting to

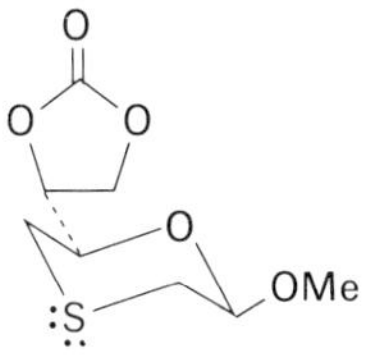

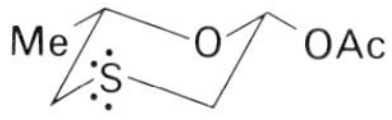

62 **63**

ascribe the observed changes from axial oxidation with **60** to equatorial oxidation with **58** to steric hindrance by the axial methoxy group in the latter, but dipole interaction may also play a role. A similar, though not as dramatic, change occurs with ozone as the oxidant, from a 50:50 equatorial

[52] A. B. Foster, T. D. Inch, M. H. Qadir, and J. M. Webber, *Chem. Commun.*, 1086 (1968).

to axial ratio with **60** to approximately a 90:10 ratio with **58**. Oxathiane **62** gave a 67:33 ratio of axial to equatorial sulfoxide with ozone.

There are other isolated examples which support the thesis that both the nature of the reagent and the structure of the substrate are important factors in determining the stereochemistry of oxidation of cyclic sulfides to sulfoxides. Oxidation of the thiacholestane (**64**) with *m*-chloroperbenzoic acid gives approximately a 2% excess of the equatorial sulfoxide.[53] On the other hand,

64

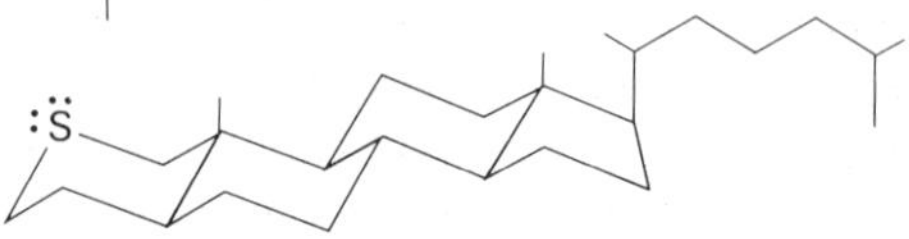

65

oxidation of **65** with perbenzoic acid gives an 88% yield of the equatorial product[54] as would be expected because of the β-angular methyl group which should greatly shield the axial electron pair on sulfur and consequently favor equatorial attack.

The bicyclic sulfide **66** gives 95% *exo* sulfoxide **67** when oxidized with *m*-chloroperbenzoic acid;[53] with *t*-butyl hypochlorite, however, the product ratio is reversed. These results are in accord with those obtained with the

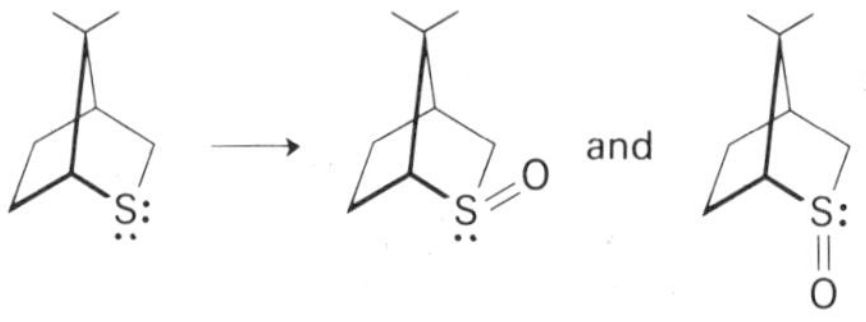

66 → **67** and **68**

4-substituted thianes (Table **8-8**). In the case of **66**, as well as **65**, one can identify with some confidence the exo and equatorial oxygenation as a formal substitution of the less hindered electron pairs of these sulfides. Mechanistically the process is better viewed as a nucleophilic attack by the less hindered electron pair of sulfur upon the oxygen of the peracid (see Sec. **8.1.2**, formula **25**). It is reasonable* to expect that the favored approach

[53] R. Nagarajan, B. H. Chollar, and R. M. Dodson, *Chem. Commun.*, 550 (1967).

[54] P. B. Sollman, R. Nagarajan, and R. M. Dodson, *Chem. Commun.*, 552 (1967).

* "Reasonable" arguments regarding the preferred conformation of the substrate and direction of peroxide attack in *acyclic* chiral sulfides led to erroneous conclusions (Sec. **8-1.3**). However, for the bicyclic skeleton of **66**, as well as the steroid nucleus of **65**, the assignment of conformation and the evaluation of the relative hindrance of the diastereotopic electron pairs on sulfur is simplified.

will be from the *exo* direction. The mechanism of the *t*-butyl hypochlorite reaction bears no resemblance to that attributed to the peracid oxidations. The best evidence available[55,56] suggests that hypochlorite oxidation involves sulfonium and sulfur dichloride intermediates whose alcoholysis, followed by elimination of isobutylene and hydrogen chloride or *t*-butyl chloride, gives the sulfoxide.[55] The introduction of oxygen by *t*-butyl hypochlorite is therefore the result of a rather complex mechanism, subject to stereochemical constraints much different than those for oxidation by the various peroxidic oxidizing agents which in general seem to operate with very similar stereoselectivities.

The influence of hydrogen bonding, dipole, and other specific interactions upon the stereochemistry of sulfide oxidations has not been carefully evaluated. Certain thioxanthrenes such as **69** show a solvent and concentration effect upon the stereochemistry of oxidation with *m*-chloroperbenzoic acid[57] to give *cis*- and *trans*-thioxanthren-9-ol 10-oxides which may be associated with variable amounts of intramolecular hydrogen bonding.

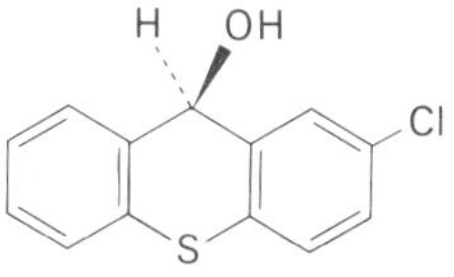

H OH Cl S **69**

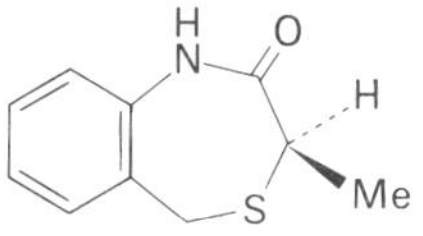

H N O H S Me **70**

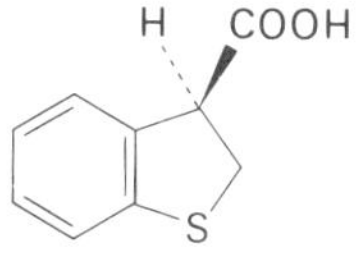

H COOH S **71**

Oxidation of **70** gave a 50% excess of sulfoxide having a *trans* S—O and methyl arrangement.[58] The structure of the *trans* product has been established by X-ray methods and thus the stereochemistry of the oxidation is firmly established. More studies of this kind are needed.

Compound **71** is an example of a system in which 1,3-induction operates. Oxidation of **71** with periodate gave *trans* sulfoxide as the only isolated product.[59a] On the other hand, with N_2O_2 (probably equilibrating conditions[49a]) a 60% excess of *cis* sulfoxide was formed. This system is also interesting in that the *cis* sulfoxide could be further oxidized to sulfone with hydrogen peroxide, but the *trans* isomer was inert, presumably due to steric hindrance. Other systems in which similar effects might operate are known[59b] and many more can be envisaged.

[55] C. Walling and M. J. Mintz, *J. Org. Chem.*, **32**, 1286 (1967).

[56] K. Mislow, T. Simmons, J. T. Melillo, and A. L. Ternay, Jr., *J. Am. Chem. Soc.*, **86**, 1452 (1964).

[57] A. L. Ternay, Jr., D. W. Chasan, and M. Sax, *J. Org. Chem.*, **32**, 2465 (1967).

[58] P. C. Thomas, I. C. Paul, T. Williams, G. Grethe, and M. Uskokovic, *J. Org. Chem.*, **34**, 365 (1969).

[59] (a) E. Jonsson, *Arkiv. Kemi*, **26**, 357 (1967). (b) S. Allenmark, *Arkiv. Kemi*, **26**, 73 (1967).

8-1.4. Synthesis of Chiral Sulfinates and Sulfinamides. Reaction of a sulfinyl chloride (**72**) with an optically active alcohol (**73**) gives unequal amounts of diastereomeric sulfinates (**74**). Frequently, (−)-menthol

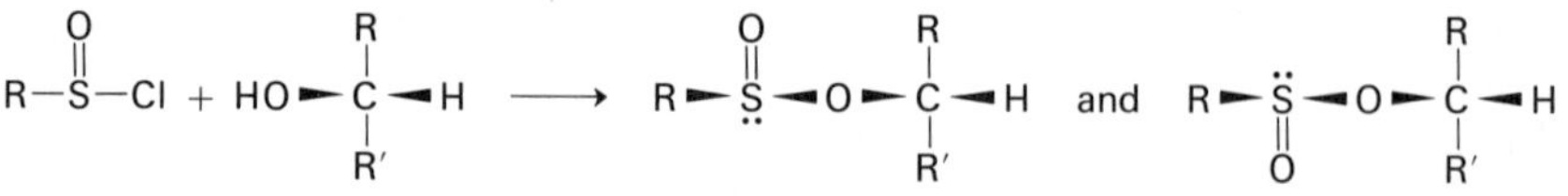

72 **73** S^*-**74** and R^*-**74**

is used as the optically active alcohol; several such asymmetric reactions are shown in Table **8-10**. With (−)-menthol, the predominant diastereomer

TABLE 8-10

Asymmetric Synthesis of Menthyl Sulfinates[a]

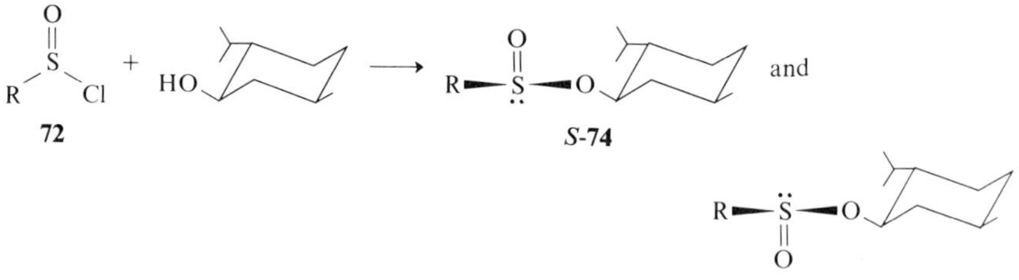

72 → *S*-**74** and *R*-**74**

No.	R =	Diastereomer ratio *R*-**74**:*S*-**74**
1	*p*-Tolyl	1.75:1
2	*p*-IC_6H_4	1.86:1
3	Benzyl	2.29:1
4	*n*-Bu	2.77:1
5	Me	1.87:1

[a] Reaction in ether at −78°; data from Reference 5f and references therein.

produced has the *R*-configuration at sulfur in the examples studied so far. Although the same epimer predominates in the hydrogen chloride-catalyzed equilibration of sulfinate from Nos. 1 and 2 in Table **8-10**, it has been shown[5f] that the product ratio in the sulfinyl chloride reaction to form the sulfinate

* Configurational designation at sulfur. This assumes that R does not take nomenclatural preference over oxygen in the sequence rule. The S=O bond is considered as a single bond for nomenclature purposes. See footnote 10 in Reference 19 for details.

is, for all practical purposes, kinetically controlled. The epimer ratio is essentially unchanged regardless of which reactant is in excess. Therefore the sulfinyl chloride, whose structure is inherently chiral, must be effectively achiral either due to rapid inversion, because of formation of an achiral intermediate such as an octahedral complex with pyridine, or due to formation of a sulfine[5f] (RCH=S=O) intermediate in certain cases. If the sulfinyl chloride structure were to be considered achiral, then this reaction would be classified as an asymmetric synthesis. But if it is a rapidly interconverting racemate, then this reaction constitutes a case of second-order asymmetric transformation (Sec. **1-4.3b**).

The reaction of a Grignard reagent with a chiral sulfinate has been shown to proceed with inversion of configuration at sulfur to give a chiral sulfoxide. It is apparent that the coupling of the synthesis of (−)-menthyl sulfinates

$$R-S(=O)-O-R^* + R'MgX \longrightarrow R-S(=O)-R' + R^*OMgX$$

with the Grignard reactions is a convenient route to chiral sulfoxides. The net effect of the sequence is the conversion of a sulfinyl chloride, by successive reactions with a chiral alcohol and achiral Grignard reagent, into a chiral sulfoxide; the chiral inducing agent is recovered in the last step. In this manner the asymmetric sulfinate synthesis has been used to correlate the configuration of several alcohols as well as to establish the configuration of sulfoxides.[60] The results of several correlations are summarized in Table **8-11**. The data indicate that alcohols represented by stereo formula **78** give an excess of *S*-(−)-**77**; therefore, an excess of the epimer of sulfinate (**76**) having the *R* configuration at sulfur[5c] must have resulted from the asymmetric synthesis step. The sensitivity of the method is illustrated by the fact that *S*-(+)-2-propanol-1-d_3 is able to induce measurable optical activity. These results indicate that the steric bulk of CH_3 is greater than CD_3 as would be expected on the basis of van der Waal radii of H versus D.

Use of the atropisomeric alcohol *S*-(+)-**79** yields an excess of *R*-(+)-**77** which implies that (+)-**79** should fit the enantiomer of stereo formula **78**. Inspection of models reveals that in *S*-**79** ring A of the biphenyl system hinders the OH more than it is hindered by ring B. When ring A is envisaged as R_L the stereo formula generated is enantiomeric with **78**. It should be noted that in this case the axial chirality of the biphenyl system is revealed rather than the chirality at an asymmetric carbon atom. The situation is similar

[60] (a) M. M. Green, M. Axelrod, and K. Mislow, *J. Am. Chem. Soc.*, **88**, 861 (1966). (b) D. J. Sandman, K. Mislow, W. P. Giddings, J. Dirlam and G. C. Hanson, *J. Am. Chem. Soc.*, **90**, 4877 (1968).

TABLE 8-11

Configurational Correlation of Alcohols via Asymmetric Synthesis of Sulfinates[60]

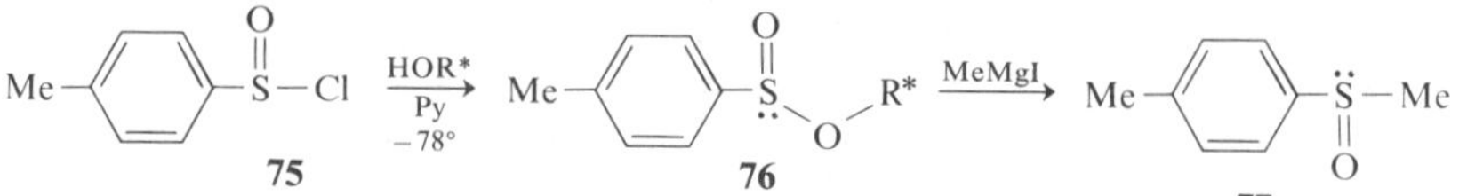

Me—C6H4—S(=O)—Cl (**75**) —HOR*, Py, −78°→ Me—C6H4—S(=O)—O—R* (**76**) —MeMgI→ Me—C6H4—S(=O)—Me (**77**)

No.	R*OH	Configuration and % e.e. of **77**[a]
1	(−)-Menthol	*S*-(−), 25.0
2	(−)-Isoborneol	*S*-(−), 50.2[b]
3	(−)-Borneol	*S*-(−), 22.7[b]
4	(+)-2-Butanol	*R*-(+), 10.0
5	(+)-3-Methyl-2-butanol	*R*-(+), 18.0
6	(+)-3, 3-Dimethyl-2-butanol	*R*-(+), 18.6
7	(+)-2-Propanol-1-d_3	*S*-(−), 0.26
8	(+)-(**79**)[c]	*R*-(+), 6.0
9	(1*R*,2*S*)-(+)-*exo*-Dehydronorborneol	*R*-(+), 4.8[b]
10	(1*R*,2*S*)-(+)-*exo*-Benznorbornenol	*R*-(+), 8.8[b]

[a] Calculated on the basis of $[\alpha]_D$ +156° (ethanol) for *R*-(+)-**77**.
[b] Corrected for the optical purity of the inducing alcohol.
[c] 4′,1″-Dimethyl-1,2,3,4-dibenz-1,3-cycloheptadiene-6-ol.

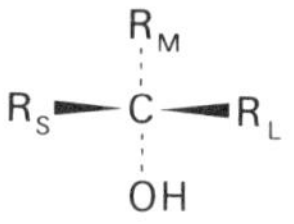

78

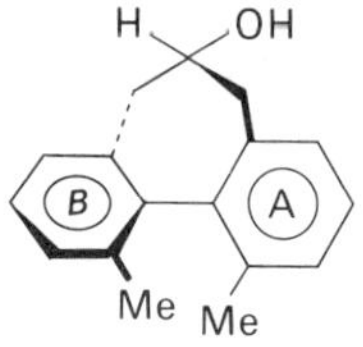

S-(+)-**79**

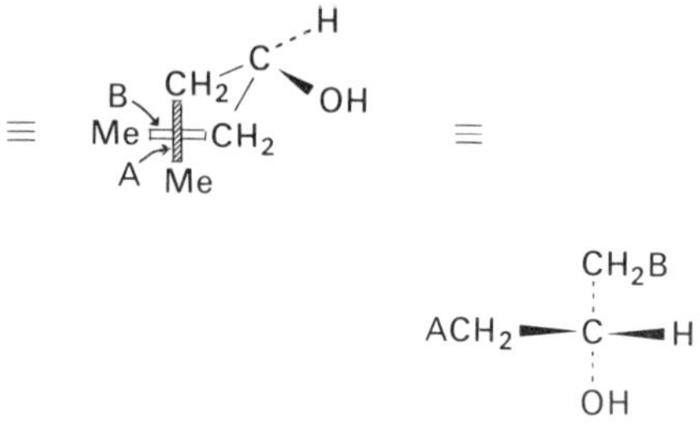

to that discussed in Chapter 2 for the atrolatic acid method of configurational correlation.

A very similar type of asymmetric synthesis takes place when benzenesulfinyl chloride (**80**) is allowed to react with optically active N-methyl-1-phenyl-2-propylamine (**81**). The product is a mixture of unequal amounts of epimeric sulfinamides (**82**) which can react with methyl lithium (methyl Grignard reagent is ineffective) to give optically active phenyl methyl sulfoxide (**83**).[61] The configuration and enantiomeric excess of the sulfoxide reflect the configuration at sulfur and the diastereomer composition of the sulfinamide precursor. It is assumed that the displacement reaction occurs

[61] J. Jacobus and K. Mislow, *Chem. Commun.*, 253 (1968).

with inversion by analogy with the similar reaction of sulfinates and Grignard reagents.* The asymmetric synthesis of **82** is very temperature sensitive,

$$\underset{\mathbf{80}}{Ph-\overset{O}{\overset{\|}{S}}-Cl} + \underset{\mathbf{81}}{PhCH_2\overset{NHMe}{\overset{|}{\overset{*}{C}}}HCH_3} \longrightarrow \underset{\mathbf{82}}{Ph-\overset{O}{\overset{\|}{\underset{*}{S}}}-\underset{Me}{\underset{|}{N}}-\overset{CH_3}{\overset{|}{\underset{*}{C}H}}-CH_2Ph} \xrightarrow{MeLi} \underset{\mathbf{83}}{Ph-\overset{O}{\overset{\|}{\underset{\cdot\cdot}{S}}}-Me}$$

increasing asymmetric bias being observed at 0° as compared to −70°. A mixture of diastereomers (**82**) prepared from *S*-(+)-**81** and **80** at 0° contained a 50% excess of the epimer having the *R* configuration at sulfur. However, at −70° equal amounts of the two epimers were obtained.

The synthetic utility of both the sulfinate and sulfinamide route to chiral sulfoxides may be extended if the intermediate diastereomers are separated by fractional crystallization prior to their reaction with a Grignard reagent or alkyl lithium compound. Andersen[5a] was the first to utilize this method for the stereospecific synthesis of optically active sulfoxides from sulfinates and extensive recent research[5] has demonstrated the generality of the reaction. Mislow and co-workers[5f] have summarized much of this work.

A recent report of separation of diastereomeric sulfite esters from the reaction of methylchlorosulfinate with *S*-(+)-diphenacyl malate[62] ($PhCOCH_2OCOCH_2\overset{*}{C}HOHCOOCH_2COPh$) opens new possibilities for asymmetric synthesis at sulfur.

8-2. Asymmetric Synthesis at Nitrogen

The reaction of imines with chiral peracids yields optically active oxaziridines.[63] Two types of optically active products have been obtained, those having both an asymmetric nitrogen and an asymmetric carbon[63a] and those whose activity is due entirely to asymmetry at nitrogen.[63b]

Numbers 1–6 in Table **8-12** illustrate the results obtained with some aldimines upon oxidation to oxaziridines with (1*S*)-(+)-monopercamphoric acid[63] (**1**). Products from Nos. 5 and 6 were separated into diastereomers indicating asymmetry at nitrogen as well as carbon. Nitrogen inversion is, in these cases, slow enough to be observable on an isolation time scale using optical rotation. Unequivocal proof of this is provided by examples Nos.

[62] M. K. Hargreaves, P. G. Modi, and J. G. Pritchard, *Chem. Commun.*, 1306 (1968).

[63] (a) D. R. Boyd, *Tetrahedron Lett.*, 4561 (1968). (b) F. Montanari, I. Moretti, and G. Torre, *Chem. Commun.*, 1694 (1968).

* There appears to be a slight loss of configurational integrity at sulfur (approximately 10%) in the conversion of **82** to **83** most likely due to partial racemization of **83**.[61]

8–11 in which there is no asymmetric carbon in the product (R′ = R″ = Ph) and the imine cannot exist in *syn* and *anti* forms.

TABLE 8-12

Asymmetric Synthesis of Oxaziridines[a]

$$\underset{\mathbf{84}}{R''(R')C{=}N{\sim}R} \xrightarrow{\text{Chiral peracid}} \underset{\mathbf{85}}{R''(R')C\overset{O}{\frown}\ddot{N}{\sim}R}$$

No.	Imine			Peracid[b]	[α] Oxaziridine (s)	
	R	R′	R″		$[\alpha]_D^{\circ c}$	$[\alpha]_{400}^{\circ}$ (neat)
1	*t*-Bu	Ph	H	(1*S*)-(+)-**1**	−12.7	—
2	*t*-Bu	α-Pyridyl	H	”	−4.3	—
3	*t*-Bu	*i*-Pr	H	”	−3.3	—
4	*t*-Bu	*p*-$NO_2C_6H_4$	H	”	−11.3 ($CHCl_3$)	—
5	*i*-Pr	*p*-$NO_2C_6H_4$	H	”	{−12.0	—
					{−1.3	—
6	Et	*p*-$NO_2C_6H_4$	H	”	{−11.0	—
					{−2.7	—
7	Pr	*i*-Bu	CH_3	”	−9.0	—
8	Me	Ph	Ph	”	−11.3 ($CHCl_3$)	−49.2
9	Me	Ph	Ph	(*S*)-(+)-**9**	—	−8.5
10	Me	Ph	Ph	(*R*)-(−)-**8**	—	+12.5
11	Me	Ph	Ph	(*S*)-(+)-**10**	—	−5.7

[a] Numbers 1–7 from Reference 63a; others from Reference 63b; Nos. 1–7 at 3°, others at −20°.
[b] See Sec. **8-1.2** for structures of peracids.
[c] Rotations were taken neat unless otherwise indicated. In Nos. 5 and 6 two diastereomers were isolated; the rotations were apparently taken in solution, but the solvent was not reported.

Maximum rotations are not known for any of the oxaziridines listed in Table **8-12**; therefore, percent asymmetric syntheses cannot be calculated. A comparison has been made[63] between the asymmetric bias of asymmetric oxidation as compared to a kinetic resolution in one of the examples studied. Reaction of 2-*n*-propyl-3-*i*-butyl-3-methyloxaziridine (**85**, R = Pr, R′ = *i*-Bu, and R″ = Me; Table **8-12**, No. 7) with brucine leads to partial conversion of the oxaziridine to the imine (**84**) with formation of brucine N-oxide.[64] The unreacted oxaziridine was optically active ($[\alpha]_D$ −3.94°). The same oxaziridine was more than twice as active (Table **8-12**, No. 7), when obtained by asymmetric oxidation of the imine with (+)-monopercamphoric acid.[63]

[64] W. D. Emmons, *J. Am. Chem. Soc.*, **79**, 5739 (1957).

Asymmetric oxidation at nitrogen with (+)-monopercamphoric acid has been used in a study of the structure of the azoxy compound formed from **86**.[65] The product is optically active ($[\alpha]_D$ + 1.2°) which is compatible with the N-oxide structure **87A** but which rules out symmetrical structures such as **87B**. Nitrogen itself is not a center of asymmetry in this example but preferential oxidation, via diastereomeric transition states, of one of the two nitrogens in **86** destroys its molecular symmetry.

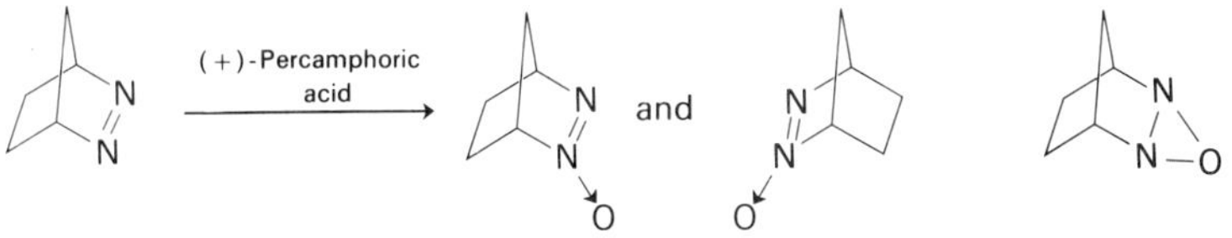

(+)-Percamphoric acid

86 **87A** and **87B**

Direct oxidation of a tertiary amine, having three different groups attached to nitrogen, with a chiral peracid is, in principle, a route to optically active amine oxides. However, in practice, several amines have been found to yield inactive amine oxides at room temperature and only very slightly active ones when the oxidation is carried out at −70°.[66] For example, N-methyl-1,2,3,4-tetrahydroquinoline [kairoline (**88**)] was allowed to react with (+)-monopercamphoric acid at −70° in chloroform-ether. The kairoline N-oxide (**89**) obtained (as the hydrochloride) had $[\alpha]_{546}$ +0.27° and $[\alpha]_{405}$ +0.54° (H_2O). The maximum reported specific rotation of kairoline N-oxide hydrochloride is $[\alpha]_{589}$ 44.2° (H_2O).[67f] Tertiary amine oxides are configurationally stable, both free and as salts.[67]

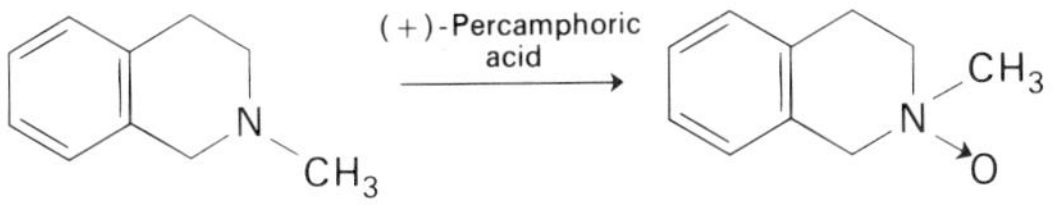

(+)-Percamphoric acid

88 (+)-**89**

Oxidation of a series of N-alkyl-N-methylanilines (R-NMePh; R = Et, *i*-Pr, and *t*-Bu) under the same conditions gave amine oxides with very low rotations ($[\alpha] < 0.8°$ from 589–405 nm); amine oxides from oxidations at room temperature showed no observable rotation.[66] The magnitude of

[65] F. D. Greene and S. S. Hecht, *Tetrahedron Lett.*, 575 (1969).

[66] J. D. Morrison and K. P. Long, unpublished observations.

[67] (a) J. Meisenheimer, *Ber.*, **41**, 3966 (1908). (b) J. Meisenheimer, *Ann. Chem.*, **385**, 117 (1911). (c) J. Meisenheimer, H. Glawe, H. Greeske, A. Schorning, and E. Vieweg, *Ann. Chem.*, **449**, 188 (1926). (d) H. S. French and C. M. Gens, *J. Am. Chem. Soc.*, **59**, 2600 (1937). (e) U. Schollkopf, U. Ludwig, M. Patsch, and W. Franken, *Ann. Chem.*, **703**, 77 (1967). (f) J. Meisenheimer and J. Dodonow, *Ann. Chem.*, **385**, 134 (1911). (g) O. Červinka and O. Kříž, *Coll. Czech. Chem. Commun.*, **31**, 1910 (1966).

rotations obtained by resolution[67] suggests that asymmetric oxidation with (+)-monopercamphoric acid produces a maximum of 1% e.e. in these examples. There is evidence[66] that other chiral peracids might give slightly higher induction, but the prognosis for significant asymmetric synthesis in such reactions is not favorable. This is not unexpected in view of the distance between chiral centers and the conformational mobility of the reactants.

The recently discovered configurational stability of N-haloaziridines[68] and N-aminoaziridines[69] opens new possibilities for the study of displacement reactions at an asymmetric nitrogen. For example, N-haloaziridines may be converted to N-amino derivatives by reaction with ammonia or lithium salts of secondary amines.[69]

The preparation of certain substituted N-haloaziridines involves asymmetric synthesis under the control of an asymmetric center in the parent compound.[68] For example, reaction of *S*-(−)-**90** with sodium hypochlorite[68d] gives a mixture of *trans*-(+)-**91**, $[\alpha]_D^{20}$ + 94° (c 0.8, *n*-nonane), 16% excess, and *cis*-(+)-**92**, $[\alpha]_D^{20}$ −81°. How one views this reaction is a matter of judgment. In one sense this is not an asymmetric synthesis, but a second-order asymmetric transformation, since the starting aziridine is a rapidly equilibrating mixture of nitrogen invertomers while chlorine substitution on nitrogen gives a stereochemically stable product. On the other hand, if one

HN — H, CH_3 — *S*-(−)-**90**; NaOCl → **91** (Cl on N, H, CH_3) and **92** (N—Cl, H, CH_3); NaOBr → **93** (Br on N, H, CH_3)

considers the average configuration at nitrogen in **90** to be planar, then dissymmetry is created by halogenation and the reaction can be considered an asymmetric synthesis. Using racemic **90**,[68a] reaction with N-chlorosuccinimide gives a 33% excess of inactive **91**.

Only *trans*-N-bromo-2-methylaziridine (**93**) $[\alpha]_D$ +497 (c 0.08, *n*-nonane) is isolated from reaction of (−)-**90** with sodium hypobromite or N-bromosuccinimide. Halogenation of (*S*)-(−)-2-*n*-propylaziridine (**94**) with N-chlorosuccinimide, sodium hypobromite, or N-bromosuccinimide likewise

[68] (a) S. J. Brois, *J. Am. Chem. Soc.*, **90**, 508 (1968). (b) J. M. Lehn and J. Wagner, *Chem. Commun.*, 148 (1968). (c) D. Felix and A. Eschenmoser, *Angew. Chem.*, **80**, 197 (1968); *Internat. Edn.*, **7**, 224 (1968). (d) R. G. Kostyanovsky, Z. E. Samojlova, and I. I. Tchervin, *Tetrahedron Lett.*, 719 (1969).

[69] S. J. Brois, *Tetrahedron Lett.*, 5997 (1968).

produces only *trans*-N-halo-2-*n*-propylaziridines.[68d] N-Chlorination of 7-azabicyclo[4.1.0]heptane[68c] (**95**) gives a 90% yield of mixed isomers of 7-aza-N-chlorobicyclo[4.1.0]heptane (**96**) in which there is a 60% excess of a more stable *exo*-chloro isomer over a labile *endo*-chloro isomer which rearranged on heating to the *exo*-form.

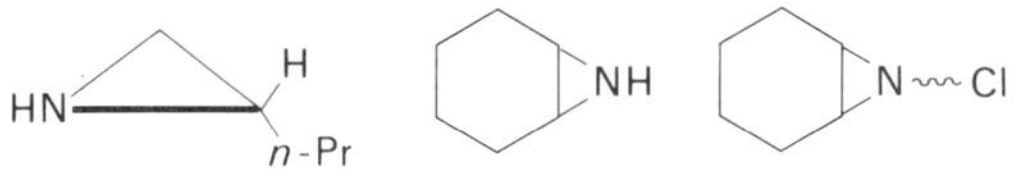

HN, H, *n*-Pr — **94**; NH — **95**; N〰Cl — **96**

8-3. Asymmetric Synthesis at Silicon

Reaction of a number of optically active N-phenyl-α-amino acids with *bis*-(N-methylacetamido)methylphenylsilane (**98**)* gives high yields of unequal amounts of diastereomeric siloxazolidones, **99**,[70] which are epimeric at silicon and which can be cleaved by successive treatment with α-naphthol and methanol to give optically active silane derivative **100** and the inducing amino acid **97**. If racemic **97** is employed, the diastereomeric siloxazolidones

RCH(NHPh)COOH (**97**) + $X_2SiMePh$ (**98**, X = NMeAc or Cl) ⟶ **99** (two epimers at Si: Si–Ph/Me and Si–Me/Ph) —1) α-Naphthol, 2) MeOH⟶ MeO(O-α-Naphthyl)Si(Me)(Ph) (**100**) + **97**

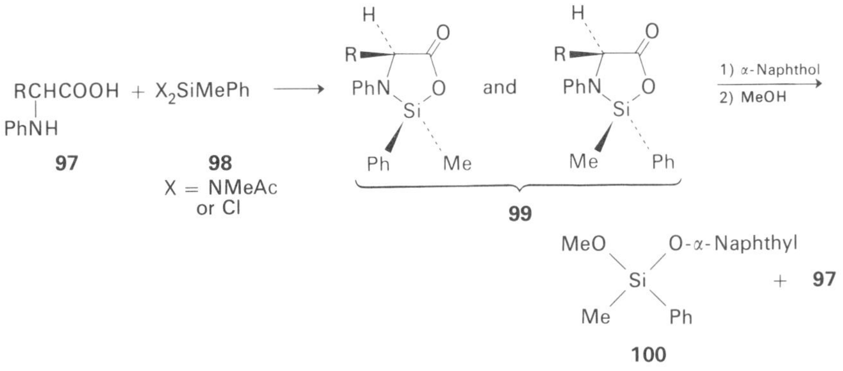

(**99**) are produced as a pair of racemates in unequal amounts. The relative amounts of the epimers from optically active amino acids or diastereomeric pairs from racemic amino acids may be assayed by nmr since the Si—Me groups have different chemical shifts.

Experimental evidence[70] suggests that the formation of **99** proceeds

* Methylphenyldichlorosilane in the presence of triethylamine also may be used, but it is less desirable because the triethylamine hydrochloride produced as a by-product interferes with the isolation of **99**.

[70] J. F. Klebe and H. Finkbeiner, *J. Am. Chem. Soc.*, **90**, 7255 (1968).

stepwise, the Si—O bond being formed first by displacement of one N-methylacetamide group followed by intramolecular displacement of the other to form the Si—N bond. Assuming that the second displacement is stereospecific and that the reaction is kinetically controlled, then the first displacement determines the diastereomer ratio. The silicon atom in **98** constitutes a prochiral center and the chiral amino acid must preferentially displace one of the two N-methylacetamido groups. The ratio of diastereomers produced at this stage is retained in **99** provided stereospecific cyclization ensues. However, it is found that when the N-phenyl derivatives are employed, the diastereomers of **99** are equilibrated after formation. When crystallization is induced by evaporation of solvent, only one diastereomer separates. After partial crystallization, equilibrium is reestablished and more of the same isomer separates on further concentration. N-Methylacetamide, present as a by-product of the reaction, is sufficiently basic to catalyze epimerization at silicon. With purified material, equilibration takes 2 days to 3 weeks in benzene or chloroform but only about 30 min in acetone or acetonitrile or in the presence of small amounts of tertiary amines. The epimerization mechanism has not been established. If during this process the silicon becomes symmetrical the rearrangement may be considered to be a thermodynamically controlled asymmetric synthesis, but other mechanistic possibilities can also be envisaged.

The phenyl group on nitrogen was found to be a key factor in the rearrangement. N-Isopropylphenylglycine, for example, reacted with **98** to give unequal amounts of diastereomers of **101**, but no rearrangement could

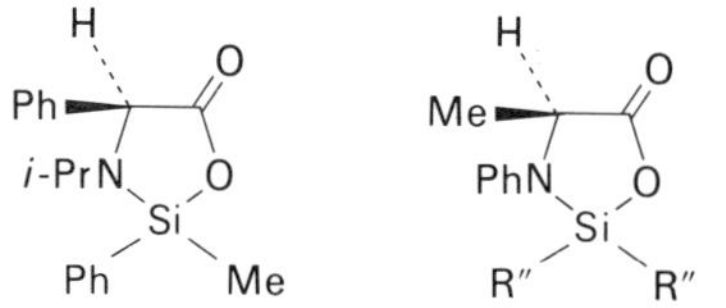

101

102 (R″ = Me)
103 (R″ = Ph)

be detected. In this case then a true asymmetric synthesis has taken place. The original paper should be consulted for interesting details concerning the rotations, nmr chemical shifts, and probable configurations of the compounds involved.

Stable nitrogen invertomers are presumably not involved, although the only hard evidence[70] that might bear on this point is that a siloxazolidone from N-phenylglycine (**99**, R = H) showed only one Si—CH_3 resonance ($\tau = 9.53$) and siloxazolidones **102** and **103** were isolated as single compounds.

8-4. Asymmetric Synthesis at Phosphorus

Reaction of strychnine with **104A** leads to methylation of the strychnine and formation of the salt **105A**.[71] In acetonitrile a dextrorotatory diastereomer of **105A** is produced in excess, whereas in methanol a levorotatory salt predominates. The diastereomeric salts can be remethylated on sulfur to give enantiomers of **106**. In some cases brucine also forms an excess of

$$\mathrm{RO{-}\overset{OMe}{\underset{OMe}{P}}{=}S} \xrightarrow{\text{Strychnine}} \left[\mathrm{RO{-}\overset{O}{\underset{OMe}{P}}{\cdots}S}\right]^{-} [\text{N-methyl-strychnine}]^{+} \longrightarrow \mathrm{RO{-}\overset{O}{\underset{OMe}{P}}{-}SMe}$$

104 **105** **106**

A, R = p-$NO_2C_6H_4$
B, R = Et
C, R = Ph

one diastereomer of **105**. Other substrates (**104B** and **104C**) have also been studied.[71]

Presumably this reaction is an example of the selection by a chiral reagent of one of two prochiral groups. Since the isolated yields of one diastereomer of **105A** was 69–86% in acetonitrile and 55–77% in methanol, there is not a simple resolution of **105A**. It is possible however that **104** and **105** are in thermodynamic equilibrium, in which case this reaction can be considered a second-order asymmetric transformation instead of an asymmetric synthesis. In either event this illustrates an interesting example of a solvent effect upon the direction of an asymmetric process.

The discovery of new routes to optically active tertiary[72,73] and secondary[74] phosphines have led to an increased number of studies on the stereochemistry of reactions at phosphorus. It is precisely because trisubstituted phosphorus compounds are configurationally stable and isolable that they are not suitable compounds for asymmetric synthesis or asymmetric transformation studies. Thus the oxidation of a tertiary phosphine to a phosphine oxide with retention of configuration is a stereospecific reaction but not an asymmetric synthesis. If a chiral oxidant is used to convert a racemic trisubstituted phosphine to the phosphine oxide, the product must be racemic *if the substrate is completely consumed.* Therefore only reactions of chiral reagents which preferentially alter or replace paired ligands at a prochiral

[71] G. Hilgetag and G. Lehmann, *J. prakt. Chem.*, [4], **9**, 3 (1959).
[72] O. Korpiun, R. A. Lewis, J. Chickos, and K. Mislow, *J. Am. Chem. Soc.*, **90**, 4842 (1968).
[73] A. Nudelman and D. J. Cram, *J. Am. Chem. Soc.*, **90**, 3869 (1968).
[74] T. L. Emmick and R. L. Letsinger, *J. Am. Chem. Soc.*, **90**, 3459 (1968).

phosphorus (for example, **104**) can qualify as asymmetric syntheses at phosphorus. This is the same situation as encountered in the silicon series as exemplified by the reactions of **98**.

1969 Addenda

Sec. 8-1.2. An interesting partial kinetic resolution of sulfoxides has been accomplished via the incomplete reduction of the racemic sulfoxide by chiral phosphonthioic acids.[75]

Sec. 8-1.3. A full account of the peroxidation of chiral esters of *o*-(methylthio)benzoic acid by perbenzoic and 2,4,6-trimethylperbenzoic acids to give chiral *o*-(methylsulfinyl)-benzoic acid (cf. Table **8-7**) has been published.[76] The maximum asymmetric synthesis was 28%, but most values were less than 10%. A transition state model for the reaction is discussed in detail.

Sec. 8-1.3a. The diastereomeric, racemic, α-methylbenzyl sulfides have been separated by GLPC and separately oxidized (H_2O_2, HOAc) to the sulfoxides. That sulfide which gave the sulfoxide showing benzylic proton equivalence in the nmr was assigned the *meso* configuration, and that which showed benzylic proton nonequivalence was assigned the *dl* configuration. Thus, the configurational assignments of the precursor sulfides and derived sulfones were defined.[77]

Sec. 8-1.3b. An excellent example of the stereoselective peroxidation of a chiral sulfide substrate is given by the oxidation of (*R*)-(−)-8-hydroxy-5-methyldihydrothiazole[2,3a]-pyridinium-3-carboxylate, prepared from L-cysteine, to give a 9:1 ratio of the *cis* versus *trans* sulfoxide (80% asymmetric peroxidation).[78] The sulfoxide could be decarboxylated under mild conditions thus removing the original chiral inducing center.

Evidence is presented for a tetracovalent intermediate in the oxidation of sulfides to sulfoxides with *t*-butylhypochlorite.[79]

Two papers dealing with asymmetric oxidation of the sulfide atom in chiral penicillin derivatives illustrate the influence of ligands at β-asymmetric centers. Oxidation of methyl phthalimidopenicillinate epimers with

[75] M. Milolajczyk and M. Para, *Chem. Commun.*, 1192 (1969).

[76] G. Barbieri, V. Davoli, I. Moretti, F. Montanari, and G. Torre, *J. Chem. Soc.* (*C*), 731 (1969).

[77] C. Y. Meyers and A. M. Malte, *J. Am. Chem. Soc.*, **91**, 2123 (1969).

[78] K. Undheim and V. Nordal, *Acta Chem. Scand.*, **23**, 1966 (1969).

[79] C. R. Johnson and J. J. Rigau, *J. Am. Chem. Soc.*, **91**, 5398 (1969).

m-chloroperbenzoic acid was shown to be under control of the 6-phthalimido group; the favored approach was from the side opposite to this bulky nitrogen substituent.[80] This conclusion was confirmed.[81] In contrast to this, it was found that oxidation of methyl 6-β-phenylacetamidopenicillinate with iodobenzenedichloride in wet pyridine gave the epimeric sulfoxides in approximately equal amounts.[81]

Sec. 8-1.4. The inversion at sulfur during the displacement of the ester moiety of a chiral sulfinate (cf. Table **8-11**) has now been established unequivocally by an X-ray absolute configuration determination.[82]

Sec. 8-2. Oxidation of the imine from benzophenone and *t*-butylamine with (+)-monopercamphoric acid gave optically active N-*t*-butyldiphenyloxaziridine ($[\alpha]_D^{25}$ − 54°, $CHCl_3$) which could be separated from the racemate by fractional crystallization to apparent optical purity ($[\alpha]_D^{25}$ − 258°, $CHCl_3$). The N-*t*-butyl compound racemized 8000 times faster than the corresponding N-methyloxaziridine at 100° in tetrachloroethylene.[83]

[80] R. D. G. Cooper, P. V. DeMarco, and D. O. Spry, *J. Am. Chem. Soc.*, **91**, 1528 (1969).

[81] D. H. R. Barton, F. Comer, and P. G. Sammes, *J. Am. Chem. Soc.*, **91**, 1529 (1969).

[82] H. Hope, U. de la Camp, G. D. Homer, A. W. Messing, and L. H. Sommer, *Angew. Chem.*, **81**, 619 (1969); *Internat. Edn.*, **8**, 612 (1969).

[83] F. Montanari, I. Moretti, and G. Torre, *Chem. Commun.*, 1086 (1969).

9

Asymmetric Rearrangement and Elimination Reactions

9-1. Intramolecular Transfer of Centrodissymmetry

Many rearrangements of the allylic type, represented by **1** → **2**, qualify as intramolecular asymmetric syntheses in which the definition is used in its broadest sense (cf. Sec. **1-2**). Since there have been several reviews which include discussions of the stereochemistry of such processes,[1–3] we shall

$$\underset{\mathbf{1}}{\mathrm{RCH{=}C(H)\text{–}\overset{*}{C}HR'\text{–}X}} \longrightarrow \underset{\mathbf{2}}{\mathrm{R\overset{*}{C}H(Y)\text{–}C(H){=}CHR'}}$$

(X and Y may or may not have the same structure)

[1] P. B. D. de la Mare, Chapter 2 in *Molecular Rearrangements*, ed. P. deMayo (New York, N.Y.: Interscience Pub., 1963), Vol. 1, pp. 27–110.

[2] H. L. Goering, *Rec. Chem. Prog.*, **21**, 109 (1960).

[3] S. J. Rhoads, Chapter 11 in *Molecular Rearrangements*, ed. P. deMayo (New York, N.Y.: Interscience Pub., 1963), Vol. I, pp. 655–706.

discuss only a relatively small number of reactions of this type which use optically active substrates and are of special interest from the standpoint of asymmetric synthesis.

9-1.1. Transfer of Centrodissymmetry from Carbon to Carbon. The *ortho* Claisen rearrangement of (*R*)-(+)-*trans*-α,γ-dimethylallyl phenyl ether (**4**) yields a mixture of chiral *cis*- and *trans*-2-(α,γ-dimethylallyl)phenols (**5**).[4–8] The geometrical configuration is largely retained,[6–8]

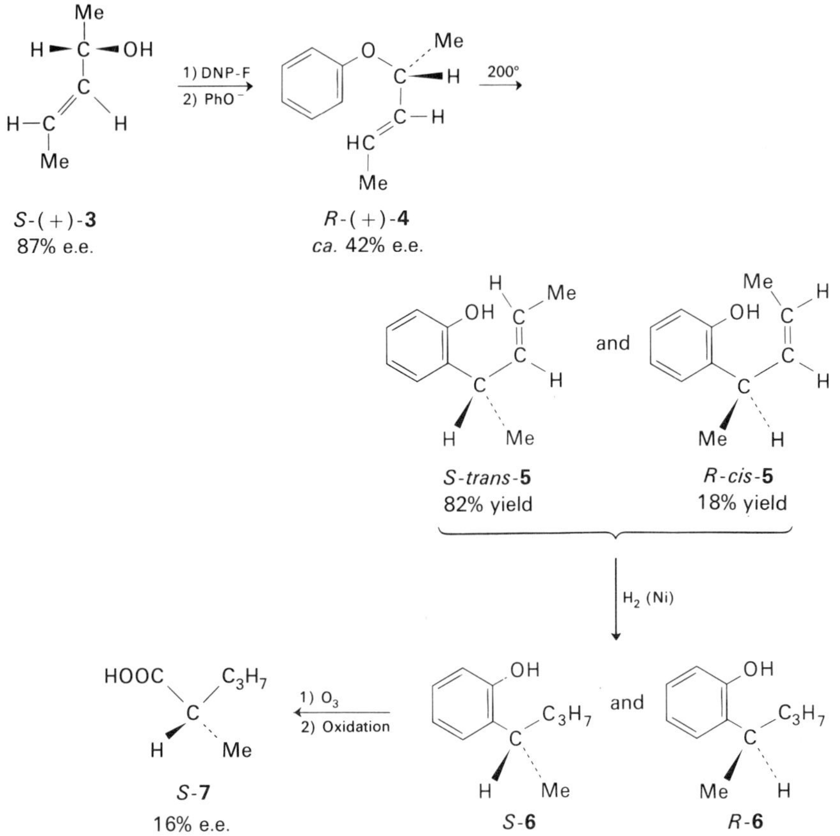

S-(+)-**3**
87% e.e.

1) DNP-F
2) PhO^-

R-(+)-**4**
ca. 42% e.e.

200°

S-*trans*-**5**
82% yield

and

R-*cis*-**5**
18% yield

H_2 (Ni)

S-**6** and *R*-**6**

1) O_3
2) Oxidation

S-**7**
16% e.e.

Fig. 9-1. The stereochemistry of the Claisen rearrangement of optically active *trans*-α,γ-dimethylallyl phenyl ether. DNP-F = 2,4-dinitrophenyl fluoride.

[4] E. R. Alexander and R. W. Kluiber, *J. Am. Chem. Soc.*, **73**, 4304 (1951).
[5] H. Hart, *J. Am. Chem. Soc.*, **76**, 4033 (1954).
[6] E. N. Marvell and J. L. Stephenson, *J. Org. Chem.*, **25**, 676 (1960).
[7] E. N. Marvell, J. L. Stephenson, and J. Ong, *J. Am. Chem. Soc.*, **87**, 1267 (1965).
[8] H. L. Goering and W. I. Kimoto, *J. Am. Chem. Soc.*, **87**, 1748 (1965).

while the configuration of the new chiral center is as shown in Fig. **9-1** for the conversion (*R*)-*trans*-**4** to (*S*)-*trans*-**5**.*

These stereochemical results are accommodated by models **8A** and **8B** which show the consequence of migration to diastereotopic faces of the double bond of *trans*-**8**. If the geometrical configuration is retained, the configuration at the chiral center is "inverted" (*R*-*trans*-**8A** → *S*-*trans*-**9A**) whereas a reversal in geometrical configuration must be accompanied by "retention" at the chiral center (*R*-*trans*-**8B** → *R*-*cis*-**9B**). The transition state for the

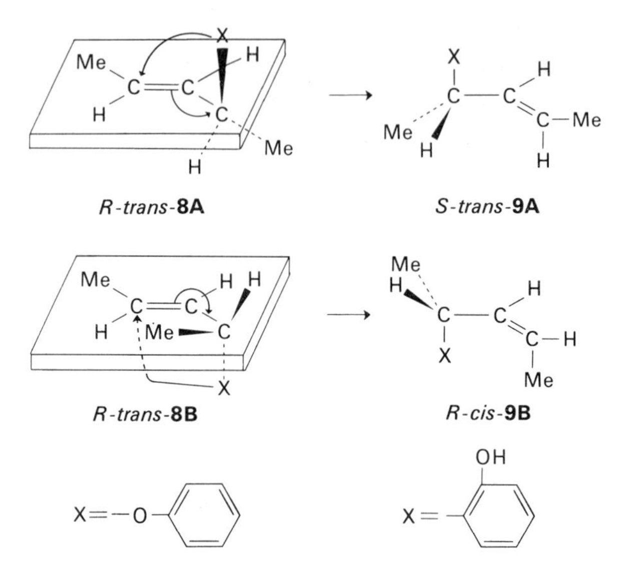

Fig. 9-2. Models which illustrate the stereochemistry of allylic rearrangement.

former process is of lower energy presumably because it involves only *trans*-butene-type interactions while the latter requires a transition state with *cis*-butene-type repulsions.[8] Since **4** is 42% enantiomerically pure and since the isomers of **5** have opposite configurations at the chiral center, the common reduction product (**6**) of the mixture can have, at best, about a 28% excess of the *S*-enantiomer, (64% of the 42% e.e.). The lower observed value (16%) results from some racemization in the rearrangement, hydrogenation, or the ozonolysis steps.

Asymmetric synthesis has also been observed in the Claisen rearrangement of vinyl ether **11**.[9] It appears that transfer of dissymmetry from **11** to

* From the first study[4] of this rearrangement using **4** which had a slight excess of the *S*-(−) enantiomer, it was erroneously concluded[5] that the reaction proceeded with predominant "retention" of configuration at the new chiral center.

[9] R. K. Hill and A. G. Edwards, *Tetrahedron Lett.*, 3239 (1964).

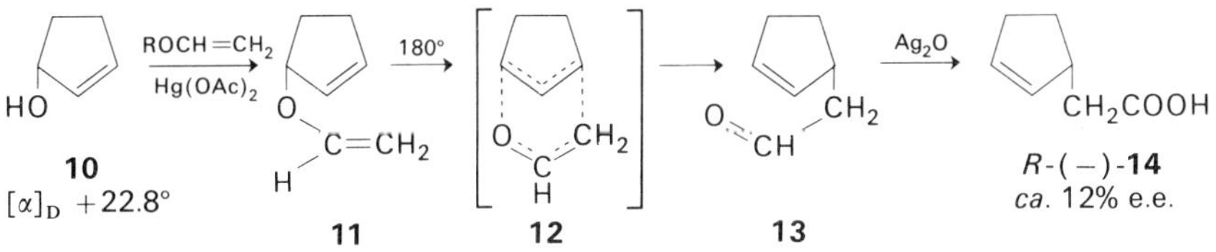

13 is complete.* This has been cited[9] as a stereochemical model for the enzymatic conversion of chorismic acid to prephenic acid.

A related thermal rearrangement is shown in the conversion of **16** to **20**.[15] The acetoacetate **16** (prepared by the reaction of diketene on 2-cyclohexenol, **15**, 8 % e.e.) rearranges via enol **17** or possibly via the enol of the ester carbonyl (**18** not shown) to give **19** which decarboxylates to **20**. Although the enantiomeric purity of the ketone (**20**, $[\alpha]_D^{25}$ + 1.43°, neat, 1 dm) has not been determined by an independent route, the stereochemistry is as shown.

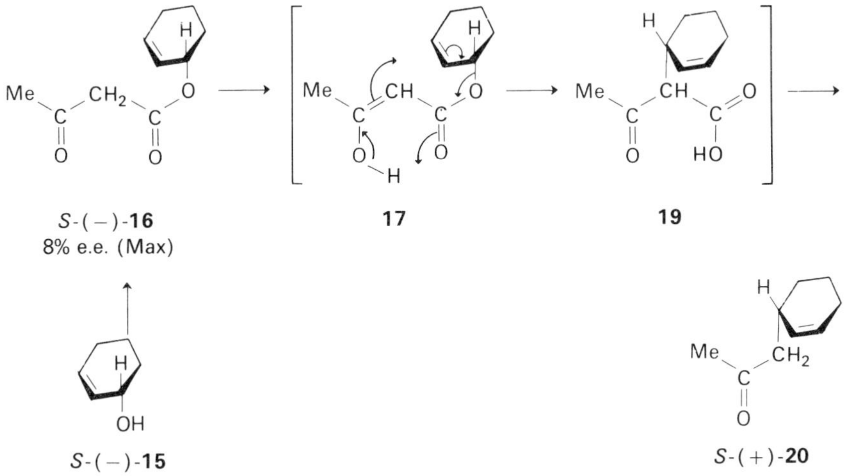

* The configuration[10,11] but not the enantiomeric purity of **10** is known but the latter can be *estimated* to be approximately 11 %.[12,13] The absolute configuration and enantiomeric purity of **13** is determined by oxidation to **14** whose configuration and enantiomeric purity have been established.[14] Based on the assumption that there is no racemization in the sequence **10** → **14**, starting material **10** with $[\alpha]_D$ +22.8° represents 12 % e.e. This is in good agreement with the independent estimate of 11 %.

[10] H. A. Vaughn, Jr., Ph.D. thesis, Columbia Univ., 1955, cited in Reference 9.

[11] Y. Sato, S. Nishioka, O. Yonemitsu, and Y. Ban, *Chem. Pharm. Bull.* (Tokyo), **11**, 829 (1963).

[12] D. B. Denney, R. Napier, and A. Cammarata, *J. Org. Chem.*, **30**, 3151 (1965).

[13] R. K. Hill and J. W. Morgan, *J. Org. Chem.*, **33**, 927 (1967).

[14] K. Mislow and I. V. Steinberg, *J. Am. Chem. Soc.*, **77**, 3807 (1955).

[15] R. K. Hill and M. E. Synerholm, *J. Org. Chem.*, **33**, 925 (1968).

Similarly, the allylic phenylurethan *S*-(−)-**21**, when heated with a catalytic amount of sodium hydride, rearranges to *R*-(+)-**24**.[16] A maximum of 65% retention of configuration was observed.

240° NaH

S-(−)-**21** **22** **23** *R*-(+)-**24** + CO_2

The allylenamine *R*-**27**, which is prepared from **25** and *R*-**26**, rearranges in high yield on heating (170–175°) in a manner analogous to the Claisen rearrangement to give an optically active mixture of *cis* and *trans*-imines, **30**.[17] The aldehydes **31** and **32** were formed in a 13:87 *cis-trans* ratio upon hydrolysis. These aldehydes were hydrogenated to the *enantiomeric* saturated aldehydes *R*-(−)-**33** and *S*-(+)-**33**, respectively. These results suggest that the transition state represented by **28A**, in which the methyl group is axial

PhCHMeCHO + PhNHC*HMeCH=CH_2 ⟶ **27** ⟶

25 *R*-(+)-**26** $[\alpha]_D$ + 2.64° ($CHCl_3$)

28A → *cis*-*S*-**30** → *cis*-*R*-**31** $\xrightarrow[(Pd\cdot C)]{H_2}$ *R*-(−)-**33**

28B → *trans*-*S*-**30** → *trans*-*S*-**32** $\xrightarrow[(Pd\cdot C)]{H_2}$ *S*-(+)-**33**

[16] M. E. Synerholm, N. W. Gilman, J. W. Morgan, and R. K. Hill, *J. Org. Chem.*, **33**, 1111 (1968).

[17] R. K. Hill and N. W. Gilman, *Tetrahedron Lett.*, 1421 (1967).

versus **28B** in which the methyl group is equatorial (preferred), accounts for the 13:87 isomer ratio. If the aldehydes **33** are enantiomerically pure (not reported), then this comprises a completely stereoselective process.

The rearrangement of *R*-(+)-**34** (95% e.e.) gives *trans*-(*S*)-(+)-**35** (91% e.e.) and *cis*-*R*-(+)-**35** (89% e.e.) in 87:13 product ratio.[18] The 87:13 ratio of products is remarkably similar to the product ratio from the amino-Claisen rearrangement[17] of **27**. This supports the view that similar transition state geometries are involved. In this case the free energy difference observed (about 2 kcal/mole) for transition states leading to *trans*-*S*-(+)-**35** and *cis*-*R*-(+)-**35** approximates the conformation free energy differences for the two chair forms of the cyclohexanes analogous to transition state models **34A** and **34B**. Chair-like transition states for thermally initiated [3,3]

R-(+)-**34** **34A** **34B** *cis*-*R*-(+)-**35** 89% e.e. *trans*-*S*-(+)-**35** 91% e.e.

sigmatropic rearrangements such as the Cope and Claisen reactions seem likely.*

9-1.2. Transfer of Centrodissymmetry from Carbon to Sulfur.

(*S*)-α-Methylallyl *p*-toluenesulfenate (**36**), which is chiral due to substitution on carbon, spontaneously rearranges to (*S*)-(−)-*trans*-crotyl *p*-tolyl sulfoxide (**38**), which is chiral due to substitution on sulfur.[22] This

[18] R. K. Hill and N. W. Gilman, *Chem. Commun.*, 619 (1967).

[19] (a) G. B. Gill, *Quart. Revs.*, **22**, 338 (1968). (b) A. Jefferson and F. Scheinmann, *Quart. Revs.*, **22**, 391 (1968).

[20] R. Hoffmann and R. B. Woodward, *Accounts Chem. Research*, **1**, 17 (1968).

[21] J. J. Vollmer and K. L. Servis, *J. Chem. Educ.*, **45**, 214 (1968).

[22] P. Bickart, F. W. Carson, J. Jacobus, E. G. Miller, and K. Mislow, *J. Am. Chem. Soc.*, **90**, 4869 (1968).

* Several reviews[19–21] of these and other rearrangements in terms of orbital symmetry arguments have been published.

product readily racemizes (half-life, 51 min in toluene at 29°); thus the observed stereoselectivity of 37% is a minimum value. It has been rigorously demonstrated that this reaction proceeds via a cyclic concerted process. The configurational and conformational equilibria involved have been considered in detail.[22]

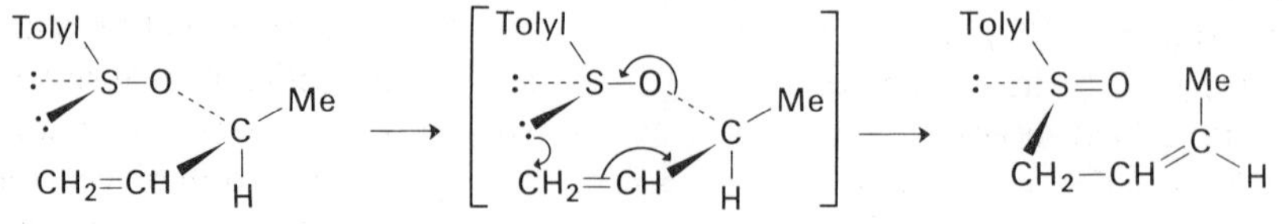

36 → **37** → **38**

9-1.3. Transfer of Centrodissymmetry from Nitrogen to Carbon.

The Stevens rearrangement of the quaternary ammonium iodide, **39**, which is chiral at nitrogen, to the amine, **42**, which is chiral at carbon, has been studied.[23] The absolute configuration of **39** was determined by diimide reduction to the known (*R*)-(+)-benzylmethylphenylpropylammonium iodide,[24] and of **42** by catalytic hydrogenation to (*R*)-(+)-2-(N-methylanilino)-1-phenylbutane which was, in turn, related to (*R*)-(−)-1-phenyl-2-butanol. Two diastereomeric transition states, **40** and **41**, appear

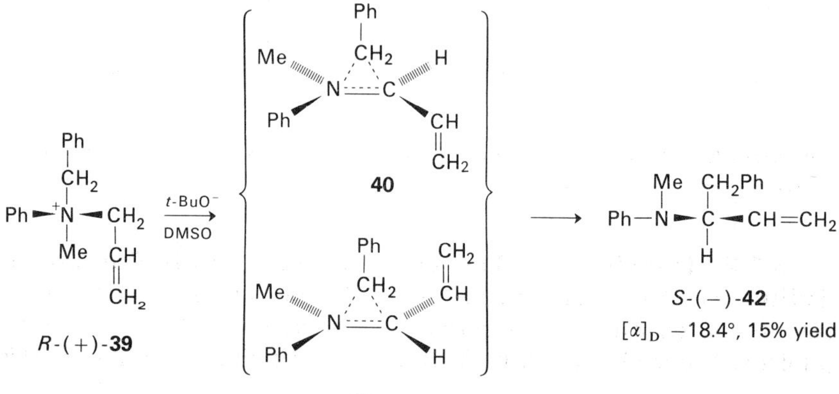

R-(+)-**39** $\xrightarrow[\text{DMSO}]{t\text{-BuO}^-}$ [**40**, **41**] → *S*-(−)-**42**

$[\alpha]_D$ −18.4°, 15% yield

reasonable for this rearrangement; from the known configurations of **39** and **42**, one deduces that **40** is of lower energy. The extent of asymmetric synthesis is probably substantial (rotation $[\alpha]_D$ −18.4°) but the maximum rotation is unknown. Preference for transition state **40** seems unexpected on steric grounds.

[23] R. K. Hill and T. H. Chan, *J. Am. Chem. Soc.*, **88**, 866 (1966).

[24] L. Horner, H. J. S. Winkler, and E. Meyer, *Tetrahedron Lett.*, 789 (1965).

9-1.4. Asymmetric Hydrogen Transfers. The base-catalyzed 1,3-tautomerism of 3-methyl-1-alkylindenes (**45**) to 3-alkyl-1-methylindenes is an ideal system for studying intramolecular asymmetric hydrogen transfers.[25–27] When R is methyl, the product of the reaction (**45C** → **46C**) is the enantiomer of the starting material. This system has been used for elegant kinetic and isotope studies by Bergson and co-workers[25,26] and by Cram and co-workers.[27] In *t*-butyl alcohol solvent, using a weak base such as triethylenediamine [1,4-diazabicyclo[2.2.2]octane, Dabco, $N(CH_2CH_2)_3N$] in the presence of the amine salt, the rearrangement of **45A** to **46A** was 96–100% stereoselective and 97–100% intramolecular.[27] This is in accord with a mechanism in which the proton is conducted from C-1 of **45** along one face of a tight ion pair to bond at C-3 of the original allylic moiety. When the hydrogen is replaced with deuterium, the transfer takes place with very

R 1 H 3 Me H —Base→ R H 3 H 1 Me

45 **46**

A, R = *t*-Bu
B, R = *i*-Pr
C, R = Me

little solvent isotope exchange. With stronger bases in ionizing solvents, racemization via proton abstraction is much faster than rearrangement. Quinine in pyridine caused (*S*)-(+)-1-methylindene to rearrange about 50% faster than the *R*-(−) enantiomer, further implicating the base as a participant via a tight ion pair.[26] Chiral α-methylbenzylamine, however, did not discriminate between the two isomers.

A related rearrangement is the base-catalyzed 1,3-hydrogen shift of aldimines.[28] When enantiomerically pure aldimine **47**, prepared by the reaction of (*S*)-(−)-α-methylbenzylamine with trimethylacetaldehyde-1-*d*, was treated with potassium *t*-butoxide in *t*-butyl alcohol, the product was essentially one isomer, namely *R*-(+)-**49**. This was shown by acid hydrolysis of **49** to (*R*)-(+)-neopentylamine-1-*d* (**50**), whose absolute configuration was

[25] (a) G. Bergson and A-M. Weidler, *Acta Chem. Scand.*, **17**, 1798 (1963). (b) A-M. Weidler, *ibid.*, **17**, 2724 (1963). (c) A-M. Weidler and G. Bergson, *ibid.*, **18**, 1487 (1964). (d) G. Bergson and A-M. Weidler, *ibid.*, **18**, 1498 (1964).

[26] L. Ohlsson, I. Wallmark, and G. Bergson, *Acta Chem. Scand.*, **20**, 750 (1966).

[27] (a) J. Almy, R. T. Uyeda, and D. J. Cram, *J. Am. Chem. Soc.*, **89**, 6768 (1967). (b) J. Almy and D. J. Cram, *J. Am. Chem. Soc.*, **91**, 4459 (1969).

[28] R. D. Guthrie, W. Meister, and D. J. Cram, *J. Am. Chem. Soc.*, **89**, 5288 (1967).

deduced to be *R* from the mechanism of the reaction[29] and application of Brewster's rules, and has been confirmed by independent synthesis.[30] The

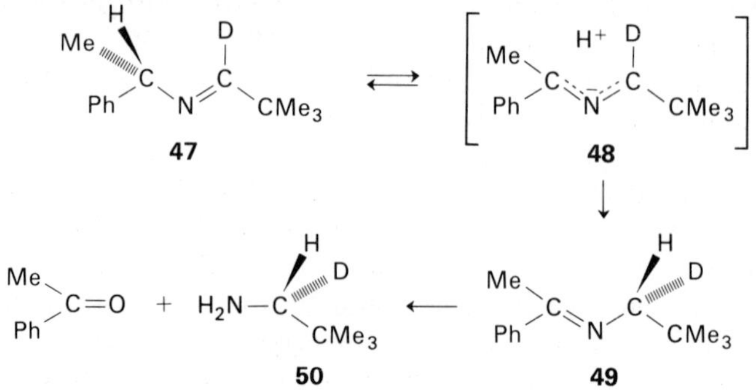

product (**50**) was essentially enantiomerically pure as shown by nmr analysis of a diastereomeric derivative.[28] This stereospecific proton transfer must involve only one face of the *aza*-allylic anion **48**. The stereospecificity is all the more surprising considering that it was shown that the rate of exchange of the hydrogen on the chiral benzylic carbon was much greater than the rate of isomerization. Therefore, although this hydrogen exchanges with the solvent from the solvent cage, it must return without inversion of the benzylic center.*

Stereoselective intramolecular hydride transfer has been observed in the polyphosphoric acid-catalyzed isomerization of **51** to **53**.[31] Based on the known configurations of **51** and **53**, transition state model **52** was proposed as the best rationalization of the preferred stereochemistry (13.6% e.e. *S*-(+)-**53** after correcting for the enantiomeric purity of (*S*)-(+)-**51**). The presumably larger groups (phenyl and methyl) at the original and incipient asymmetric centers are visualized as being oriented "equatorially" in a chair-like cyclohexane conformation.

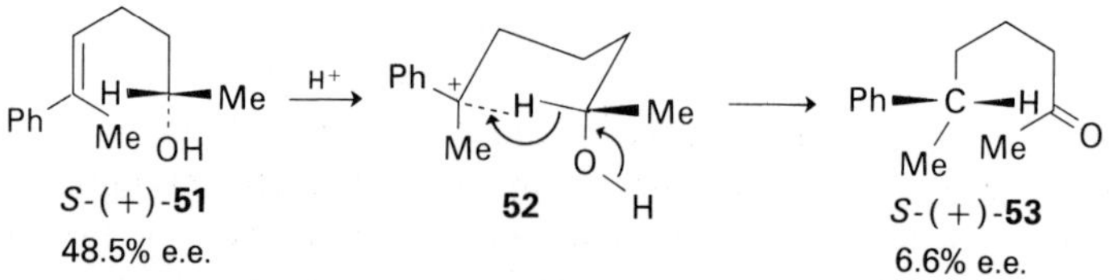

S-(+)-**51**
48.5% e.e.

52

S-(+)-**53**
6.6% e.e.

[29] D. J. Cram, *Fundamentals of Carbanion Chemistry* (New York: Academic Press, 1965).

[30] G. Solladié and H. S. Mosher, unpublished results from the S_N2 displacement by azide ion on (*S*)-neopentyl-1-*d* tosylate in hexamethylphosphoramide (HMPA) solvent followed by lithium aluminum hydride reduction.

[31] R. K. Hill and R. M. Carlson, *J. Am. Chem. Soc.*, **87**, 2772 (1965).

* The base-catalyzed exchange of tricyclo[4.3.1]deca-2,4,7-triene is characterized by a remarkable stereospecific abstraction of a proton and, similar to the case above, by stereospecific reprotonation, apparently as a result of special stereoelectronic effects. [P. Radlick and W. Rosen, *J. Am. Chem. Soc.*, **89**, 5308 (1967).]

54
(−)-Thebaine
($5R,9R,13S$)
1) PhMgBr
2) H_2O
55
(Biphenyl-R,9-R)
and
56
(Biphenyl-R,9-S)
57
(Biphenyl-R)
$CH{=}CHPh$
$CH{=}CH_2$
PhMgBr
58
59
60
MeO
HO
MgO
NMe
N—Me
Ph

9-2. The Interconversion of Centrodissymmetry and Molecular Dissymmetry

9-2.1. Optically Active Biphenyls. The classical example of the conversion of carbon centrodissymmetry to molecular dissymmetry of the biphenyl type is the reaction of thebaine (**54**) with phenyl Grignard reagent to give unequal amounts of diastereomeric phenyldihydrothebaines (**55** and **56**). These two substituted biphenyls are epimeric at the asymmetric benzylic carbon atom but have the same axial chirality as shown by degradation to the same biphenyl, **57**, upon elimination of the chiral center at C-9. This unusual reaction was discovered by Freund;[32] subsequent investigations[33] revealed the constitution of these products and thus the complexity of the reaction. Based on the established absolute configuration of (−)-thebaine and on an analysis of the course of the reaction, Berson and Greenbaum[33] concluded that the configuration of the biphenyl moiety in **55** and **56** as well as **57** must be *R* as shown.* Development of a biphenyl with *S* chirality is prevented by intolerable ring strain in the "intermediate" corresponding to **59** which would be required to give the opposite biphenyl chirality. These conversions, in addition to the Meerwein-Ponndorf asymmetric reductions (Sec. **5-1.7**) and the Prelog atrolactic asymmetric synthesis (Sec. **2-2**, Table **2-5**), illustrate the use of asymmetric synthesis to deduce absolute configurations in the biphenyl series.

Related to the thebaine reaction is the conversion of the alkaloid caranine (**61**) to a chiral biphenyl (**63**, $[\alpha]_D$ −27°) via Hofmann degradation of its methiodide (**62**) under mild conditions.[34] The configuration of **61** is known

MeI → ; t-BuO⁻ / DMSO →

61 **62** **63**

[32] M. Freund, *Ber.*, **38**, 3234 (1905).

[33] J. A. Berson and M. A. Greenbaum, *J. Am. Chem. Soc.*, **80**, 445 (1958) and references therein.

[34] E. W. Warnhoff and S. V. Lopez, *Tetrahedron Lett.*, 2723 (1967).

* The designation of the chiral symbol *R* to the biphenyl portion of **55** and **56** is based on the revised method of analysis [R. S. Cahn, C. Ingold, and V. Prelog, *Angew. Chem. Internat. Edn.*, **5**, 385 (1966)] according to which "the fiducial groups shall be the pairs nearest together of groups directly bonded to atoms on the axis, that lie one pair in each of the planes of atoms which intersect along the axis." The previous assignment was *S* when the original method for the specification of biphenyl chirality was used as outlined in E. Eliel, *Stereochemistry of Carbon Compounds* (New York: McGraw-Hill Book Co., 1962), p. 171.

and that of **63** assigned on the basis of optical rotatory dispersion data. Apparently the C-11 hydrogen, which is retained as an *ortho*-hydrogen in the incipient biphenyl system, retards rotation of the aryl group past the nitrogen bridge. The product (**63**) racemizes readily on heating, and Hofmann degradation at 120° gives inactive compound.

Symmetrization of all four asymmetric centers accompanies the conversion of the enol ether of chaparrol (**64**) to the chiral *o*-bridged biphenyl **65** via intermediates resulting from a reverse aldol condensation and dehydration sequences.[35] Again, as in the previous example, a distortion

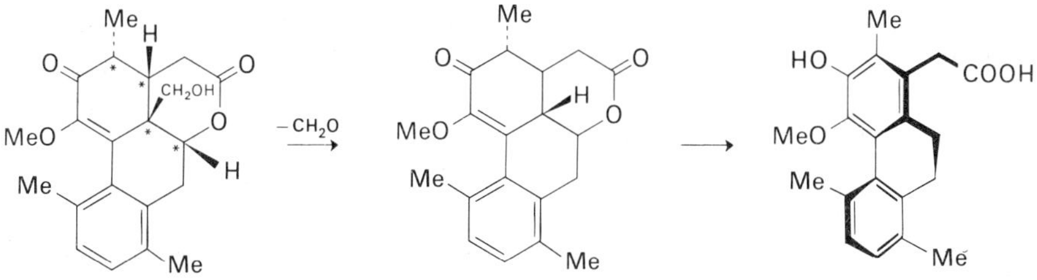

64 $\xrightarrow{-CH_2O}$ ⟶ *S*-**65**

present in the starting material as the result of conformational demands within the centrodissymmetric framework is retained in the product as molecular dissymmetry. A less complex transfer of a center of asymmetry to an axis of dissymmetry was attempted by Berson and Brown[36] and has been discussed in Sec. **1-4.4c** and Sec. **10-4.1.**

Transfer of asymmetry in the other direction, i.e., from atropisomerism to centrodissymmetry, has been achieved in the Stevens rearrangement of **66**.[37] The *S*-(+) enantiomer of **66**, which possesses only axial chirality, rearranges in the presence of phenyllithium to a mixture of **67A** and **67B**. These epimers have opposite biphenyl chiralities, but the configuration at C-9 is the same and is believed to be *S*.[37] The nearly exclusive production of *S*

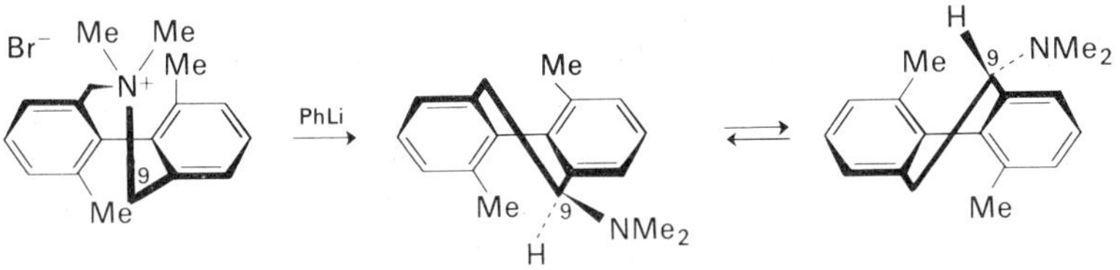

(+)-**66** *S*-(Biphenyl) $\xrightarrow{PhLi}$ (+)-**67A** *S*-(Biphenyl) *S*-(C-9) ⇄ (−)-**67B** *R*-(Biphenyl) *S*-(C-9)

[35] (a) T. R. Hollands, P. deMayo, and M. Nisbet, *Can. J. Chem.*, **43**, 2996 (1965). (b) T. R. Hollands, P. deMayo, M. Nisbet, and P. Crabbé, *ibid.*, **43**, 3008 (1965).

[36] J. A. Berson and E. Brown, *J. Am. Chem. Soc.*, **77**, 450 (1955).

[37] H. Joshua, R. Gans, and K. Mislow, *J. Am. Chem. Soc.*, **90**, 4884 (1968).

chirality at the asymmetric carbon atom demands a concerted or tight ion pair mechanism for the rearrangement.[37] The mutarotation of epimers **67A** and **67B** has been carefully examined and both crystalline isomers have been isolated. They have almost equal but opposite rotations which are similar in magnitude to that of 9,10-dihydro-4,5-dimethylphenanthrene in which the dimethylamino group of **67** is replaced by hydrogen. It is apparent that the major contribution to the optical rotation of these compounds is the biphenyl chirality; the saturated asymmetric center makes only a minor contribution to overall optical rotation.

9-2.2. Chiral Allenes. Stereoselective syntheses have played a more important role in the preparation of optically active allenes than have classical resolution procedures.[38] Furthermore, asymmetric reactions have played an important role in establishing the configurations of key allene compounds. One of the early stereoselective syntheses involved the dehydration of a racemic allylic alcohol in the presence of a chiral acid catalyst.[39] The alcohol, **68**, in which the geometry around the double bond is unknown, gave upon treatment with *d*-camphorsulfonic acid a 90% yield of *dl* allene **69** and a 1.3% yield of the enantiomerically pure crystalline (+) isomer **69A**.

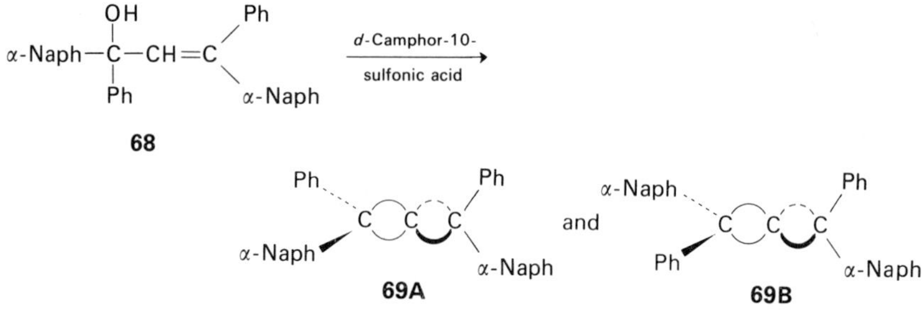

α-Naph—C(OH)(Ph)—CH=C(Ph)(α-Naph) —(*d*-Camphor-10-sulfonic acid)→

68

(Ph)(α-Naph)C=C=C(Ph)(α-Naph) and (α-Naph)(Ph)C=C=C(Ph)(α-Naph)

69A **69B**

Rotation of the reaction mixture indicated an asymmetric synthesis of 4–5%. Use of *l*-camphor-10-sulfonic acid gave enantiomer **69B** in excess; *d*-bromocamphor-10-sulfonic acid was not as effective. Since the starting material was racemic, this might constitute a kinetic resolution but this cannot be the entire answer since the reaction was carried to completion.

Dehydration of the acetylenic diol **70**, catalyzed by *d*-camphorsulfonic acid gives, instead of the initially reported[39b] 1,5-diphenyl-1,5-di-*t*-butyl-pentatetraene, the chiral indenoallene **71** of relatively low optical rotation[39c] ($[\alpha]_D$ + 11°).

[38] D. R. Taylor, *Chem. Revs.*, **67**, 317 (1967).

[39] (a) P. Maitland and W. H. Mills, *Nature*, **B5**, 994 (1935); *J. Chem. Soc.*, 987 (1936). (b) M. Nakagawa, K. Shingū, and K. Naemura, *Tetrahedron Lett.*, 802 (1961). (c) R. Kuhn and B. Schulz, *Angew. Chem.*, **74**, 292 (1962).

A case of basic chiral catalysis for the synthesis of an allene is the quinine-on-alumina catalyzed rearrangement of the acetylene **72** to the allene **73**.[40]

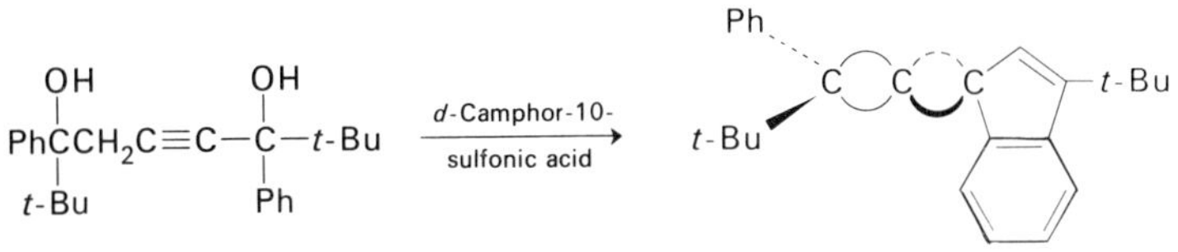

PhC(OH)(t-Bu)CH₂C≡C—C(OH)(Ph)—t-Bu —(d-Camphor-10-sulfonic acid)→

70 **71**

The solution developed a maximum rotation of $[\alpha]_D$ +8.57° but it was not possible to purify the product. Neither the maximum rotation nor absolute

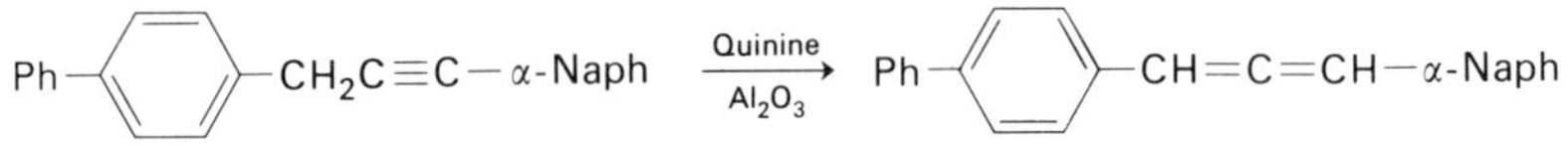

Ph—C₆H₄—CH₂C≡C—α-Naph —(Quinine, Al_2O_3)→ Ph—C₆H₄—CH=C=CH—α-Naph

72 **73**

configuration of **73** is known. Brucine-on-alumina catalyzed the formation of the (−) enantiomer in excess. The triple bond position isomer of **72** underwent the same stereoselective rearrangement but did not develop as high rotation. 1,3-Diphenylpropyne in like manner gave 1,3-diphenylallene of low activity when rearranged with either brucine- or quinine-on-alumina. Thus the use of a chiral catalyst in the formation of chiral allenes, although of historical and theoretical interest, seems of very limited practical utility.

The conversion of a centrodissymmetric compound to a chiral allene is observed on the treatment of *S*-(+)-**74** with thionyl chloride to produce *R*-(+)-**75**.[41] The configuration of **74** was deduced using the Prelog atrolactic acid method (cf. Table **2-6**) and that of **75** was inferred by the probable S_Ni' reaction mechanism. Analogously, the configuration of allene **77** was

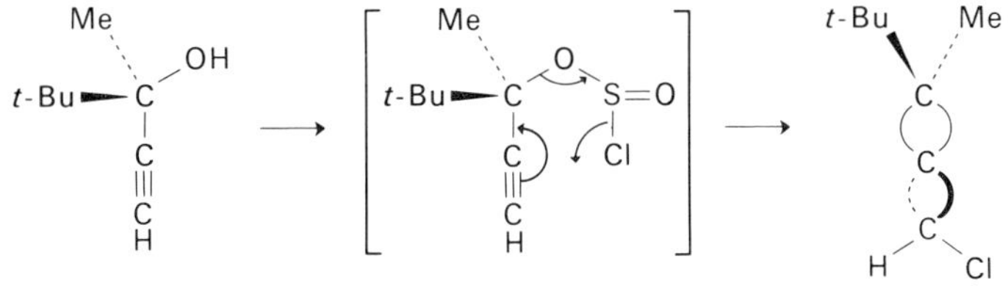

S-(+)-**74** *R*-(+)-**75**

assigned on the basis of its preparation from *R*-(+)-**76** via pyrolysis of its isobutyl acetal which was assumed to proceed by a concerted mechanism requiring that the allene have the *R* configuration.[42] The same kind of

[40] T. L. Jacobs and D. Dankner, *J. Org. Chem.*, **22**, 1424 (1957).

[41] R. J. D. Evans, S. R. Landor, and R. Taylor Smith, *J. Chem. Soc.*, 1506 (1963); R. J. D. Evans and S. R. Landor, *J. Chem. Soc.*, 2553 (1965).

[42] E. R. H. Jones, J. D. Loder, and M. C. Whiting, *Proc. Chem. Soc.*, 180 (1960).

Claisen-type rearrangement has also been used to establish the stereochemistry of the reduction of some achiral enynols **78** to allenic alcohols **79**

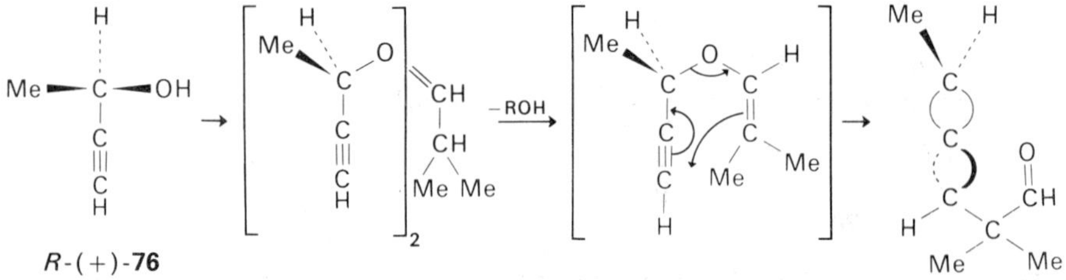

R-(+)-**76** → [ ]$_2$ $\xrightarrow{-ROH}$ [ ] → *R*-(−)-**77**

(Table **9-1**) using LAH modified by treating with menthol or a sugar derivative[43] (cf. Sec. **5-3**).* The *S* configuration was assigned to (+)-3,4-hexadiene-

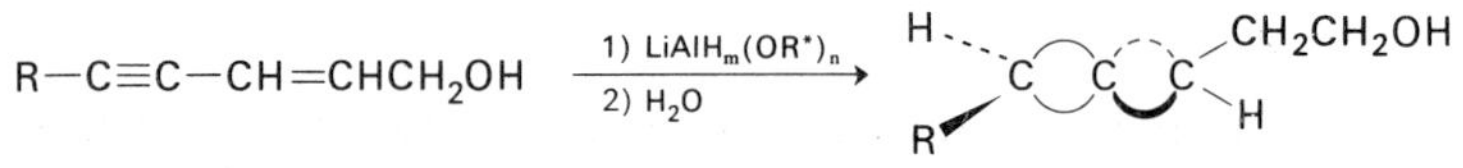

$$R{-}C{\equiv}C{-}CH{=}CHCH_2OH \xrightarrow[2)\ H_2O]{1)\ LiAlH_m(OR^*)_n} R(H)C{=}C{=}C(H)CH_2CH_2OH$$

78 **79**

1-ol (**79**, R = Me, Table **9-1**, No. 1) based on its stereoselective synthesis from *S*-(+)-**80** using the LAH-menthol reagents, by a process assumed to involve the following cyclic mechanism as shown in **80** → **79**. Assuming a

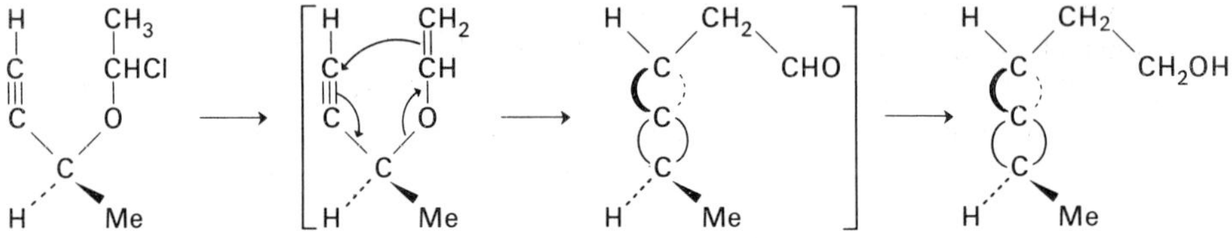

S-(+)-**80** → [ ] → *S*-(+)-**79**

uniform stereochemical course for these reductions by the chiral LAH-sugar complex, all the (−) allenes in Table **9-1** have been assigned the *R* configuration. The last two examples are the naturally occurring marasin and 9-methylmarasin whose maximum rotations are $[\alpha]_D^{20}$ +385° and +340° (neat), respectively. The asymmetric syntheses in these cases can therefore be calculated to be 6.8 and 3.3 %, respectively. The stereoselectivity

[43] (a) R. J. D. Evans, S. R. Landor, and J. P. Regan, *Chem. Commun.*, 397 (1965). (b) S. R. Landor, B. J. Miller, J. P. Regan, and A. R. Tatchell, *Chem. Commun.*, 585 (1966).

* A process somewhat related to these is the reaction of the chiral alkynol, $HC{\equiv}C{-}\overset{*}{C}HOHPh$, with a chlorophosphine to give an allenic phosphine oxide, $OPR_2CH{=}CHPh$; stereochemical details however are unavailable [A. Sevin and W. Chodkiewicz, *Tetrahedron Lett.*, 2975 (1967)].

of the reaction is not high but the convenience of this method for correlating configurations commends itself to much wider study and verification.

TABLE 9-1

Asymmetric Conversion of Alkenynols to Chiral Allenic Alcohols by Use of LAH-Chiral Carbinol Reagents[a]

$$\underset{\mathbf{78}}{RC{\equiv}C{-}CH{=}CHCH_2OH} \xrightarrow[2)\ H_2O]{1)\ LiAlH_m(OR^*)_n} \underset{\mathbf{79}}{(H)(R)C{=}C{=}C(CH_2CH_2OH)(H)}$$

No.	R	R* = (−)-Menthyl [α]D °	R* = (−)-Menthyl Configuration	3-O-Benzyl-1,2-cyclo-hexylidene-α-D-glucosyl [α]D °	3-O-Benzyl-1,2-cyclo-hexylidene-α-D-glucosyl Configuration	% e.e.
1	Me	+8.3	*S*	−10	*R*	—
2	Et	+2.3	*S*	−8.9	*R*	—
3	*n*-Bu	—	—	−7.4	*R*	—
4	*t*-Bu	—	—	−12.5	*R*	—
5	Ph	—	—	−3.1	*R*	—
6	HC≡CC≡C—[c,d]	—	—	−26.6	*R*	6.8
7	$CH_3C{\equiv}CC{\equiv}C{-}$[c,e]	—	—	−11.3	*R*	3.3

Allenic alcohols, **79**

[a] Data from Reference 43a unless otherwise noted.

[b] Based on the established configuration of the first example, it is proposed that all the (−)-allenes prepared by this method have the *R* configuration and the (+)-allenes the *S* configuration.

[c] Data from Reference 43b.

[d] Marasin.

[e] 9-Methylmarasin.

An asymmetric synthesis of allenes,[44] which permits recovery of the inducing reagent is the reaction of a phosphonium ylide **81A**, containing a chiral ester moiety, with an achiral acid chloride (**81B**) to produce a phosphonium salt (**81C**) which, in the presence of excess **81A**, eliminates triphenylphosphine oxide to yield a mixture of diastereomeric allenic esters **81E**. Hydrolysis regenerates the original chiral alcohol moiety of the ester [R*OH where R* is either (−)-menthyl or (−)-*sec*-octyl] and gives optically active allenic acid (**81F**), $[\alpha]_D^{22}$ −2.8 to −5.1° where R is either *n*-butyl or *n*-amyl. The maximum rotations of the products are unknown; however, from the known rotation of other aliphatic allenes it is probable that these reactions represent asymmetric syntheses of 5–10% at best. The

[44] (a) I. Tömösközi and H. J. Bestmann, *Tetrahedron Lett.*, 1293 (1964). (b) I. Tömösközi and G. Janzso, *Chem. and Ind.*, 2055 (1962).

absolute configuration of the predominant product is presumed by these authors to be *S*, based on the following analysis of the reaction mechanism.

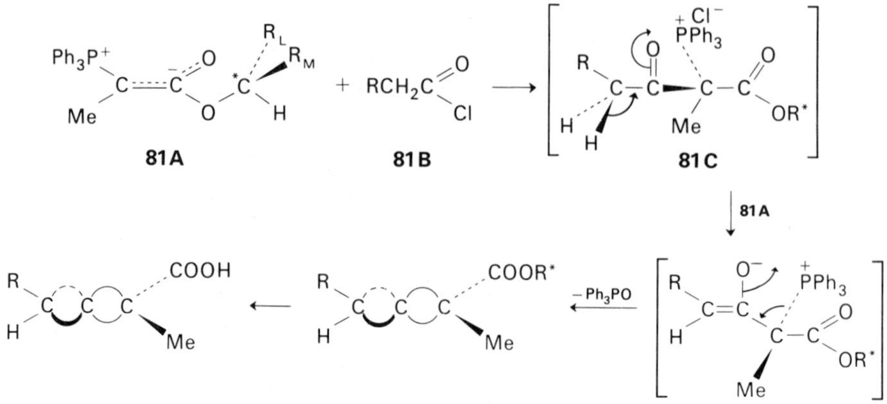

Ph3P+ C=C O Me O C* RL RM H **81A** + RCH2COCl **81B** → [81C] ↓ 81A [81D] —Ph3PO→ **81E** ← **81F**

Fig. 9-3. Asymmetric synthesis of allenic acids;[44] R* represents either (−)-menthyl or (−)-*sec*-octyl; R is either *n*-butyl or *n*-amyl.

It is assumed via a Prelog-Cram type model that the original acylation (**81A** → **81C**) proceeds in the conformation shown with the favored approach of the reagent from the least hindered side (from above the plane of the page in **81A**). It is further assumed that the proton elimination (**81C** → **81D**) is most favorable as indicated in the conformation shown to give the enol with the R group and oxygen *cis* to each other. This enolate then eliminates triphenylphosphine oxide in the manner shown to give **81E** during which the centro-asymmetry, at the carbon α to phosphorus, is transferred to the axial asymmetry of the allenic system. It would seem desirable to have many more examples in this series before judging the value of this method for the correlation of allenic configurations.

Achiral 4-substituted cyclohexanones **82A** react with the chiral menthyl phosphono ester **82B** to give the cyclohexylideneacetic acid derivative **82C** after hydrolysis.[44b] These products owe their chirality to axial dissymmetry as do the allenes.[44a] The stereoselectivity was found to be 51% for the

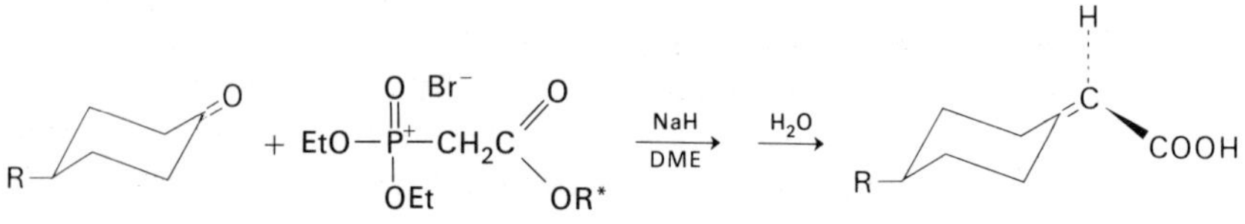

82A + EtO–P(O)(OEt)+–CH2C(O)OR* Br− **82B** —NaH/DME→ —H2O→ **82C**

R = Me or *t*-Bu, R* = (−)-menthyl, DME = 1,2-dimethoxyethane.

4-methyl derivative but it could not be determined for the 4-*t*-butyl derivative because of racemization upon hydrolysis of the intermediate.

An alternative related allene asymmetric synthesis[45] introduces the asymmetric carbon α to phosphorus by use of a chiral acid chloride as shown in Fig. **9-4**. The course of the reaction can be viewed as follows: When the

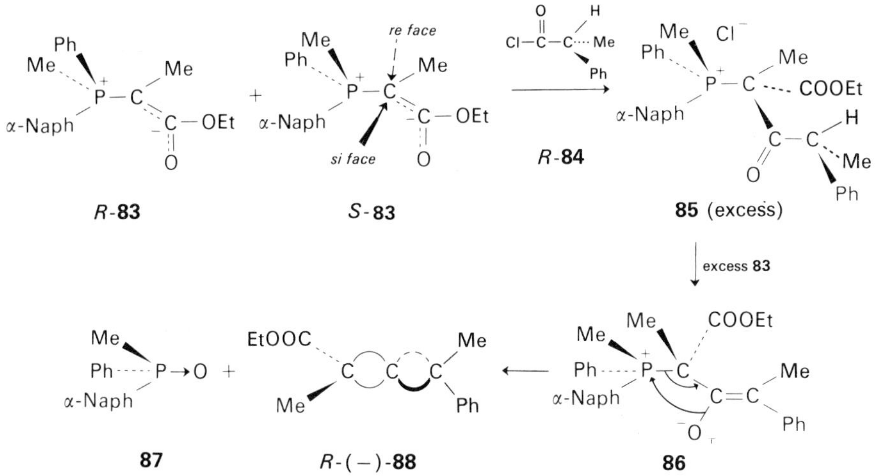

Fig. 9-4. Asymmetric synthesis of allene carboxylic ester and simultaneous partial kinetic resolution to produce a chiral phosphine oxide.[45]

racemic ylide *RS*-**83** reacts with (*R*)-(−)-hydratropic acid chloride (**84**), attack is most rapid on the *si* face of *S*-**83**. This produces **85** in excess and leaves the unreacted ylide **83** enriched in the *R* isomer. Removal of a proton from **85** by excess **83** destroys the original asymmetry of the acid chloride moiety generating **86** which, via stereospecific elimination of **87**, produces *R*-(−)-**88** (39% e.e.). Using (*S*)-hydratropic acid chloride (*S*-**84**), *S*-(+)-**88** is produced (41% e.e.). Using (*R*)- and (*S*)-α-phenylbutyric acid chlorides instead of (*R*)- and (*S*)-hydratropic acid chlorides gives the homologs of *R*-(−)-**88** and *S*-(+)-**88** (72% and 48% e.e., respectively).* When **83** reacts with **85**, it is converted to a phosphonium salt which can be hydrolyzed *with inversion* to give **87** which is enriched in the *S* isomer. This unique reaction sequence is therefore both an asymmetric synthesis and a kinetic resolution.

[45] H. J. Bestmann and I. Tömösközi, *Tetrahedron*, **24**, 3299 (1968).

* This latter value was reported[45] as 77.5% but there appears to be a mistake. One explanation for the difference between these two values is that the extent of kinetic resolution in forming **85**, which may vary from reaction to reaction, determines the stereoselectivity in forming allene **88**.

A general method for generating optically active allene hydrocarbons (**92**) from olefins (**89**) involves the ring opening of diazocyclopropanes arising from base-catalyzed decomposition of chiral N-nitroso-N-cyclopropylcarbamates (**91**).[46] Only an estimate of the extent of asymmetric transfer

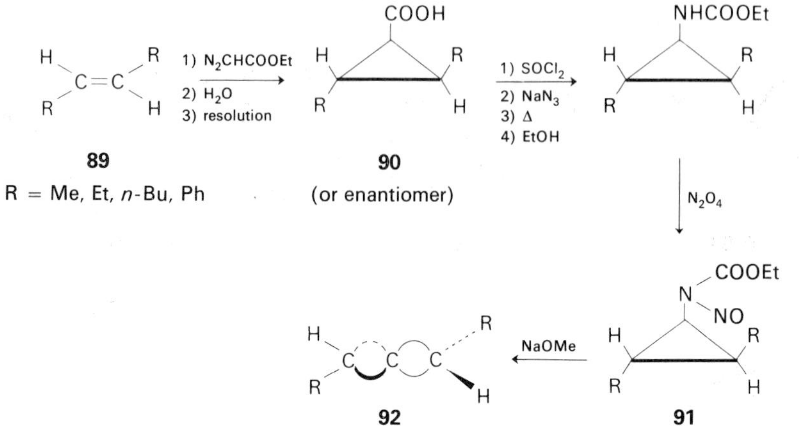

Fig. 9-5. Synthesis of optically active allenes from chiral cyclopropanecarboxylic acids. The configurations in this figure are written to represent the conversion of 2*S*,3*S*-cyclopropanecarboxylic acids to (*S*)-allenes, **92** (where R is a normal alkyl group), in spite of the fact that the configuration of the isomer used and formed was not always known.[46] Assuming the correctness of Lowe's generalization [*Chem. Commun.*, 411 (1965) cf. also 47a], those allenes, **92**, with a positive rotation would be *S*.

from the cyclopropane to the allene can be made since the enantiomeric purities of the cyclopropylcarbamates (**91**) and allenes (**92**) are not known with certainty. It has been estimated[47] that 2,3-pentadiene (**92**, R = Me) has a maximum specific rotation of about 174° (ethanol). If one disregards the different solvents used for the rotations, the 2,3-pentadiene, $[\alpha]_D$ −25.3° (CCl_4), obtained by this method is approximately 15% enantiomerically pure. If the resolution of the *trans*-2,3-dimethylcyclopropanecarboxylic acid (**90**, R = Me) is complete, the stereoselectivity is 15%.

The 1,3-diphenylallene, obtained from 2*S*,3*S*-(−)-**90** (R = Ph), of known configuration, and presumably completely resolved, contained at best a 40% excess of the *R*-(+)-**92** isomer. Two crops of crystals were obtained ($[\alpha]_D$ +405° and 459°) which when combined and recrystallized three times

[46] J. M. Walbrick, J. W. Wilson, Jr., and W. M. Jones, *J. Am. Chem. Soc.*, **90**, 2895 (1968).

[47] (a) J. H. Brewster, *Topics in Stereochemistry*, ed., N. L. Allinger, and E. L. Eliel (New York: John Wiley & Sons, Inc., 1967), Vol. 2, pp. 33–39. (b) W. L. Waters, W. S. Linn, and M. C. Caserio, *J. Am. Chem. Soc.*, **90**, 6741 (1968).

from hexane gave crystals ($[\alpha]_D$ +1020°) which may possibly represent the pure *R*-(+) isomer of **92**. The initial crystals may have undergone some enrichment by fractionation from the racemate and thus the 40% asymmetric synthesis is a maximum value.

Ring opening of the precursor diazocyclopropane or cyclopropylidene, **93**, with conrotation in one direction, **93A**, involves increased phenyl interactions and produces *S*-(−)-**92** while conrotation in the opposite direction, **93B**, minimizes phenyl interactions and leads to the production of excess *R*-(+)-**92** as observed. Disrotation (not shown) would pass through a planar stage giving a racemate.

93A → *S*-(−)-**92**

93B —Favored→ *R*-(+)-**92**

Reduction of the bicyclic *gem*-dibromides (**94**) with chromous (+)-tartrate or an *n*-butyllithium complex of (−)-sparteine gave cyclic allenes (**95**) that had low specific rotations.[48] Diastereomers **94A** (*cis* ring junction) and **94B** (*trans* ring junction) gave enantiomeric allenes in excess with both chromous (+)-tartrate (**95A**, $[\alpha]_D$ + 0.43; **95B**, $[\alpha]_D$ − 0.44) and the *n*-butyllithium-(−)-sparteine complex (**95A**, $[\alpha]_D$ +1.4; **95B**, $[\alpha]_D$ −2.8; **95C**, $[\alpha]_D$ + 1.9). The observation that asymmetric synthesis occurs supports a mechanism for the debromination that involves chiral carbenoid intermediates (see also Sec. **6-3.1**). Conrotation around the C_α—C_β and C_α—$C_{\beta'}$

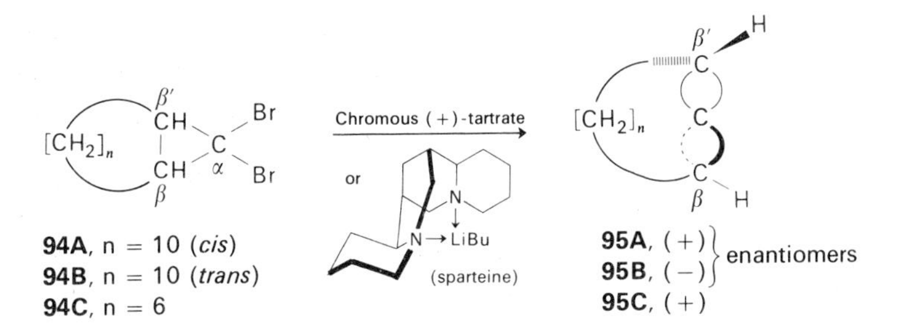

[48] H. Nozaki, T. Aratani, and R. Noyori, *Tetrahedron Lett.*, 2087 (1968).

bonds of **94**, as the C_β—$C_{\beta'}$ bond breaks, yields enantiomeric allenes depending upon the direction of the conrotation. The fact that enantiomeric products result from diastereomers **94A** and **94B** is in consonance with this mechanism.

Examples of the conversion of allenic axial dissymmetry to centrodissymmetry are less common than asymmetric transfers in the reverse direction which have just been discussed. To the extent that the mechanism of such reactions and the configurations of the products are known, these transformations may be used to deduce the configuration of allenes. The first published report of such a conversion is the Diels–Alder reaction of (−)-pentadiendioic acid [(−)-glutinic acid, **96**] with cyclopentadiene.[49] Two products, **97** and **98**, were formed in roughly equal amounts, and their structures were established by the formation of iodolactone **99** from **97** and diacid **100** from **98**

HOOCCH=C=CHCOOH + (cyclopentadiene) ⟶ (+)-**97** and (−)-**98**

(−)-**96**

using racemic materials. The absolute configurations were established by their conversion to norcamphor (**101**); **97** gave (+)-**101** and **98** gave (−)-**101**.

99 ← I_2, Na_2CO_3 — (+)-**97** — 1) H_2 (Pd·C) 2) O_3, −70° 3) HCl, Δ → (+)-**101**

100 ← H_2 (PtO_2) — (−)-**98** → (−)-**101**

Four products can be expected, a priori, starting with one enantiomer of **96**. We need to consider the reaction at only one double bond since in any one isomer the double bonds are stereochemically equivalent. The cyclopentadiene can add to (*R*)-glutinic acid (*R*-**96**) either from above (**A** or **A**′) or below (**B** or **B**′) the plane of the double bond and the cyclopentadiene can be oriented either for maximum overlap with the carboxyl group of the reacting double bond (**A**, **B**) according to the *Alder–Stein* rules or for a minimum overlap (**A**′ or **B**′). In each case, attack from above as shown

[49] W. C. Agosta, *J. Am. Chem. Soc.*, **86**, 2638 (1964).

(**A** or **A**′) involves approach of the cyclopentadiene from the side of the upright hydrogen on the adjacent double bond. This should be of lower energy than the approach from below (**B** or **B**′) due to greater steric interaction with the bulky carboxyl group. If the absolute configuration of (−)-**96** is *R* as depicted in Fig. **9-6**, the attack according to **A** will result in (+)-**97** and that according to **A**′ will result in (−)-**98**. If the configuration of

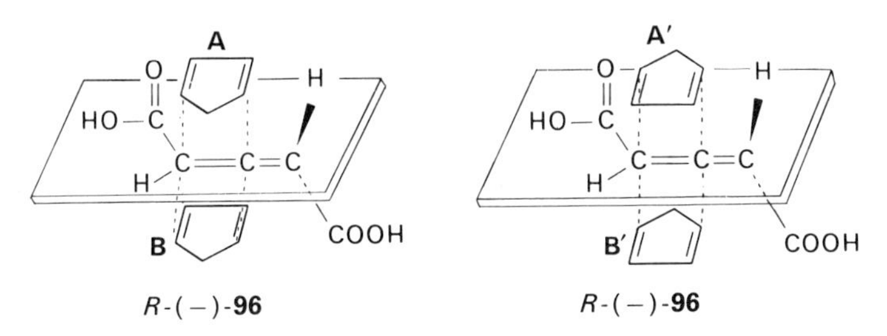

Fig. 9-6. Models for Diels–Alder addition to (−)-(*R*)-glutinic acid (**96**). There is a mirror-image set of reactions for *S*-**96** not shown.

(−)-**96** were *S* (not shown), then attack from the less hindered direction, with and without maximum overlap, would give (−)-**97** and (+)-**98**, respectively. Attack from the more hindered side of *S*-**96** would yield a pair of diastereomers (not shown) that are the enantiomers of those that would result from approach in the more hindered direction (below, **B** and **B**′) on *R*-**96**. Compounds having the latter structures were not found, and it seems eminently reasonable to assume that the controlling factor is the lower energy for approach from the side of the out-of-plane hydrogen rather than the side of the out-of-plane carboxyl. Therefore, since (−)-**96** gave (+)-**97** and (−)-**98** the configuration of (−)-**96** must be *R* as shown in Fig. **9-6**.

Stereoselective brominations have been used for the correlation of allenic and centrodissymmetric systems. A series of optically active allenic

S-(+)-**102** $\xrightarrow{Br_2}$ **103** $\xrightarrow{KMnO_4}$ **104** $\longrightarrow$ **105**

R = H, Me, Et, *i*-Pr, *t*-Bu

acids have been converted to optically active α-hydroxy acids by the sequence of reactions shown in **102** → **105**.[50] The carboxyl group is ideally positioned for intramolecular participation in the bromination and its orientation with respect to the β,γ-double bond is clearly revealed by the configuration of the products. The process is reported to approach 100% stereoselectivity for R = H, Me, Et and presumably also for R = *i*-Pr and *t*-Bu, although maximum rotations are not known with certainty in the latter two cases. Essentially the same type of configurational correlation scheme was used earlier by Gianni and Kuivila[51] who converted *R*-(+)-**106** to *R*-(+)-**108**.

106 →(Br_2) **107** →(LAH) **108**

Chiral allenic hydrocarbons also undergo stereospecific reactions which lead to chiral centrodissymmetric products.[47b] Optically active **109** [believed to contain a 9% e.e. of the *R*-(−)-isomer] was prepared via kinetic resolution by incomplete hydroboration of racemic material with "di-3-pinanylborane." Oxymercuration gave more of **112** than **113** presumably because of

R-(−)-**109** (~9% e.e.) ⇌ ($Hg(OAc)_2$, MeOH) [**110** and **111**]

110 → *S*-*trans*-**112** (83%) → *S*-(+)-**114** (7% e.e.)

111 → *R*-*cis*-**113** (17%)

[50] K. Shingū, S. Hagishita, and M. Nakagawa, *Tetrahedron Lett.*, 4371 (1967).

[51] M. H. Gianni, Ph.D. thesis, University of New Hampshire (1960) with H. Kuivila; *Chem. Abst.*, **55**, 16500h (1961).

greater hindrance to methanol attack on the mercurinum ion, **111**, from the side of the methyl group on the adjacent double bond, than on mercurinium ion **110** from the side of the hydrogen. A sample of **112** was converted to the chloromercuri derivative, recrystallized to enrich the amount of *trans*-isomer present, and demercurated to **114** [7% e.e., *S*-(+)].

Likewise, *R*-(−)-**109** was brominated in methanol to a mixture of *S*-*trans*-**115** (83%) and the *R*-*cis* isomer (17%). Bromination in carbon tetrachloride or ether gave mainly *S*-*trans*-**116** (83% versus 17% *R*-*cis* isomer). Iodine and

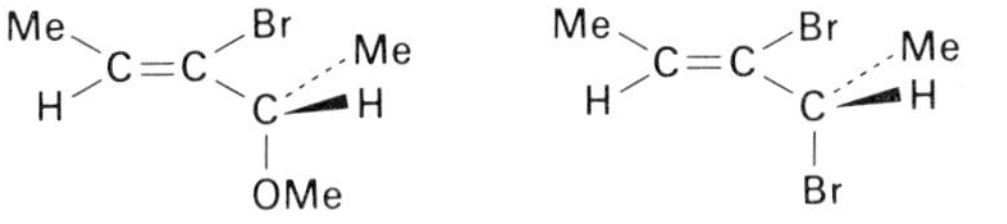

S-*trans*-**115** *S*-*trans*-**116**

methanol produced an 80% excess of the iodo isomer corresponding to **115**. It was concluded[47b] that oxymercuration, bromination, and iodination of allenes are all predominantly *trans* additions and probably highly stereoselective.

Similarly, hemihydrogenation of *R*-(−)-**117** (R = Me, $[\alpha]_D$ − 29.4°, probably 100% e.e.) gave a mixture of products from which *R*-(−)-*cis*-**118** was isolated and, in turn, hydrogenated to *R*-(−)-**119** ($[\alpha]_D$ − 7.4°, approximately 95% e.e.).[52] Under the conditions of the hydrogenation of **117** it

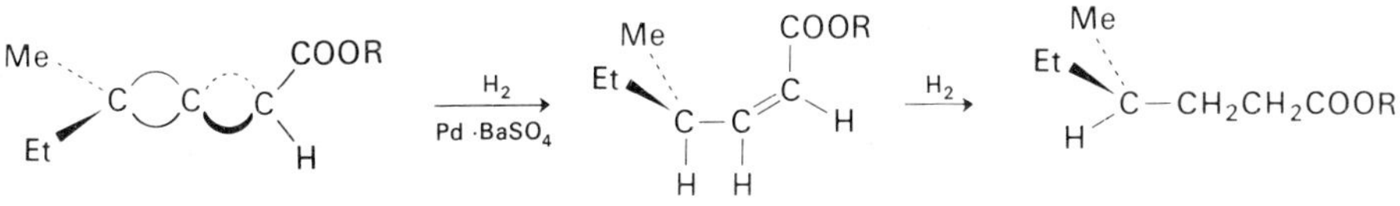

R-(−)-**117** *R*-(−)-**118** *R*-(−)-**119**

was found that **118** was partially converted to the *trans* isomer; smaller amounts of inactive *cis* and *trans* isomers from hydrogenation of the α,β-double bond of **117** and some fully hydrogenated product were also formed. The sequence shown involves *cis* addition of hydrogen to the β,γ-double bond of **117** and is the preferred reaction pathway.

9-2.3. Spiranes, Alkylidenecycloalkanes, and Related Cases.

Although several optically active spiro compounds are known,[53] some having been prepared in very low enantiomeric purity by asymmetric

[52] L. Crombie and P. A. Jenkins, *Chem. Commun.*, 870 (1967).

[53] E. L. Eliel, *Stereochemistry of Carbon Compounds* (New York: McGraw-Hill Book Co., 1962), p. 310.

catalysis,[54] attempts to interconvert spirodissymmetry and centrodissymmetry[53] were largely unsuccessful until Krow and Hill[55] reported the synthesis of an optically active spiro compound (**122**) in a manner that reveals its absolute configuration. Pyrolysis of the ammonium salt of *R*-(−)-**120** gave centrodissymmetric (−)-**121** which was then converted to *S*-(−)-**122** having only axial chirality. The configuration of the starting material (**120**) was established by its conversion to **125** which, in turn, was related to *R*-(+)-**127** of known absolute configuration.

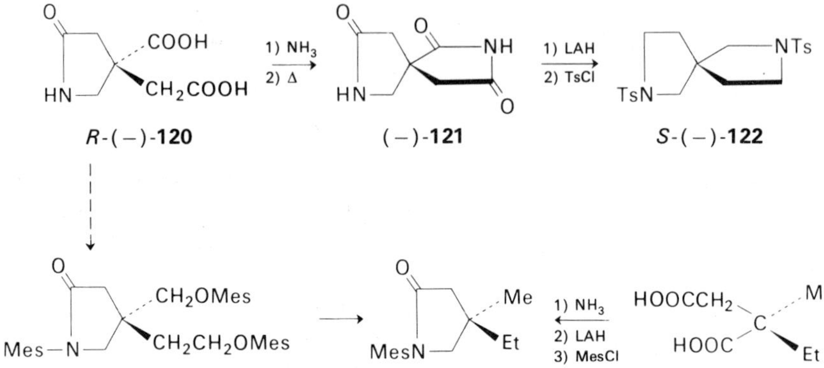

Fig. 9-7. Stereospecific synthesis of a spiro compound.[54] (Mes = CH_3SO_2—; Ts = *p*-$CH_3C_6H_4SO_2$—.)

Although this illustrates the conversion of a centrodissymmetric to axially dissymmetric system (**121** → **122**), the reaction has none of the usual characteristics of an asymmetric synthesis. The substituents on the skeleton are removed without altering the spirane nucleus responsible for the chirality of both starting material and products; no competing diastereomeric transition states are involved.

Certain cycloalkylidenes have only axial chirality and both their synthesis from and conversion to centrodissymmetric compounds have been studied. An example not unlike the spirane case (**121** → **122**) involves the *R*-(−)-2-benzylidene-5-methylcyclohexanone **128A**, prepared from benzaldehyde and *R*-(+)-3-methylcyclohexanone, which was reduced with "$AlCl_2H$" (3:1 $AlCl_3$—LAH) to *S*-(+)-1-benzylidene-4-methylcyclohexane (**129**) plus some *R*-(+)-1-benzyl-4-methylcyclohexene (**130**).[56] Photochemical isomerization

[54] A. A. Ponomarev and V. V. Zelenkova, *Dokl. Akad. Nauk SSSR*, **87**, 423 (1952); *Chem. Abstr*., **48**, 663d (1954).

[55] G. Krow and R. K. Hill, *Chem. Commun*., 430 (1968).

[56] J. H. Brewster and J. E. Privett, *J. Am. Chem. Soc*., **88**, 1419 (1966).

of **128A** to **128B** followed by reduction gave racemic **129**. The transfer of centro- to axial-dissymmetry most likely fails in the conversion of **128B** to

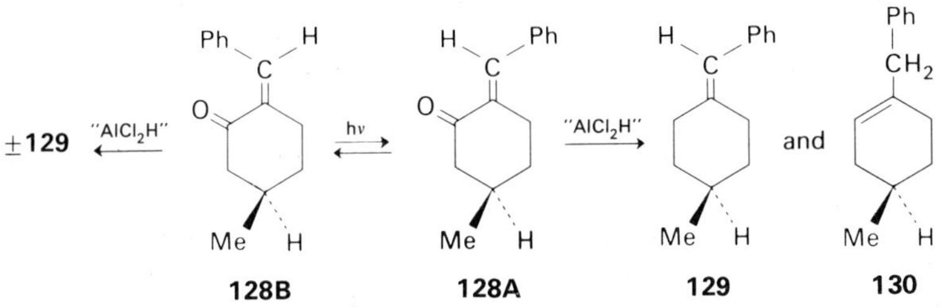

±129 "$AlCl_2H$" ← Ph C H O Me H **128B** $h\nu$ ⇄ H C Ph O Me H **128A** "$AlCl_2H$" → H C Ph Me H **129** and Ph CH_2 Me H **130**

129 because rotation of the benzylidene substituent to relieve the steric crowding of the phenyl group can occur within the lifetime of an intermediate delocalized carbonium ion.[56]

Another interconversion of an axially dissymmetric alkylidenecyclohexane (**131**) to compounds with centrodissymmetry, which was accomplished many years ago,[57] has now been analyzed stereochemically.[56,58] 4-Methylcyclohexylideneacetic acid (**131**) was resolved. Each isomer was

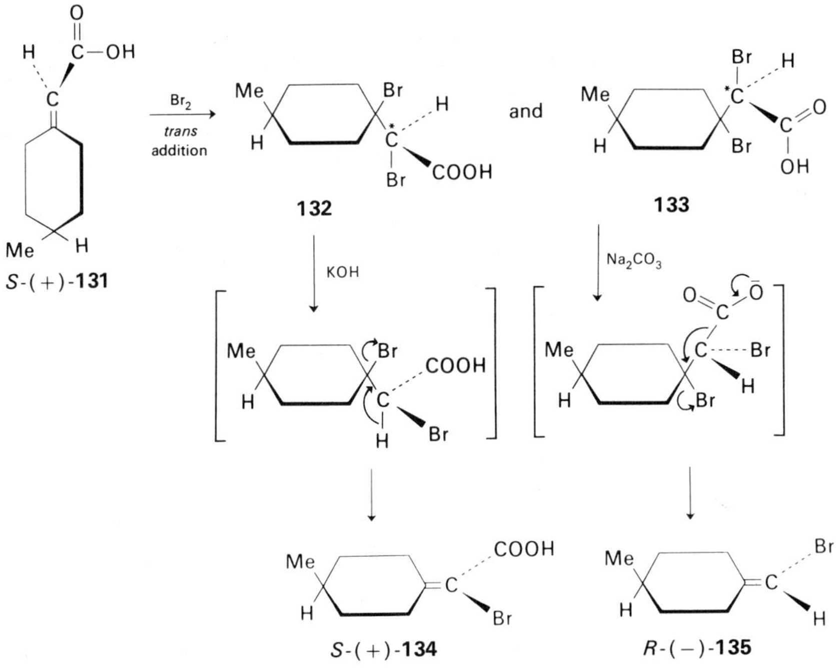

O H C—OH C Me H *S*-(+)-**131**

Br_2 → *trans* addition

Me H Br H C* Br COOH **132** and Me H Br H C* Br C=O OH **133**

KOH

Me H Br C COOH H Br

Na_2CO_3

Me H O=C O⁻ C Br H Br

Me H C COOH Br *S*-(+)-**134**

Me H C Br H *R*-(−)-**135**

[57] W. H. Perkin and W. J. Pope, *J. Chem. Soc.*, **99**, 1510 (1911).
[58] H. Gerlach, *Helv. Chim. Acta*, **49**, 1291 (1966).

carried through the indicated conversions represented on page 399 for one enantiomer. The axially dissymmetric compound, *S*-(+)-**131**, was brominated to the centrodissymmetric diastereomers **132** and **133**. The former was dehydrobrominated (KOH) to the axially dissymmetric *S*-(+)-**134** and the latter decarboxylatively debrominated to give the axially dissymmetric *R*-(−)-**135**. The configurations shown are based on models for the correlation of chirality and rotation[56] in axially dissymmetric alkylidene cyclohexanes.

A detailed and elegant study[58] of a similar transformation from axial- to centro-dissymmetry reveals that **136** is hydrogenated to an approximately 2:1 mixture of **137** and **138**. The configurations of derivatives of these products (**139** and **140**) are related to those of alcohols obtained by asymmetric reduction of *cis* and *trans* **141** with the bromomagnesium salt of (1*R*,2*R*,4*R*)-isoborneol-2-*d* (**142**). Knowing the stereochemistry of **141** (*cis* and *trans*, respectively) and assuming that both of these aldehydes are reduced by **142** to carbinols having primarily the *S* configuration (Sec. **5-1.4**, Table **5-2**) enables the stereochemistry of the carbinols (**143** and **144**) to be defined. The reactions used to convert **143** to **140** involves one inversion (azide displacement) and therefore the configurations of **143** and **140** are as shown. Curtius rearrangement (retention) on the azide of **138**, followed by treatment with phthalic anhydride, leads to *R*-(−)-**140**, the same product obtained from **143**. Similar treatment of *trans*-**141** gives *trans*-*R*-(−)-**139**, the diastereomer of *trans*-*S*-(+)-**139**. Making the reasonable assumption that hydrogenation of (+)-**136** involves *cis* addition, the preceding evidence enables the configurations in Fig. **9-8** to be assigned as shown, with a high level of confidence.

A reaction which might be mistaken for a transformation from axial to centrodissymmetry is the Beckmann rearrangement of **145** to **146**,[59] and ultimately to **147** of known configuration. The precursor pyridone of **145** is a *meso* compound (C_s point group) but conversion into the oxime introduces an element of chirality (**145** belongs to point group C_1) which depends upon the

HON H H Ph N Ph Me — 1) $PhSO_2Cl$ 2) NaOH → O H N H H Ph N Ph Me — HCl → H_2N CH_2 H H—N Ph Me

(+)-*syn*-*R*-**145** **146** **147**

[59] (a) G. G. Lyle and E. T. Pelosi, *J. Am. Chem. Soc.*, **88**, 5276 (1966). (b) R. E. Lyle and G. G. Lyle, *J. Org. Chem.*, **24**, 1679 (1959). See R. S. Cahn, C. Ingold, and V. Prelog, *Angew. Chem. Internat. Edn.*, **5**, 385 (1966), page 399, for similar examples which do not fall within the scope of axial chirality.

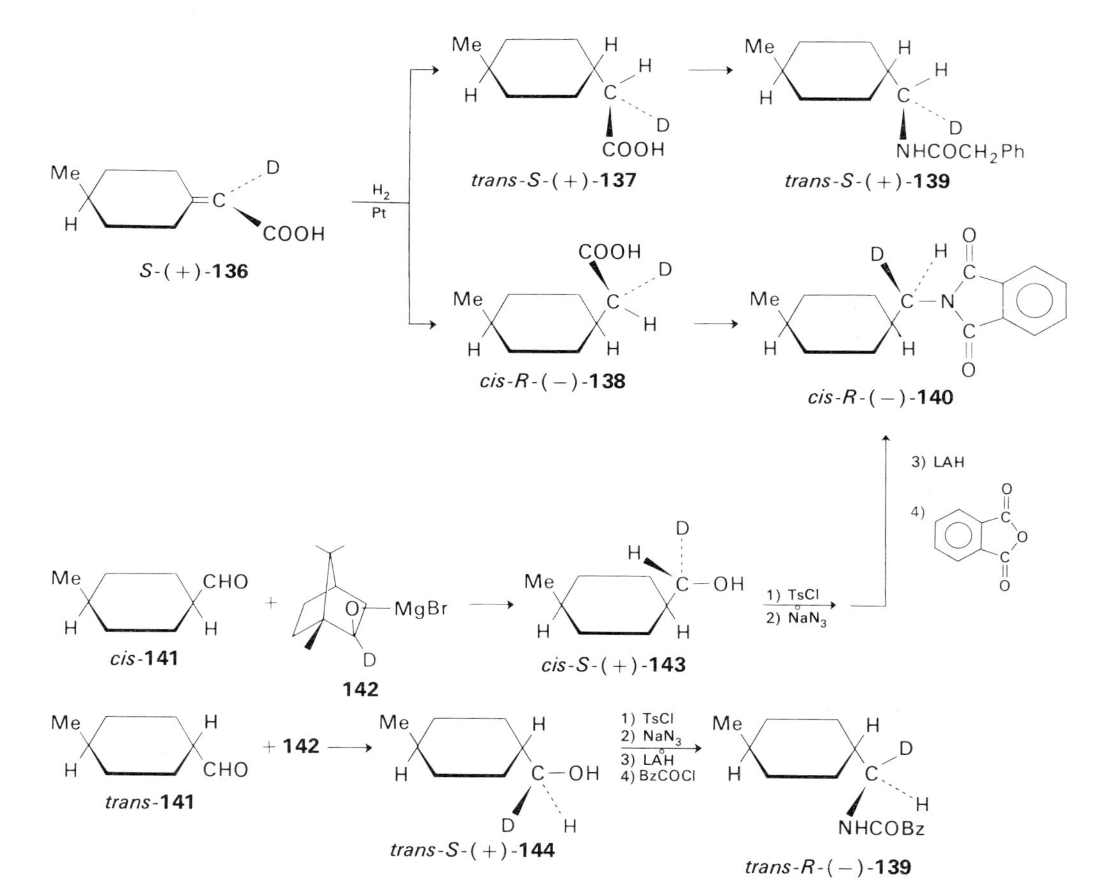

Fig. 9-8. Conversion of axially dissymmetric alkylidenecyclohexane to centrodissymmetric derivatives and proof of configurations. Bz represents the benzyl group.

configuration of the oxime, *i.e.*, whether the OH is on the side of the *R* or *S* asymmetric carbon adjacent to the ring carbon atom. However, the Beckmann rearrangement product, **146**, is a classical case of one enantiomer of a set of two diastereomeric pairs; its chirality is due to the asymmetric carbon atoms already present as such in the precursor oxime.

Chiral *trans*-cyclooctene (C_2 point group) has been converted to a centrodissymmetric compound* in the course of determining its absolute configuration.[60] Although it has been obtained by asymmetric synthesis (Sec. **9-3**), for this purpose, it was resolved by crystallization of diastereomeric *trans*-dichloro-(*trans*-cyclooctene)-(α-methylbenzylamine)-platinum(II) complexes.[61] The levorotatory enantiomer (**148**) was converted to (1*S*,2*S*)-(+)-1,2-dimethoxycyclooctane (**149**) which in turn was synthesized from (2*R*,3*R*)-(+)-tartaric acid (**150**) in a manner which established its configuration. The conformation of **148** shown may not be the only one of low energy; in fact, it has been suggested[62,63] that hindered rotation about the C-5/C-6 bond introduces a second element of dissymmetry which could lead to the existence of a pair of isomers with an appreciable barrier to interconversion† for the *trans* compound. That such conformational

1) OsO_4
2) CH_2N_2, HBF_4

15 steps

148
trans-R-(−)-Cyclooctene

149

150
(2*R*,3*R*)-(+)-Tartaric acid

diastereomers could be isolated at room temperature seems very unlikely. In any event, the configurational correlation scheme can only reveal the chirality resulting from the conformation at the double bond. However, the meaning of activation parameters for the racemization of optically active *trans*-cyclooctene[64] is clouded[62] by these additional conformation possibilities.

[60] A. C. Cope and A. S. Mehta, *J. Am. Chem. Soc.*, **86**, 5626 (1964).

[61] A. C. Cope, C. R. Ganellin, H. W. Johnson, Jr., T. V. Van Auken, and H. J. S. Winkler, *J. Am. Chem. Soc.*, **85**, 3276 (1963).

[62] G. Binsch and J. D. Roberts, *J. Amer. Chem. Soc.*, **87**, 5157 (1965).

[63] (a) A. Moscowitz and K. Mislow, *J. Am. Chem. Soc.*, **84**, 4605 (1962). (b) M. Yaris, A. Moscowitz, and R. S. Berry, *J. Chem. Phys.*, **49**, 3150 (1968).

[64] A. C. Cope and B. A. Pawson, *J. Am. Chem. Soc.*, **87**, 3649 (1965).

* See Sec. **9-3** for another such example (*R*-**148** → **154D** → **154E**).

† It has been pointed out[63] that the optical rotation of *trans*-cyclooctene is due primarily to the dissymmetry associated with the twisted double bond, but initially[63a] the absolute configuration of (−)-*trans*-cyclooctene was erroneously predicted using theoretical calculations thought to be appropriate for such a case. This assignment was subsequently reversed.[63b]

9-3. Asymmetric Eliminations

Several reactions are known in which a chiral group is eliminated with the formation of an optically active alkene. This is a form of self-immolative asymmetric synthesis. An example[65] that proceeds with a reasonably high conservation of chirality is the Hofmann elimination reaction on (−)-**151** which gives (+)-*cis,trans*-1,5-cyclooctadiene (**152**, 80% e.e.*). Both isomers (**A** and **B**) were studied. The absolute configurations of (−)-**151** and (+)-**152** are not known. However, it is possible to predict the stereochemical relation-

1) AgOH
2) Δ
and
I^- $\overset{+}{N}Me_3$

151
151A, $[\alpha]_D$ −14.92°
151B, $[\alpha]_D$ +14.30°

R-**152** — $[\alpha]_D$ +121.3°
S-**152** — $[\alpha]_D$ −120.5°

ship between **151** and **152** by considering the probable mechanism of the elimination and conformational models. It has been demonstrated recently that Hofmann elimination on N,N,N-trimethylcyclooctylammonium hydroxide gives a mixture of *cis*- and *trans*-cyclooctene; the *trans*-cyclooctene is formed by a 100% *syn* mechanism.[67] Assuming the same is true of the quaternary hydroxide from **151**, it would appear from an inspection of models that loss of the *pro-R* hydrogen from the *S* enantiomer at C-8 should be favored to produce *S*-**152**. Loss of the *pro-S* hydrogen from the same *S*-enantiomer, again by a *syn* mechanism, will give the inactive *cis,cis*-1,5-cyclooctadiene. Thus if *S*-**151** represents the (−) enantiomer, then *S*-**152** is predicted to be the (+) enantiomer or vice versa. The lack of 100% enantiomeric purity in the product could be the result of some stereo-nonselective *syn* or *anti* elimination, partial racemization of either products or starting materials under reaction conditions, or the use of less than optically pure starting material.

In another study on the asymmetric Hofmann elimination reaction,[68] pyrolysis of the cyclooctyl substituted, chiral quaternary base from (−)-**153**

[65] A. C. Cope, C. F. Howell, and A. Knowles, *J. Am. Chem. Soc.*, **84**, 3190 (1962).
[66] A. C. Cope, J. K. Hecht, H. W. Johnson, Jr., H. Keller, and H. J. S. Winkler, *J. Am. Chem. Soc.*, **88**, 761 (1966).
[67] J. L. Coke and M. C. Mourning, *J. Am. Chem. Soc.*, **90**, 5561 (1968).
[68] A. C. Cope, W. R. Funke, and F. N. Jones, *J. Am. Chem. Soc.*, **88**, 4693 (1966).

* The maximum specific rotation for **152** of −152° was obtained by resolution[66] via (+)-*trans*-dichloro-(*cis-trans*-1,5-cyclooctadiene)-(α-methylbenzylamine)-platinum(II); it is possible that resolution was incomplete (Reference 66, Footnote 9).

(99.6% e.e., configuration unknown) was found to give an 89% yield of mixed *trans*- and *cis*-cyclooctenes in a 60:40 ratio. The *trans* isomer was found to possess slight optical activity [*R*-(−)-**148**, 1.4% e.e. max]. When

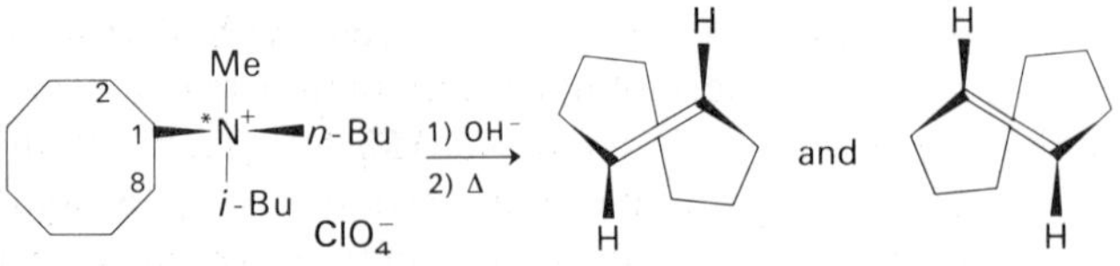

(−)-**153** *R*-(−)-**148** *S*-(+)-**148**

trans-cyclooctene was produced by treating (−)-**153** with potassium amide in liquid ammonia at −40°, the enantiomer was formed in excess [*S*-(+)-**148**, 2.4–2.7% e.e.]. For both of these reactions, enantiomeric starting materials gave products of opposite chirality.

In this reaction there is a transfer of nitrogen centroasymmetry to the C_2 dissymmetry of *trans*-cyclooctene. Elimination presumably occurs via an E_2-*syn* mechanism,[67] with hydrogen abstraction at C-2 or C-8 yielding enantiomeric products. The two *syn-alpha* hydrogens are diastereotopic and thus the transition states for the elimination of one versus the other will not be identical and the enantiomers of **148** will not be produced in exactly the same amount. It is futile to attempt to analyze in detail the factors which lead to the small extents of asymmetric bias in these reactions and which lead to the reversal of stereoselectivity under different reaction conditions. The difference between −1.4 and +2.7% asymmetric synthesis is less than 0.1 kcal/mole and the change in solvent, temperature, and reagent could easily account for an effect of this magnitude.

An attempted preparation of optically active *trans*-cyclooctene and *trans*-cyclodecene by pyrolysis of chiral cyclooctyl and cyclodecyl hydratropates[69] gave inactive cycloalkene in both cases. These results are to be expected, since at the temperature of the reaction the products are racemized.[70]

Elegant direct syntheses of optically active *trans*-cyclooctene (nearly 100% e.e.) have been accomplished via the centrodissymmetric compounds **154A** and **154C**.[71] Thionocarbonate **154A** ($[\alpha]_D^{21}$ +17.2°, $CHCl_3$) was prepared from (+)-*trans*-1,2-cyclooctanediol ($[\alpha]_D^{22}$ +16.9°, absolute EtOH, resolved via the strychnine salt of the half-phthalate) by treatment with N,N′-thiocarbonyl-diimidazole. Reaction of (+)-**154A** with triisooctyl phosphosphite at 130° (using a nitrogen sweep to remove **148** as formed in order to prevent racemization) gave an 84% yield of R-(−)-*trans*-cyclooctene

[69] O. Červinka, J. Budilova, and M. Daněček, *Coll. Czech. Chem. Commun.*, **32**, 2381 (1967).

[70] (a) A. C. Cope and B. A. Pawson, *J. Am. Chem. Soc.*, **87**, 3649 (1965). (b) A. C. Cope, K. Banholzer, H. Keller, B. A. Pawson, J. J. Whang, and H. J. S. Winkler, *ibid.*, **87**, 3644 (1965).

[71] E. J. Corey and J. I. Shulman, *Tetrahedron Lett.*, 3655 (1968).

[*R*-(−)-**148**, 99% *trans*, $[\alpha]_D$ −423°, enantiomerically pure based on $[\alpha]_D$ max from resolution via a platinum-α-methylbenzylamine complex[61]].

S-(+)-**148** $\xrightarrow{CH_2N_2}$ **154D** $\xrightarrow[-N_2]{h\nu}$ **154E**

R-(−)-**148**

154A, X = O, Y = S
154B, X = S, Y = NH
154C, X = S, Y = S

From (+)-**154A**: 100% e.e., *R*-(−)-**148**
From (−)-**154C**: 96% e.e., *S*-(+)-**148**

Reaction of *cis*-cyclooctene with thiocyanogen and then hydrobromic acid yielded the imino dithiocarbonate **154B**. This was resolved via the (−)-1-phenylethanesulfonic acid salt to give (+)-**154B** which was converted to (−)-**154C** with hydrogen sulfide in ethanol. Treatment of (−)-**154C** with 1,3-dibenzyl-2-methyl-1,3,2-diazophospholidine at 30° gave *S*-(+)-**148** (99% *trans*, 96% e.e.).

The optically active *trans*-cyclooctene obtained by the procedures above has been converted to centrodissymmetric derivatives[71] via reaction with OsO_4 and diazomethane (see p. 402, **148** → **149**). From *R*-(−)-**148** there was obtained (−)-**154D** which on photolysis eliminated nitrogen to yield (+)-**154E** containing only 4% of the *cis*-fused isomer. The enantiomeric purity of **154D** and **154E** prepared in this way is not known, but it is believed[71] to be nearly as high as that of the *trans*-cyclooctene from which they are prepared.

An attempted asymmetric synthesis of tropidine from hyoscyamine[72a] failed because the hyoscyamine racemizes below the temperature at which tropidine was produced by elimination. However, pyrolyses of chiral esters of 4-methylcyclohexyl hydrotropates[72] were successful. All four diastereomers of **155** gave optically active 4-methylcyclohexene, but the enantiomeric excess was less than 1% in every case (Fig. **9-9**). The configuration of the 4-methylcyclohexene is known[72b] and transition state topologies were proposed for correlating this with that of the ester being pyrolyzed. However the established asymmetric center is remote from the site which is responsible for generating the new chiral unit and the several conformational possibilities render such models very speculative, especially when the asymmetric bias of the reaction is so low.

[72] (a) S. I. Goldberg and F. L. Lam, *J. Org. Chem.*, **31**, 2336 (1966); (b) *ibid.*, **31**, 240 (1966); (c) *Tetrahedron Lett.*, 1893 (1964).

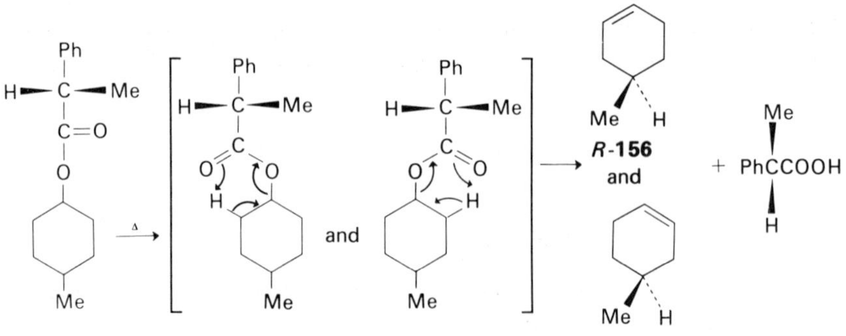

4-Methylcyclohexanol	4-Methylcyclohexene	Hydratropic acid
cis	*S*-(−) 0.24 ± 0.08% e.e.	*R*
cis	*R*-(+) 0.87 ± 0.08% e.e.	*S*
trans	*R*-(+) 0.54 ± 0.41% e.e.	*R*
trans	*S*-(−) 0.41 ± 0.14% e.e.	*S*

Fig. 9-9. Asymmetric pyrolysis of 4-methylcyclohexyl hydratropates [illustrated with *S*-(+)-hydratropic acid].

Much higher asymmetric syntheses are observed when chiral 4-methylcyclohexyl *p*-tolyl sulfoxides[73] are pyrolyzed. The *trans* sulfoxide having the *R* configuration at sulfur (*R*-(+)-**157**) gives *R*-(+)-4-methylcyclohexene (*R*-(+)-**156**), while *S*-(−)-**157** gives an excess of *S*-(−)-**156**. Extents of asymmetric synthesis in these eliminations range up to 70% in contrast to those for the pyrolysis of the comparable chiral esters (Fig. **9-9**). This is undoubtedly a result of the fact that the chiral center on sulfur is intimately involved in the transition state. Elimination of *p*-tolylsulfenic acid takes place via a cyclic concerted mechanism, and removal of hydrogen from either C-2 or C-6 can conceivably occur.[73] It is presumably the *syn* axial hydrogen (rather than the equatorial hydrogen) which is transferred but this has not been rigorously established by isotope studies. The configurations of starting materials and products are correctly correlated by a model as shown in Fig. **9-10** in which the preferred configuration for the *R* isomer is the one having the nonbonding electron pair on sulfur compressed against the hydrogens at C-2, rather than the more bulky *p*-tolyl group compressed against the hydrogens at C-6.

In the pyrolysis of certain steroidal sulfoxides the substrate has multiple chiral centers and therefore the asymmetric synthesis is not sharply defined. Nevertheless the same factors are involved from a mechanistic standpoint.

[73] S. I. Goldberg and M. S. Sahli, *J. Org. Chem.*, **32**, 2059 (1967).

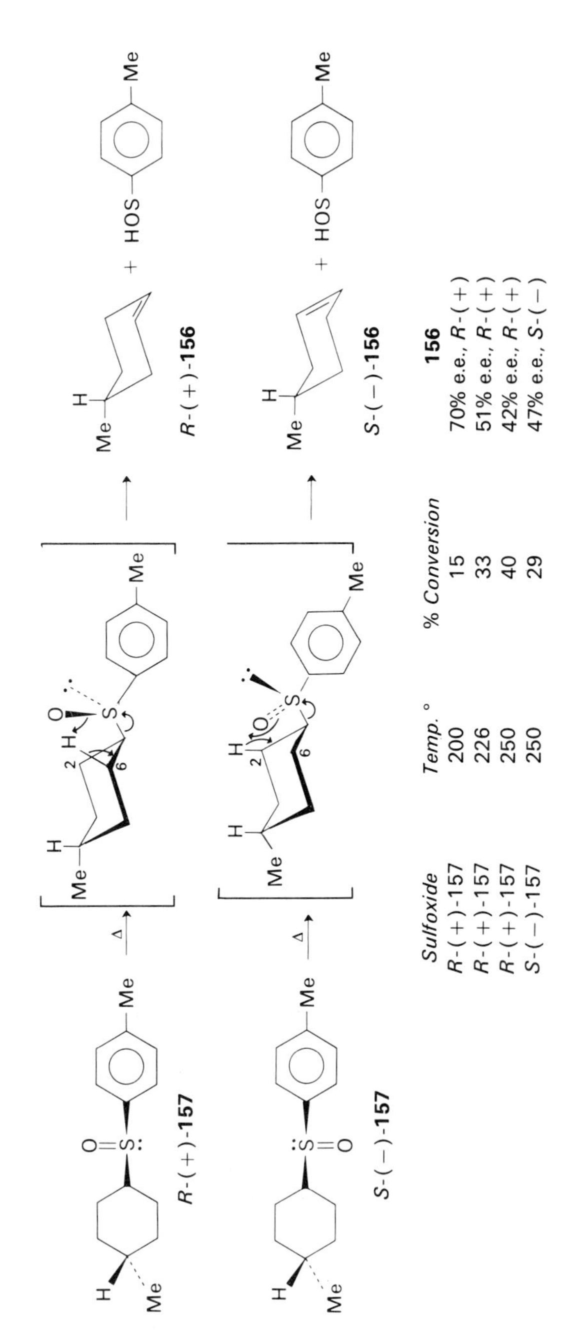

Sulfoxide	*Temp.* °	*% Conversion*	**156**
R-(+)-157	200	15	70% e.e., *R*-(+)
R-(+)-157	226	33	51% e.e., *R*-(+)
R-(+)-157	250	40	42% e.e., *R*-(+)
S-(−)-157	250	29	47% e.e., *S*-(−)

Fig. 9-10. Pyrolysis of chiral 4-methylcyclohexyl *p*-tolyl sulfoxides.

For instance,[74] (*S*)-4β-phenylsulfinylcholestane (**158**) readily undergoes pyrolytic elimination to 3-cholestene (**159**) in refluxing benzene. The oxygen of the phenylsulfinyl group can remove the 3β-hydrogen without forcing the phenyl into serious compression with the steroid moiety. However in the *R*-4β-

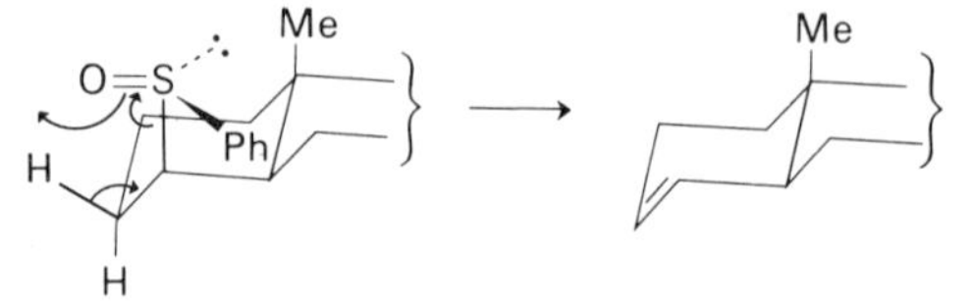

158 **159**

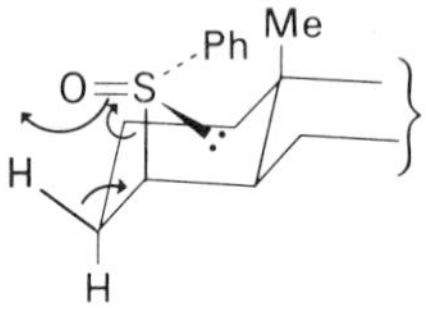

160 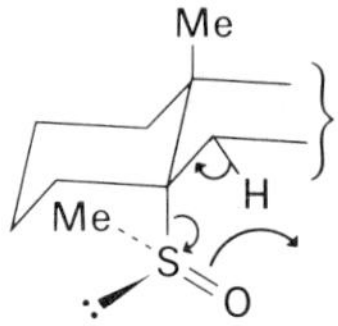 *R*-**161A** 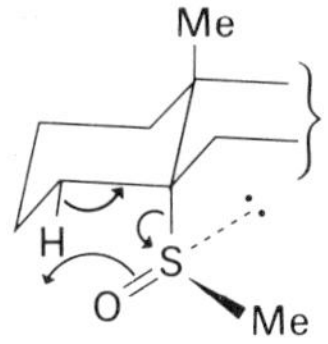 *R*-**161B**

isomer (**160**), in order that the oxygen may be properly oriented for transfer of the 3-β-hydrogen, the phenyl ring is forced into severe 1,3-diaxial interactions with both the angular methyl group and the 2β-hydrogen. This isomer is inert to refluxing benzene in agreement with this analysis. Comparable results were obtained with *R*- and *S*-4β-methylsulfinylcholestanes.[74b] Similar arguments explain the preferential formation of 4-cholestene over 5-cholestene when *R*-**161** is pyrolyzed.[74b] A transition state that retains the features of **161B** would be expected to be more favorable than one resembling **161A** because of the methyl interactions with the ring in *R*-**161A** but not in *R*-**161B**. The pyrolysis results for a number of other steroidal sulfoxides have been rationalized in the same way.[74]

Optically active 4-methylcyclohexylamine oxides have also been pyrolyzed to optically active 4-methylcyclohexene. Heating (+)-**162A** at 110° gave a 60% yield of *R*-(+)-**156** [approximately 30% e.e.], while (−)-**162A** gave *S*-(−)-**156** in excess.[75] If the configuration of the 4-methylcyclohexyl portion is *cis* as tentatively assigned, then (+)-**162A** probably has the *S* configuration at nitrogen as shown. This would be in agreement with the more detailed and definitive study of Goldberg and Lam[76] concerning the stereochemistry of pyrolytic elimination from four isomeric N-methyl-N-neopentyl-4-methylcyclohexylamine oxides (for instance **162B**, cf. Table

[74] (a) D. N. Jones and M. J. Green, *J. Chem. Soc.*, (*C*), 532 (1967). (b) D. N. Jones, M. J. Green, M. A. Seed, and R. D. Whitehouse, *J. Chem. Soc.*, (*C*), 1362 (1968).

[75] G. Berti and G. Bellucci, *Tetrahedron Lett.*, 3853 (1964).

[76] S. I. Goldberg and F. L. Lam, *J. Am. Chem. Soc.*, **91**, 5113 (1969).

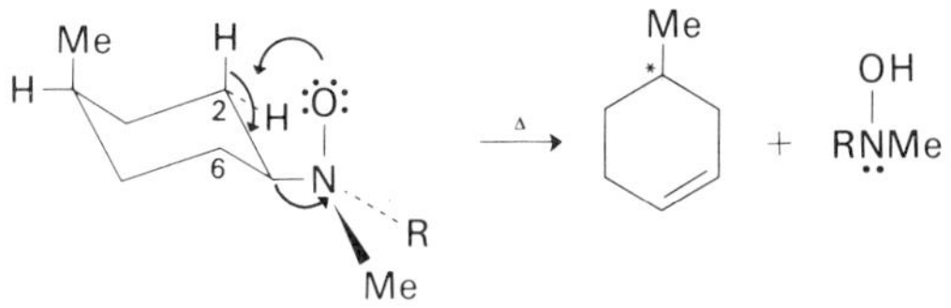

162A R = Ph
162B R = Neopentyl

R-(+)-**156**

9-2). The situation is exactly analogous to that encountered in the sulfoxide eliminations. The transition state for *syn* elimination from **162** is of lower energy if the oxygen "swings right" to remove a proton from C-2, since this involves compression of methyl with the ring; whereas to "swing left" for removal of a proton from C-6 would push phenyl (**162A**) or neopentyl

TABLE 9-2

Assignment of Absolute Configuration at Nitrogen in N-Methyl-N-neopentyl-4-methylcyclohexylamine Oxides[76] (162B)[a]

No.	Pyrolyzed amine oxide	Assigned config-uration	Pyrolysis product, 120° 4-methylcyclohexene configuration	% e.e.
1	(−)-*cis*	*S*	*S*-(−)	9.4 ± 0.6
2	(−)-*trans*	*S*	*R*-(+)	1.9 ± 0.2
3	(+)-*cis*	*R*	*R*-(+)	7.2 ± 0.4
4	(+)-*trans*	*R*	*S*-(−)	5.4 ± 0.4

[a] The % e.e. was slightly higher at 85°. Minimum values of % e.e. since enantiomeric purities of the starting amine oxides are not known.

(**162B**) against a ring methylene. Based upon this analysis, the known configuration of the 4-methylcyclohexene produced on pyrolysis can be used to establish the configuration at nitrogen of the amine oxide (Table **9-2**).

1969 Addenda

Sec. 9-2.2. An added example of change from axial dissymmetry to centrodissymmetry is the ethoxymercuration of optically active 1,2-cyclononadiene ($[\alpha]_D^{25}$ − 81.8°, max. 48% e.e.) to give optically active 3-ethoxycyclononene ($[\alpha]_D$ + 15.3°) of unreported configuration and optical purity.[77]

[77] R. D. Bach, *J. Am. Chem. Soc.*, **91**, 1771 (1969).

Both enantiomers of di-*t*-butylallene have been obtained by treatment of resolved propargylic sulfonates with lithium aluminum hydride which results in the concerted insertion of a hydride ion and displacement of the sulfonate group.[78]

Thermal dimerization of racemic 1,2-cyclononadiene leads to three products. Two result from combination of enantiomeric products and the third from the combination of allenes of the same configuration. Comparison of these product ratios from those obtained using partially resolved 1,2-cyclononadiene allows the conclusion that the substrate reacts one-fourth as fast with itself as it does with its enantiomer.[79]

An additional example of the interconversion of centrodissymmetry to biphenyl dissymmetry (cf. the conversion of thebaine, **54**, to phenyldihydrothebaines, **55** and **56**) is found in the rearrangement of the alkaloid schelhammeridine.[80]

Sec. 9-2.3. A case of the transfer of axial chirality of a spiro compound to centro chirality is given by the Stevens rearrangement (cf. **9-1.3**) of a spiro quaternary ammonium hydroxide (possessing C_2 symmetry), by heating with sodium hydroxide in diglyme, to give a chiral, bridged, tertiary amine.[81]

An attempt to prepare optically active spiro[3.3]hepta-1,5-diene by Hofmann elimination on the quaternary base from optically active 2,6-*bis*-dimethylaminospiro[3.3]heptane dimethiodide was unsuccessful.[82]

The chiral Wittig reagent made from (*S*)-(+)-benzylmethylphenyl-*n*-propylphosphonium bromide was treated with 4-methylcyclohexanone to give (*S*)-(+)-benzylidene-4-methylcyclohexane (43% e.e.). Seven other chiral benzylidene derivatives were prepared from other prochiral cycloalkanones.[83]

Sec. 9-3. The configurations at sulfur of eight 6- and 7-methylsulfinyl-5-α-cholestanes[84] and eighteen 6-alkylsulfinyl-5-α-cholestanes[85] were assigned on the basis of their pyrolysis patterns. Chirality at sulfur was found to influence the ratio of C-7 methylsulfinyl epimers resulting from base catalyzed equilibration. The assignment of configuration at sulfur did not correlate in every case with an empirical rule relating configuration and ord.

[78] W. T. Borden and E. J. Corey, *Tetrahedron Lett.*, 313 (1969).
[79] W. R. Moore, R. D. Bach, and T. M. Ozretich, *J. Am. Chem. Soc.*, **91**, 5918 (1969).
[80] S. R. Johns, J. A. Lamberton, A. A. Sioumis, and H. Suares, *Chem. Commun.*, 646 (1969).
[81] J. H. Brewster and R. S. Jones, Jr., *J. Org. Chem.*, **34**, 354 (1969).
[82] J. P. M. Houbiers, L. A. Hulshof, and Hans Wynberg, *Chem. Commun.*, 91 (1969).
[83] H. J. Bestmann and J. Lienert, *Angew. Chem.*, **81**, 751 (1969); *Internat. Edn.*, **8**, 763 (1969).
[84] D. N. Jones, M. J. Green, and R. D. Whitehouse, *J. Chem. Soc.* (*C*), 1166 (1969).
[85] D. N. Jones, D. Mundy, and R. D. Whitehouse, *J. Chem. Soc.* (*C*), 1668 (1969).

10

Miscellaneous Topics in Asymmetric Synthesis

Several subjects and several special examples which do not lend themselves to treatment in the previous sections have been collected here under the headings of (1) Asymmetric Reactions in Chiral Media, (2) Asymmetric Synthesis by Substitution at a Prochiral Ligand, (3) Asymmetric Polymer Syntheses, and (4) Absolute Asymmetric Reactions.

10-1. Asymmetric Reactions in Chiral Media

A prochiral substrate and an achiral reagent may react to give a chiral product if the reaction is conducted in a chiral environment. There have been many attempts to effect asymmetric syntheses using optically active additives that do not function as catalysts but only enter into the reaction as intermediate complexes or solvates. It seems self-evident that the more intimately the chiral additive or solvent is involved in the transition state, the more likely the possibility for a significant asymmetric bias to the reaction. Many early experiments were unsuccessful for lack of appreciation of this principle[1] which, however, was recognized by Bredig and Balcom[2] who

[1] P. D. Ritchie [*Asymmetric Synthesis and Asymmetric Induction* (London: Oxford University Press, 1933), pp. 78, 79] has reviewed these early experiments.

[2] G. Bredig and R. W. Balcom, *Ber.*, **41**, 740 (1908).

TABLE 10-1

Electrophilic Substitutions in Chiral (2*R*,3*R*)-(+)-Dimethyoxybutane[a]

$$\text{R-Metal} + \text{substrate} \xrightarrow{(+)\text{-DMB}^{a}} \xrightarrow{H_2O} \text{product}$$

No.	R-Metal	Substrate		Product	$[\alpha]_D^{\circ}$	Ref.
1	EtCHMeMgCl	+ PhN=C=O	$\xrightarrow{b}\longrightarrow$	EtCHMeCOOH	+0.29[c]	3
2		+ PhN=C=O	$\xrightarrow{d}\longrightarrow$	EtCHMeCOOH	+0.18[c]	3
3		+ PhN=C=O	$\xrightarrow{e}\longrightarrow$	EtCHMeCOOH	+0.30[c]	3
4		+ PhN=C=O	$\xrightarrow{f}\longrightarrow$	EtCHMeCOOH	+0.07[c]	3
5		+ CO_2	$\longrightarrow\longrightarrow$	EtCHMeCOOH	−0.21[c]	3
6		+ H_2CO	$\longrightarrow\longrightarrow$	$EtCHMeCH_2OH$	0.0	3
7		+ $HgCl_2$	$\longrightarrow\longrightarrow$	$(EtCHMe)_2Hg$	+0.20	3
8		+ PhCODur[g]	$\longrightarrow\longrightarrow$	EtCHMe CODur[g]	+0.44[h]	3
9	EtCHMeLi	+ CO_2	$\longrightarrow\longrightarrow$	EtCHMeCOOH	−0.16[c]	4
10	PhCHMeMgCl	+ PhN=C=O	$\xrightarrow{b}\longrightarrow$	PhCHMeCOOH	−0.67[h,i]	3
11		+ PhN=C=O	$\xrightarrow{d}\longrightarrow$	PhCHMeCOOH	+2.20[j]	3
12	PhCHMeMgCl	+ PhCHMeCl	$\longrightarrow\longrightarrow$	PhCHMeCHMePh	0.0	3
13	PhCMeEtMgCl	+ EtOH	$\longrightarrow\longrightarrow$	PhCHMeEt	0.0	5
14	$PhCH_2CHPhMgCl$	+ PhN=C=O	$\longrightarrow\longrightarrow$	$PhCH_2CHPhCOOH$	+3.16[k]	6
15	PhCH(MgCl)COOMgCl	+ $PhCH_2Cl$	$\xrightarrow{l}\longrightarrow$	$PhCH_2CHPhCOOH$	+1.52[k]	3
16	PhCH(MgCl)COONa	+ MeCOMe	$\xrightarrow{l}\longrightarrow$	$PhCHOHCMe_2COOH$	+2.94	3
17	PhCHNaCHNaPh	+ Me_2SO_4	$\longrightarrow\longrightarrow$	PhCHMeCHMePh	0.0[m]	7
18	PhCHLiCHLiPh	+ CO_2	$\longrightarrow\longrightarrow$	HOOCCHPhCHPhCOOH	5.4[n]	7
19	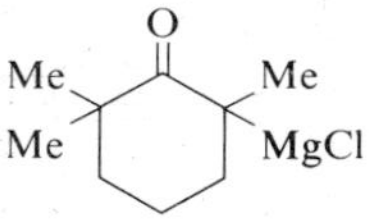	+ EtOH	$\longrightarrow\longrightarrow$	R-H[o]	+0.31[p]	4
20		+ CO_2	$\longrightarrow\longrightarrow$	RCOOMe[o]	+5.83	4
21		+ O_2	$\longrightarrow\longrightarrow$	ROH[o]	+0.57	4

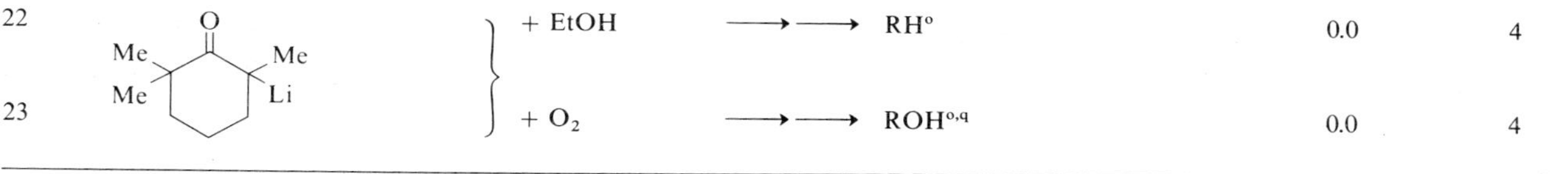

[a] Chiral additive is (+)-DMB [(2*R*,3*R*)-dimethoxybutane] in benzene unless otherwise noted.
[b] Grignard added to phenylisocyanate at 0°. The anilide product was hydrolyzed to the acid.
[c] As the anilide in acetone.
[d] Phenylisocyanate added to Grignard at 0°.
[e] Grignard added to phenylisocyanate in toluene at −70°.
[f] Phenylisocyanate added to Grignard in toluene at −70°.
[g] Dur is the 2,3,5,6-tetramethylphenyl, duryl, radical.
[h] Benzene solvent.
[i] $[\alpha]_D$ max 76.5° ($CHCl_3$); 81.1° (EtOH).
[j] Acetic acid solvent.
[k] Acetic acid solvent, $[\alpha]_D$ max 94°.
[l] Heterogeneous reaction mixture.
[m] Mixture of *threo* and *meso* products; *threo* was inactive.
[n] The *threo* acid was converted to the methyl ester without crystallization.
[o] R is the 2-(2,6,6-trimethylcyclohexanone)yl radical.
[p] Neat, *ca.* 2.5 % e.e., $[\alpha]_D$ max 12.36° (neat).
[q] Isolated as the semicarbazone.

[3] H. L. Cohen and G. F. Wright, *J. Org. Chem.*, **18**, 432 (1953).
[4] K. R. Bharucha, H. L. Cohen, and G. F. Wright, *J. Org. Chem.*, **19**, 1097 (1954).
[5] N. Allentoff and G. F. Wright, *Can. J. Chem.*, **35**, 900 (1957).
[6] N. Allentoff and G. F. Wright, *J. Org. Chem.*, **22**, 1 (1957).
[7] A. G. Brook, H. L. Cohen, and G. F. Wright, *J. Org. Chem.*, **18**, 447 (1953).

found that enantiomeric camphor carboxylic acids decarboxylated at the same rate in the hydrocarbon solvent (−)-limonene but at different rates in (−)-nicotine. In this latter case, of course, there is salt formation and the reaction may be considered as the decomposition of diastereomeric salts.

Most of the successful experiments with chiral solvents involve organometallic reagents which coordinate with the solvent. In these cases there is no sharp boundary between a classical chiral reagent such as the LAH · quinine reagents discussed in Chapter **5** and one that is chiral due to coordination of solvent, e.g., the Simmons–Smith reagent in menthol solvent (Table **6-7**). Two main types of asymmetric reactions in chiral media have been studied: those involving electrophilic substitution at the carbon-metal bond and those involving nucleophilic addition to carbonyl groups. Examples of the first type are summarized in Table **10-1**.

10-1.1. Electrophilic Substitution in Chiral Media. In a sense, none of these reactions represent a true asymmetric synthesis since these reagents which have a metal attached to a chiral center are racemic. For simple compounds such as these, however, the carbon-metal bond is not stereochemically stable and the enantiomers are in rapid equilibrium (**1A** ⇆ **1B**). The reactions listed in Table **10-1** were carried out in the presence of (2*R*, 3*R*)-(+)-dimethoxybutane, (+)-DMB, which forms a solvate with these Grignard reagents as shown in **2**. In such a complex the magnesium becomes a chiral

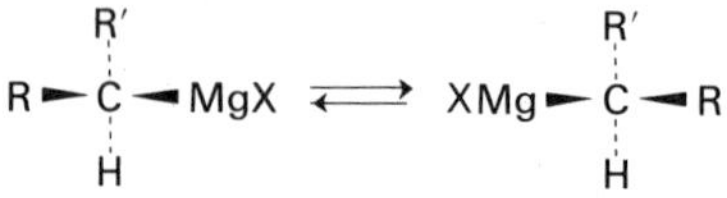

R–C(R′)(H)–MgX ⇄ XMg–C(R′)(H)–R

1A **1B**

R″(X)Mg←O(Me)–C(Me)(H)–C(H)(Me)–O(Me)→

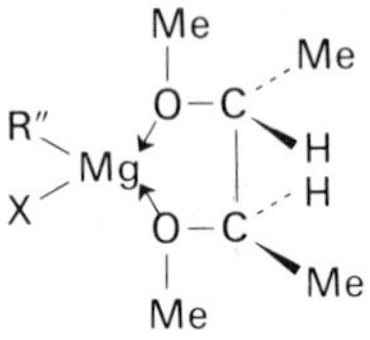

2

center and **2** must exist in diastereomeric forms. When the R″ group of **2** can exist in two forms as seen in **1A** ⇆ **1B**, this complex represents four stereochemically distinct solvated species, not necessarily present in equal amounts, each of which can react with a prochiral substrate at different rates.

The most apparent conclusion from the data in Table **10-1** is that the extent of induced stereoselectivity is generally very low,* at best perhaps 2–3%. If the stereoselectivities are this low with solvents which complex

* Maximum rotations of many of the products were unknown. Rotation of the same products were often taken in different solvents and direct comparison of such results is not possible. Thus in Table 10-1 it is not certain that the product from No. 11, $[\alpha]_D$ −0.67° (C_6H_6) is even of opposite configuration from that of run No. 12, $[\alpha]_D$ +2.20° (HOAc).

with the reagent, it is very likely that they will be much lower with solvents which are not thus involved in the transition state.

10-1.2. Carbonyl Additions in Chiral Media. Examples of the addition of organometallic reagents to carbonyl compounds in the presence of chiral solvent or other chiral additive are summarized in Tables **10-2** (ketones) and **10-3** (aldehydes). The stereoselectivities are somewhat higher than encountered in the examples in Table **10-1** but still are generally low (less than 5% e.e.) with the exception of the addition of phenyl Grignard reagent to 2-butanone in (+)-DMB (Table **10-2**, No. 2, 17% e.e.) and the addition of

TABLE 10-2

Additions to Ketones in Chiral Media

$$\text{R-Metal} + \text{R}'\text{—}\overset{\overset{\displaystyle \text{O}}{\|}}{\text{C}}\text{—R}'' \xrightarrow{\text{Chiral solvent}} \xrightarrow{\text{Hydrolysis}} \text{R—}\underset{\underset{\displaystyle \text{R}'}{|}}{\overset{\overset{\displaystyle \text{OH}}{|}}{\text{C}}}\text{—R}''$$

No.	R-Metal	R′	R″	Chiral solvent[a]	Product	Ref.
1	EtMgCl	Ph	COOEt	(+)-DMB[b,c]	5% e.e., *S*-(+)	3
2	PhMgBr	Me	Et	(+)-DMB[c]	17% e.e., *R*-(+)	6, 8
3		Me	Et	*D*-(−)-APME[c]	2.3% e.e., *R*-(+)	6
4		Me	Et	*D*-(+)-MHME[c]	4.3% e.e., *S*-(−)	6
5	EtMgCl	Me	Ph	(+)-DMB[c]	1% e.e., *R*-(+)	6
6	EtMgBr	Me	Ph	(+)-DMB[c]	3.5% e.e., *R*-(+)	6
7	EtMgI	Me	Ph	(+)-DMB[c]	2.5% e.e., *R*-(+)	6
8	PhMgBr	Me	Et	(−)-MME[c]	0	6
9	EtMgBr	Me	Ph	Et-O-CH_2CHMeEt[c,d]	0	8
10		Ph	COOEt	(−)-Sparteine[c,e]	18% e.e., *S*-(+)	9
11		Ph	Me	(−)-Sparteine[e,f]	0	9
12	*n*-BuLi	Ph	COOEt	(−)-Sparteine[e,f,g]	$[\alpha]_D$ +6.7[h,i], *S*-(+)	9
13		Ph	Me	(−)-Sparteine[e,f,g]	$[\alpha]_D$ −1.6[h,i], *S*-(+)	9

[a] (+)-DMB is (2*R*,3*R*)-dimethyoxybutane; APME is arabitol pentamethyl ether; MHME is mannitol hexamethyl ether; MME is methyl (−)-menthyl ether.
[b] Hetereogeneous reaction.
[c] Benzene cosolvent.
[d] (*S*)-(+)-1-Ethoxy-2-methylbutane.
[e] See structural drawing page 393.
[f] Toluene cosolvent, −70°.
[g] Hexane solution, −70°.
[h] $CHCl_3$ solution.
[i] Comparison rotation data for enantiomerically pure products were not given.

[8] C. Blomberg and J. Coops, *Rec. Trav. Chim.*, **83**, 1083 (1964).
[9] H. Nozaki, T. Aratani, and T. Toraya, *Tetrahedron Lett.*, 4097 (1968).

TABLE 10-3

Additions to Aldehydes in Chiral Media

$$\text{R-Metal} + \text{R}'\text{—}\overset{\text{O}}{\overset{\|}{\text{C}}}\text{—H} \xrightarrow{\text{Chiral solvent}} \xrightarrow{\text{H}_2\text{O}} \text{R}'\text{—}\overset{\text{OH}}{\overset{|}{\text{C}}}\text{H—R}$$

No.	R-Metal	R′	Chiral solvent[a]	Product	Ref.
1	*n*-BuLi	Ph	(−)-Sparteine[b,c]	6% e.e., *R*-(+)	9
2	EtMgBr	Ph	(−)-Sparteine[c,d]	$[\alpha]_D^{20}$ +9.4°, *R*[e]	9
3	MeMgCl	Ph	(+)-DMB[a]	1.3% e.e., *S*-(−)	10[f]
4	Me_2Mg	Ph	(+)-DMB[a]	4.8% e.e., *S*-(−)	10[f]
5	MeMgI	Ph	"Me_2Bornyl amine"[g]	0.0	11, 12
6	PhMgBr	Me	"Me_2Bornyl amine"[g]	0.0	11, 12
7	EtMgBr	Ph	*S*-(+)-MeO-R*[h]	≪1% e.e., *R*-(+)	8
8		Ph	*S*-(+)-EtO-R*[h]	≪1% e.e., *R*-(+)	8
9		Ph	*S*-(+)-*i*-BuO-R*[h]	≪1% e.e., *R*-(+)	8
10		Ph	(+)-DMB[a]	*ca.* 2.4% e.e., *S*-(−)	8
11	PhMgBr	Et	*S*-(+)-EtO-R*[h]	<1% e.e., *S*-(−)	8
12		Et	(+)-DMB[a]	*ca.* 1.3% e.e., *R*-(+)	8
13	*i*-BuMgBr	Me	*S*-(+)-MeO-R*[h]	*ca.* 1% e.e., *R*-(+)	8
14		Me	*S*-(+)-EtO-R*[h]	*ca.* 1.4% e.e., *R*-(+)	8
15		Me	*S*-(+)-*n*-OctylO-R*[h]	*ca.* 1.9% e.e., *R*-(+)	8
16		Me	MeO-(−)-Menthyl	<1% e.e., *R*-(+)	8
17		Me	(+)-DMB[a]	*ca.* 3% e.e., *S*-(−)	8
18	*n*-PrMgBr	Et	MeO-(−)-Menthyl	≪1% e.e., *S*-(+)	8

[a] (+)-DMB = (2*R*,3*R*)-dimethoxybutane.
[b] Hexane cosolvent, −70°.
[c] See page 393 for (−)-sparteine structure.
[d] Toluene cosolvent, −70°.
[e] $CHCl_3$ solution, % e.e. not reported.
[f] The $[\alpha]_D$ max assumed for the product in this paper is one-fourth the actual value. The corrected values are reported here.
[g] Diastereomer composition and enantiomeric purity are not known. Reference 11 reported optically active products but this was not confirmed in Reference 12.
[h] R* is $-CH_2\overset{*}{C}HMeEt$ from (*S*)-(−)-2-methyl-1-butanol.

ethyl Grignard reagent to ethyl benzoylformate in a sparteine-benzene solution (No. 10, 18% e.e.). It seems generally true that the monodentate chiral ethers give low asymmetric inductions as do the polydentate ethers such as mannitol hexamethyl ether (MHME) and arabitol pentamethyl ether

[10] W. French and G. F. Wright, *Can. J. Chem.*, **42**, 2474 (1964).
[11] M. Bettie and E. Lucchi, *Boll. sci. facolta chim. ind. Bologna*, **1–2**, 2–5 (1940); *Chem. Abst.*, **34**, 2354 (1940).
[12] D. S. Tarbell and M. C. Paulson, *J. Am. Chem. Soc.*, **64**, 2842 (1942).

(APME) [cf. also dimethylisosorbide DMI (7)] while the didentate ether, DMB, and the didentate amine, sparteine, give the highest asymmetric inductions.

The general interpretation of addition to aldehydes in chiral media (Table **10-3**) is clouded by the observation[13] that salts of racemic secondary alcohols undergo asymmetric equilibration via Meerwein–Ponndorf–Verley reduction of ketones in chiral solvents. For example, the reaction of the bromomagnesium salt of racemic methylphenylcarbinol with acetophenone in (+)-2,3-dimethoxybutane gives a heterogeneous mixture from which optically active (*S*)-(−)-methylphenylcarbinol (39% e.e.) is isolated upon hydrolysis. Such a postaddition asymmetric equilibrium could contribute to the observed optical activity of some of the products listed in Table **10-3**.

10-1.3. Asymmetric Reductions in Chiral Media. Several asymmetric reductions using achiral Grignard reagents in chiral media have been carried out. The reaction of racemic menthone (**3**) with *i*-propylmagnesium bromide in (+)-DMB-benzene[5] gave the results shown in

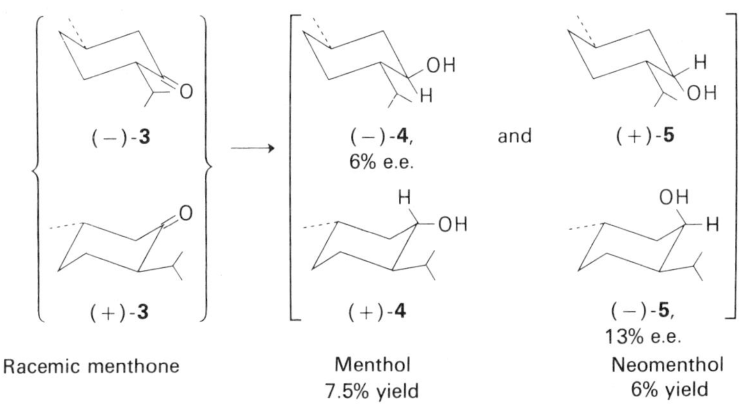

Fig. 10-1. Reaction of isopropyl Grignard reagent with racemic menthone in (+)-2,3-dimethoxybutane-benzene solvent.

Fig. **10-1**. The reduction yield was about 14% with a little more menthol (7.5% yield) being isolated than neomenthol (6% yield). However, reduction via equatorial attack to produce neomenthol gave higher stereoselectivities (13% e.e.) than via axial attack to give menthol (6% e.e.).* The major reaction was enolization to produce menthone (64% yield, racemic) and isomenthone (6% yield, 6% e.e., *levo* isomer).

[13] J. D. Morrison and R. W. Ridgway, *Tetrahedron Lett.*, 573 (1969).

* Asymmetric postreduction Meerwein–Ponndorf–Verley equilibration in the chiral solvent [(−)-**4** ⇄ (−)-**3** ⇄ (+)-**5** versus (+)-**4** ⇄ (+)-**3** ⇄ (−)-**5**] cannot be excluded as a factor contributing to these differences.

Methyl *t*-butyl ketone (**6A**) has been reduced by *t*-butyl-, isobutyl-, and *n*-propyl-magnesium chlorides in the presence of the chiral ether dimethylisosorbide (**7**, DMI) in THF[14] with asymmetric reductions from 0 to 2.3%.

$$\text{R—CO—}t\text{-Bu} + \text{R'MgCl} \xrightarrow{\mathbf{7},\ \text{DMI}} \xrightarrow{H_2O} \text{R—}\overset{*}{\text{C}}\text{H(OH)—}t\text{-Bu}$$

6A, R = Me **8A**, R = Me
6B, R = Ph **8B**, R = Ph

The stereoselectivity increased as the molar ratio of DMI and Grignard increased from 1:1 to 5:1. The reductions of phenyl *t*-butyl ketone (**6B**) with isopropyl magnesium chloride in DMI (1:1) gave less than 0.2% stereoselectivity.

Reduction of phenyl isopropyl ketone (**10**) with two equivalents of the racemic Grignard reagent from 2-phenylbutyl chloride (**9**) in (+)-DMB gave phenylisopropylcarbinol [**11**, 8.4% e.e., *S*-(−)-isomer].[15] Furthermore, the 2-phenylbutane (**12**) resulting from hydrolysis of the unreacted, excess Grignard reagent was optically active [4.9% e.e., *R*-(−)-isomer]. These results indicate that the solvate of *S*-**9** reduces phenyl isopropyl ketone

$$\text{PhCHEtCH}_2\text{MgCl} \xrightarrow{(+)\text{-DMB}} \xrightarrow[\mathbf{10}\ (1\ \text{mole})]{\text{PhCO-}i\text{-Pr}} \text{Ph—C(OH)(H)—}i\text{-Pr} + \text{H—C(Ph)(Et)—CH}_3$$

RS-**9** (2 moles) *S*-(−)-**11** 8.4% e.e. *R*-(−)-**12** 4.9% e.e.

faster than the diastereomeric solvate. Therefore this reaction is not only an asymmetric reduction of the ketone but it is also a kinetic resolution of the Grignard reagent. Postreductive asymmetric equilibrium probably is not a crucial factor in this case, since equilibration of the bromomagnesium salt of racemic **11** in (+)-DMB led to 2.1% e.e. of (+)-**11** after 5 days.

A final example of an asymmetric reaction in a chiral medium is that of the electrolytic reduction of acetophenone at a mercury cathode in the presence of (+)-ephedrine hydrochloride to give (*S*)-(−)-methylphenylcarbinol (4.6% e.e.); reduction in the presence of (−)-ephedrine gave (*R*)-(+)-carbinol (4.2% e.e.).[16] These results should be compared with those for the electrolytic reduction of coumarins (cf. Fig. **6-35**).

[14] F. H. Walker, M.S. thesis, Stanford Univ., 1959.
[15] J. D. Morrison and R. W. Ridgway, *Tetrahedron Lett.*, 569 (1969).
[16] L. Horner and D. Degner, *Tetrahedron Lett.*, 5889 (1968).

It is significant that in all the cases cited in this section there is no example of a successful asymmetric reaction conducted in a nonpolar chiral solvent.

10-2. Asymmetric Synthesis by Substitution at a Prochiral Ligand

The majority of asymmetric syntheses involve the stereoselective addition of a reagent to either one or the other prochiral faces of a double bond (C=C, C=O, and C=N). The number of asymmetric syntheses resulting from preferential reaction at either one or the other prochiral ligands of a molecule is much smaller but perhaps even more interesting.

The symmetry situation we are considering in this section is typified by the alkaloid-catalyzed decarboxylation of methylethylmalonic acid (**13**). When brucine is added to an aqueous solution of **13** which is then evaporated to dryness and the residue decarboxylated at 170°, optically active methylethylacetic acid [**14**, 10% e.e., *S*-(+)-isomer] is formed.[17–19] This was

pro-R → COOH (on C); Me–C(COOH)(Et)–COOH ← *pro-S*

13 $\xrightarrow[170°,\ -CO_2]{\text{Brucine}}$ Me–C(COOH)(H)–Et, *S*-(+)-**14**, 55% and Me–C(H)(COOH)–Et, *R*-(−)-**14**, 45%

claimed to be the first asymmetric synthesis but the interpretation of the reaction has occasioned much controversy,* for although the acid is achiral, the salts are diastereomeric and are in equilibrium with each other. It is hardly profitable to pursue this historically interesting example but a mechanistically clearer example is the reaction of β-phenylglutaric anhydride (**15**) with *R*-(−)-α-methylbenzylamine[20] (**16**), via diastereomeric transition states, to give an unequal mixture of the half amides **17** in 95% yield in 60:40 ratio.†

[17] W. Marckwald, *Ber.*, **37**, 349 (1904).
[18] S. Tijmstra Bz., *Ber.*, **38**, 2165 (1905).
[19] E. Erlenmeyer and F. Landsberger, *Biochem. Z.*, **64**, 366 (1914).
[20] P. S. Schwartz and H. E. Carter, *Proc. Nat'l. Acad. Sci., U.S.*, **40**, 499 (1954).

* See Reference 1, p. 411.

† The absolute configuration at the β-position in the glutaramide is unknown; thus the ratio may be reversed from that shown. If the 5% unaccounted for were all the minor isomer, the asymmetric bias would be 57:43 instead of 60:40 or if the material unaccounted for were all the major component, the ratio would be 62:38. The true stereoselectivity must fall within these limits.

15 + R-16 → 17A, 60% + 17B, 40% (95%)

Reaction of the same glutaric anhydride **15** with (−)-menthol (**18**) gave the diastereomeric monomenthyl esters **19** (95.9% yield) in a ratio of 54:46.[21,22] It is possible in this case that the reaction is thermodynamically controlled instead of kinetically controlled. The isolated diastereomers were separately converted to the enantiomeric monopyrrolidine amides **20** thus regenerating the inducing (−)-menthol and achieving an asymmetric synthesis in the Marckwald sense.

15 + R*OH (18) → 19A → 20A + R*OH; 19B → 20B + R*OH

A further study of this type is the chymotrypsin-catalyzed hydrolysis of a group of unnatural substrates for this enzyme, namely malonic and glutaric esters **21** which are achiral but possess prochiral ligands.[23–26] Both the acylamino substituted malonate (**21A**) and glutarate (**21B**) were hydrolyzed,

[21] R. Altschul, P. Berstein, and S. G. Cohen, *J. Am. Chem. Soc.*, **78**, 5091 (1956).
[22] P. J. Scheuer and S. G. Cohen, *J. Am. Chem. Soc.*, **80**, 4933 (1958).
[23] S. G. Cohen and L. H. Klee, *J. Am. Chem. Soc.*, **82**, 6038 (1960).
[24] S. G. Cohen and E. Khedouri, *ibid.*, **83**, 1093 (1961).
[25] S. G. Cohen, Y. Sprinzak, and E. Khedouri, *ibid.*, **83**, 4225 (1961).
[26] S. G. Cohen and E. Khedouri, *ibid.*, **83**, 4228 (1961).

$$\underset{\textbf{21}}{EtOOC(CH_2)_nCH(R)(CH_2)_nCOOEt} \xrightarrow[\text{Chymotrypsin}]{H_2O} \underset{\textbf{22}}{EtOOC(CH_2)_n\overset{*}{C}H(R)(CH)_nCOOH}$$

21A, n = 0, R = NHAc	Probably 100% e.e.
21B, n = 1, R = NHAc	Probably 100% e.e.
21C, n = 0, R = OH	Zero% e.e.
21D, n = 1, R = OH	100% e.e., *R* isomer

probably with 100% stereoselectivity, while the hydroxy substituted malonate **(21C)** gave racemic product in contrast to the hydroxy glutarate **(21D)** which was hydrolyzed with 100% stereoselectivity.

Although not an asymmetric synthesis, the same principles are involved in the now classic experiments which have shown that enzymes can distinguish between the two enantiotopic CH_2COOH groups of citric acid. There have been several demonstrations of this* but perhaps the clearest[27] is the preparation of asymmetrically labeled radioactive citric acid-C^{14}, **24**, from enantiomerically pure (−)-γ-chloro-β-carboxy-β-hydroxybutyric acid $[\alpha]_D^{25}$ − 44.9°, **23**, by treatment with radioactive cyanide. One isotopic enantiomer of citric acid upon enzyme degradation gave α-ketoglutaric acid, **25**, which

$$\underset{(-)\text{-}\textbf{23}}{ClCH_2\overset{*}{C}(OH)(COOH)CH_2COOH} \xrightarrow[2)\ H_2O]{1)\ ^{14}CN^-} \underset{\textbf{24}}{\overset{pro\text{-}S}{HOOC^{14}CH_2}C(OH)(COOH)\overset{pro\text{-}R}{CH_2COOH}} \xrightarrow[\text{2) Isocitric dehydrogenase}]{\text{1) Aconitase}}$$

$$\underset{\textbf{25}}{HOOC^{14}CH_2CH_2C(=O)COOH} + CO_2 \longrightarrow \underset{\textbf{26}}{HOOC^{14}CH_2CH_2COOH}$$

was further degraded to carbon dioxide and radioactive succinic acid, **26**. The radioactivity of the succinic acid was, within experimental error, the same as that of the precursor acid, **24**; none of the label was found in the carbon dioxide. Although the experiment was not done, if labeled citric acid had been made from (+)-**23** and had been treated similarly, the succinic acid would have been unlabeled and all the radioactivity would have been found in the carbon dioxide. It has been shown that aconitase is active only with respect to the *pro-R* CH_2COOH group.

* This work has been reviewed by H. R. Levy, P. Talalay, and B. Vennesland, in *Progress in Stereochemistry*, Vol. 3, eds., P. B. D. de la Mare and W. Klyne (London : Butterworth & Co., Ltd., 1962), pp. 320–335.

[27] P. E. Wilcox, C. Heidelberger, and V. R. Potter, *J. Am. Chem. Soc.*, **72**, 5019 (1950).

4-Methylcyclohexanone (**27**) has been reported to react with optically active 2-octylnitrite in the presence of sodium ethoxide to give the sodium salt of oxime **28** (15% yield) which was optically active. Dextro nitrite gave levo salt (α_D^{25} $-12.2°$)* and levo nitrite gave dextro salt (α_D^{25} $+17.55$ and $16.51°$).*[28] The product racemized in 10–12 hr. The mechanism for involvement of the chiral reagent in this reaction is obscure. The facile racemization of a compound with structure **28** is difficult to rationalize and these results should be viewed with reservation until these points have been satisfactorily resolved.

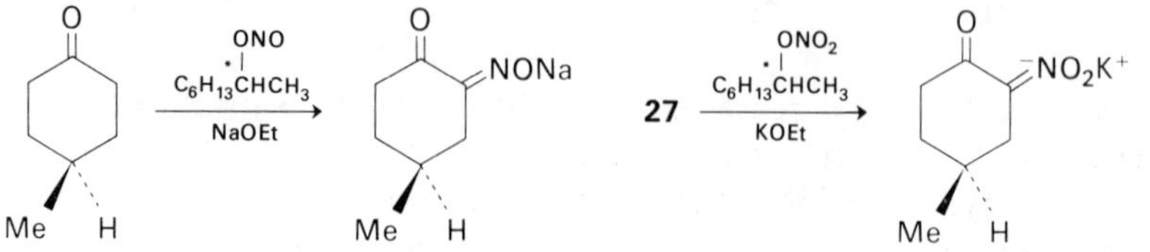

The reaction of 4-methylcyclohexanone (**27**) with (+)-2-octyl nitrate in the presence of potassium ethoxide reportedly produced the potassium salt of 2-nitro-4-methylcyclohexanone (**29**, $[\alpha]_D^{25}$ $+9.02°$, solvent and concentration unspecified, 34% yield) which was racemized in approximately 1 hr.[29] Again it is difficult to account for the facile racemization and to rationalize the mechanism and asymmetric reaction. These studies were prompted by earlier accounts which claimed that chiral 2-nitrooctane retained its stereochemistry as the anion. It has been shown[30] that these erroneous results arose from optically active impurities.

A carefully documented example of asymmetric substitution at a prochiral ligand is the reaction of *t*-butylhydroperoxide (**31**) with cyclohexene (**30**) in the presence of copper (+)-α-ethyl camphorate (**32**).[31] The resulting diastereomeric esters (**34**) are hydrolyzed in 84% yield to cyclohex-2-en-1-ol (**35**) which is optically active[32] (6.6% excess *S* isomer). Copper acetyltartrate was equally satisfactory as a chiral catalyst in the one example reported. Cyclopentene, cyclooctene, and bicyclo[3.2.1]octene-2 gave optically active products ($[\alpha]_D$ $-13.8°$, $+5.8°$ and $-5.2°$, respectively; configurations and maximum rotations are unknown) but cyclododecene and 1-octene gave inactive products.

* The solid sodium salt was precipitated with ether. The solvent for the rotations and the concentration were unspecified.

[28] M. Pezold and R. L. Shriner, *J. Am. Chem. Soc.*, **54**, 4707 (1932).
[29] R. L. Shriner and E. A. Parker, *J. Am. Chem. Soc.*, **55**, 766 (1933).
[30] N. Kornblum, J. T. Patton, and J. B. Nordmann, *J. Am. Chem. Soc.*, **70**, 746 (1948).
[31] D. D. Denney, R. Napier, and A. Cammarata, *J. Org. Chem.*, **30**, 3151 (1965).
[32] R. K. Hill and J. W. Morgan, *J. Org. Chem.*, **33**, 927 (1968).

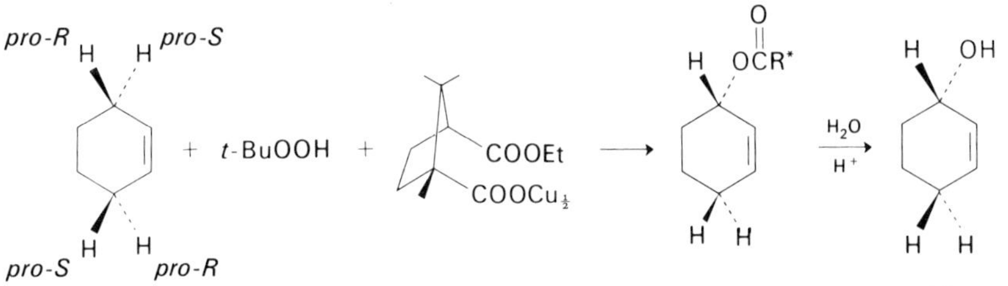

pro-R H H pro-S + t-BuOOH + COOEt COOCu½ → H OCR* H₂O/H⁺ → H OH
pro-S H H pro-R

30 **31** **32** **34** **35**

Some of the asymmetric eliminations studied by Goldberg and co-workers and discussed in Sec. **9-1** can be considered examples of asymmetric substitution at a prochiral ligand and should be referred to at this point.

The net effect of treating various carbohydrate derived diethylthioacetals (**36**), with mercuric acetate is the preferential replacement of one of two prochiral thioethyl groups with the establishment of a new chiral center in the product.[33,34] The inducing chiral moiety cannot be removed easily in these sugar derivatives and thus a demonstration of an asymmetric synthesis in the Marckwald sense is not easily accomplished. It seems most likely that the mechanism involves the asymmetric addition by the neighboring acetyl group to the carbon–sulfonium double bond (**38** → **39**) with displacement by

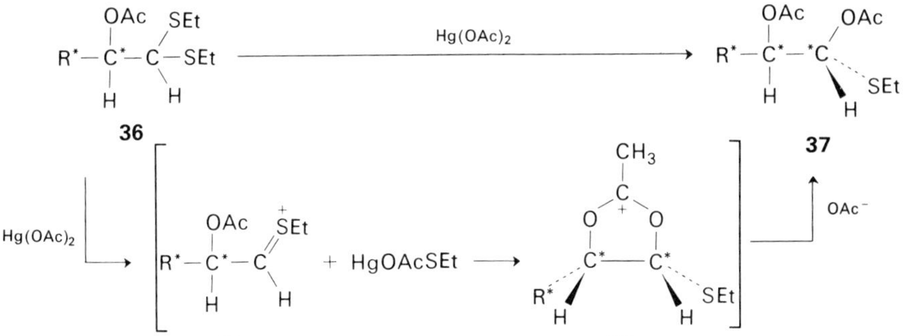

R*—C*(OAc)(H)—C(SEt)(H)—SEt (**36**) —$Hg(OAc)_2$→ R*—C*(OAc)(H)—*C(OAc)(H)(SEt) (**37**)

$Hg(OAc)_2$ → [R*—C*(OAc)(H)—C(=$\overset{+}{S}$Et)(H) (**38**) + HgOAcSEt → cyclic acetoxonium ion with CH_3, R*, H, SEt (**39**)] —OAc^-→ **37**

acetate ion to give the product. The preferential methylation of strychnine by one of the two methyl groups in prochiral dimethyl *p*-nitrophenyl thiophosphate described in Sec. **8-4** is another similar example.

A most interesting use of an asymmetric reaction at enantiotopic ligands is shown in the conversion of the achiral cyclopenta-1,3-diketone **40** by reaction with (*R*,*R*)-tartramic hydrazide (**41**) to diastereomeric hydrazones **42A** and **42B**, the latter of which is a precursor for hormones of the

[33] E. P. Painter, *Can. J. Chem.*, **42**, 2018 (1964).
[34] E. P. Painter, *Tetrahedron*, **21**, 1337 (1965).

estradiol series.[35] Hydrazones **42A** and **42B** either are produced at different rates and/or are in equilibrium with each other. Recrystallization of the initial precipitate (84% yield) gave purified **42B** (75% yield). This process is

40 + **41** ⟶ **42A** and **42B** ⟶ **43** ⟶ **44**

4-O-Methyl-8-isoestradiol

not a simple resolution although it could be a case of resolution of stereochemically equilibrating diastereomers (second-order asymmetric transformation, Sec. **1-4.3**). Ring closure of **42B** with hydrogen chloride in dioxane gave the ketone **43** which on hydrogenation and sodium borohydride reduction was converted into the estradiol derivative **44** belonging to the natural stereochemical series.

[35] R. Bucourt, L. Nédélec, J-C. Gasc, and J. Weill-Raynal, *Bull. Soc. Chim. Fr.*, 561 (1967).

10-3. Asymmetric Polymer Synthesis

The subject of asymmetric polymer synthesis has become extensive and specialized; although it properly belongs in the present treatment of asymmetric organic reactions, we shall not attempt to cover it here. This topic has been reviewed recently from various viewpoints[36–41] and thus the information is generally available in convenient form.

The various types of asymmetric synthesis which have been applied to polymerizations can be classified as (a) polymerization of a prochiral monomer in the presence of a chiral catalyst; (b) the polymerization of a chiral monomer whose chiral inducing unit can be cleaved from the polymer; (c) polymerizations carried out in a chiral solvent, solid matrix, or conceivably by a chiral physical force.

An example of the first of these is the polymerization of 1,3-pentadiene (**45**) in the presence of triethylaluminum and titanium menthoxide to give the *cis*-1,4-isotactic polymer[42] **46** ($[\alpha]_D$ $-22.3°$). By the action of diethyl

$$CH_2{=}CH{-}CH{=}CH(Me) \xrightarrow[\text{Ti (menthoxide)}_4]{AlEt_3} \text{---}[CH_2CH{=}CHCH(Me)]_n\text{---}$$

45 → **46**

45 $\xrightarrow[\text{Ti (menthoxide)}_4]{AlEt_2Cl}$ **47** and **48** etc.

aluminum chloride and titanium menthoxide on the same monomer, it was possible to obtain a mixture of cyclic trimers (10–15% yield) which were

[36] R. C. Schulz and E. Kaiser, *Adv. Polym. Sci.*, **4**, 236–315 (1965).

[37] P. Pino, *Adv. Polym. Sci.*, **4**, 393–456 (1965).

[38] M. Goodman, *Topics in Stereochemistry*, Vol. 2, eds. N. L. Allinger and E. L. Eliel (New York: Interscience Pub., 1967), pp. 73–156.

[39] G. Natta and F. Danusso, *Stereoregular Polymers and Stereospecific Polymerization* (New York–London: Pergamon Press, 1967).

[40] (a) E. I. Klabunovskii, *Stereospecific Catalysis* (Moscow: Nauka, 1968). (b) E. I. Klabunovskii, *Russian Chem. Rev.*, **37**, 969–984 (1968) in English; *Uspekki Khimii*, **39**, 2192 (1968).

[41] A. D. Ketley, ed., *The Stereochemistry of Macromolecules*, Vol. 3 (London: Arnold, 1968).

[42] G. Natta, L. Porri, and S. Valenti, *Makromol. Chem.*, **67**, 225 (1963).

shown to be optically active ($[\alpha]_D$ −8.6, −4.4°).[43] This mixture contained both trimethyl *trans-trans-trans* (**47**) and *trans-trans-cis* (**48**) cyclododecatriene isomers.

The second type is illustrated by the polymerization of the menthyl ester of sorbic acid (**49**) with butyl lithium. The menthyl group was first removed from the polymer by hydrolysis and then ozonolysis of the polymer (**50**) gave methylsuccinic acid (**51**) with a 6% excess of the *R* isomer.[44]

$$2n\ \overset{\text{COOR}^*}{\text{HC}}=\text{CH}-\text{CH}=\overset{\text{Me}}{\text{CH}} \xrightarrow{\text{LiBu}} \xrightarrow[-\text{R}^*\text{OH}]{\text{H}_2\text{O}} \cdots \left[{}^*\overset{\text{COOH}}{\text{CH}}\text{CH}=\text{CH}^*\overset{\text{Me}}{\text{CH}}-{}^*\overset{\text{COOH}}{\text{CH}}\text{CH}=\text{CH}^*\overset{\text{Me}}{\text{CH}} \right]_n \cdots$$

49 **50**

$$\xrightarrow{\text{O}_3} \text{HOOCCH}_2{}^*\overset{\text{Me}}{\text{CH}}\text{COOH}$$

51

The third type is illustrated by the ingenious experiment in which 1,3-pentadiene (**45**) is polymerized by γ irradiation while trapped in the chiral matrix of solid, optically active, (*R*)-(−)-perhydrotriphenylene (**52**) ($[\alpha]_D^{25}$ −93°).[45] The isotactic *trans*-1,4-polypentadiene (**53**) shows a rotation of $[\alpha]_D^{25}$ +2.5 ± 0.3° ($CHCl_3$).

Me —γ rays→ [Me Me Me]$_n$

45 **52** **53**

10-4. Absolute Asymmetric Reactions

The subject of absolute asymmetric synthesis has been introduced in Sec. **1-4.4c**. Because of its relationship to the development of optically active compounds in living systems, it has held a position of high interest for chemists since the process was first conceived by Emil Fischer. Currently the problem associated with the origin of chirality in nature is not that of discovering a process whereby such chirality could arise but rather it is

[43] J. Furukawa, T. Kakuzen, H. Morikawa, R. Yamamoto, and O. Okuno, *Bull. Chem. Soc. Japan*, **41**, 155 (1968).

[44] M. Farina, M. Modena, and W. Ghizzoni, *Rend. Acc. Naz. Lincei*, **32** [8], 91 (1962); *Chem. Abst.*, **57**, 15320 (1962).

[45] M. Farina, G. Audisio, and G. Natta, *J. Am. Chem. Soc.*, **89**, 5071 (1967).

that of trying to evaluate which of several alternatives seems most credible. Since this subject has been examined many times[46–53] and will be the subject of a definitive forthcoming review,[54] we shall limit ourselves to a discussion of the chemical aspects of absolute asymmetric synthesis as we have defined it (Sec. **1-4.4c**). We shall now consider those processes which produce a chiral product by the action of some dissymmetric physical force. These clearly can be divided into two types: (a) those which start with a racemate and achieve preferential destruction or alteration of one enantiomer, and (b) those which bring about the conversion of an achiral substrate or substrates into a chiral molecule.

Since the first of these types starts with a racemate, individual molecules of which are chiral, such examples do not represent asymmetric syntheses by our definition, but instead they represent a kinetic resolution. However, we must briefly consider certain pertinent reactions of this first class because of their relationship to absolute asymmetric syntheses.

10-4.1. Absolute Kinetic Resolutions. A necessary but not sufficient condition for an absolute asymmetric synthesis is that the chiral physical force must be directly involved in the reaction under study; that is, the reaction will not proceed (or will proceed at a different rate) without the physical force. By about 1900 this principle was generally appreciated,[55] but prior to this time there had been several futile attempts to obtain absolute asymmetric syntheses which failed because this necessary condition was not met. For instance, experiments conducted in the presence of a magnetic field alone were destined to failure since such a field is achiral, having a plane of symmetry perpendicular to the lines of force.[56] Other experiments failed which did use a chiral physical force but one not directly involved in the chemical process being studied. An obvious example of this was the addition

[46] See Chapter 1, References 3, 19, 21, 92, 93, and 95.

[47] H. Eyring, L. L. Jones, and H. D. Spikes, *Horizons in Biochemistry*, M. Kasha and B. Pullman, eds. (New York: Academic Press, 1962), pp. 242–250.

[48] E. Havinga, *Biochem. Biophys. Acta*, **13**, 171 (1954).

[49] A. Amariglo, H. Amariglo, and X. Duval, *Ann. Chim.*, **3**, 5–25 (1968).

[50] J. Keosian, *The Origin of Life*, 2nd ed. (New York: Reinhold Book Corp., 1964).

[51] A. I. Oparin, *Origin of Life*, 2nd ed. (New York: Dover Pub. Inc., 1953).

[52] G. Wald, *Proc. Natl. Acad. Sci., U.S.*, **52**, 595–611 (1964).

[53] E. I. Klabunovskii, *Absolute Asymmetric Synthesis and Asymmetric Catalysis in Origins of Life on Earth*, Rept. Int'l. Symp., Moscow (New York: Pergamon Press, 1959), pp. 158–168.

[54] W. A. Bonner, *Origins of Molecular Chirality*, in *Current Aspects of Exobiology*, C. Ponnamperuma, ed. (Amsterdam: North-Holland Pub. Co., *in press*).

[55] A. Byk, *Z. physikal. Chem.*, **49**, 641 (1904); *Ber.*, **37**, 4696 (1904).

[56] Several such experiments are as follows: the addition of bromine to stilbene, methyl cinnamate, and methyl fumarate, and the reduction of benzoylformic acid in a magnetic field. These theoretically hopeless experiments have been reviewed in Reference 1, Chapter 4.

of hydrocyanic acid to acetaldehyde in the presence of circularly polarized light.[57–60]

It seems likely that the attempted absolute asymmetric synthesis by the addition of hydrogen bromide or bromine to the indane-1,3-dione **54** to give the lactone **55** failed because the light used for the irradiation of the mixture was not crucially involved in the reaction.[61]

$CH_2CH{=}CH_2$, COOEt; $h\nu$, +HBr(A) or Br_2(B); CH_2R, H, O

54

55A, R = H
55B, R = Br

The first generally accepted report of the production of an optically active product brought about by a chiral physical force is that of the irradiation of racemic ethyl α-bromopropionate with circularly polarized light (2800 A). The results are much more convincing when racemic N,N-dimethyl α-azidopropionamide is similarly photolyzed.[62] This example, which is further described in Sec. **1-4.4c**, now serves as the classic case of an absolute asymmetric kinetic resolution. The extent of asymmetric resolution is appreciable (2% e.e.); the rotations are far from marginal (α_D −1.04°), enantiomers are obtained in excess by use of right versus left circularly polarized light, and the asymmetric decomposition is based on the known difference of absorption by the (+) and (−) isomers of circularly polarized light of a wavelength which causes the chemical decomposition.

Another example is the photochemical decomposition of racemic humulene nitrosite[63] (**57**) which is blue and which is formed from the achiral sesquiterpene humulene (**56**) by the addition of nitrogen trioxide. Irradiation by right circularly polarized light (4358 A) causes the rotation of a solution

[57] J. Pirak, *Biochem. Z.*, **130**, 76 (1922).

[58] It appears that van't Hoff was the first to point out that circularly polarized light might be used to achieve an absolute asymmetric synthesis [van't Hoff, *Die Lagerung der Atome und Raume*, 2nd ed. (1894), p. 30].

[59] A. Cotton first proposed that circularly polarized light might be used to achieve an absolute asymmetric kinetic resolution by the preferential destruction of one isomer of a racemic pair. [A. Cotton, *J. Chim. Phys.*, **7**, 81 (1909).]

[60] A. Cotton, *Trans. Faraday Soc.*, **26**, 377 (1930).

[61] D. Râdulescu and V. Moga, *Bul. Soc. Chim. Romania*, **1**, [2] (1939); *Chem. Abst.*, **37**, 4070 (1943). These authors reported that the product did show a slight but disappointingly low rotation but H. Pracejus was unable to confirm this [*Fortschr. Chem. Forsch.*, **8**, 541 (1967)].

[62] (a) W. Kuhn and E. Braun, *Naturwiss.*, **17**, 227 (1929). (b) W. Kuhn and E. Knopf, *Z. Phys. Chem.*, **7B**, 292 (1930).

[63] S. Mitchell, *J. Chem. Soc.*, 1829 (1930).

of **57** to change from 0 to a maximum of +0.30° (or −0.30° in the case of left circularly polarized light) over a 36-hr period and then go to 0 after a total of 64 hr. The structure of humulene was not known at the time of these

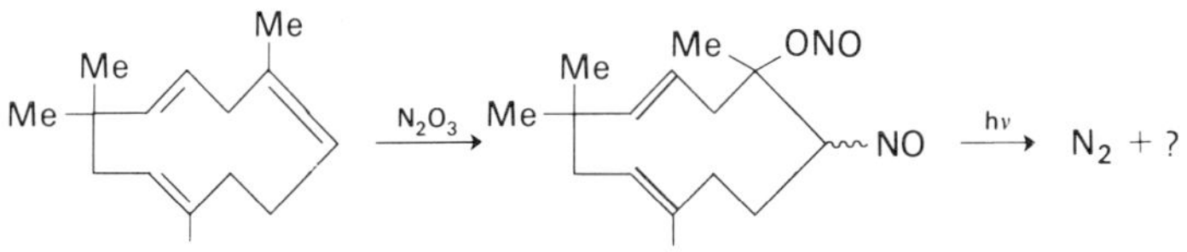
Me
Me
Me
N_2O_3
Me
Me
Me
ONO
NO
hν
N_2 + ?

56 **57**

experiments; details such as the nature of the products and the circular dichroism curves were not investigated.

An ingeniously conceived absolute asymmetric decomposition is illustrated by the irradiation with elliptically polarized light (3600 A) of the racemic dihydropyridine compound **58** to give the substituted 4-*o*-nitrosophenylpyridine derivative **59** which possesses biphenyl-type molecular asymmetry.[64] After one-third of the racemate of **58** had been decomposed, the residue from the reaction mixture had an observed rotation of α_D −0.022 ± 0.006°.

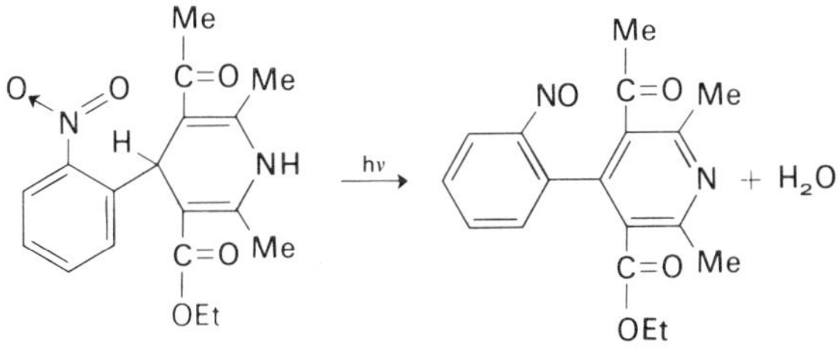
Me
C=O Me
O
N
O
H
NH
C=O Me
OEt
hν
Me
NO C=O Me
N + H_2O
C=O Me
OEt

58 **59**

It is unknown whether this rotation is due to undecomposed starting material **58** or to product **59** or both. Maximum rotations and stereochemical stabilities of pure **58** and **59** were not experimentally investigated.

In general it seems likely that higher stereoselectivities by absolute asymmetric decomposition can be achieved under proper circumstances. However, the soundness of the principles involved have been established and there is little activity in this field at the present time.

High energy electrons resulting from β-decay of radioactive elements (and from cosmic rays) possess a helicity which is passed on to the γ radiation (Bremsstrahlung), released as these high energy electrons slow down. As a result of the nonconservation of parity, all such radiation caused by β-decay on the earth (and that which is caused by cosmic rays from our galaxy) is

[64] J. A. Berson and E. Brown, *J. Am. Chem. Soc.*, 77, 450 (1955).

slightly circularly polarized and thus constitutes a universal chiral physical force which is theoretically available for the initiation of chemical reactions.[65]

After several unsuccessful experiments by Vester and Ulbricht,[66] it has been claimed by Garay[67] that D-tyrosine is destroyed faster than L-tyrosine by the decomposition of strontium-90 over a period of 18 months. The relative rates of decomposition were followed by observing the uv spectral changes. Similar experiments in which *dl*-alanine, *dl*-tryptophan, and *dl*-tyrosine were exposed to phosphorus-32 decay gave negative results after a few months. Thus generation of *optical activity* by irradiation of racemic samples was not demonstrated. It was demonstrated that there was no decomposition in the absence of strontium-90. This may be the definitive experiment which links natural radioactive decay with the production of optically active compounds in nature. Further development of these experiments will be watched with much interest.

10-4.2. Absolute Asymmetric Synthesis. Following are the known studies in which attempts have been made to convert a prochiral substrate into chiral product under the influence of a chiral physical force.

Davis and Heggie[68] studied the addition of both chlorine and bromine to trinitrostilbene (**60**) in the presence of circularly polarized light (3600–4500 A). Five experiments on the addition of chlorine were described. The initial observed rotation of zero rose to a maximum of about α_D $+0.030 \pm 0.003°$ over a period of about 45 min and then diminished to zero. Both trinitrostilbene dibromide (**61**, X = Br) and dichloride (**61**, X = Cl) decompose spontaneously, more rapidly in the light than dark, and no record was published of an attempt to isolate active material.

C_6H_5–CH=CH–$C_6H_2(NO_2)_3$ (**60**) $\xrightarrow[h\nu]{X_2}$ C_6H_5–CHXCHX–$C_6H_2(NO_2)_3$ (**61**)

The photochemically catalyzed addition of anhydrous hydrogen peroxide to neat diethyl fumarate (**62**) was studied[69] using left circularly polarized light (2535–2539 A). A maximum observed rotation of $+0.073 \pm 0.008°$ *

[65] T. L. V. Ulbricht, *Quart. Rev.*, **13**, 48 (1959).

[66] (a) F. Vester, T. L. V. Ulbricht, and H. Krauch, *Naturwiss.*, **46**, 68 (1959). (b) T. L. V. Ulbricht and F. Vester, *Tetrahedron*, **18**, 629 (1962).

[67] A. S. Garay, *Nature*, **219**, 338 (1968).

[68] T. L. Davis and R. Heggie, *J. Am. Chem. Soc.*, **57**, 377, 1622 (1935).

[69] T. L. Davis and J. Ackerman, *J. Am. Chem. Soc.*, **67**, 486 (1945).

* Using a 1 dm center-filled 0.2 ml polarimeter tube and a Schmidt and Haensch polarimeter which read to 0.01°.

$$\underset{\mathbf{62}}{\text{(H)(EtOOC)C=C(COOEt)(H)}} \xrightarrow[h\nu]{H_2O_2} \underset{\mathbf{63}}{\text{EtOOC–CHOH–CHOH–COOEt}}$$

developed in one experiment after 119 hrs. A second experiment was not nearly so convincing; only one reading after 187 hours, +0.030°, was substantially beyond the experimental error from zero. An attempt to induce the formation of the enantiomer by use of oppositely circularly polarized light was unsuccessful. It was determined that the nickel mirror used for the circular polarization of the plane polarized light had badly deteriorated by this time. There was no isolation of active material in these experiments, although it was shown that a 6.6% yield of *dl* tartaric acid (**63**) was formed under the same conditions using nonpolarized radiation. One must assume that the role of the radiation in this reaction is to convert the hydrogen peroxide to hydroxyl radicals. Since a hydroxyl radical is achiral, the rationale for success of this experiment is not apparent and these results should be considered with reservation until confirmed.

Similar considerations apply to the experiments reported on the gas phase addition of chlorine to propene and to butadiene[70] (and of hydrogen chloride to 2-pentene) in the presence of circularly polarized light. The observed rotations were reported not to exceed 0.04–0.05° in a decimeter tube.

An absolute asymmetric synthesis of a triaryl chloromethane has been reported by Karagunis and Drikos[71] using a solution of preformed phenyl-*p*-biphenyl-α-naphthyl free radicals which was irradiated with circularly polarized light (4350 A) during reaction with chlorine. The apparatus was so designed that rotations were taken at the D line of sodium while the reaction progressed. Observed rotations rose from 0 to a maximum of +0.07° with right circularly polarized light and to about −0.06° with left circularly polarized light after about 1 hr, and then they declined. A control experiment with nonpolarized light gave negative results. These results were interpreted in terms of circularly polarized light inducing chirality in a nonplanar triaryl free radical. No optically active material was isolated.

TrioxalatochromateIII (**64**) undergoes both thermal and photochemical interconversion between (+) and (−) forms. Upon irradiation with right

[70] M. Betti and E. Lucchi, *Atti X^e congr. intern. chim.*, **2**, 112 (1938); *Chem. Abst.*, **33**, 7273 (1939).

[71] G. Karagunis and G. Drikos, *Praktika* (*Akad. Athenon*), **9**, 177 (1934); *Chem. Abst.*, **30**, 4486 (1936); *Naturwiss.*, **21**, 607 (1933); *Nature*, **132**, 354 (1933); *Z. phys. Chem.*, **B26**, 428–38 (1934).

circularly polarized light (5460 A), a steady state equilibrium is established,[72] (**64A** ⇆ **64B**) with α_D + 0.06° at 1.4°C and α_D + 0.03° at 13.3°C; left polarized light gives an equilibrium with the (−) form predominating to the same extent. Di-oxalatodiaquochromateIII, *cis* and *trans*, and di-μ-hydroxytetra-oxalatodichromateIII give similar results. Although a racemic substrate was

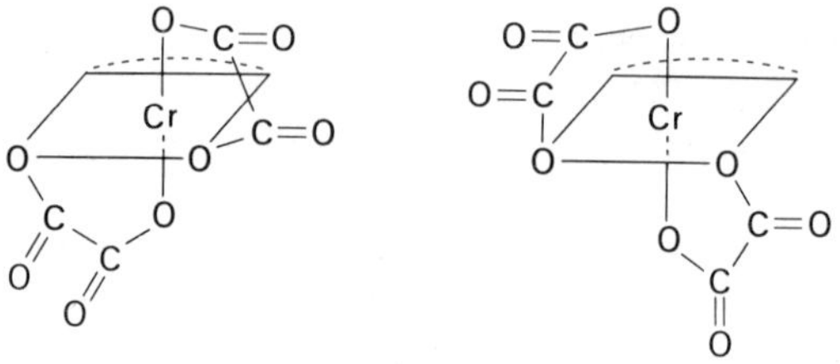

64A **64B**

used, one isomer is converted into another and there is a real generation of chirality in the system which is not simply the destruction of one enantiomer. In this sense a true absolute asymmetric synthesis has been achieved.

1969 Addenda

Sec. 10-1. (+)-Methyl-2-methylbutyl ether has been used as a chiral solvent to induce asymmetry in the *threo* pinacol resulting from the reduction of phenyl ferrocenyl ketone, thereby distinguishing between the *dl* and *meso* forms.[73]

Treatment of isopropylferrocene with *n*-butyllithium in the presence of (−)-sparteine resulted in the formation of chiral 3-isopropyl-1,1′-dilithio-ferrocene which was subsequently converted to optically active derivatives having about 3% optical purity. Treatment of *S*-(+)-1-ferrocenylmethyl-2-methylpiperidine with *n*-butyllithium gave what was judged to be 100% asymmetric lithiation.[74]

Sec. 10-1.1. Bosnich[75,76] claimed an asymmetric synthesis in chiral 2,3-butanediol solvent in which the *d* and *l* forms of racemic *cis*-dichloro*bis*-(ethylenediamine)-cobalt(III) tetraphenylborate underwent asymmetric equilibrium. It has been shown in a closely related study that optically active 1,2-propanediol solvent is stereospecifically substituted as a ligand in

[72] K. L. Stevenson and J. F. Verdieck, *J. Am. Chem. Soc.*, **90**, 2974 (1968).
[73] S. I. Goldberg and W. D. Bailey, *J. Am. Chem. Soc.*, **91**, 5685 (1969).
[74] T. Aratani, T. Gonda, and H. Nozaki, *Tetrahedron Lett.*, 2265 (1969).
[75] B. Bosnich, *J. Am. Chem. Soc.*, **89**, 6143 (1967).
[76] B. Bosnich and D. W. Watts, *ibid.*, **90**, 6228 (1968).

place of the chloride ions of the substrate complex and that an asymmetric equilibration by virtue of the optically active solvent did not take place.[77]

Sec. 10-1.2. A very effective asymmetric reaction which may be formally classified as a carbonyl addition in a chiral media is the addition of Grignard reagents to various carbonyl compounds in the presence of the chiral sugar derivative, 1,2:5,6-di-O-isopropylidene-α-D-glucofuranose.[78] The stereoselectivities were highest for the addition of methylmagnesium bromide to phenyl cyclohexyl ketone (70% e.e.) and lowest for the addition of phenylmagnesium bromide to propionaldehyde (9% e.e.). The sugar derivative has a free hydroxyl group which reacts with the Grignard reagent to give an alcoholate which most likely complexes with the excess reagent. Thus this reaction is more properly considered to be one with a chirally modified reagent rather than one in a chiral solvent.

Sec. 10-2. Heterogeneous asymmetric syntheses in general have not been treated except for a few representative examples involving carbon–carbon double bonds (cf. **6-7.3**) and asymmetric amino acid syntheses (cf. **7-21**). Recent reports[79,80] of the Raney nickel catalyzed reduction of carbonyl compounds to secondary carbinols in the presence of chiral additives such as (*R*)-tartaric acid, (*S*)-glutamic acid, and (*S*)-alanine can be considered as asymmetric syntheses in a chiral media (cf. **10-1**). Stereoselectivities from 4–27% e.e. were reported for ethyl acetoacetate, while those for ethyl pyruvate and methyl ethyl ketone ranged from 0–11% e.e. *Meso*-α,α'-diaminopimelic acid has been studied as a substrate for asymmetric synthesis by substitution at a prochiral ligand.[81]

Several enzymatic reactions which distinguish between the *pro-R* and *pro-S* hydrogens in glycine have been characterized. Serine hydroxymethylase promotes exchange of the *pro-S* hydrogen in D_2O or T_2O and in the presence of formaldehyde catalyzes the production of L-serine.[82]

Photochemical catalyzed alkylations of di- and polypeptides have also been shown to result in preferential replacement (up to 40%) of the diasteriotopic hydrogens of the glycine units.[83–85]

[77] A. A. Smith and R. A. Haines, *ibid.*, **91**, 6280 (1969).

[78] T. D. Inch, G. J. Lewis, G. L. Sainsbury, and D. J. Sellers, *Tetrahedron Lett.*, 3657 (1969).

[79] Y. Izumi, M. Imaida, T. Harada, T. Tanabe, S. Yajima, and T. Ninomiya, *Bull. Chem. Soc. Jap.*, **42**, 241 (1969).

[80] Y. Izumi, S. Tatsumi, and M. Imaida, *Bull. Chem. Soc. Jap.*, **42**, 2373 (1969).

[81] J. van Heijenoort, E. Bricas, and C. Nicot, *Bull. Soc. Chim. Fr.*, 2743 (1969).

[82] M. Akhtar and P. M. Jordan, *Tetrahedron Lett.*, 875 (1969).

[83] D. Elad and J. Sperling, *Chem. Commun.*, 234 (1969).

[84] D. Elad and J. Sperling, *J. Chem. Soc.* (*C*), 1579 (1969).

[85] J. Sperling, *J. Am. Chem. Soc.*, **91**, 5389 (1969).

Chlorination of optically active benzyl methyl sulfoxide with iodobenzene dichloride preferentially replaced one of the diastereotopic hydrogens to give an α-chlorobenzyl methyl sulfoxide. This sulfoxide was oxidized to an optically active α-chlorobenzyl methyl sulfone in which the chirality at sulfur had been removed.[86] The chlorination preferentially replaced the opposite hydrogen from that exchanged in NaOD-D_2O, and the assignment of configuration was based on the reported stereochemistry of the exchange reaction [S. Wolfe and A. Rauk, *Chem. Commun.*, 778 (1966)]. It has now been shown that the earlier assignment was incorrect.[87] The conclusion that diastereotopic hydrogens are involved is correct but the previous prochiral assignment of the respective benzylic protons should be reversed. Other papers on stereoselective H-D exchange in sulfoxides have appeared.[88,89]

Sec. 10-4.2. A special category of absolute asymmetric synthesis is illustrated by the reaction of liquid (or gaseous) bromine with molecularly achiral *trans*-4,4′-dimethyl-chalcone in the form of chiral crystals, to give the dibromo addition product, 6% e.e.[90]

[86] M. Cinquini, S. Colonna, and F. Montanari, *Chem. Commun.*, 607 (1969).
[87] J. E. Baldwin, R. E. Hackler, and R. M. Scott, *Chem. Commun.*, 1415 (1969).
[88] R. R. Fraser and F. J. Schuber, *Chem. Commun.*, 397 (1969).
[89] B. J. Hutchinson, K. K. Andersen, and A. R. Katritzky, *J. Am. Chem. Soc.*, **91**, 3839 (1969).
[90] K. Penzien and G. M. J. Schmidt, *Angew. Chem.*, **81**, 576 (1969); *Internat. Edn.*, **8**, 608 (1969).

Author index

Subject index